Grundlehren Text Editions

Springer
Berlin
Heidelberg
New York
Hong Kong
London
Milan
Paris
Tokyo

Nicole Berline
Ezra Getzler
Michèle Vergne

Heat Kernels and Dirac Operators

Springer

Nicole Berline
Michèle Vergne
École Polytechnique
Centre de Mathématiques
91128 Palaiseau CX
France
e-mail: nicole.berline@math.polytechnique.fr
michele.vergne@math.polytechnique.fr

Ezra Getzler
Northwestern University
Department of Mathematics
Sheridan Road 2033
60208-2730 Evanston, IL
U.S.A.
e-mail: getzler@northwestern.edu

Cataloging-in-Publication Data applied for

A catalog record for this book is available from the Library of Congress.

Bibliographic information published by Die Deutsche Bibliothek.
Die Deutsche Bibliothek lists this publication in the Deutsche Nationalbibliografie; detailed bibliographic data is available in the Internet at http://dnb.ddb.de.

Mathematics Subject Classification (2000): 58J20, 58J35, 58A10, 53C05

ISSN 1618-2685
ISBN 3-540-20062-2 Springer-Verlag Berlin Heidelberg New York
(Original edition as Vol. 298 of series *Grundlehren der mathematischen Wissenschaften*, 1992)

Springer-Verlag is a part of Springer Science+Business Media

springeronline.com

Printed in Germany

Cover design: Erich Kirchner, Heidelberg
Typeset by the authors using a Springer TEX macro package
Printed on acid-free paper SPIN 10816225 41/3142/YL - 5 4 3 2 1 0

Preface

This book, which began as a seminar in 1985 at MIT, contains complete proofs of the local index theorem for Dirac operators using the heat kernel approach, together with its generalizations to equivariant Dirac operators and families of Dirac operators, as well as background material on superconnections and equivariant differential forms.

Since the publication of the first edition, the subjects treated here have continued to find new applications. Equivariant cohomology plays an important role in the study of symplectic reduction, and Bismut superconnections and the local index theorem for families have had many applications, through the construction of higher analytic torsion forms and currents. (For a survey of some of these developments, we recommend reading Bismut's talk at the Berlin International Congress of Mathematicians, reference [33].)

Although this book lacks some of the usual attributes of a textbook (such as exercises), it has been widely used in advanced courses in differential geometry; for many of the topics discussed here, there are no other treatments available in monograph form. Because of the continuing demand from students for the book, we were very pleased when our editor Catriona Byrne at Springer Verlag proposed reissuing it in the series "Grundlehren Text Editions." The proofs in this book remain among the simplest available, and we have decided to retain them without any change in the new edition.

We have not attempted to give a definitive bibliography of this very large subject, but have only tried to draw attention to the articles that have influenced us.

We would like to take the opportunity to thank the other participants in the MIT seminar, especially Martin Andler and Varghese Mathai, for their spirited participation. Discussions with many other people have been important to us, among whom we would like to single out Jean-Michel Bismut, Dan Freed and Dan Quillen. Finally, we are pleased to be able to thank all of those people who read all or part of the book as it developed and who made many comments which were crucial in improving the book, both mathematically and stylistically, especially Jean-François Burnol, Michel Duflo, Sylvie Paycha, Christophe Soulé, and Shlomo Sternberg. We also thank the referee for suggestions which have improved the exposition.

To all of the following institutes and funds, we would like to express our gratitude: the Centre for Mathematical Analysis of the ANU, the ENS-Paris, the Harvard Society of Fellows, the IHES, MIT, and the Université de Paris-Sud. We also received some assistance from the CNRS, the NSF, and the Sloan Foundation.

We wish to thank readers of the first edition who were kind enough to send us corrections: any remaining errors are of course our own.

Paris and Chicago,
September, 2003

N. Berline,
E. Getzler,
M. Vergne

Contents

Introduction

Dirac operators on Riemannian manifolds, which were introduced in the articles of Atiyah and Singer [13] and Lichnerowicz [81], are of fundamental importance in differential geometry: they occur in situations such as Hodge theory, gauge theory, and geometric quantization, to name just a few examples. Most first-order linear differential operators of geometric origin are Dirac operators.

After Atiyah and Singer's fundamental work on the index for general elliptic operators, methods based on the heat kernel were applied to prove the Atiyah-Singer Index Theorem in the special case of Dirac operators, by Patodi [90], Gilkey [65] and Atiyah-Bott-Patodi [8]. In recent years, new insights into the local index theorem of Patodi and Gilkey have emerged, which have simplified the proofs of their results, and permitted the extension of the local index theorem to other situations. Thus, we felt it worthwhile to write a book in which the Atiyah-Singer Index Theorem for Dirac operators on compact Riemannian manifolds and its more recent generalizations would receive elementary proofs. Many of the theorems which we discuss are due to J.-M. Bismut, although we have replaced his use of probability theory by classical asymptotic expansion methods.

Our book is based on a simple principle, which we learned from D. Quillen: Dirac operators are a quantization of the theory of connections, and the supertrace of the heat kernel of the square of a Dirac operator is the quantization of the Chern character of the corresponding connection. From this point of view, the index theorem for Dirac operators is a statement about the relationship between the heat kernel of the square of a Dirac operator and the Chern character of the associated connection. This relationship holds at the level of differential forms and not just in cohomology, and leads us to think of index theory and heat kernels as a quantization of Chern-Weil theory.

Following the approach suggested by Atiyah-Bott and McKean-Singer, and pursued by Patodi and Gilkey, the main technique used in the book is an explicit geometric construction of the kernel of the heat operator e^{-tD^2} associated to the square of a Dirac operator D. The importance of the heat kernel is that it interpolates between the identity operator, when $t = 0$, and the projection onto the kernel of the Dirac operator D, when $t = \infty$. However, we will study the heat kernel, and more particularly its restriction to the diagonal, in its own right, and not only as a tool in understanding the kernel of D.

Lastly, we attempt to express all of our constructions in such a way that they generalize easily to the equivariant setting, in which a compact Lie group G acts on the manifold and leaves the Dirac operator invariant.

We will consider the most general type of Dirac operators, associated to a Clifford module over a manifold, to avoid restricting ourselves to manifolds with spin structures. We will also work within Quillen's theory of superconnections, since this is conceptually simple, and is needed for the formulation of the local family index theorem of Bismut in Chapters 9 and 10.

We will now give a rapid account of some of the main results discussed in our book. Dirac operators on a compact Riemannian manifold M are closely related to the Clifford algebra bundle. The Clifford algebra $C_x(M)$ at the point $x \in M$ is the associative complex algebra generated by cotangent vectors $\alpha \in T_x^*M$ with relations

$$\alpha_1 \cdot \alpha_2 + \alpha_2 \cdot \alpha_1 = -2(\alpha_1, \alpha_2),$$

where (α_1, α_2) is the Riemannian metric on T_x^*M. If e_i is an orthonormal basis of T_xM with dual basis e^i, then this amounts to saying that $C_x(M)$ is generated by elements c^i subject to the relations

$$(c^i)^2 = -1, \text{ and } c^i c^j + c^j c^i = 0 \text{ for } i \neq j.$$

The Clifford algebra $C_x(M)$ is a deformation of the exterior algebra ΛT_x^*M, and there is a canonical bijection $\sigma_x : C_x(M) \to \Lambda T_x^*M$, the symbol map, defined by the formula

$$\sigma_x(c^{i_1} \dots c^{i_j}) = e^{i_1} \wedge \dots \wedge e^{i_j}.$$

The inverse of this map is denoted by $\mathbf{c}_x : \Lambda T_x^*M \to C_x(M)$.

Let $\mathscr{E}$ be a complex $\mathbb{Z}_2$-graded bundle on M, that is, $\mathscr{E} = \mathscr{E}^+ \oplus \mathscr{E}^-$. We say that $\mathscr{E}$ is a bundle of Clifford modules, or just a Clifford module, if there is a bundle map $c : T^*M \to \operatorname{End}(\mathscr{E})$ such that

1. $c(\alpha_1)c(\alpha_2) + c(\alpha_2)c(\alpha_1) = -2(\alpha_1, \alpha_2)$, and
2. $c(\alpha)$ swaps the bundles $\mathscr{E}^+$ and $\mathscr{E}^-$.

That is, $\mathscr{E}_x$ is a $\mathbb{Z}_2$-graded module for the algebra $C_x(M)$. If M is even-dimensional, the Clifford algebra $C_x(M)$ is simple, and we obtain the decomposition

$$\operatorname{End}(\mathscr{E}) \cong C(M) \otimes \operatorname{End}_{C(M)}(\mathscr{E}).$$

From now on, most of our considerations only apply to even-dimensional oriented manifolds. If M is a spin manifold, that is, a Riemannian manifold satisfying a certain topological condition, there is a Clifford module $\mathscr{S}$, known as the spinor bundle, such that $\operatorname{End}(\mathscr{S}) \cong C(M)$. On such a manifold, any Clifford module may be written as a twisted spinor bundle $\mathscr{W} \otimes \mathscr{S}$, with $\mathscr{W} = \operatorname{Hom}_{C(M)}(\mathscr{S}, \mathscr{E})$. Let $\Gamma_M \in \Gamma(M, \mathbb{C}(M))$ be the chirality operator in $C(M)$, given by the formula

$$\Gamma_M = i^{\dim(M)/2} c^1 \dots c^n,$$

so that $\Gamma_M^2 = 1$.

If $\mathscr{E}$ is a vector bundle on M, let $\Gamma(M,\mathscr{E})$ be the space of smooth sections of $\mathscr{E}$, and let $\mathscr{A}(M,\mathscr{E}) = \Gamma(M, \Lambda T^*M \otimes \mathscr{E})$ be the space of differential forms on M with values in $\mathscr{E}$. We make the obvious, but crucial, remark that if $\mathscr{E}$ is a Clifford module, then by the symbol map, the space of sections

$$\Gamma(M, \mathrm{End}(\mathscr{E})) \cong \Gamma(M, C(M) \otimes \mathrm{End}_{C(M)}(\mathscr{E}))$$

is isomorphic to the space of bundle-valued differential forms

$$\mathscr{A}(M, \mathrm{End}_{C(M)}(\mathscr{E})) \cong \Gamma(M, \Lambda T^*M \otimes \mathrm{End}_{C(M)}(\mathscr{E})).$$

Thus, a section $k \in \Gamma(M, \mathrm{End}(\mathscr{E}))$ corresponds to a differential form $\sigma(k)$ with values in $\mathrm{End}_{C(M)}(\mathscr{E})$. If M is a spin manifold and $\mathscr{E} = \mathscr{W} \otimes \mathscr{S}$ is a twisted spinor bundle, $\sigma(k)$ is a differential form with values in $\mathrm{End}(\mathscr{W})$.

A Clifford connection on a Clifford module $\mathscr{E}$ is a connection $\nabla^{\mathscr{E}}$ on $\mathscr{E}$ satisfying the formula

$$[\nabla^{\mathscr{E}}_X, c(\alpha)] = c(\nabla_X \alpha),$$

where α is a one-form on M, X is a vector field, and $\nabla_X \alpha$ is the Levi-Civita derivative of α.

The Dirac operator D associated to the Clifford connection $\nabla^{\mathscr{E}}$ is the composition of arrows

$$\Gamma(M,\mathscr{E}) \xrightarrow{\nabla^{\mathscr{E}}} \Gamma(M, T^*M \otimes \mathscr{E}) \xrightarrow{c} \Gamma(M,\mathscr{E}).$$

With respect to a local frame e^i of T^*M, D may be written

$$\mathsf{D} = \sum_i c^i \nabla^{\mathscr{E}}_{e_i}.$$

A number of classical first-order elliptic differential operators are Dirac operators associated to a Clifford connection. Let us list three examples:

1. The exterior bundle ΛT^*M is a Clifford bundle with Clifford action by the one-fom $\alpha \in \Gamma(M, T^*M)$ defined by the formula
$$c(\alpha) = \varepsilon(\alpha) - \varepsilon(\alpha)^*;$$
here $\varepsilon(\alpha) : \Gamma(M, \Lambda^\bullet T^*M) \to \Gamma(M, \Lambda^{\bullet+1} T^*M)$ is the operation of exterior multiplication by α. The Levi-Civita connection on ΛT^*M is a Clifford connection. The associated Dirac operator is $d + d^*$, where d is the exterior differential operator. The kernel of this operator is just the space of harmonic forms on M, which by Hodge's Theorem is isomorphic to the de Rham cohomology $H^\bullet(M)$ of M.
2. If M is a complex manifold and $\mathscr{W}$ is a Hermitian bundle over M, the bundle $\Lambda(T^{0,1}M)^* \otimes \mathscr{W}$ is a Clifford module, with $\alpha = \alpha_{1,0} + \alpha_{0,1} \in \mathscr{A}^{1,0}(M) \oplus \mathscr{A}^{0,1}(M)$ acting by

$$c(\alpha) = \sqrt{2}(\varepsilon(\alpha_{0,1}) - \varepsilon(\overline{\alpha_{1,0}})^*).$$

The Levi-Civita connection on ΛT^*M preserves $\Lambda(T^{0,1}M)^*$ and defines a Clifford connection if M is Kähler, and the Dirac operator associated to this connection is $\sqrt{2}(\bar\partial + \bar\partial^*)$. If $\mathscr{W}$ is a holomorphic vector bundle with its canonical connection, the tensor product of this connection with the Levi-Civita connection on $\Lambda(T^{0,1}M)^*$ is a Clifford connection with associated Dirac operator $\sqrt{2}(\bar\partial + \bar\partial^*)$. The kernel of this operator is the space of harmonic forms on M lying in $\mathscr{A}^{0,\bullet}(M,\mathscr{W})$, which by Dolbeault's theorem is isomorphic to the sheaf cohomology $H^\bullet(M,\mathscr{W})$.

3. If M is a spin manifold, its spinor bundle $\mathscr{S}$ is a Clifford module, the Levi-Civita connection is a Clifford connection, and the associated Dirac operator is known simply as **the** Dirac operator. Its kernel is the space of harmonic spinors.

Thus, we see from these examples that the kernel of a Dirac operator often has a topological, or at least geometric, significance.

The heat kernel $\langle x \mid e^{-t\mathsf{D}^2} \mid y\rangle \in \mathrm{Hom}(\mathscr{E}_y, \mathscr{E}_x)$ of the square of the Dirac operator D is the kernel of the heat semigroup $e^{-t\mathsf{D}^2}$, that is,

$$(e^{-t\mathsf{D}^2}s)(x) = \int_M \langle x \mid e^{-t\mathsf{D}^2} \mid y\rangle s(y)\,|dy| \quad \text{for all } s \in \Gamma(M,\mathscr{E}),$$

where $|dy|$ is the Riemannian measure on M. The following properties of the heat kernel are proved in Chapter 2:

1. it is smooth;
2. the fact that smooth kernels are trace-class, from which we see that the kernel of D is finite-dimensional;
3. the uniform convergence of $\langle x \mid e^{-t\mathsf{D}^2} \mid y\rangle$ to the kernel of the projection onto $\ker(\mathsf{D})$ as $t \to \infty$;
4. the existence of an asymptotic expansion for $\langle x \mid e^{-t\mathsf{D}^2} \mid y\rangle$ at small t, (where $\dim(M) = n = 2\ell$),

$$\langle x \mid e^{-t\Delta^2} \mid y\rangle \sim (4\pi t)^{-\ell} e^{-d(x,y)^2/4t} \sum_{i=0}^{\infty} t^i f_i(x,y),$$

where f_i is a sequence of smooth kernels for the bundle $\mathscr{E}$ given by local functions of the curvature of $\nabla^{\mathscr{E}}$ and the Riemannian curvature of M, and $d(x,y)$ is the geodesic distance between x and y.

Note that the restriction to the diagonal $x \mapsto \langle x \mid e^{-t\mathsf{D}^2} \mid x\rangle$ is a section of $\mathrm{End}(\mathscr{E})$. The central object of our study will be the behaviour at small time of the differential form

$$\sigma(\langle x \mid e^{-t\mathsf{D}^2} \mid x\rangle) \in \mathscr{A}(M, \mathrm{End}_{C(M)}(\mathscr{E})),$$

obtained by taking the image under the symbol map σ of $\langle x \mid e^{-t\mathsf{D}^2} \mid x\rangle$.

Let us describe the differential forms which enter in the study of the asymptotic expansion of $\langle x \mid e^{-t\mathsf{D}^2} \mid x\rangle$. If $\mathfrak{g}$ is a unimodular Lie algebra, let $j_{\mathfrak{g}}(X)$ be the analytic function on $\mathfrak{g}$ defined by the formula

$$j_{\mathfrak{g}}(X) = \det\left(\frac{\sinh(\operatorname{ad}X/2)}{\operatorname{ad}X/2}\right);$$

it is the Jacobian of the exponential map $\exp : \mathfrak{g} \to G$. We define $j_{\mathfrak{g}}^{1/2}$ in a neighbourhood of $0 \in \mathfrak{g}$ to be the square-root of $j_{\mathfrak{g}}$ such that $j_{\mathfrak{g}}^{1/2}(0) = 1$.

Let $R \in \mathcal{A}^2(M, \mathfrak{so}(TM))$ be the Riemannian curvature of M. Choose a local orthonormal frame e_i of TM, and consider the matrix R with two-form coefficients,

$$\mathsf{R}_{ij} = (Re_i, e_j) \in \mathcal{A}^2(M).$$

Then $\left(\frac{\sinh(\mathsf{R}/2)}{\mathsf{R}/2}\right)$ is a matrix with even degree differential form coefficients, and since the determinant is invariant under conjugation by invertible matrices,

$$j(\mathsf{R}) = \det\left(\frac{\sinh(\mathsf{R}/2)}{\mathsf{R}/2}\right)$$

is an element of $\mathcal{A}(M)$ independent of the frame of TM used in its definition. Note that the zero-form component of $j(\mathsf{R})$ equals 1. Thus, we can define the $\hat{A}$-genus $\hat{A}(M)$ of the manifold M by

$$\hat{A}(M) = j(\mathsf{R})^{-1/2} = \det{}^{1/2}\left(\frac{R/2}{\sinh(R/2)}\right) \in \mathcal{A}(M);$$

it is a closed differential form whose cohomology class is independent of the metric on M. It is a fascinating puzzle that the function $j_{\mathfrak{g}}^{-1/2}$ occurs in a basic formula of representation theory for Lie groups, the Kirillov character formula, while its cousin, the $\hat{A}$-genus, plays a similar role in a basic formula of differential geometry, the index theorem for Dirac operators. Understanding the relationship between these two objects is one of the aims of this book.

Note that our normalization of the $\hat{A}$-genus, and of other characteristic classes, is not the same as that preferred by topologists, who multiply the $2k$-degree component by $(-2\pi i)^k$, so that it lies in $H^{2k}(M, \mathbb{Q})$. We prefer to leave out these powers of $-2\pi i$, since it is in this form that they will arise in the proof of the local index theorem.

Let $\mathcal{E}$ be a Clifford module on M, with Clifford connection $\mathcal{E}$ and curvature $F^{\mathcal{E}}$. The twisting curvature $F^{\mathcal{E}/S}$ of $\mathcal{E}$ is defined by the formula

$$F^{\mathcal{E}/S} = F^{\mathcal{E}} - R^{\mathcal{E}} \in \mathcal{A}^2(M, \operatorname{End}_{C(M)}(\mathcal{E})),$$

where

$$R^{\mathcal{E}}(e_i, e_j) = \tfrac{1}{2} \sum_{k<l} (R(e_i, e_j)e_k, e_l) c^k c^l.$$

If M is a spin manifold with spinor bundle $\mathcal{S}$ and $\mathcal{E} = \mathcal{W} \otimes \mathcal{S}$, $F^{\mathcal{E}/S}$ is the curvature of the bundle $\mathcal{W}$.

For $a \in \Gamma(M, C(M)) \cong \Gamma(M, \Lambda T^*M)$, we denote the k-form component of $\sigma(a)$ by $\sigma_k(a)$.

The first four chapters lead up to the proof of the following theorem, which calculates the leading order term, in a certain sense, of the heat kernel of a Dirac operator restricted to the diagonal.

Theorem A. *Consider the asymptotic expansion of* $\langle x \mid e^{-t\mathsf{D}^2} \mid x\rangle$ *at small times* t,

$$\langle x \mid e^{-t\mathsf{D}^2} \mid x\rangle \sim (4\pi t)^{-\ell} \sum_{i=0}^{\infty} t^i k_i(x)$$

with coefficients $k_i \in \Gamma(M, C(M) \otimes \mathrm{End}_{C(M)}(\mathscr{E}))$. *Then*

1. $\sigma_j(k_i) = 0$ *for* $j > 2i$
2. *Let* $\sigma(k) = \sum_{i=0}^{\ell} \sigma_{2i}(k_i) \in \mathscr{A}(M, \mathrm{End}_{C(M)}(\mathscr{E}))$. *Then*

$$\sigma(k) = \det{}^{1/2}\Big(\frac{R/2}{\sinh(R/2)}\Big) \exp(-F^{\mathscr{E}/S}).$$

In Chapter 4, we give a proof of Theorem A which relies on an approximation of D^2 by a harmonic oscillator, which is easily derived from Lichnerowicz's formula for the square of the Dirac operator, and properties of the normal coordinate system.

Since the zero-form piece of the $\hat{A}$-genus equals 1, we recover Weyl's formula

$$\lim_{t\to 0} (4\pi t)^{\ell} \langle x \mid e^{-t\mathsf{D}^2} \mid x\rangle = \mathrm{rk}(\mathscr{E});$$

in this sense, Theorem A is a refinement of Weyl's formula for the square of a Dirac operator.

Define the index of D to be the integer

$$\mathrm{ind}(\mathsf{D}) = \dim(\ker(\mathsf{D}^+)) - \dim(\ker(\mathsf{D}^-)),$$

where $\mathsf{D}^{\pm}$ is the restriction of D to $\Gamma(M, \mathscr{E}^{\pm})$. For example, $\mathrm{ind}(d + d^*)$ is the Euler characteristic

$$\mathrm{Eul}(M) = \sum_{i=0}^{n} (-1)^i \dim(H^i(M))$$

of the manifold M, while $\mathrm{ind}(\bar\partial + \bar\partial^*)$ is the Euler characteristic of the sheaf of holomorphic sections

$$\mathrm{Eul}(M, \mathscr{W}) = \sum_{i=0}^{\ell} (-1)^i \dim(H^i(M, \mathscr{W})).$$

These indexes are particular cases of the Atiyah-Singer Index Theorem, and are given by well-known formulas, respectively the Gauss-Bonnet-Chern theorem and the Riemann-Roch-Hirzebruch theorem; we will show in Section 4.1 how these formulas follow from Theorem A.

In Chapter 3, we establish some well-known properties of the index, such as its homotopy invariance, using the formula of McKean and Singer. If E is a $\mathbb{Z}_2$-graded vector space and A is an endomorphism on E, we define the supertrace $\mathrm{Str}(A)$ of A to be the trace of the operator ΓA, where $\Gamma \in \mathrm{End}(E)$ is the chirality operator, which equals ± 1 on $E^{\pm}$. The McKean-Singer formula says that for each $t > 0$, the index of D equals

$$\operatorname{ind}(\mathsf{D}) = \operatorname{Str}(e^{-t\mathsf{D}^2}) = \int_M \operatorname{Str}\langle x \mid e^{-t\mathsf{D}^2} \mid x\rangle \, |dx|.$$

This formula shows that the index of D is given in terms of the restriction to the diagonal of the heat kernel of D^2 at arbitrarily small times, which we know by the asymptotic expansion of the heat kernel to be given by local formulas in the curvature of the connection $\nabla^{\mathscr{E}}$ and the Riemannian curvature of M. Using the McKean-Singer formula, we see that

$$\operatorname{ind}(\mathsf{D}) = (4\pi)^{-n/2} \int_M \operatorname{Str}(k_{n/2}(x))\, dx.$$

From Theorem A, we see that

$$\sigma(k_{n/2}(x))_{[n]} \in \mathscr{A}^n(M, \operatorname{End}_{C(M)}(\mathscr{E})) \cong \Gamma(M, \operatorname{End}_{C(M)}(\mathscr{E}))$$

equals the n-form component of

$$\det{}^{1/2}\Big(\frac{R/2}{\sinh(R/2)}\Big)\exp(-F^{\mathscr{E}/S}) \in \mathscr{A}^n(M, \operatorname{End}_{C(M)}(\mathscr{E})).$$

The differential form

$$\operatorname{ch}(\mathscr{E}/S) = 2^{-n/2}\operatorname{Str}(\Gamma_M \cdot \exp(-F^{\mathscr{E}/S}))$$

is a generalization of the Chern character for Clifford modules, called the relative Chern character; when M is a spin manifold and $\mathscr{E} = \mathscr{W} \otimes \mathscr{S}$, the form $\operatorname{ch}(\mathscr{E}/S)$ equals the Chern character $\operatorname{ch}(\mathscr{W})$ of the twisting bundle $\mathscr{W}$. In this way, we obtain the formula

$$\operatorname{ind}(\mathsf{D}) = (2\pi i)^{-n/2} \int_M \hat{A}(M)\operatorname{ch}(\mathscr{E}/S).$$

This is the Atiyah-Singer index theorem for the Dirac operator D. Thus, we see that our main theorem is at the same time a generalization of Weyl's formula for the leading term of the heat kernel, and of the Atiyah-Singer index theorem. As conjectured by McKean-Singer, we establish that

$$\lim_{t\to 0} \operatorname{Str}\langle x \mid e^{-t\mathsf{D}^2} \mid x\rangle$$

exists and is given by the above integrand. This result, due to Patodi [90] and Gilkey [65], is known as the local index theorem. It is very important, since it generalizes to certain situations inaccessible to the global index theorem of Atiyah and Singer, such as when the manifold M is no longer closed. One important example is the signature theorem for manifolds with boundary of Atiyah-Patodi-Singer [11]. As another example, it is possible in the case of infinite covering spaces to prove a local index theorem which gives the "index per unit volume" of the Dirac operator in question (Connes and Moscovici [52]).

Note that the proofs of Patodi and Gilkey of the local index theorem required the calculation of the index of certain Dirac operators on homogeneous spaces. In the proof which we present, we are able to calculate the index density by analysis of Lichnerowicz's formula for the square of the Dirac operator, which was proved in [81].

In Chapter 5, we give another proof of Theorem A. This proof, while similar in some respects to the first, uses an integral expression for the heat kernel of D^2 in terms of the scalar heat kernel of the orthonormal frame bundle $\mathrm{SO}(M)$; the Jacobian of the exponential map on $\mathrm{SO}(M)$ leads to the appearance of the $\hat{A}$-genus. Once more, Lichnerowicz's formula is fundamental to the proof.

The second topic of this book is an equivariant version of the results of Chapter 4. Suppose that a compact Lie group G acts on the manifold M and on the bundle $\mathscr{E}$, and that the operator D is invariant under this action (for this to be possible, G must act on M by isometries). In this situation, the action of the group G leaves the kernel of D invariant; in other words, $\ker(\mathsf{D})$ is a representation of G. Thus, the index of D may be generalized to a character-valued index, defined by the formula

$$\operatorname{ind}_G(\gamma, \mathsf{D}) = \operatorname{Tr}(\gamma, \ker(\mathsf{D}^+)) - \operatorname{Tr}(\gamma, \ker(\mathsf{D}^-)).$$

The analogue of the McKean-Singer formula holds here:

$$\operatorname{ind}_G(\gamma, \mathsf{D}) = \int_M \operatorname{Str}\langle x \mid \gamma_{\mathscr{E}} \cdot e^{-t\mathsf{D}^2} \mid x\rangle \, |dx|,$$

where $\gamma_{\mathscr{E}}$ is the action of $\gamma \in G$ on the space of sections of the vector bundle $\mathscr{E}$. In Chapter 6, we prove a formula, due to Gilkey [66], for the distribution

$$\lim_{t \to 0} \operatorname{Str}\langle x \mid \gamma_{\mathscr{E}} \cdot e^{-t\mathsf{D}^2} \mid x\rangle.$$

It is a product of the delta-function along the fixed point set M^γ of the action of γ on M and a function involving the curvature of M^γ, the curvature of $\mathscr{E}$, and the curvature of the normal bundle $\mathscr{N}$ of M^γ. Integrating over M, we obtain an expression for $\operatorname{ind}_G(\gamma, \mathsf{D})$ as an integral over M^γ, which is a theorem due to Atiyah, Segal and Singer [12], [15]. The proof that we give is modelled on that of Chapter 5, and is hardly more difficult.

The Weyl character formula for representations of compact Lie groups is a special case of the equivariant index theorem for Dirac operators. By the Borel-Weil-Bott theorem, a finite-dimensional representation of G may be realized as the kernel of a Dirac operator on the flag variety G/T of G, where T is a maximal torus of G. If g is a regular element of G, then the fixed-point set $(G/T)^g$ is in one-to-one correspondence with the Weyl group, and the fixed point formula for the index may be identified with the Weyl character formula, as is explained by Atiyah-Bott [6].

Generalizing ideas of Bott [43] and Baum-Cheeger [18], we will show how the fixed point formula for the equivariant index $\operatorname{ind}_G(e^X, D)$ may be rewritten as an integral over the whole manifold. To do this, we introduce the notion of equivariant differential forms in Chapter 7; this is of independent interest. Let M be a manifold with an action of a compact Lie group G; if X is in the Lie algebra $\mathfrak{g}$ of G, let X_M be the corresponding vector field on M. Let $\mathbb{C}[\mathfrak{g}]$ denote the algebra of complex valued polynomial functions on $\mathfrak{g}$. Let $\mathcal{A}_G(M)$ be the algebra of G-invariant elements of $\mathbb{C}[\mathfrak{g}] \otimes \mathcal{A}(M)$, graded by

$$\deg(P \otimes \alpha) = 2\deg(P) + \deg(\alpha)$$

for $P \in \mathbb{C}[\mathfrak{g}]$ and $\alpha \in \mathcal{A}(M)$. We define the equivariant exterior differential $d_{\mathfrak{g}}$ on $\mathbb{C}[\mathfrak{g}] \otimes \mathcal{A}(M)$ by

$$(d_{\mathfrak{g}}\alpha)(X) = d(\alpha(X)) - \iota(X)(\alpha(X)),$$

where $\iota(X)$ denotes contraction by the vector field X_M. Thus, $d_{\mathfrak{g}}$ increases by one the total degree on $\mathbb{C}[\mathfrak{g}] \otimes \mathcal{A}(M)$, and preserves $\mathcal{A}_G(M)$. Cartan's homotopy formula implies that $d_{\mathfrak{g}}^2$ vanishes on elements of $\mathcal{A}_G(M)$; thus $(\mathcal{A}_G(M), d_{\mathfrak{g}})$ is a complex, called the complex of equivariant differential forms; of course, if $G = \{1\}$, this is the de Rham complex. Its cohomology $H_G^\bullet(M)$ is called the equivariant cohomology of the manifold with coefficients in $\mathbb{R}$; it was proved by H. Cartan [47] that this is the real cohomology of the homotopy quotient $M \times_G EG$. The Chern-Weil construction of differential forms associated to a vector bundle with connection may be generalized to this equivariant setting.

Integration over M defines a map

$$\int_M : \mathcal{A}_G(M) \to \mathbb{C}[\mathfrak{g}]^G$$

by the formula $(\int_M \alpha)(X) = \int_M \alpha(X)$; this map vanishes on equivariantly exact forms $d_{\mathfrak{g}}\alpha$, and hence defines a homomorphism from $H_G^\bullet(M)$ to $\mathbb{C}[\mathfrak{g}]^G$. An important property of equivariant differential forms is the localization formula, a generalization of a formula of Bott, which expresses the integral of an equivariantly closed differential form evaluated at $X \in \mathfrak{g}$ as an integral over the zero set of the vector field X_M. This formula has many applications, especially when the vector field X_M has discrete zeroes.

Using the localization theorem, we can rewrite the formula for the equivariant index as an integral over the entire manifold, at least in a neighbourhood of the identity in G.

Theorem B. *The equivariant index is given by the formula, for $X \in \mathfrak{g}$ sufficiently small,*

$$\operatorname{ind}_G(e^{-X}, D) = (2\pi i)^{-n/2} \int_M \hat{A}_{\mathfrak{g}}(X, M) \operatorname{ch}(X, \mathcal{E}/S),$$

where $\operatorname{ch}_{\mathfrak{g}}(X, \mathcal{E}/S) \in \mathcal{A}_G(M)$ is a closed equivariant differential form which specializes to the relative Chern character at $X = 0$, and $\hat{A}_{\mathfrak{g}}(X, M) \in \mathcal{A}_G(M)$ similarly specializes at $X = 0$ to the $\hat{A}$-genus.

Thus, the integrand is a $d_{\mathfrak{g}}$-closed equivariant differential form depending analytically on X for X near $0 \in \mathfrak{g}$, and which coincides with $\hat{A}(M)\,\mathrm{ch}(\mathscr{E}/S)$ at $X=0$.

Let G be a compact Lie group, and let T be a finite dimensional irreducible representation of G. As we will see in Chapter 8, Kirillov's formula for the characters of the representation T of G is a special case of the above equivariant index theorem: this formula says that there is an orbit M of the coadjoint representation of G on $\mathfrak{g}^*$ such that

$$\mathrm{Tr}(T(e^X)) = j_{\mathfrak{g}}^{-1/2}(X) \int_M e^{i\langle f,X\rangle}\, d\beta(f),$$

where $d\beta(f)$ is the Liouville measure on M.

This formula, which unlike the Weyl character formula has a generalization to any Lie group, inspired our expression for the equivariant index in terms of equivariant differential forms. Note the similarity with the equivariant index theorem: the equivariant $\hat{A}$-genus corresponds to the factor $j_{\mathfrak{g}}^{-1/2}(X)$, while the equivariant Chern character corresponds to the measure $e^{i\langle f,X\rangle}\, d\beta(f)$.

The final major topic of the book is the local index theorem for families of Dirac operators. This theorem, which is due to Bismut, is a relative version of the local index theorem, in which we consider a fibre bundle $\pi : M \to B$ and a smooth family of Dirac operators D^z, one for each fibre $M_z = \pi^{-1}(z)$, $(z \in B)$. The kernels of the Dirac operators D^z now fit together to form a vector bundle, if they are of constant dimension. The index of a single operator D, $\mathrm{ind}(\mathsf{D}) = \dim(\ker(\mathsf{D}^+)) - \dim(\ker(\mathsf{D}^-))$, is now replaced by the notion of the index bundle, which is the difference bundle

$$\mathrm{ind}(\mathsf{D}) = [\ker(\mathsf{D}^+)] - [\ker(\mathsf{D}^-)];$$

following Atiyah-Singer, we define in Chapter 9 a smooth index bundle even if the dimension of the kernel $\ker(\mathsf{D}^z)$ is not constant, but it is no longer canonical. The index theorem of Atiyah-Singer [16] for a family gives a formula for the Chern character of this index bundle.

It turns out that the formulation of a local version of this family index theorem requires the introduction of generalization of connections due to Quillen, known as superconnections. Let $\mathscr{H} = \mathscr{H}^+ \oplus \mathscr{H}^-$ be a $\mathbb{Z}_2$-graded vector bundle. A superconnection on $\mathscr{H}$ is an operator acting on $\mathscr{A}(M,\mathscr{H})$ of the form

$$\mathbb{A} = \mathbb{A}_{[0]} + \mathbb{A}_{[1]} + \mathbb{A}_{[2]} + \dots,$$

where $\mathbb{A}_{[1]}$ is a connection preserving the $\mathbb{Z}_2$-grading on $\mathscr{H}$, and $\mathbb{A}_{[i]}$ for $i \neq 1$ are i-form valued sections of $\mathrm{End}(\mathscr{H})$ which exchange $\mathscr{H}^+$ and $\mathscr{H}^-$ for i even and preserves them for i odd; usually, $\mathbb{A}_{[0]} \in \Gamma(M,\mathrm{End}(\mathscr{H}))$ is supposed to be self-adjoint as well. The curvature of a superconnection $\mathbb{A}^2$ is its square, which is an element of $\mathscr{A}(M,\mathrm{End}(\mathscr{H}))$ of the form $\mathbb{A}^2 = \mathbb{A}_{[0]}^2 + \dots$. Quillen showed that as a de Rham cohomology class, the Chern character of the difference bundle $[\mathscr{H}^+] - [\mathscr{H}^-]$ is equal to the Chern character of $\mathbb{A}$, defined by the supertrace

$$\mathrm{ch}(\mathbb{A}) = \mathrm{Str}\big(e^{-\mathbb{A}^2}\big) = \mathrm{Str}\big(e^{-\mathbb{A}_{[0]}^2+\dots}\big),$$

which is a closed even degree differential form whose cohomology class is independent of $\mathbb{A}$. Superconnections play a role in two situations: to define the Chern character on non-compact manifolds, where $\mathbb{A}_{[0]}$ grows at infinity, thus allowing $\mathrm{ch}(\mathbb{A})$ to represent an integrable cohomology class, and for infinite dimensional vector bundles, where $\mathbb{A}_{[0]}$ is such that $e^{-\mathbb{A}_{[0]}^2}$ is trace-class. We explain the extension of the Chern-Weil construction of characteristic forms to the case of superconnections in Chapter 1.

Let D be an odd endomorphism of a finite-dimensional superbundle $\mathscr{H}$ with components $\mathsf{D}^{\pm} : \mathscr{H}^{\pm} \to \mathscr{H}^{\mp}$, such that $\ker(\mathsf{D})$ has constant rank, and let $\mathbb{A}$ be a superconnection on $\mathscr{H}$ with zero-degree term $\mathbb{A}_{[0]}$ equal to D. At the level of cohomology,

$$\mathrm{ch}(\mathbb{A}) = \mathrm{ch}(\ker(\mathsf{D}^+)) - \mathrm{ch}(\ker(\mathsf{D}^-)).$$

In Chapter 9, we prove a refinement of this equation which holds at the level of differential forms.

If $t > 0$, we may define the rescaled superconnection on $\mathscr{H}$, by the formula

$$\mathbb{A}_t = t^{1/2}\mathsf{D} + \mathbb{A}_{[1]} + t^{-1/2}\mathbb{A}_{[2]} + \dots,$$

with Chern character

$$\mathrm{ch}(\mathbb{A}_t) = \mathrm{Str}(e^{-\mathbb{A}_t^2}).$$

We define a connection on the bundle $\ker(\mathsf{D})$ by the formula

$$\nabla_0 = P_0 \cdot \mathbb{A}_{[1]} \cdot P_0,$$

where P_0 is the projection from $\mathscr{H}$ onto its sub-bundle $\ker(\mathsf{D})$; thus,

$$\mathrm{ch}(\nabla_0) = \mathrm{ch}(\ker(\mathsf{D}^+)) - \mathrm{ch}(\ker(\mathsf{D}^-)).$$

Intuitively speaking, as t tends to infinity, the supertrace

$$\mathrm{ch}(\mathbb{A}_t) = \mathrm{Str}(e^{-\mathbb{A}_t^2}) = \mathrm{Str}(e^{-t\mathsf{D}^2 + O(t^{1/2})})$$

is pushed onto the sub-bundle $\ker(\mathsf{D})$. In fact, we can prove the following formula,

$$\lim_{t\to\infty} \mathrm{ch}(\mathbb{A}_t) = \mathrm{ch}(\nabla_0).$$

Let $M \to B$ be a fibre bundle with compact Riemannian fibres of dimension n, and let $\mathscr{E}$ be a bundle of modules for the vertical Clifford algebras $C_x(M/B)$ associated to the vertical tangent spaces of $M \to B$. Bismut showed that the above picture can be extended to an infinite-dimensional setting when D is a smooth family of Dirac operators on the fibres $\mathscr{E}_z \to M_z$, where $\mathscr{E}_z$ and M_z are the fibres over $z \in B$ of $\mathscr{E}$ and M. He introduced the infinite-dimensional bundle $\pi_*\mathscr{E}$ whose fibre at $z \in B$ is the space of sections of $\mathscr{E}$ over the fibre M_z,

$$(\pi_*\mathscr{E})_z = \Gamma(M_z, \mathscr{E}_z).$$

Given a connection on the fibre bundle $M \to B$, that is, a choice of horizontal tangent bundle in TM transverse to the vertical tangent bundle, we may define a unitary connection on $\pi_*\mathscr{E}$, whose formula involves the connection on $\mathscr{E}$ and the mean curvature of the fibres M_z. If we form the superconnection $\mathbb{A}_t = t^{1/2}\mathsf{D} + \nabla^{\pi_*\mathscr{E}}$, then the heat kernel of $\mathbb{A}_t^2$ has a supertrace in $\mathscr{A}(B)$, since

$$\mathbb{A}_t^2 = t\mathsf{D}^2 + t^{1/2}\nabla^{\pi_*\mathscr{E}}\mathsf{D} + (\nabla^{\pi_*\mathscr{E}})^2$$

is a perturbation of the operator $t\mathsf{D}^2$. In this way, we may define $\mathrm{ch}(\mathbb{A}_t) \in \mathscr{A}(B)$.

Assume that $\ker(\mathsf{D})$ has constant rank. The projection $P_0 \cdot \nabla^{\pi_*\mathscr{E}} \cdot P_0$ of the connection $\nabla^{\pi_*\mathscr{E}}$ defines a connection ∇_0 on the finite dimensional bundle $\ker(\mathsf{D})$. We obtain the following transgression formula:

$$\mathrm{ch}(\mathbb{A}_t) - \mathrm{ch}(\nabla_0) = d\left(\int_t^\infty \mathrm{Str}\left(\frac{d\mathbb{A}_s}{ds}e^{-\mathbb{A}_s^2}\right)ds\right).$$

In general, it is impossible to take the limit $t \to 0$ in this formula. However, Bismut showed that this limit exists if a certain term $\mathbb{A}_{[2]}$ is added to $\mathbb{A}$, proportional to Clifford multiplication by the curvature of the fibre bundle $M \to B$; the curvature of this superconnection satisfies a formula very similar to the Lichnerowicz formula for the square of a Dirac operator. By methods almost identical to those of Chapter 4, we prove the following theorem in Chapter 10.

Theorem C. *Let $R^{M/B} \in \mathscr{A}^2(M, \mathfrak{so}(T(M/B)))$ be the curvature of the vertical tangent bundle of M/B, with $\hat{A}$-genus $\hat{A}(M/B)$. Then, if $\mathbb{A}$ is the Bismut superconnection,*

$$\lim_{t\to 0}\mathrm{ch}(\mathbb{A}_t) = (2\pi i)^{-n/2}\int_{M/B}\hat{A}(M/B)\,\mathrm{ch}(\mathscr{E}/S) \in \mathscr{A}(B).$$

This result has turned out to be of interest to mathematicians working in such varied areas as string theory, Arakelov theory, and the theory of moduli spaces of Yang-Mills fields. Interestingly, when the fibre bundle $M \to B$ has a compact structure group G and all of the data are compatible with G, the above theorem is equivalent to the Kirillov formula for the equivariant index.

The book is not necessarily meant to be read sequentially, and consists of four groups of chapters:

1. Chapters 1, which gives various preliminary results in differential geometry, and Chapter 7, on equivariant differential forms, do not depend on any other chapters.
2. Chapters 2, 3 and 4 introduce the main ideas of the book, and take the reader through the main properties of Dirac operators, culminating in the local index theorem.
3. Chapters 5, 6 and 8 are on the equivariant index theorem, and may be read after the first four chapters, although Chapter 7 is needed in Chapter 8.
4. Chapters 9 and 10 are on the family index theorem, and can be read after the first four chapters, except for Sections 9.4 and 10.7, which have Chapter 8 as a prerequisite.

1

Background on Differential Geometry

In this chapter, we will survey some of the notions of differential geometry that are used in this book: principal bundles and their associated vector bundles, connections and superconnections, and characteristic classes. For further details about these topics, we refer the reader to the bibliographic notes at the end of the chapter.

1.1 Fibre Bundles and Connections

Definition 1.1. *Let $\pi : \mathcal{E} \to M$ be a smooth map from a manifold $\mathcal{E}$ to a manifold M. We say that $(\mathcal{E}, \pi)$ is a* **fibre bundle** *with typical fibre E over M if there is a covering of M by open sets U_i and diffeomorphisms $\phi_i : \pi^{-1}(U_i) \to U_i \times E$, such that $\pi : \pi^{-1}(U_i) \to U_i$ is the composition of ϕ_i with projection onto the first factor U_i in $U_i \times E$. The space $\mathcal{E}$ is called the total space of the fibre bundle, and M is called the base.*

It follows from this definition that $\pi^{-1}(x)$ is diffeomorphic to E for all $x \in M$; we will call E "the" fibre of $\mathcal{E}$. In this book, we will often make use of the convenient convention that the bundles denoted $\mathcal{E}$, $\mathcal{V}$, $\mathcal{W}$, and so on, have typical fibres E, V, W, etc.

Consider the diffeomorphism $\phi_j \circ \phi_i^{-1}$ of $(U_i \cap U_j) \times E$ which we obtain from the definition of a fibre bundle. It is a map from $U_i \cap U_j$ to the group of diffeomorphisms $\mathrm{Diff}(E)$ of the fibre E.

Definition 1.2. *A fibre bundle $\pi : \mathcal{E} \to M$ is a* **vector bundle** *if its typical fibre is a vector space E, and if the diffeomorphisms ϕ_i may be chosen in such a way that the diffeomorphisms $\phi_j \circ \phi_i^{-1} : \{x\} \times E \to \{x\} \times E$ are invertible linear maps of E for all $x \in U_i \cap U_j$.*

A **section** s of a vector bundle $\mathcal{E}$ over M is a map $s : M \to \mathcal{E}$ such that $\pi s(x) = x$ for all $x \in M$. In this book, we will use the word smooth to mean infinitely differentiable. The space of all smooth sections is denoted by $\Gamma(M, \mathcal{E})$, while the space of all compactly supported smooth sections (those equal to 0 outside a compact subset of M) is denoted by $\Gamma_c(M,\mathscr{E})$.

If $\phi : M_1 \to M_2$ is a smooth map of manifolds and $\mathscr{E}$ is a vector bundle on M_2, we denote by $\phi^*\mathscr{E}$ the vector bundle over M_1 obtained by pulling back $\mathscr{E}$: it is the smooth vector bundle given by restricting the vector bundle $M_1 \times \mathscr{E}$ over $M_1 \times M_2$ to the graph $\Gamma(\phi) = \{(x_1,\phi(x_1)) \in M_1 \times M_2\}$ of ϕ, so that

$$\phi^*\mathscr{E} = \{(x,v) \in M_1 \times \mathscr{E} \mid v \in \mathscr{E}_{\phi(x)}\}.$$

There is induced a pull-back map $\phi^* : \Gamma(M_2,\mathscr{E}) \to \Gamma(M_1,\phi^*\mathscr{E})$.

Definition 1.3. *A* **principal bundle** *P with structure group G is a fibre bundle P with a right action of a Lie group G on the fibres, that is, $(p\cdot g)\cdot h = p\cdot(gh)$ for $p \in P$, $g,h \in G$ such that*

$$\pi(pg) = \pi(p) \quad \text{for all } p \in P \text{ and } g \in G,$$

and such that the action of G is free and transitive on the fibres. The fibres of the bundle P are diffeomorphic to G itself, and so the base M may be identified with the quotient manifold P/G.

A **vertical vector** on a fibre bundle $\mathscr{E}$ with base M is a tangent vector on $\mathscr{E}$ which is tangential to the fibres: that is, $X(\pi^*f) = 0$ for any $f \in C^\infty(M)$, where $\pi^*f = f \circ \pi$. We denote the bundle of vertical tangent vectors by $V\mathscr{E}$; it is a subbundle of the tangent bundle $T\mathscr{E}$. The space of vertical tangent vectors to a point p in a principal bundle can be canonically identified with $\mathfrak{g}$, the Lie algebra of G, by using the derivative of the right action of G on P. If $X \in \mathfrak{g}$, we denote by X_P the vector field such that if $f \in C^\infty(P)$,

$$(X_P f)(p) = \frac{d}{d\varepsilon}\Big|_{\varepsilon=0} f(p\exp\varepsilon X).$$

Thus, we obtain a map from $P \times \mathfrak{g}$ to TP, whose image at p is just the vertical subspace V_pP. The vector field X_P will often be simply denoted by X.

If P is a principal bundle with structure group G and E is a left G-space, with action of G on it given by the homomorphism

$$\rho : G \to \mathrm{Diff}(E),$$

we can form a fibre bundle over M with E as fibre, called the **associated bundle**, by forming the fibred product $P \times_G E$, defined by taking the product $P \times E$ and dividing it by the equivalence relation

$$(p\cdot g, f) \sim (p, \rho(g)f),$$

for all $p \in P$, $g \in G$ and $f \in E$. In particular, if we take for E a vector space which carries a linear representation of G, we obtain a vector bundle on M.

Every vector bundle with N-dimensional fibres on M is an associated bundle for a principal bundle on M with structure group $\mathrm{GL}(N)$, called the bundle of frames. In this book, we will use vector spaces over both $\mathbb{R}$ and $\mathbb{C}$, and when it is clear from the context, we will not explicitly mention which field we take homomorphisms or tensor products over. Thus, when we write here $\mathrm{GL}(N)$, we mean either $\mathrm{GL}(N,\mathbb{R})$ or $\mathrm{GL}(N,\mathbb{C})$, depending on the context.

Proposition 1.4. *If $\mathscr{E}$ is a vector bundle over M with fibre $\mathbb{R}^N$, let $\mathrm{GL}(\mathscr{E})$ be the bundle whose fibre over the point $x \in M$ is the space of all invertible linear maps from $\mathbb{R}^N$ to the fibre $\mathscr{E}_x$. Then $\mathrm{GL}(\mathscr{E})$ is a principal bundle for the group $\mathrm{GL}(N,\mathbb{R})$, under the action*

$$(p\cdot g)(v) = p(g\cdot v),$$

where $p : \mathbb{R}^N \to \mathscr{E}_x$, $g \in \mathrm{GL}(N,\mathbb{R})$ and $v \in \mathbb{R}^N$, and $\mathscr{E}$ is naturally isomorphic to $\mathrm{GL}(\mathscr{E}) \times_{\mathrm{GL}(N,\mathbb{R})} \mathbb{R}^N$. The same result holds when $\mathbb{R}$ is replaced by $\mathbb{C}$. We call $\mathrm{GL}(\mathscr{E})$ the **frame bundle** *of $\mathscr{E}$.*

Proof. Since locally, $\mathscr{E}$ has the form $U \times \mathbb{R}^N$, it follows that locally, $\mathrm{GL}(\mathscr{E})$ has the form $U \times \mathrm{GL}(N,\mathbb{R})$, proving that P is a principal bundle.

The isomorphism from $\mathrm{GL}(\mathscr{E}) \times_{\mathrm{GL}(N,\mathbb{R})} \mathbb{R}^N$ to $\mathscr{E}$ is given by mapping the equivalence class of $(p,v) \in \mathrm{GL}(\mathscr{E})_x \times \mathbb{R}^N$ to $p(v) \in \mathscr{E}_x$; it is easily seen that this map is equivariant under the action of $\mathrm{GL}(N,\mathbb{R})$, that is, $(p\cdot g, v)$ maps to the same point as $(p, g\cdot v)$. □

The formation of an associated bundle is functorial with respect to morphisms between G-spaces; in other words, we could consider it as a map from the category of G-spaces to the category of fibre bundles on the base. Indeed, if $\kappa : E_1 \to E_2$ is a G-map from one G-space to another, then it induces a map from $P \times_G E_1$ to $P \times_G E_2$. This map is constructed by the commutative diagram

$$\begin{array}{ccc} P \times E_1 & \xrightarrow{1\times\kappa} & P \times E_2 \\ \downarrow & & \downarrow \\ P \times_G E_1 & \longrightarrow & P \times_G E_2 \end{array}$$

Definition 1.5. *Let $\pi : \mathscr{E} \to M$ be a fibre bundle and let G be a Lie group. We say that $\mathscr{E}$ is an* **G-equivariant bundle** *if G acts smoothly on the left of both $\mathscr{E}$ and M in a compatible fashion, that is,*

$$\gamma\cdot\pi = \pi\cdot\gamma \quad \textit{for all } \gamma \in G.$$

If $\mathscr{E}$ is a vector bundle, we require in addition that the action $\gamma^{\mathscr{E}} : \mathscr{E}_x \to \mathscr{E}_{\gamma\cdot x}$ is linear.

The group G acts on the space of sections $\Gamma(M,\mathscr{E})$ of an G-equivariant bundle by the formula

$$(\gamma\cdot s)(x) = \gamma^{\mathscr{E}}\cdot s(\gamma^{-1}\cdot x). \tag{1.1}$$

Let $\mathfrak{g}$ be the Lie algebra of G. We denote by $\mathscr{L}^{\mathscr{E}}(X)$, $X \in \mathfrak{g}$, the corresponding infinitesimal action on $\Gamma(M,\mathscr{E})$,

$$\mathscr{L}^{\mathscr{E}}(X)s = \frac{d}{d\varepsilon}\Big|_{\varepsilon=0} \exp(\varepsilon X)\cdot s, \tag{1.2}$$

which is called the **Lie derivative**. Clearly, $\mathscr{L}^{\mathscr{E}}(X)$ is a first-order differential operator on $\Gamma(M,\mathscr{E})$.

Let TM be the tangent bundle of M. A section $X \in \Gamma(M,TM)$ is called a vector field on M. If $\phi : M_1 \to M_2$ is a smooth map, it induces a map $\phi_* : TM_1 \to TM_2$, in such a way that

$$\phi_*(v) \in T_{\phi(x)}M_2 \quad \text{if } v \in T_xM_1.$$

In particular, diffeomorphisms ϕ of M induce a diffeomorphism ϕ_* of TM. Informally, TM is a $\mathrm{Diff}(M)$-equivariant vector bundle.

Let $\mathrm{GL}(M) = \mathrm{GL}(TM)$ be the frame bundle obtained by applying the construction of Proposition 1.4 to the tangent bundle TM; it has structure group $\mathrm{GL}(n)$ with $n = \dim(M)$. From this principal bundle, we can construct a vector bundle on M corresponding to any representation E of $\mathrm{GL}(n)$; these vector bundles are called **tensor bundles**. We see from this that any tensor bundle carries a natural action of $\mathrm{Diff}(M)$, making it into a $\mathrm{Diff}(M)$-equivariant vector bundle over M.

If $\mathscr{V}$ is a tensor bundle, the Lie derivative $\mathscr{L}^{\mathscr{V}}(X)$ by a vector field X on M is the first-order differential operator defined on $\Gamma(M,\mathscr{V})$ by the formula

$$\mathscr{L}^{\mathscr{V}}(X)s = \frac{d}{dt}\Big|_{t=0} \phi_t \cdot s,$$

where ϕ_t is the family of diffeomorphisms generated by X. For example, if $\mathscr{V} = TM$, then we obtain the **Lie bracket** $\mathscr{L}(X)Y = [X,Y]$.

One example of a tensor bundle is obtained by taking the exterior algebra $\Lambda(\mathbb{R}^n)^*$; this leads to the bundle of exterior differentials ΛT^*M. The space of sections $\Gamma(M,\Lambda T^*M)$ of the bundle of exterior differentials is called the space of **differential forms** $\mathscr{A}(M)$; it is an algebra graded by

$$\mathscr{A}^i(M) = \Gamma(M,\Lambda^i T^*M).$$

Similarly, we denote by $\mathscr{A}_c(M)$ the algebra $\Gamma_c(M,\Lambda T^*M)$ of compactly supported differential forms. If α is a differential form on M, we will denote by $\alpha_{[i]}$ its component in $\mathscr{A}^i(M)$. If $\alpha \in \mathscr{A}^i(M)$, we will write $|\alpha|$ for the degree i of α; such an α is called **homogeneous**. The space $\mathscr{A}(M)$ is a super-commutative algebra (see Section 3 for the definition of this term), that is,

$$\alpha \wedge \beta = (-1)^{|\alpha|\cdot|\beta|}\beta \wedge \alpha.$$

We denote by d the **exterior differential**; it is the unique operator on $\mathscr{A}(M)$ such that

1. $d : \mathscr{A}^\bullet(M) \to \mathscr{A}^{\bullet+1}(M)$ satisfies $d^2 = 0$;
2. if $f \in C^\infty(M)$, then $df \in \mathscr{A}^1(M)$ is the one-form such that $(df)(X) = X(f)$ for a vector field X;

3. (Leibniz's rule) d is a derivation, that is, if α and β are homogeneous differential forms on M, then

$$d(\alpha\wedge\beta)=(d\alpha)\wedge\beta+(-1)^{|\alpha|}\alpha\wedge(d\beta).$$

If $\alpha\in\mathscr{A}^k(M)$ and $X_0,\ldots,X_k\in\Gamma(M,TM)$, then

$$(d\alpha)(X_0,\ldots,X_k)=\sum_{i=0}^{k}(-1)^i X_i(\alpha(X_0,\ldots,\widehat{X}_i,\ldots,X_k))$$
$$+\sum_{0\le i<j\le k}(-1)^{i+j}\alpha([X_i,X_j],X_0,\ldots,\widehat{X}_i,\ldots,\widehat{X}_j,\ldots,X_k). \quad (1.3)$$

Definition 1.6. *If V is a vector space, the* **contraction** *operator $\iota(v):\Lambda V^*\to\Lambda V^*$, ($v\in V$), is the unique operator such that*

1. *$\iota(v)\alpha=\alpha(v)$ if $\alpha\in V^*$;*
2. *$\iota(v)(\alpha\wedge\beta)=(\iota(v)\alpha)\wedge\beta+(-1)^{|\alpha|}\alpha\wedge(\iota(v)\beta)$ if α and β are homogeneous elements of ΛV^*.*

The **exterior** *operator $\varepsilon(\alpha):\Lambda V^*\to\Lambda V^*$, is the operation of left exterior multiplication by $\alpha\in V^*$.*

We will denote the operation of **contraction** with a vector field X by

$$\iota(X):\mathscr{A}^\bullet(M)\to\mathscr{A}^{\bullet-1}(M).$$

Similarly, on $\mathscr{A}(M,\mathscr{E})$, $\iota(X)$ is defined in such a way that for elements of the form αs, $\alpha\in\mathscr{A}(M)$ and $s\in\Gamma(M,\mathscr{E})$,

$$\iota(X)(\alpha s)=(\iota(X)\alpha)s.$$

On differential forms, the Lie derivative satisfies the two properties,

$$\mathscr{L}(X)d=d\mathscr{L}(X),$$
$$\mathscr{L}(X)(\iota(Y)\alpha)=\iota([X,Y])\alpha+\iota(Y)(\mathscr{L}(X)\alpha),$$

from which follows easily E. Cartan's homotopy formula:

$$\mathscr{L}(X)=d\cdot\iota(X)+\iota(X)\cdot d. \quad (1.4)$$

Note that if α is an n-form, where n is the dimension of M, then this formula implies that $\mathscr{L}(fX)\alpha=\mathscr{L}(X)(f\alpha)$ for all $f\in C^\infty(M)$.

If $\phi:M_1\to M_2$ is a smooth map of manifolds, we denote by $\phi^*:\mathscr{A}^\bullet(M_2)\to\mathscr{A}^\bullet(M_1)$ the pull-back of differential forms; it commutes with exterior product and with d, and hence defines a homomorphism of differential graded algebras.

Since $d^2=0$, we can form the cohomology of the complex $(\mathscr{A}(M),d)$, which is

$$H^i(\mathscr{A}(M),d)=\ker(d|_{\mathscr{A}^i(M)})/\operatorname{im}(d|_{\mathscr{A}^{i-1}(M)}).$$

We denote $H^i(\mathscr{A}(M),d)$ by $H^i(M)$, and call $H^\bullet(M)$ the **de Rham cohomology** groups of M. If $d\alpha = 0$, we say that α is **closed**, and we write its class in $H^\bullet(M)$ as $[\alpha]$. If $\alpha = d\beta$ for some β, we say that α is **exact**; exact forms represent zero in $H^\bullet(M)$. The graded vector space $\sum_{i=0}^n H^i(\mathscr{A}(M),d)$ is a graded algebra, with product defined by

$$[\alpha_1] \wedge [\alpha_2] = [\alpha_1 \wedge \alpha_2] \quad \text{for closed forms } \alpha_1 \text{ and } \alpha_2.$$

This definition depends on the fact that if α_1 and α_2 are closed and one of them is exact, then so is $\alpha_1 \wedge \alpha_2$; for example, if $\alpha_1 = d\beta$, then

$$\alpha_1 \wedge \alpha_2 = d(\beta \wedge \alpha_2).$$

If M is a non-compact manifold, for example, a vector bundle, the definition of the de Rham cohomology may be repeated with $\mathscr{A}(M)$ replaced by the differential forms of compact support $\mathscr{A}_c(M)$. The cohomology of the complex $(\mathscr{A}_c^\bullet(M),d)$ is denoted $H_c^\bullet(M)$ and called the de Rham cohomology with compact supports.

Let P be a pricipal bundle with structure group G. There is a representation of the sections of an associated vector bundle $P \times_G E$ as functions on the corresponding principal bundle that is extremely useful in doing calculations in a coordinate free way.

Proposition 1.7. *Let $C^\infty(P,E)^G$ denote the space of equivariant maps from P to E, that is, those maps $s : P \to E$ that satisfy $s(p \cdot g) = \rho(g^{-1})s(p)$. There is a natural isomorphism between $\Gamma(M, P \times_G E)$ and $C^\infty(P,E)^G$, given by sending $s \in C^\infty(P,E)^G$ to s_M defined by*

$$s_M(x) = [p, s(p)];$$

here p is any element of $\pi^{-1}(x)$ and $[p,s(p)]$ is the element of $P \times_G E$ corresponding to $(p,s(p)) \in P \times E$.

Proof. We first note that the definition of s_M is unambiguous; if, for some point $x \in M$, we had chosen some lifting $p \cdot g$ instead of $p \in \pi^{-1}(x)$, then $[p,s(p)]$ would have been replaced by

$$[p \cdot g, s(p \cdot g)] = [p \cdot g, \rho(g^{-1})s(p)] = [p, s(p)].$$

To complete the proof, we only have to show that any section of $P \times_G E$ is of the form s_M for some $s \in C^\infty(P,E)^G$. To do this, we merely define $s(p)$ to equal the unique $v \in E$ such that (p,v) is a representative of $s_M(x)$. □

Observe the following infinitesimal version of equivariance: a function s in $C^\infty(P,E)^G$ satisfies the formula

$$(X_P \cdot s)(x) + \rho(X)s(x) = 0, \quad \text{for } X \in \mathfrak{g}, \tag{1.5}$$

where we also denote by ρ the differential of the representation ρ.

If $\mathscr{E}$ is a vector bundle on M, let $\mathscr{A}(M,\mathscr{E})$ denote the space of differential forms on M with values in $\mathscr{E}$, in other words,

$$\mathscr{A}^k(M,\mathscr{E}) = \Gamma(M, \Lambda^k T^*M \otimes \mathscr{E}).$$

The differential forms will be taken over the field of real numbers unless either they take values in a complex bundle or unless explicitly stated otherwise. If $\mathscr{E}$ is the trivial bundle $\mathscr{E} = M \times E$, we may write $\mathscr{A}^k(M,E)$ instead of $\mathscr{A}^k(M,\mathscr{E})$. When $P \times_G E$ is an associated bundle on M, we will describe $\mathscr{A}(M, P \times_G E)$ as a subspace of the space of differential forms on P with values in E.

The group G acts on $\mathscr{A}(P,E) = \mathscr{A}(P) \otimes E$ by the formula

$$g \cdot (\beta \otimes e) = g \cdot \beta \otimes g \cdot e.$$

We denote by $\mathscr{A}(P,E)^G \subset \mathscr{A}(P) \otimes E$ the space of invariant forms. Elements of $\mathscr{A}(P,E)^G$ satisfy the following infinitesimal form of invariance:

$$\mathscr{L}(X)\alpha + \rho(X)\alpha = 0,$$

for all $X \in \mathfrak{g}$ and $\alpha \in \mathscr{A}(P,E)^G$.

Definition 1.8. *A* **horizontal differential form** *on a fibre bundle $\mathscr{E}$ is a differential form α such that $\iota(X)\alpha = 0$ for all vertical vector fields X.*

A **basic differential form** *on a principal bundle P with structure group G, taking values in the representation (E, ρ) of G, is an invariant and horizontal differential form, that is, a form $\alpha \in \mathscr{A}(P,E)$ which satisfies*

1. *$g \cdot \alpha = \alpha$, $g \in G$;*
2. *$\iota(X)\alpha = 0$ for any vertical vector field X on P.*

The space of all basic forms with values in E is denoted $\mathscr{A}(P,E)_{\mathrm{bas}}$.

We will not prove the following proposition, since its proof is similar to that of Proposition 1.7.

Proposition 1.9. *If $\alpha \in \mathscr{A}^q(P,E)_{\mathrm{bas}}$, define $\alpha_M \in \mathscr{A}^q(M, P \times_G E)$ by*

$$\alpha_M(\pi_* X_1, \dots, \pi_* X_q)(x) = [p, \alpha(X_1, \dots, X_q)(p)],$$

where $p \in P$ is any point such that $\pi(p) = x$, and $X_i \in T_p P$. Then α_M is well-defined, and the map $\alpha \mapsto \alpha_M$ from $\mathscr{A}(P,E)_{\mathrm{bas}}$ to $\mathscr{A}(M, P \times_G E)$ is an isomorphism.

Let $\mathscr{E}$ be a fibre bundle with base M. The quotient of the tangent bundle $T\mathscr{E}$ by its subbundle $V\mathscr{E}$ is isomorphic to the pull-back $\pi^* TM$ of the tangent bundle of the base to $\mathscr{E}$; that is, we have a short exact sequence of vector bundles

$$0 \to V\mathscr{E} \to T\mathscr{E} \to \pi^* TM \to 0.$$

There is no canonical choice of splitting of this short exact sequence, but it is important to have such a splitting in many situations. Such a splitting is called a **connection**.

Definition 1.10. *Let $\pi : \mathscr{E} \to M$ be a fibre bundle. A* **connection one-form** *is a one-form $\omega \in \mathscr{A}^1(\mathscr{E}, V\mathscr{E})$ such that $\iota(X)\omega = X$ for any vertical vector field on $\mathscr{E}$. The kernel of the one-form ω is called the* **horizontal bundle**, *and written $H\mathscr{E}$; it is isomorphic to π^*TM.*

If X is a vector field on the base M, we denote by $X_{\mathscr{E}}$ the section of $H\mathscr{E}$ such that $\pi_* X_{\mathscr{E}} = X$; we call $X_{\mathscr{E}}$ the **horizontal lift** *of X with respect to the connection determined by ω.*

Note that we can equally well specify a connection by the horizontal subbundle $H\mathscr{E}$ or by the connection form ω. In many situations, however, the connection form is more convenient.

An important object associated to a connection on a fibre bundle $\mathscr{E}$ is its **curvature** Ω, which is a section of the bundle $\Lambda^2\pi^*T^*M \otimes V\mathscr{E}$: if X and Y are vector fields on M and if $X_{\mathscr{E}}$ and $Y_{\mathscr{E}}$ are their horizontal lifts, then $\Omega(X,Y)$ is the vertical vector field on $\mathscr{E}$ defined by the formula

$$\Omega(X,Y) = [X,Y]_{\mathscr{E}} - [X_{\mathscr{E}}, Y_{\mathscr{E}}]. \tag{1.6}$$

Note that this tensor is local on M, in the sense that for any $f \in C^\infty(M)$

$$\Omega(fX,Y) = (\pi^* f)\,\Omega(X,Y);$$

this follows immediately from the fact that $X_{\mathscr{E}}(\pi^* f) = \pi^*(Xf)$.

If P is a principal bundle, we only consider connections on it which are stable under the action of the structure group G by multiplication on the right. This condition is most conveniently characterized in terms of the connection form. Observe that, by the identification of VP with the trivial bundle $P \times \mathfrak{g}$, a connection form on P will be an element of $\mathscr{A}^1(P, \mathfrak{g})$.

Definition 1.11. *A* **connection one-form** *on a principal bundle P is an invariant $\mathfrak{g}$-valued one-form $\omega \in \mathscr{A}^1(P,\mathfrak{g})^G$ such that $\iota(X_P)\omega = X$ if $X \in \mathfrak{g}$.*

A differential form $\alpha \in \mathscr{A}(P,\mathfrak{g})^G$ satisfies the formula

$$\mathscr{L}(X)\alpha + [X,\alpha] = 0 \quad \text{for all } X \in \mathfrak{g}. \tag{1.7}$$

Proposition 1.12. *The space of connection one-forms on a principal bundle P is an affine space modelled on the space of one-forms $\mathscr{A}^1(M, P \times_G \mathfrak{g})$.*

Proof. Once a connection form ω_0 is chosen, all others are described uniquely in the form $\omega_0 + \alpha$ for some $\alpha \in \mathscr{A}^1(P,\mathfrak{g})_{\mathrm{bas}}$. The result now follows from Proposition 1.9. □

The **curvature** Ω of the connection associated to ω is the element of $\mathscr{A}^2(P,\mathfrak{g})_{\mathrm{bas}}$ defined by

$$\Omega(X,Y)_P = h[hX,hY] - [hX,hY].$$

Here, we have denoted by h the projection onto the horizontal sub-bundle of TP, and X and Y are two vector fields on P. In stating the following proposition, we use the Lie bracket on $\mathscr{A}(P,\mathfrak{g})$ defined by

$$[\alpha\otimes A_1,\beta\otimes A_2]=(\alpha\wedge\beta)\otimes[A_1,A_2],$$

for α and β in $\mathscr{A}(P)$, and A_1 and A_2 in $\mathfrak{g}$. This makes $\mathscr{A}(P,\mathfrak{g})$ into a Lie superalgebra, as we will see in the Section 1.3. If α is a $\mathfrak{g}$-valued one-form and X_1 and X_2 are two tangent vectors, then $[\alpha,\alpha](X_1,X_2)=2[\alpha(X_1),\alpha(X_2)]$.

Proposition 1.13. *The curvature form Ω satisfies the formula*

$$\Omega=d\omega+\tfrac{1}{2}[\omega,\omega].$$

Proof. If $X\in\mathfrak{g}$, then

$$\iota(X)(d\omega+\tfrac{1}{2}[\omega,\omega])=\mathscr{L}(X)\omega-d(\iota(X)\omega)+[X,\omega]=0$$

by (1.7), since $d(\iota(X)\omega)=dX=0$. Thus, the differential form $d\omega+\frac{1}{2}[\omega,\omega]$ is seen to be horizontal; as it is clearly invariant, it is seen to be basic. To complete the proof, we need only evaluate the two-form $d\omega+\frac{1}{2}[\omega,\omega]$ on the horizontal vectors Y and Z; it is easy to see that we obtain $-\iota([Y,Z])\omega$, which is exactly equal to $\Omega(Y,Z)$. □

The choice of a connection on a principal bundle P determines connections on all associated bundles $\mathscr{E}=P\times_G E$, by defining $H\mathscr{E}$ to be the image of HP under the projection $P\times E\to\mathscr{E}$. We will be especially interested in this construction for associated vector bundles, since it provides an analogue of the exterior differential on spaces of twisted differential forms $\mathscr{A}(M,\mathscr{E})$, called a covariant derivative.

Definition 1.14. *If $\mathscr{E}$ is a vector bundle over a manifold M, a* **covariant derivative** *on $\mathscr{E}$ is a differential operator*

$$\nabla:\Gamma(M,\mathscr{E})\to\Gamma(M,T^*M\otimes\mathscr{E})$$

which satisfies Leibniz's rule; that is, if $s\in\Gamma(M,\mathscr{E})$ and $f\in C^\infty(M)$, then

$$\nabla(fs)=df\otimes s+f\nabla s.$$

Note that a covariant derivative extends in a unique way to a map

$$\nabla:\mathscr{A}^\bullet(M,\mathscr{E})\to\mathscr{A}^{\bullet+1}(M,\mathscr{E})$$

that satisfies Leibniz's rule: if $\alpha\in\mathscr{A}^k(M)$ and $\theta\in\mathscr{A}(M,\mathscr{E})$, then

$$\nabla(\alpha\wedge\theta)=d\alpha\wedge\theta+(-1)^k\alpha\wedge\nabla\theta.$$

This formula is the starting point for the definition of a superconnection, which generalizes the notion of a covariant derivative and forms the topic of Section 3.

If X is a vector field on M, we will denote by ∇_X the differential operator $\iota(X)\nabla$, which is called the covariant derivative by the vector field X. It depends linearly on X, $\nabla_{fX}=f\nabla_X$, while the commutator $[\nabla_X,f]$ of ∇_X and the operator of multiplication by a smooth function f is multiplication by $X\cdot f$.

We see from the definition of a connection that if $\alpha \in \mathscr{A}^k(M,\mathscr{E})$ and $X_0,\ldots,X_k \in \Gamma(M,TM)$, then

$$(\nabla\alpha)(X_0,\ldots,X_k) = \sum_{i=0}^{k}(-1)^i \nabla_{X_i}(\alpha(X_0,\ldots,\widehat{X}_i,\ldots,X_k)) \\ + \sum_{0\le i<j\le k}(-1)^{i+j}\alpha([X_i,X_j],X_0,\ldots,\widehat{X}_i,\ldots,\widehat{X}_j,\ldots,X_k). \quad (1.8)$$

Two covariant derivatives ∇_0 and ∇_1 on a vector bundle $\mathscr{E}$ differ by a one-form $\omega \in \mathscr{A}^1(M,\mathrm{End}(\mathscr{E}))$:

$$\nabla_1 s = \nabla_0 s + \omega\cdot s \quad \text{for } s \in \Gamma(M,\mathscr{E}).$$

If $\mathscr{E} = M\times E$ is a trivial vector bundle, then the exterior differential d is a covariant derivative on $\mathscr{E}$, so any covariant derivative on $M\times E$ may be written

$$\nabla s = ds + \omega\cdot s, \quad (1.9)$$

for some $\omega \in \mathscr{A}^1(M,\mathrm{End}(\mathscr{E})) = \mathscr{A}^1(M)\otimes\mathrm{End}(E)$. We say that ω is the connection one-form of ∇ in the given trivialization of $\mathscr{E}$.

The **curvature** of a covariant derivative ∇ is the $\mathrm{End}(\mathscr{E})$-valued two-form on M defined by

$$F(X,Y) = [\nabla_X,\nabla_Y] - \nabla_{[X,Y]},$$

for two vector fields X and Y. It is easy to check, using Leibniz's rule, that $F(X,Y)$ really is a local operator, in other words, that it commutes with multiplication by a smooth function f:

$$[F(X,Y),f] = [\nabla_X, Yf] + [Xf,\nabla_Y] - [X,Y]f = 0.$$

Proposition 1.15. *The operator* $\nabla^2 : \mathscr{A}^\bullet(M,\mathscr{E}) \to \mathscr{A}^{\bullet+2}(M,\mathscr{E})$ *is given by the action of* $F \in \mathscr{A}^2(M,\mathrm{End}(\mathscr{E}))$ *on* $\mathscr{A}(M,\mathscr{E})$.

Proof. If $\mathscr{E}$ is a trivial bundle and $\nabla = d+\omega$, then F is the $\mathrm{End}(E)$-valued two-form on M given by the formula

$$F = d\omega + \omega\wedge\omega.$$

On the other hand, the square of the operator $d+\omega$ is also given by the action of $d\omega+\omega\wedge\omega$. □

We will study the curvature thoroughly later, in the more general context of superconnections.

If $\mathscr{E}$ is an associated bundle $\mathscr{E} = P\times_G E$, then a covariant derivative ∇ may be formed from a connection one-form ω on P, by means of the diagram

$$\begin{array}{ccc} C^\infty(P,E)^G & \xrightarrow{d+\rho(\omega)} & \mathscr{A}^1(P,E)_{\mathrm{bas}} \\ \downarrow & & \downarrow \\ \Gamma(M,\mathscr{E}) & \xrightarrow{\nabla} & \mathscr{A}^1(M,\mathscr{E}) \end{array}$$

To show this, we must check that if s is an equivariant map from P to E, then $ds+\rho(\omega)s$ is basic. But choosing $X \in \mathfrak{g}$, we see that

$$\iota(X)(ds+\rho(\omega)s) = X\cdot s+\rho(X)s,$$

which vanishes, since $s \in C^\infty(P,E)^G$. Likewise, the one-form $ds+\rho(\omega)s$ lies in $\mathscr{A}^1(P,E)^G$. It is easy to verify that ∇ is indeed a covariant derivative.

There is another formula for the covariant derivative when written on the principal bundle in terms of the projection onto the horizontal subspaces of TP:

$$\nabla s = P_{HM}\circ ds \quad \text{for } s \in C^\infty(P,E)^G.$$

Indeed, if X_H is a horizontal tangent vector, then

$$\iota(X_H)(d+\rho(\omega))s = \iota(X_H)ds$$

since X_H is horizontal, while if X_V is a vertical tangent vector,

$$\iota(X_V)(d+\rho(\omega))s = 0,$$

since $s \in C^\infty(P,E)^G$. Thus, if $s_P \in C^\infty(P,E)^G$ corresponds to the section $s \in \Gamma(M,\mathscr{E})$, and $X \in \Gamma(M,TM)$ is a vector field with horizontal lift X_P, then for any $p \in P$ above x,

$$(\nabla_X s)(x) = [p,(X_P s_P)(p)].$$

Proposition 1.16. *There is a one-to-one correspondence between connections on the frame bundle* $\mathrm{GL}(\mathscr{E})$ *and covariant derivatives on the vector bundle* $\mathscr{E}$.

Proof. If ∇ is a covariant derivative on $\mathscr{E}$, we may construct a one-form $\omega \in \mathscr{A}^1(P,\mathrm{End}(E))^G$ by

$$\omega(X)s(p) = (\nabla_{\pi_* X}s)(p) - (X\cdot s)(p) \quad \text{for } s \in C^\infty(P,E)^G, X \in T_pP.$$

This defines ω uniquely, since for any $(p,v) \in P\times E$, there exists an $s \in C^\infty(P,E)^G$ such that $s(p) = v$. The one-form ω satisfies

$$\iota(X)\omega = \rho(X) \quad \text{for all } X \in \mathfrak{g},$$

and hence is a connection form. □

As an example of this result, let $\mathscr{E} = M\times E$ be a trivial vector bundle on M with connection $\nabla = d+\omega$. The frame bundle $\mathrm{GL}(\mathscr{E})$ of $\mathscr{E}$ is equal to $\pi : M\times \mathrm{GL}(E) \to M$. Let $g^{-1}dg$ be the Maurer-Cartan one-form in $\mathscr{A}^1(\mathrm{GL}(E),\mathrm{End}(E))$ defined using the tautological map

$$g \in C^\infty(\mathrm{GL}(E),\mathrm{End}(E)),$$

which sends an element of the group $\mathrm{GL}(E)$ to the corresponding linear map in $\mathrm{End}(E)$. Then the connection one-form on $\mathrm{GL}(\mathscr{E})$ associated to ∇ is

$$\omega_{\mathrm{GL}(\mathscr{E})} = g^{-1} \cdot \pi^* \omega \cdot g + g^{-1} dg.$$

In this way, we see that the two different notions of connections, as a connection on a principal bundle and as a covariant derivative on sections of a vector bundle, are two faces of a single object.

If the covariant derivative has been defined in terms of a connection ω on a principal bundle P, then we obtain a representation of its curvature as a basic two-form Ω on P with values in $\mathfrak{g}$: $F = \rho(\Omega)$, with $\Omega = d\omega + \frac{1}{2}[\omega, \omega]$, since for $s \in C^\infty(P,E)^G$,

$$\begin{aligned} F(X,Y)s &= ([X_P, Y_P] - [X,Y]_P)s = -\Omega(X,Y)s \\ &= \rho(\Omega(X,Y))s. \end{aligned}$$

The advantage of constructing a covariant derivative from a connection on a principal bundle is that this provides a compatible family of covariant derivatives on all vector bundles associated to P simultaneously, in a way which is compatible with the natural linear maps between these bundles, and with the tensor product of bundles. Thus, if we are given a G-equivariant map $\kappa : V \to W$ and $s \in \Gamma(M, P \times_G V)$, then we may form the bundle maps $\kappa : P \times_G V \to P \times_G W$ and $1 \otimes \kappa : T^*M \otimes P \times_G V \to T^*M \otimes P \times_G W$, and then

$$\nabla \kappa(s) = (1 \otimes \kappa)(\nabla s) \in \mathscr{A}^1(M, P \times_G W).$$

We now give some ways of contructing new covariant derivatives out of old. If ∇ is a covariant derivative on $\mathscr{E}$, we obtain a covariant derivative on the dual bundle $\mathscr{E}^*$, by the following formula, for $s \in \Gamma(M, \mathscr{E})$ and $t \in \Gamma(M, \mathscr{E}^*)$:

$$d\langle s,t\rangle = \langle \nabla s, t\rangle + \langle s, \nabla t\rangle,$$

that is, if $X \in \Gamma(M, TM)$ is a vector field on M,

$$X\langle s,t\rangle = \langle \nabla_X s, t\rangle + \langle s, \nabla_X t\rangle.$$

Here, we use the notation $\langle s,t\rangle$ to denote the pairing between $\mathscr{E}$ and $\mathscr{E}^*$.

If $\nabla^{\mathscr{E}_i}$ is a covariant derivative on $\mathscr{E}_i$ for $i = 1, 2$, we obtain a covariant derivative $\nabla^{\mathscr{E}_1 \otimes \mathscr{E}_2}$ on the tensor product by

$$\nabla^{\mathscr{E}_1 \otimes \mathscr{E}_2}(s_1 \otimes s_2) = \nabla^{\mathscr{E}_1} s_1 \otimes s_2 + s_1 \otimes \nabla^{\mathscr{E}_2} s_2$$

for $s_i \in \Gamma(M, \mathscr{E}_i)$. We refer to $\nabla^{\mathscr{E}_1 \otimes \mathscr{E}_2}$ as the tensor product covariant derivative, and write

$$\nabla^{\mathscr{E}_1 \otimes \mathscr{E}_2} = \nabla^{\mathscr{E}_1} \otimes 1 + 1 \otimes \nabla^{\mathscr{E}_2}.$$

Let M be a manifold on which a compact Lie group G acts smoothly, and let $\mathscr{E}$ be an equivariant bundle over M. We say that a covariant derivative $\nabla^{\mathscr{E}}$ is **invariant** if it commutes with the action of G,

$$g^{T^*M \otimes \mathscr{E}} \cdot \nabla = \nabla \cdot g^{\mathscr{E}}$$

for all $g \in G$. If G is compact, there will exist such connections, since we may average with respect to the Haar measure on G,

$$\int_G \left(g^{T^*M\otimes\mathscr{E}}\right)^{-1} \cdot \nabla \cdot g^{\mathscr{E}}\, dg \tag{1.10}$$

is an invariant covariant derivative.

Let $\phi : N \to M$ be a smooth map, and let $\mathscr{E}$ be a vector bundle on M. If $\nabla^{\mathscr{E}}$ is a covariant derivative on $\mathscr{E}$, then the formula

$$\nabla^{\phi^*\mathscr{E}}(f\phi^*s) = df \otimes \phi^*s + f\phi^*(\nabla^{\mathscr{E}}s)$$

for $f \in C^\infty(N)$ and $s \in \Gamma(M, \mathscr{E})$, defines a covariant derivative $\nabla^{\phi^*\mathscr{E}}$ on the bundle $\phi^*\mathscr{E}$, which we call the pull-back of the covariant derivative $\nabla^{\mathscr{E}}$.

In particular, if $\gamma(t) : \mathbb{R} \to M$ is a smooth curve in M with tangent vector $\dot\gamma(t) \in T_{\gamma(t)}M$, and if $s : \mathbb{R} \to \gamma^*\mathscr{E}$ is smooth, then $\nabla^{\mathscr{E}}_{\dot\gamma(t)}s(t)$ is defined to be the covariant derivative $\nabla^{\gamma^*\mathscr{E}}(s)(t)$. If $\mathscr{E}$ is the trivial bundle $M \times E$ and $\nabla^{\mathscr{E}} = d + \omega$, we may write $s(t) = (\gamma(t), f(t))$ with $f : \mathbb{R} \to E$, and

$$\nabla^{\mathscr{E}}_{\dot\gamma(t)}s(t) = \left(\gamma(t), \frac{df(t)}{dt} + \omega(\dot\gamma(t))f(t)\right).$$

The **parallel transport** map along $\gamma(t)$,

$$\tau_\gamma(t) \in \operatorname{Hom}(\mathscr{E}_{\gamma(0)}, \mathscr{E}_{\gamma(t)})$$

is defined by solving the ordinary differential equation

$$\nabla^{\mathscr{E}}_{\dot\gamma(t)}\tau_\gamma(t) = 0, \tag{1.11}$$

with initial condition $\tau_\gamma(0) = I$.

Suppose U is an open ball in $\mathbb{R}^n$ and $\mathscr{E}$ is a vector bundle over U with connection ∇. If $\mathbf{x}^i$ are the coordinates on U and let ∂_i are the corresponding partial derivatives, let

$$\mathscr{R} = \sum_i \mathbf{x}^i \partial_i$$

be the radial vector field on U. The following result follows from the theory of ordinary differential equations.

Proposition 1.17. *Parallel transport along rays $t \mapsto tv$, $v \in U$, gives a smooth trivialization of $\mathscr{E}$ over U, by identifying the fibres $\mathscr{E}_v$ with $E = \mathscr{E}_0$.*

In terms of this trivialization, the connection one-form on $\mathscr{E}$ is the $\operatorname{End}(E)$-valued one-form $\omega = \sum_i \omega_i d\mathbf{x}^i$ such that $\nabla = d + \omega$, and the curvature of $\mathscr{E}$ is the two-form F with values in $\operatorname{End}(E)$ defined by the equation $F = d\omega + \omega \wedge \omega$. By the definition of this frame of $\mathscr{E}$, we have $\iota(\mathscr{R})\omega = 0$. Thus the following formula for the radial derivative of the connection form ω holds:

$$\mathscr{L}(\mathscr{R})\omega = [\iota(\mathscr{R}), d]\omega = \iota(\mathscr{R})(d\omega + \omega\wedge\omega) = \iota(\mathscr{R})F. \tag{1.12}$$

A useful property of this framing of $\mathscr{E}$ is that all of the coefficients of the Taylor expansion of ω_i at 0 are determined by the coefficients of the Taylor expansion of F at 0.

Proposition 1.18. *The Taylor expansion of ω_i at $x_0 = 0$ has the following form:*

$$\omega_i(\mathbf{x}) \sim -\tfrac{1}{2}\sum_j F(\partial_i, \partial_j)_{x_0}\mathbf{x}^j + \sum_{|\alpha|\geq 2} \partial^\alpha \omega_i(x_0)\frac{\mathbf{x}^\alpha}{\alpha!}.$$

Proof. Expanding the Taylor's series of both sides of (1.12), we obtain

$$\sum_\alpha (|\alpha|+1)\partial^\alpha \omega_\ell(x_0)\mathbf{x}^\alpha/\alpha! = \sum_{\alpha,k} \partial^\alpha F(\partial_k, \partial_\ell)_{x_0}\mathbf{x}^k\mathbf{x}^\alpha/\alpha!$$

By equating coefficients of $\mathbf{x}^\alpha$ on both sides, we see from this formula that

$$\partial_j\omega_i(x_0) = -\tfrac{1}{2}F(\partial_i, \partial_j)_{x_0}.$$

Furthermore, it follows that the Taylor coefficients of ω_i at x_0 to order m only depend on those of $F(\partial_i, \partial_j)$ to order $m-1$. □

A **metric** on a vector bundle is a smooth family of non-degenerate inner products on the fibres of $\mathscr{E}$. If $\mathscr{E}$ is a real bundle of rank N, then this amounts to requiring that $\mathscr{E}$ is associated to a $\mathrm{O}(N)$-bundle P with fibre the standard representation $\mathbb{R}^N$. In fact, the principal bundle may be taken equal to $\mathrm{O}(\mathscr{E})$, the sub-bundle of $\mathrm{GL}(\mathscr{E})$ consisting of orthonormal frames. (There are corresponding statements with $\mathbb{R}$ replaced by $\mathbb{C}$ and $\mathrm{O}(N)$ replaced by $\mathrm{U}(N)$.) A real vector bundle with metric will be called a **Euclidean** vector bundle, while a complex vector bundle with metric will be called a **Hermitian** vector bundle.

We say that a covariant derivative ∇ preserves a metric if it satisfies the formula

$$d(s_1, s_2) = (\nabla s_1, s_2) + (s_1, \nabla s_2) \quad \text{for } s_i \in \Gamma(M, \mathscr{E}).$$

If ∇ preserves a metric on $\mathscr{E}$, it is easy to see that its curvature $F_x(X,Y)$ lies in the Lie algebra $\mathfrak{so}(\mathscr{E}_x)$ of skew-adjoint endomorphisms of $\mathscr{E}_x$.

The covariant derivative ∇ preserves a metric on $\mathscr{E}$ if and only if the parallel transport map also preserves this metric, since if $s \in \mathscr{E}_{\gamma(0)}$, we have

$$\frac{d}{dt}|\tau_\gamma(t)s|^2 = 2\left(\tau_\gamma(t)s, \nabla^{\mathscr{E}}_{\dot\gamma(t)}\tau_\gamma(t)s\right) = 0.$$

This shows that the trivialization of Proposition 1.17 is compatible with the metrics on the fibres of $\mathscr{E}$. Thus, if e_a is an orthonormal basis of $E = \mathscr{E}_0$, we obtain smooth sections, which we also denote by e_a, which are orthonormal at each point of U; these sections make up an **orthonormal frame**.

Let $\pi : \mathscr{E} \to M$ be a vector bundle on M. A covariant derivative $\nabla^{\mathscr{E}}$ on $\mathscr{E}$ determines a connection on $\mathrm{GL}(\mathscr{E})$, and hence a connection on the total space $\mathscr{E}$, that is, a splitting

$$T\mathscr{E} = V\mathscr{E} \oplus H\mathscr{E}.$$

The bundle $V\mathscr{E}$ is isomorphic to $\pi^*\mathscr{E}$. We denote by $\theta \in \mathscr{A}^1(\mathscr{E}, V\mathscr{E})$ the connection one-form of this connection.

Definition 1.19. *If $\mathscr{E}$ is a vector bundle, its* **tautological section** $\mathbf{x} \in \Gamma(\mathscr{E}, \pi^*\mathscr{E})$ *is the smooth section which to a point $v \in \mathscr{E}$ assigns the point $(v, v) \in \pi^*\mathscr{E}$.*

Proposition 1.20. *Let $\mathscr{E}$ be a vector bundle with covariant derivative $\nabla^{\mathscr{E}}$. If $\nabla^{\pi^*\mathscr{E}}$ denotes the pull-back of the covariant derivative on $\mathscr{E}$ to $\pi^*\mathscr{E}$, then θ, thought of as an element of $\mathscr{A}^1(\mathscr{E}, \pi^*\mathscr{E})$, equals $\nabla^{\pi^*\mathscr{E}}\mathbf{x}$.*

Proof. Locally, $\mathscr{E}$ may be trivialized; thus, we assume that $\mathscr{E} = M \times E$, with covariant derivative $\nabla = d + \omega$. Denote a point of $\mathscr{E}$ by $(x, v) \in M \times E$; then $\mathbf{x}(x, v) = v$, and $\nabla^{\pi^*\mathscr{E}}\mathbf{x} = dv + \omega v$. Thus

$$\nabla^{\pi^*\mathscr{E}}_{(X,V)}\mathbf{x} = V + \omega(X)v.$$

The frame bundle $\mathrm{GL}(\mathscr{E})$ has tangent vectors $(X, A) \in T_{(x,g)}\mathrm{GL}(\mathscr{E}) = T_xM \times \mathfrak{gl}(E)$ at the point $(x, g) \in \mathrm{GL}(\mathscr{E}) = M \times \mathrm{GL}(E)$. The horizontal tangent space at (x, g) is

$$H_{(x,g)}\mathrm{GL}(\mathscr{E}) = \{(X, -\omega(X)) \mid X \in T_xM\}.$$

Under the map $((x, g), v) \mapsto (x, gv)$ from $\mathrm{GL}(\mathscr{E}) \times E$ to $\mathscr{E}$, the image of $H_{(x,e)}\mathrm{GL}(\mathscr{E})$ is

$$H_{(x,v)}\mathscr{E} = \{(X, -\omega(X)v) \mid X \in T_xM\}.$$

Since $\nabla^{\pi^*\mathscr{E}}\mathbf{x}$ vanishes on this space and satisfies

$$\nabla^{\pi^*\mathscr{E}}_{(0,V)}\mathbf{x} = V,$$

we see that it equals θ. □

We will now study connections on the tangent bundle TM: these are often called **affine connections**.

Definition 1.21. *The* **fundamental one-form** θ *on M is the element of $\mathscr{A}^1(M, TM)$ such that $\iota(X)\theta = X$ for $X \in TM$. If ∇ is a connection on TM, then the* **torsion** *of ∇ is the differential form $T = \nabla\theta \in \mathscr{A}^2(M, TM)$. If $T = 0$, we say that the connection ∇ is* **torsion-free**.

If X and Y are two vector fields on M, then $T(X, Y) \in \Gamma(M, TM)$ is given by the explicit formula

$$T(X, Y) = \nabla_X Y - \nabla_Y X - [X, Y].$$

A connection on TM gives rise to a covariant derivative on any vector bundle which is associated to $\mathrm{GL}(M)$, in particular the exterior bundle ΛT^*M; by naturality, this connection satisfies

$$\nabla_X(\alpha \wedge \beta) = \nabla_X \alpha \wedge \beta + \alpha \wedge \nabla_X \beta$$

for any homogeneous differential forms α and β, and vector field X. If α is a k-form and $X, Y_1, \ldots, Y_k$ are vector fields, we see that

$$X \cdot \alpha(Y_1,\ldots,Y_k) = (\nabla_X \alpha)(Y_1,\ldots,Y_k) + \sum_{i=1}^{k} \alpha(Y_1,\ldots,\nabla_X Y_i,\ldots,Y_k).$$

In statement of the following proposition, we use the exterior product map

$$\varepsilon : T^*M \otimes \Lambda^\bullet T^*M \to \Lambda^{\bullet+1} T^*M.$$

Proposition 1.22. *If ∇ is a torsion-free connection on TM, the exterior differential is equal to the composition*

$$\Gamma(M,\Lambda T^*M) \xrightarrow{\nabla} \Gamma(M,T^*M \otimes \Lambda T^*M) \xrightarrow{\varepsilon} \Gamma(M,\Lambda^{\bullet+1}T^*M).$$

Proof. Let us denote the operator defined above by D. Using the fact that ∇ satisfies Leibniz's rule, we see that D does too:

$$D(\alpha\beta) = D\alpha \wedge \beta + (-1)^{|\alpha|} \alpha \wedge D\beta.$$

(Note that this part of the proof is independent of whether ∇ is torsion-free or not.)

Given Leibniz's rule, it clearly suffices to show that D agrees with d on functions (which is clear), and one-forms. The proof now rests upon the following formula: for any $f \in C^\infty(M)$,

$$D^2 f = D(df) = -\langle T, df \rangle \in \mathscr{A}^2(M),$$

where $T \in \mathscr{A}^2(M,TM)$ is the torsion of the affine connection ∇. In particular, if T vanishes, then D agrees with d on one-forms. To prove this, we choose a local coordinate system on M, and let dx^i and $X_i = \partial/\partial x^i$ be the corresponding frames of the cotangent and tangent bundles. If we denote the covariant derivative in the direction X_i by ∇_i, D is given by the formula

$$D = \sum_i \varepsilon(dx^i)\nabla_i.$$

Thus, $D(df)$ equals

$$\sum_{ij} \varepsilon(dx^i)\nabla_i(\partial_j f\, dx^j) = \sum_{ij} \varepsilon(dx^i)\partial_j f \nabla_i dx^j.$$

But by Leibniz's rule,

$$0 = \nabla_i \langle dx^j, X_k \rangle = \langle \nabla_i dx^j, X_k \rangle + \langle dx^j, \nabla_i X_k \rangle$$

so that

$$\begin{aligned} D(df) &= -\sum_{ijk} \varepsilon(dx^i dx^k)\partial_j f \langle dx^j, \nabla_i X_k \rangle \\ &= -\sum_{i<k} \varepsilon(dx^i dx^k)\langle df, T(X_i,X_k)\rangle. \end{aligned}$$

□

Let M be a manifold of dimension n. Another example of a vector bundle associated to the principal bundle $\mathrm{GL}(M)$ is the bundle, denoted $|\Lambda_M|^s$, of **s-densities** on M, where s is any real number. (If the manifold M is clear from the context, we will write simply $|\Lambda|^s$.) This is the bundle associated to the character $A \mapsto |\det(A)|^{-s}$ of $\mathrm{GL}(n,\mathbb{R})$. An element of the fibre $|\Lambda|^s_x$ is a function $\psi : \Lambda^n T_x M - \{0\} \to \mathbb{R}$ such that $\psi(\lambda X) = |\lambda|^s \psi(X)$ for $\lambda \neq 0$. If U is an open subset of $\mathbb{R}^n$, we denote by $|d\mathbf{x}|^s$ the s-density such that

$$|d\mathbf{x}|^s(\partial_1 \wedge \dots \partial_n) = 1,$$

where ∂_i is the vector field $\partial/\partial\mathbf{x}^i$ corresponding to differentiation in the ith coordinate direction. Using a partition of unity, we see that we can always construct a nowhere-vanishing section of $|\Lambda|^s$, so that $|\Lambda|^s$ is a trivializable bundle.

The importance of the one-density bundle, or **density bundle** as it is usually known, comes from the following result, which follows immediately from the change of variables formula for integrals in several variables.

Proposition 1.23. *Let M be a manifold of dimension n. There is a unique linear form, the* **integral**, *denoted by*

$$\int_M : \Gamma_c(M, |\Lambda|) \to \mathbb{R}$$

which is invariant under diffeomorphisms, and agrees in local coordinates with the Lebesgue integral:

$$\int_M f(x)\,|dx| = \int_{\mathbb{R}^n} f(x)\,dx_1 \dots dx_n.$$

For any vector field X on M and α any compactly supported C^1 density, we see that

$$\int_M \mathcal{L}(X)\alpha = 0. \tag{1.13}$$

Since $\mathcal{L}(fX)\alpha = f\mathcal{L}(X)\alpha + X(f)\alpha$ if $\alpha \in \Gamma(M, |\Lambda|)$, it is easy to see that if $\beta \in \Gamma(M, |\Lambda|^{1/2})$ is a half-density on M,

$$\mathcal{L}(fX)\beta = f\mathcal{L}(X)\beta + \tfrac{1}{2}X(f)\beta. \tag{1.14}$$

The bundle of densities $|\Lambda|$ is very closely related to the bundle of **volume forms** $\Lambda^n T^*M$; the first corresponds to the character $|\det(A)|^{-1}$ of $\mathrm{GL}(n)$ and the second to the character $\det(A)^{-1}$. Unlike $|\Lambda|$, the line-bundle $\Lambda^n T^*M$ is trivializable if and only if M is orientable. If M is a connected orientable manifold, its frame bundle $\mathrm{GL}(M)$ has two components; an **orientation** of M is the choice of one component, which is a principal bundle for the group $\mathrm{GL}^+(n)$ of matrices with positive determinant. We say that frames in the chosen component $\mathrm{GL}^+(M)$ are oriented. An orientation of a manifold M is equivalent to the choice of an isomorphism between the bundles $\Lambda^n T^*M$ and $|\Lambda_M|$. If $\nu \in \mathcal{A}^n(M)$ is a volume form on M, the corresponding density $|\nu|$ is such that $|\nu|(X) = \nu(X)$ if $X \in \Lambda^n T_x M$ is oriented.

If M is an n-dimensional oriented manifold, we define the integral of a compactly supported differential form $\alpha \in \mathcal{A}_c(M)$ to be the integral of the density corresponding to $\alpha_{[n]}$:

$$\int_M \alpha = \int_M |\alpha_{[n]}|.$$

There is a fibred version of this integral. Let $\pi : M \to B$ be a fibre bundle with n-dimensional fibre, such that both M and B are oriented. If $\alpha \in \mathscr{A}_c^k(M)$ is a compactly-supported differential form on M, its integral over the fibres of $M \to B$ is the differential form $\int_{M/B} \alpha \in \mathscr{A}^{k-n}(B)$ such that

$$\int_B \Big(\int_{M/B} \alpha \Big) \wedge \beta = \int_M \alpha \wedge \pi^* \beta, \tag{1.15}$$

for all differential forms β on the base B. We sometimes write $\pi_* \alpha$ instead of $\int_{M/B} \alpha$. It follows easily from (1.15) that

$$\pi_*(\alpha \wedge \pi^* \beta) = \pi_* \alpha \wedge \beta \tag{1.16}$$

for all $\alpha \in \mathscr{A}_c(M)$ and $\beta \in \mathscr{A}(B)$.

The integral over the fibres commutes with the exterior differential, in the sense that

$$d_B \int_{M/B} \alpha = (-1)^n \int_{M/B} d_M \alpha. \tag{1.17}$$

This formula shows that the integral over the fibres induces a map

$$\int_{M/B} : H_c^k(M) \to H_c^{k-n}(B)$$

on de Rham cohomology.

The following result is an example of a transgression formula: it says that the cohomology class of the integral over the fibres does not change if the map π is deformed smoothly.

Proposition 1.24. *Let $\pi : M \times \mathbb{R} \to B$ be a smooth map such that $\pi^t = \pi|_{M \times \{t\}} : M \to B$ is a fibre bundle for each t. Then there is a smooth t-dependent map $\tilde{\pi}_*^t : \mathscr{A}_c^k(M) \to \mathscr{A}_c^{k-n-1}(B)$ such that*

$$\frac{d(\pi_*^t \alpha)}{dt} = (-1)^n \tilde{\pi}_*^t (d\alpha) - d(\tilde{\pi}_*^t \alpha).$$

Proof. Think of π as a fibre bundle map $M \times \mathbb{R} \to B \times \mathbb{R}$ which preserves the t-coordinate. Let $q : M \times \mathbb{R} \to M$ be the projection which forgets the t-coordinate. Given $\alpha \in \mathscr{A}^k(M)$, the differential form $\pi_* q^* \alpha \in \mathscr{A}^{k-n}(B \times \mathbb{R})$ may be decomposed as follows:

$$\pi_* q^* \alpha = \pi_*^t \alpha + \tilde{\pi}_*^t \alpha \wedge dt,$$

where this formula serves as the definition of $\tilde{\pi}_*^t \alpha$. Applying this formula to $d\alpha$, we see that on the one hand,

$$\pi_* q^* d\alpha = \pi_*^t d\alpha + dt \wedge \tilde{\pi}_*^t d\alpha,$$

while on the other hand,

$$\begin{aligned}\pi_* q^* d\alpha &= \pi_* d q^* \alpha \\ &= (-1)^n d\left(\pi^t_* \alpha + \tilde{\pi}^t_* \alpha \wedge dt\right) \\ &= (-1)^n d_B \pi^t_* \alpha + (-1)^n \left(\frac{d\pi^t_* \alpha}{dt} + d_B \tilde{\pi}^t_* \alpha\right) \wedge dt.\end{aligned}$$

Equating coefficients of dt, the proposition follows. □

Corollary 1.25. *The map induced by π^t_* on de Rham cohomology with compact support is independent of t.*

1.2 Riemannian Manifolds

If M is an n-dimensional manifold, then a **Riemannian structure** on M is a metric on the tangent bundle TM. Using the Riemannian structure, we can construct the **orthonormal frame bundle**, which is denoted by $\mathrm{O}(M)$:

$$\mathrm{O}(M) = \{(x,(e_1,\dots,e_n)) \mid e_i \text{ form an orthonormal frame of } T_xM\}$$

An orthonormal frame $p \in \mathrm{O}(M)$ over $x \in M$ may be considered to be an isometry $u = (u_1,\dots,u_n) \in \mathbb{R}^n \mapsto \sum u_i e_i \in T_xM$ from $\mathbb{R}^n$ to T_xM. We will denote by $\pi : \mathrm{O}(M) \to M$ the projection map. The bundle $\mathrm{O}(M)$ is an $\mathrm{O}(n)$-principal bundle, and the tangent bundle of M is an associated bundle of $\mathrm{O}(M)$, corresponding to the standard representation $\mathbb{R}^n$ of $\mathrm{O}(n)$:

$$TM \cong \mathrm{O}(M) \times_{\mathrm{O}(n)} \mathbb{R}^n.$$

If, in addition, M is oriented, then the bundle of oriented orthonormal frames is a $\mathrm{SO}(n)$-principal bundle which we denote by $\mathrm{SO}(M)$.

The tangent bundle TM of a Riemannian manifold M has a canonical covariant derivative ∇, called the **Levi-Civita connection**. This is the unique covariant derivative which preserves the Riemannian metric,

$$d(X,Y) = (\nabla X, Y) + (X, \nabla Y),$$

for all vector fields X and Y on M, and which is torsion-free,

$$\nabla_X Y - \nabla_Y X = [X,Y].$$

It is easily verified from the above two formulas that the Levi-Civita derivative $\nabla_X Y$ is given by the formula

$$\begin{aligned}2(\nabla_X Y, Z) = {} & ([X,Y],Z) - ([Y,Z],X) + ([Z,X],Y) \\ & + X(Y,Z) + Y(Z,X) - Z(X,Y).\end{aligned} \tag{1.18}$$

Denote by $\mathfrak{so}(TM)$ the bundle $\mathrm{O}(M) \times_{\mathrm{O}(n)} \mathfrak{so}(n)$, whose fibre at $x \in M$ is the Lie algebra consisting of antisymmetric endomorphisms of the inner product space T_xM.

The curvature of the Levi-Civita connection is an $\mathfrak{so}(TM)$-valued two-form on M, called the **Riemannian curvature**, which we will denote by $R \in \mathscr{A}^2(M,\mathfrak{so}(TM))$, that is, if X, Y and Z are vector fields on M,

$$R(X,Y)Z = \nabla_X\nabla_Y Z - \nabla_Y\nabla_X Z - \nabla_{[X,Y]}Z.$$

If X and Y are two vector fields on M, we will write $R(X,Y)$ for the section of $\mathfrak{so}(TM)$ given by contracting the two-form R with X and Y, and (RX,Y) for the two-form obtained by contracting the curvature in its $\mathfrak{so}(TM)$ part.

For future reference, we list some well-known properties of the Riemannian curvature.

Proposition 1.26. *Let X, Y, Z and W be vector fields on M.*

1. $R(X,Y) = -R(Y,X)$
2. $(R(W,X)Y,Z) + (Y,R(W,X)Z) = 0$
3. $R(X,Y)Z + R(Y,Z)X + R(Z,X)Y = 0$
4. $(R(W,X)Y,Z) = (R(Y,Z)W,X)$

Proof. (1) says that R is a two-form, and (2) that $R_x(W,X) \in \mathfrak{so}(T_xM)$.

Let us prove (3), which holds for the curvature of any torsion-free connection ∇ on TM. If R is the curvature tensor of ∇,

$$\nabla_X\nabla_Y Z - \nabla_X\nabla_Z Y = \nabla_X[Y,Z],$$

and similarly if we cycle the three vectors X, Y and Z. In this way, we obtain three equations which, when added together, give

$$\begin{aligned}\nabla_X\nabla_Y Z + \nabla_Y\nabla_Z X + \nabla_Z\nabla_X Y - \nabla_Y\nabla_X Z - \nabla_Z\nabla_Y X - \nabla_X\nabla_Z Y \\ = \nabla_X[Y,Z] + \nabla_Y[Z,X] + \nabla_Z[X,Y].\end{aligned}$$

Clearly, this is equivalent to (3).

The last property is a purely algebraic consequence of the first three, which we leave as an exercise. □

When we are given a frame X_i for the tangent bundle with dual frame X^i, we can write the components of the curvature matrix in the form $R^i_{jkl} = \langle R(X_k,X_l)X_j,X^i\rangle$; here, the i and j indices refer to the fact that R takes values in $\mathrm{End}(TM)$, and the k and l indices are the two-form indices. Using the metric, define

$$R_{ijkl} = (R(X_k,X_l)X_j,X_i). \tag{1.19}$$

The components R_{ijkl} satisfy $R_{ijkl} = R_{klij}$. If e_i is an orthonormal frame of TM with dual frame e^i, the two-form (Re_i,e_j) is given by the formula

$$(Re_i,e_j) = -\sum_{k<l} R_{ijkl}e^k \wedge e^l. \tag{1.20}$$

From the curvature tensor, we may form a symmetric tensor

$$\mathrm{Ric}_{ij} = \sum_k R_{ikjk} \in \Gamma(M, S^2(TM)),$$

called the **Ricci curvature**, and its trace the scalar $r_M = \sum_{lm} R_{lmlm}$ called the **scalar curvature** of M.

We denote by $|dx| \in \Gamma(M, |\Lambda|)$ the **Riemannian density** associated to the Riemannian metric on M, which is such that

$$|dx|(e_1 \wedge \ldots \wedge e_n) = 1$$

when e_i is an orthonormal frame of TM. The existence of $|dx|$ corresponds to the fact that the character $|\det|$ of $\mathrm{GL}(n)$ used to define the line bundle $|\Lambda|$ is trivial on the structure group $\mathrm{O}(n)$, so that the density bundle is canonically isomorphic to $M \times \mathbb{R}$. It follows that $|dx|$ is covariant constant under the Levi-Civita derivative. We will often write dx instead of $|dx|$, although this notation should be restricted to oriented manifolds.

If f is a smooth function on a Riemannian manifold M, the **gradient** of M, denoted $\operatorname{grad} f$, is the vector field on M dual to df, that is, for every vector field X on M,

$$(\operatorname{grad} f, X) = X \cdot f.$$

By a smooth path $x_t : [0,1] \to M$ on a Riemannian manifold, we mean the restriction to $[0,1]$ of a smooth map $x_t : (-\varepsilon, 1+\varepsilon) \to M$ for some $\varepsilon > 0$. If x_t is such a path, we define its length by the formula

$$L(x) = \int_0^1 \sqrt{(\dot{x}_t, \dot{x}_t)}\, dt,$$

where by $\dot{x}_t$ we mean the tangent vector in T_{x_t} to the path x_t at time t. Define the Riemannian distance between two points $d(x_0, x_1)$ to be the infimum of $L(x_t)$ over all smooth paths between them.

A smooth path x_t is called a **geodesic** if it satisfies the ordinary differential equation

$$\nabla_{\dot{x}_t} \dot{x}_t = 0.$$

Like any second-order ordinary differential equation, the equation for a geodesic has a unique solution for t small, given the starting point x_0 and the derivative $\mathbf{x} = \dot{x}_0 \in T_{x_0}M$ at the starting point. The end point x_1 of the resulting curve is called the **exponential** of the vector $\mathbf{x}$ at x_0 and is written $\exp_{x_0} \mathbf{x}$. It is defined if $\mathbf{x}$ is small enough. (This terminology arises because on a compact Lie group with the left and right invariant metric, the geodesics turn out to be the same as the one-parameter subgroups and we recover the exponential map of the Lie group.)

The derivative $d\exp_{x_0}$ of the exponential map at x_0 itself is just the identity map of $T_{x_0}M$, so that by the inverse-function theorem, the exponential map is a diffeomorphism from a small ball around zero in $T_{x_0}M$ to a neighbourhood of x_0 in M. (The radius of the largest ball in $T_{x_0}M$ for which this is true is called the **injectivity radius** at x_0.) Thus the exponential map defines a system of coordinates around x_0, called the **normal coordinate** system (this is also known as the canonical coordinate system). The important property of this coordinate system is that the rays emanating from the origin in $T_{x_0}M$ map onto geodesics in M. We will use letters of the form $\mathbf{x}$ to denote coordinates in the normal coordinate system: thus, $\mathbf{x} \in T_{x_0}M$ represents the coordinates of the point $\exp_{x_0} \mathbf{x}$.

We will now collect some formulas for the pull-back of the metric and covariant derivatives by the exponential map. In particular we will see that the ray from 0 to $\mathbf{x}$ is the curve of shortest length between x_0 and $\exp_{x_0} \mathbf{x}$ when $\mathbf{x}$ is small.

If we apply the construction of Proposition 1.17 to the tangent bundle with its Levi-Civita connection, we obtain a smooth trivialization of TM over a normal coordinate chart centred at a point $x_0 \in M$. Choose an orthonormal frame of the tangent space $T_{x_0}M$, with respect to which the coordinate functions are $\mathbf{x}^i$, and the corresponding partial derivatives are ∂_i; thus, the vectors ∂_i are orthonormal at x_0, although not in general elsewhere. In this coordinate system, the point x_0 has coordinates $\mathbf{x} = 0$. The radial vector field $\mathscr{R}$ is the vector field defined by the formula $\mathscr{R} = \sum_{i=1}^n \mathbf{x}^i \partial_i$. Let e_i be the orthonormal frame of TM obtained from the smooth trivialization of TM in this coordinate patch; thus, $e_i = \partial_i + O(|\mathbf{x}|)$ and $\nabla_{\mathscr{R}} e_i = 0$.

Proposition 1.27. *In the normal coordinate system, the following formulas hold:*

1. $\nabla_{\mathscr{R}} \mathscr{R} = \mathscr{R}$
2. $\mathscr{R} = \sum_i \mathbf{x}^i e_i$, *and thus* $(\mathscr{R}, \mathscr{R}) = |\mathbf{x}|^2$
3. $(\mathscr{R}, \partial_i) = \mathbf{x}^i$
4. $d(x_0, \exp_{x_0} \mathbf{x}) = |\mathbf{x}|$

Proof. (1) In the normal coordinate system, the curve $\mathbf{x}_t = t\mathbf{x}$ is a geodesic. Since $\mathscr{R}(\mathbf{x}_t) = t\dot{\mathbf{x}}_t$, the geodesic equation gives

$$\nabla_{\mathscr{R}} \mathscr{R} = t \nabla_{\dot{\mathbf{x}}_t} (t\dot{\mathbf{x}}_t) = t\dot{\mathbf{x}}_t = \mathscr{R}.$$

(2) If we differentiate the function $(\mathscr{R}, e_i)$ with respect to the vector field $\mathscr{R}$, we obtain

$$\mathscr{R}(\mathscr{R}, e_i) = (\nabla_{\mathscr{R}} \mathscr{R}, e_i) + (\mathscr{R}, \nabla_{\mathscr{R}} e_i) = (\mathscr{R}, e_i).$$

In other words, $(\mathscr{R}, e_i)$ is homogeneous of order one. Since

$$(\mathscr{R}, e_i) = \sum_j \mathbf{x}^j (\partial_j, e_i) = \mathbf{x}^i + O(|\mathbf{x}|^2),$$

the result follows from these two equations.

(3) The function $(\mathscr{R}, \partial_i)$ is equal to $\sum_j \mathbf{x}^j (\partial_i, \partial_j)$, so it must equal $\mathbf{x}^i + O(|\mathbf{x}|^2)$. On the other hand,

$$\mathscr{R}(\mathscr{R}, \partial_i) = (\nabla_{\mathscr{R}} \mathscr{R}, \partial_i) + (\mathscr{R}, \nabla_{\mathscr{R}} \partial_i).$$

Since the Levi-Civita connection is torsion-free and $[\mathscr{R}, \partial_i] = -\partial_i$, it follows that

$$\begin{aligned}(\mathscr{R}, \nabla_{\mathscr{R}} \partial_i) &= (\mathscr{R}, \nabla_{\partial_i} \mathscr{R}) + (\mathscr{R}, [\mathscr{R}, \partial_i]) \\ &= \tfrac{1}{2} \partial_i |\mathscr{R}|^2 - (\mathscr{R}, \partial_i).\end{aligned}$$

If we assemble all of the terms that we have obtained, we see that

$$\mathcal{R}(\mathcal{R},\partial_i) = \tfrac{1}{2}\partial_i|\mathcal{R}|^2 = \mathbf{x}^i,$$

from which the result follows.

(4) The equation (3) shows that $\mathcal{R}$ is orthogonal for the Riemannian metric to the vector fields $\mathbf{x}^i\partial_j - \mathbf{x}^j\partial_i$. If $\mathbf{x}_t$, $0 \le t \le 1$, is any curve from 0 to $\mathbf{x}$, decompose its tangent vector $\dot{\mathbf{x}}_t$ into the sum of a multiple of $\mathcal{R}$ and a vector lying in the span of the vector fields $\mathbf{x}^i\partial_j - \mathbf{x}^j\partial_i$ tangent to the sphere. It follows that

$$|\dot{\mathbf{x}}_t| \ge \left|\frac{d|\mathbf{x}_t|}{dt}\right|;$$

thus the length of the curve $\mathbf{x}_t$ is greater than or equal to $|\mathbf{x}|$. Since the length of the curve $\mathbf{x}_t = t\mathbf{x}$ is equal to $|\mathbf{x}|$, the formula follows. □

A property of the normal coordinate system is that near the origin, the coefficients of the metric $(\partial_i, \partial_j)_{\mathbf{x}}$ agree with the Kronecker delta δ_{ij} up to second order. Although we will not use the full force of the next proposition in the rest of this book, we give its proof because of its applications to the theory of the heat kernel.

Proposition 1.28. *The Taylor expansion of the functions $g_{ij}(\mathbf{x})$, defined by $(\partial_i, \partial_j)_{\mathbf{x}}$, in normal coordinates at x_0 has the following form:*

$$g_{ij}(\mathbf{x}) \sim \delta_{ij} - \tfrac{1}{3}\sum_{kl} R_{ikjl}(x_0)\mathbf{x}^k\mathbf{x}^l + \sum_{|\alpha|\ge 3}(\partial^\alpha g_{ij})(x_0)\frac{\mathbf{x}^\alpha}{\alpha!}.$$

Proof. To prove this result, we will make use of two distinct frames of the tangent bundle TM in a normal coordinate chart around a point x_0, the first being given by the vectors ∂_i (which is *not* an orthonormal frame in general), and the second by the orthonormal frame e_i obtained by radial parallel transport of the orthonormal frame $(\partial_i)_{x_0}$ of $T_{x_0}M$. The matrix relating these two frames is the matrix θ_i^j such that

$$\partial_i = \sum_j \theta_i^j e_j.$$

With respect to the frame e_j, the fundamental one-form

$$\theta = \sum_i d\mathbf{x}^i\partial_i \in \mathcal{A}^1(M,TM)$$

equals $\sum\theta^j e_j$, where $\theta^j = \sum_i \theta_i^j d\mathbf{x}^i$. Thus the $\mathbb{R}^n$-valued one-form $\theta = (\theta^j)$ satisfies the structure equation

$$d\theta + \omega\wedge\theta = 0,$$

where ω is the $\operatorname{End}(\mathbb{R}^n)$-valued one-form of the Levi-Civita connection in the frame e_i. Observe that

$$\iota_{\mathcal{R}}\theta = \mathcal{R} = (\mathbf{x}^1,\dots,\mathbf{x}^n),$$

since $\mathscr{R} = \sum \mathbf{x}^i e_i$, as follows from Proposition 1.27(2).

Let $\Omega = d\omega + \omega \wedge \omega$ be the curvature of M in this coordinate system; as in Proposition 1.18, we see that $\mathscr{L}(\mathscr{R})\omega = \iota(\mathscr{R})\Omega$. Substituting the formulas $\iota_{\mathscr{R}}\omega = 0$ and $(\mathscr{L}(\mathscr{R}) - 1)\mathbf{x} = 0$, into the identity $\iota_{\mathscr{R}}(d\theta + \omega \wedge \theta) = 0$, we obtain

$$(\mathscr{L}(\mathscr{R}) - 1)\mathscr{L}(\mathscr{R})\theta = (\mathscr{L}(\mathscr{R})\omega)\mathbf{x} = (\iota(\mathscr{R})\Omega)\mathbf{x},$$

where we think of the curvature Ω as a matrix of two-forms, so that $\Omega\mathbf{x}$ is a vector of two-forms. Using the formula $\mathscr{R} = \sum \mathbf{x}^i e_i$ once more gives

$$\iota(\partial_k)(\mathscr{L}(\mathscr{R}) - 1)\mathscr{L}(\mathscr{R})\theta^a = \sum_{ij} \mathbf{x}^i \mathbf{x}^j (\Omega(\partial_i, \partial_k)\partial_j, e_a). \tag{1.21}$$

Now e_j is expressed as a function of the ∂_i by the inverse of the matrix θ^i_j, and $\theta^j_i(x_0) = \delta_{ij}$. Thus this equation determines the Taylor expansion of θ^a_b to order m in terms of the Taylor expansion of the curvature coefficients $R_{ijkl} = (\Omega(\partial_k, \partial_l)\partial_j, \partial_i)_{\mathbf{x}}$ to order $m-2$. In particular, the linear term is zero, and the quadratic piece may be calculated from (1.21)

$$\theta^a_b = \delta_{ab} - \frac{1}{6} \sum_{i,j} \mathbf{x}^i \mathbf{x}^j R_{jaib} + O(|\mathbf{x}|^3).$$

Since the metric on M is expressed in terms of the one-form θ by the formula

$$g_{ij}(\mathbf{x}) = \sum_k \theta^k_i \theta^k_j,$$

the Taylor coefficients of g_{ij} may be expressed in terms of those of θ. □

The differential $d_{\mathbf{x}} \exp_{x_0}$ of the exponential map $\exp_{x_0} : T_{x_0}M \to M$ at the point $\mathbf{x} \in T_{x_0}M$ is a map from $T_{x_0}M$ to T_xM, where $x = \exp_{x_0} \mathbf{x}$. Since $T_{x_0}M$ and T_xM are metric spaces, the Jacobian

$$j(\mathbf{x}) = |\det(d_{\mathbf{x}} \exp_{x_0})|$$

is well defined. By definition, if e_i is an orthonormal frame of T_xM and $\partial_i = \sum_j \theta^j_i e_j$, then $j(\mathbf{x}) = |\det(\theta^j_i)|$, so that

$$j(\mathbf{x}) = |\det(g_{ij}(\mathbf{x}))|^{1/2}. \tag{1.22}$$

The frames ∂_i and e_i of the tangent bundle TM that were introduced at the start of the above proof will be used constantly throughout this book, and the reader must be very careful to distinguish them. The first frame is frequently useful because it satisfies the formula $[\partial_i, \partial_j] = 0$, while the frame e_i possesses the advantage of being orthonormal at each point in the normal coordinate chart. In fact, we will only use the following two consequences of Propositions 1.18 and 1.28 in the rest of this book:

$$\omega_i(\mathbf{x}) = -\tfrac{1}{2} \sum_j F(\partial_i, \partial_j)_{x_0} \mathbf{x}^j + O(|\mathbf{x}|^2); \tag{1.23}$$

$$g_{ij}(\mathbf{x}) = 1 + O(|\mathbf{x}|^2). \tag{1.24}$$

1.3 Superspaces

Although this is not a book on supergeometry, we will constantly use the terminology of super-objects. Recall that a **superspace** E is a $\mathbb{Z}_2$-graded vector space

$$E = E^+ \oplus E^-,$$

and that a **superalgebra** is an algebra A whose underlying vector space is a superspace, and whose product respects the $\mathbb{Z}_2$-grading; in other words, $A^i \cdot A^j \subset A^{i+j}$. (We will denote the two elements of $\mathbb{Z}_2$ by $+$ and $-$, although we should really denote them by $0 \mod 2$ and $1 \mod 2$.) An exterior algebra is an example of a superalgebra, with the $\mathbb{Z}_2$-grading

$$\Lambda^\pm E = \sum_{(-1)^i = \pm 1} \Lambda^i E.$$

An ungraded vector space E is implicitly $\mathbb{Z}_2$-graded with $E^+ = E$ and $E^- = 0$. The algebra of endomorphisms $\mathrm{End}(E)$ of a superspace is a superalgebra, when graded in the usual way:

$$\begin{aligned}\mathrm{End}^+(E) &= \mathrm{Hom}(E^+,E^+) \oplus \mathrm{Hom}(E^-,E^-),\\ \mathrm{End}^-(E) &= \mathrm{Hom}(E^+,E^-) \oplus \mathrm{Hom}(E^-,E^+).\end{aligned}$$

Definition 1.29. *A* **superbundle** *on a manifold M is a bundle $\mathcal{E} = \mathcal{E}^+ \oplus \mathcal{E}^-$, where $\mathcal{E}^+$ and $\mathcal{E}^-$ are two vector bundles on M. Thus, the fibres of $\mathcal{E}$ are superspaces.*

If $\mathcal{E}$ is a vector bundle, we will identify it with the superbundle such that $\mathcal{E}^+ = \mathcal{E}$ and $\mathcal{E}^- = 0$.

There is a general principle in dealing with elements of a superspace, that whenever a formula involves commuting an element a past another b, one must insert a sign $(-1)^{|a|\cdot|b|}$, where $|a|$ is the parity of a, which equals 0 or 1 according to whether the degree of a is even or odd. Thus, the **supercommutator** of a pair of odd parity elements of a superalgebra is actually their anticommutator, due to the extra minus sign. We will use the same bracket notation for this supercommutator as is usually used for the commutator:

$$[a,b] = ab - (-1)^{|a|\cdot|b|} ba$$

This bracket satisfies the axioms of a **Lie superalgebra**:

1. $[a,b] + (-1)^{|a|\cdot|b|}[b,a] = 0,$
2. $[a,[b,c]] = [[a,b],c] + (-1)^{|a|\cdot|b|}[b,[a,c]].$

We say that a superalgebra is **super-commutative** if its superbracket vanishes identically: an example is the exterior algebra ΛE. The following definition gives the extension of the notion of a trace to the superalgebra setting.

Definition 1.30. *A* **supertrace** *on a superalgebra A is a linear form ϕ on A satisfying $\phi([a,b]) = 0$.*

On the superalgebra $\operatorname{End}(E)$, there is a canonical linear form given by the formula

$$\operatorname{Str}(a) = \begin{cases} \operatorname{Tr}_{E^+}(a) - \operatorname{Tr}_{E^-}(a) & \text{if } a \text{ is even,} \\ 0 & \text{if } a \text{ is odd.} \end{cases}$$

Proposition 1.31. *The linear form* Str *defined above is a supertrace on* $\operatorname{End}(E)$.

Proof. We must verify that $\operatorname{Str}[a,b] = 0$. If a and b have opposite parity, this is clear, since $[a,b]$ is then odd in parity and hence $\operatorname{Str}[a,b] = 0$.

If $a = \begin{pmatrix} a^+ & 0 \\ 0 & a^- \end{pmatrix}$ and $b = \begin{pmatrix} b^+ & 0 \\ 0 & b^- \end{pmatrix}$ are both even, then

$$[a,b] = \begin{pmatrix} [a^+,b^+] & 0 \\ 0 & [a^-,b^-] \end{pmatrix}$$

has vanishing supertrace, since

$$\operatorname{Tr}_{E^+}[a^+,b^+] = \operatorname{Tr}_{E^-}[a^-,b^-] = 0.$$

If $a = \begin{pmatrix} 0 & a^- \\ a^+ & 0 \end{pmatrix}$ and $b = \begin{pmatrix} 0 & b^- \\ b^+ & 0 \end{pmatrix}$ are both odd, then

$$[a,b] = \begin{pmatrix} a^-b^+ + b^-a^+ & 0 \\ 0 & a^+b^- + b^+a^- \end{pmatrix}$$

has supertrace

$$\operatorname{Str}[a,b] = \operatorname{Tr}_{E^+}(a^-b^+ + b^-a^+) - \operatorname{Tr}_{E^-}(a^+b^- + b^+a^-) = 0. \qquad \square$$

If $E = E^+ \oplus E^-$ and $F = F^+ \oplus F^-$ are two superspaces, then their **tensor product** $E \otimes F$ is the superspace with underlying vector space $E \otimes F$ and grading

$$(E \otimes F)^+ = E^+ \otimes F^+ \oplus E^- \otimes F^-,$$
$$(E \otimes F)^- = E^+ \otimes F^- \oplus E^- \otimes F^+.$$

If A and B are superalgebras, then their tensor product $A \otimes B$ is the superalgebra whose underlying space is the $\mathbb{Z}_2$-graded tensor product of A and B, and whose product is defined by the rule

$$(a_1 \otimes b_1) \cdot (a_2 \otimes b_2) = (-1)^{|b_1| \cdot |a_2|} a_1 a_2 \otimes b_1 b_2.$$

(This product is imposed by the sign rule.) Note that this $\mathbb{Z}_2$-graded product is not the same as the usual product on $A \otimes B$, so is this algebra is sometimes denoted $A \hat{\otimes} B$; however, we will never make use of the ungraded tensor product, so there is no ambiguity.

Definition 1.32. *If A is a supercommutative algebra and E is a superspace, we extend the supertrace on* $\operatorname{End}(E)$ *to a map*

$$\operatorname{Str} : A \otimes \operatorname{End}(E) \to A,$$

by the formula $\operatorname{Str}(a \otimes M) = a \operatorname{Str}(M)$ *for $a \in A$ and $M \in \operatorname{End}(E)$.*

This extension of the supertrace still vanishes on supercommutators in $A \otimes \mathrm{End}(E)$, since

$$[a \otimes M, b \otimes N] = (-1)^{|M| \cdot |b|} ab \otimes [M,N] \tag{1.25}$$

when A is supercommutative.

If E_1 and E_2 are two superspaces, we may identify $\mathrm{End}(E_1 \otimes E_2)$ with $\mathrm{End}(E_1) \otimes \mathrm{End}(E_2)$, the action being as follows:

$$(a_1 \otimes a_2)(e_1 \otimes e_2) = (-1)^{|a_2| \cdot |e_1|}(a_1 e_1) \otimes (a_2 e_2),$$

where $a_i \in \mathrm{End}(E_i)$ and $e_i \in E_i$.

If E is a superspace, such that $\dim(E^{\pm}) = m_{\pm}$, we define its **determinant line** to be the one-dimensional vector space

$$\det(E) = (\Lambda^{m_+} E^+)^{-1} \otimes \Lambda^{m_-} E^-.$$

Here, we make use of the commonly-used notation L^{-1} for L^* when L is a one-dimensional vector space. The vector space $\det(E)$ is one-dimensional, and if $E^+ = E^-$, then $\det(E)$ is canonically isomorphic to $\mathbb{R}$.

Definition 1.33. *If $\mathscr{E} = \mathscr{E}^+ \oplus \mathscr{E}^-$ is a superbundle, then its* **determinant line bundle** $\det(\mathscr{E})$ *is the bundle*

$$\det(\mathscr{E}) = (\Lambda^{m_+} \mathscr{E}^+)^{-1} \otimes (\Lambda^{m_-} \mathscr{E}^-),$$

where $m_{\pm} = \mathrm{rk}(\mathscr{E}^{\pm})$.

For example, if M is an n-dimensional manifold, $\det(TM)$ is the volume-form bundle $\Lambda^n T^*M$.

Definition 1.34. *A* **Hermitian superspace** *is a superspace E such that both E^+ and E^- are complex vector spaces with Hermitian structures.*

If E is a Hermitian superspace, we say that $u \in \mathrm{End}^-(E)$ is odd self-adjoint if it has the form

$$u = \begin{pmatrix} 0 & u^- \\ u^+ & 0 \end{pmatrix},$$

where $u^+ : E^+ \to E^-$, and u^- is the adjoint of u^+.

Let V be a real vector space with basis e_i. We will denote $\iota_k = \iota(e_k)$ and $\varepsilon^k = \varepsilon(e^k)$. We will identify $\Lambda^k V^*$ with $(\Lambda^k V)^*$ by setting $\langle e^I, e_J \rangle = \delta^I_J$, where

$$I = \{1 \le i^1 < \cdots < i^k \le \dim(V)\} \quad \text{and}$$
$$J = \{1 \le j_1 < \cdots < j_k \le \dim(V)\}$$

are multi-indices, that is, subsets of $\{1, \ldots, \dim V\}$, and $e^I = e^{i_1} \ldots e^{i_k}$, $e_J = e_{j_1} \ldots e_{j_k}$. (Here, as occasionally in the rest of this book, we write vw instead of $v \wedge w$ for the product in an exterior algebra.) Using this identification, it is easy to see that $\varepsilon(\alpha)^* = \iota(\alpha) \in \mathrm{End}(\Lambda V)$ and that $\iota(v)^* = \varepsilon(v) \in \mathrm{End}(\Lambda V)$.

If $A \in \mathrm{End}(V)$, we denote by $\lambda(A)$ the unique derivation of the superalgebra ΛV which coincides with A on $V \subset \Lambda V$; it is given by the explicit formula

$$\lambda(A) = \sum_{jk} \langle e^j, Ae_k \rangle \varepsilon_j \iota^k. \tag{1.26}$$

A non-zero linear map $T : \Lambda V \to \mathbb{R}$ which vanishes on $\Lambda^k V$ for $k < \dim(V)$ is called a **Berezin integral**. Let us explain why such a linear map is called an integral. If V is a real vector space, the superalgebra ΛV^* is the algebra of polynomial functions on the purely odd superspace with $E^+ = 0$, $E^- = V$. If e_i is a basis for V with dual basis e^i, the elements $e^i \in \Lambda V^*$ play the role of coordinate function on V. From this point of view, the contraction $\iota(v) : \Lambda V^* \to \Lambda V^*$, $(v \in V)$, is the operation of differentiation in the direction v. We now see that a Berezin integral is an analogue of the Lebesgue integral: a Berezin integral $T : \Lambda V^* \to \mathbb{R}$ is a linear form on the function space ΛV^* which vanishes on "partial derivatives":

$$T \cdot \iota(v)\alpha = 0 \quad \text{for all } v \in V \text{ and } \alpha \in \Lambda V^*. \tag{1.27}$$

If V is an oriented Euclidean vector space, there is a canonical Berezin integral, defined by projecting $\alpha \in \Lambda V$ onto the component of the monomial $e_1 \wedge \ldots \wedge e_n$; here n is the dimension of V, and e_i form an oriented orthonormal basis of V. We will denote this Berezin integral by T:

$$T(e_I) = \begin{cases} 1, & |I| = n, \\ 0, & \text{otherwise.} \end{cases} \tag{1.28}$$

If $\alpha \in \Lambda V$, we will often denote $T(\alpha)$ by $\alpha_{[n]}$, although strictly speaking $\alpha_{[n]}$ is an element of $\Lambda^n V$ and not of $\mathbb{R}$. If $A \in \Lambda^2 V$, its exponential in the algebra ΛV will be denoted by $\exp_\Lambda A$.

Definition 1.35. *The* **Pfaffian** *of an element $A \in \Lambda^2 V$ is the number*

$$\mathrm{Pf}_\Lambda(A) = T(\exp_\Lambda A)$$

The Pfaffian of an element $A \in \mathfrak{so}(V)$ is the number

$$\mathrm{Pf}(A) = T\Big(\exp_\Lambda \sum_{i<j} \langle Ae_i, e_j \rangle e_i \wedge e_j\Big).$$

If $V = \mathbb{R}^2$ with orthonormal basis $\{e_1, e_2\}$, and if

$$A = \begin{pmatrix} 0 & -\theta \\ \theta & 0 \end{pmatrix},$$

then the Pfaffian of A is θ.

The Pfaffian vanishes if the dimension of V is odd, while if it is even, $\mathrm{Pf}(A)$ is a polynomial of homogeneous order $n/2$ in the components of A. If the orientation of V is reversed, it changes sign.

Proposition 1.36. *The Pfaffian of an antisymmetric linear map is a square root of the determinant:*

$$\mathrm{Pf}(A)^2 = \det(A).$$

Proof. By the spectral theorem, we can choose an oriented basis e_j of V such that there are real numbers c_j, $1 \leq j \leq n/2$ for which

$$Ae_{2j-1} = c_j e_{2j}$$
$$Ae_{2j} = -c_j e_{2j-1}.$$

In this way, we reduce the proof to the case in which V is the vector space $\mathbb{R}^2$, and $Ae_1 = \theta e_2, Ae_2 = -\theta e_1$. In this case, $\mathrm{Pf}(A) = \theta$, while the determinant of $A = \left(\begin{smallmatrix} 0 & -\theta \\ \theta & 0 \end{smallmatrix}\right)$ is θ^2. □

We will frequently use the notation

$$\det{}^{1/2}(A) = \mathrm{Pf}(A). \tag{1.29}$$

However, the choice of sign in the square root $\det^{1/2}(A)$ depends on the choice of orientation of E.

1.4 Superconnections

In this section, we describe Quillen's concept of a superconnection. If M is a manifold and $\mathscr{E} = \mathscr{E}^+ \oplus \mathscr{E}^-$ is a superbundle on M, let $\mathscr{A}(M,\mathscr{E})$ be the space of $\mathscr{E}$-valued differential forms on M. This space has a $\mathbb{Z}$-grading given by the degree of a differential form, and we will denote the component of exterior degree i of $\alpha \in \mathscr{A}(M,\mathscr{E})$ by $\alpha_{[i]}$. In addition, we have the total $\mathbb{Z}_2$-grading, which we will denote by

$$\mathscr{A}(M,\mathscr{E}) = \mathscr{A}^+(M,\mathscr{E}) \oplus \mathscr{A}^-(M,\mathscr{E}),$$

and which is defined by

$$\mathscr{A}^\pm(M,\mathscr{E}) = \sum_i \mathscr{A}^{2i}(M,\mathscr{E}^\pm) \oplus \sum_i \mathscr{A}^{2i+1}(M,\mathscr{E}^\mp);$$

for example, the space of sections of $\mathscr{E}^\pm$ is contained in $\mathscr{A}^\pm(M,\mathscr{E})$.

If $\mathfrak{g}$ is a bundle of Lie superalgebras on M, then $\mathscr{A}(M,\mathfrak{g})$ is a Lie superalgebra with respect to the Lie superbracket defined by

$$[\alpha_1 \otimes X_1, \alpha_2 \otimes X_2] = (-1)^{|X_1|\cdot|\alpha_2|}(\alpha_1 \wedge \alpha_2) \otimes [X_1, X_2].$$

Likewise, if $\mathscr{E}$ is a superbundle of modules for $\mathfrak{g}$ with respect to an action ρ, then $\mathscr{A}(M,\mathscr{E})$ is a supermodule for $\mathscr{A}(M,\mathfrak{g})$, with respect to the action

$$\rho(\alpha \otimes X)(\beta \otimes v) = (-1)^{|X|\cdot|\beta|}(\alpha \wedge \beta) \otimes (\rho(X)v).$$

In particular, this construction may be applied to the bundle of Lie superalgebras $\mathrm{End}(\mathscr{E})$, where $\mathscr{E}$ is a superbundle on M, since $\Lambda T^*M \otimes \mathscr{E}$ is a bundle of modules for the superalgebra $\Lambda T^*M \otimes \mathrm{End}(\mathscr{E})$; we see that $\mathscr{A}(M, \mathrm{End}(\mathscr{E}))$ is a Lie superalgebra, which has $\mathscr{A}(M,\mathscr{E})$ as a supermodule. We will frequently use the fact that any differential operator on $\mathscr{A}(M,\mathscr{E})$ which supercommutes with the action of $\mathscr{A}(M)$ is given by the action of an element of $\mathscr{A}(M, \mathrm{End}(\mathscr{E}))$; such an operator will be called local. (This is consistent with the "super" point of view, that $\mathscr{A}(M)$ is the algebra of functions on a supermanifold fibred over M, and that elements of $\mathscr{A}(M, \mathrm{End}(\mathscr{E}))$ are the zeroth order differential operators on this supermanifold.)

In defining a superconnection, Quillen abstracted the main properties of a covariant derivative, namely, that it is an odd operator satisfying Leibniz's rule. It turns out that many of the results which hold for ordinary connections continue to hold for superconnections.

Definition 1.37. *If $\mathscr{E}$ is a bundle of superspaces over a manifold M, then a* **superconnection** *on $\mathscr{E}$ is an odd-parity first-order differential operator*

$$\mathbb{A} : \mathscr{A}^{\pm}(M,\mathscr{E}) \to \mathscr{A}^{\mp}(M,\mathscr{E})$$

which satisfies Leibniz's rule in the $\mathbb{Z}_2$-graded sense: if $\alpha \in \mathscr{A}(M)$ and $\theta \in \mathscr{A}(M,\mathscr{E})$, then

$$\mathbb{A}(\alpha \wedge \theta) = d\alpha \wedge \theta + (-1)^{|\alpha|} \alpha \wedge \mathbb{A}\theta.$$

Let $\mathbb{A}$ be a superconnection on $\mathscr{E}$. The operator $\mathbb{A}$ may be extended to act on the space $\mathscr{A}(M, \mathrm{End}(\mathscr{E}))$ in a way consistent with Leibniz's rule:

$$\mathbb{A}\alpha = [\mathbb{A}, \alpha] \quad \text{for } \alpha \in \mathscr{A}(M, \mathrm{End}(\mathscr{E})).$$

In order to check that $\mathbb{A}\alpha$, as defined by this formula, is an element of $\mathscr{A}(M, \mathrm{End}(\mathscr{E}))$, we need only observe that the operator $[\mathbb{A}, \alpha]$ commutes with exterior multiplication by any differential form $\beta \in \mathscr{A}(M)$.

Define the **curvature** of a superconnection $\mathbb{A}$ to be the operator $\mathbb{A}^2$ on $\mathscr{A}(M,\mathscr{E})$.

Proposition 1.38. *The curvature F is a local operator, and hence is given by the action of a differential form $F \in \mathscr{A}(M, \mathrm{End}(\mathscr{E}))$, which has total degree even, and satisfies the Bianchi identity $\mathbb{A}F = 0$.*

Proof. The operator $\mathbb{A}^2$ is local because it supercommutes with exterior multiplication by any $\alpha \in \mathscr{A}(M)$:

$$[\mathbb{A}^2, \varepsilon(\alpha)] = [\mathbb{A}, [\mathbb{A}, \varepsilon(\alpha)]] = \varepsilon(d^2\alpha) = 0.$$

Thus, $\mathbb{A}^2$ is given by the action of a differential form $F \in \mathscr{A}(M, \mathrm{End}(\mathscr{E}))$. The Bianchi identity is just another way of writing the obvious identity $[\mathbb{A}, \mathbb{A}^2] = 0$. □

It is clear that in the case when the superconnection $\mathscr{E}$ is a connection, the above definition of curvature coincides with that of Section 1.

A superconnection is entirely determined by its restriction to $\Gamma(M,\mathscr{E})$, which may be any operator $\mathbb{A} : \Gamma(M,\mathscr{E}^\pm) \to \mathscr{A}^\mp(M,\mathscr{E})$ that satisfies

$$\mathbb{A}(fs) = df\cdot s + f\mathbb{A}s \quad \text{for all } f\in C^\infty(M),\, s\in\Gamma(M,\mathscr{E}).$$

Indeed, if we define $\mathbb{A}(\alpha\otimes s) = d\alpha s + (-1)^{|\alpha|}\alpha\mathbb{A}s$ for $\alpha\in\mathscr{A}(M)$ and $s\in\Gamma(M,\mathscr{E})$, this gives the extension of $\mathbb{A}$ to all of $\mathscr{A}(M,\mathscr{E})$.

In order to better understand what a superconnection consists of, we can break it up into its homogeneous components $\mathbb{A}_{[i]}$, which map $\Gamma(M,\mathscr{E})$ to $\mathscr{A}^i(M,\mathscr{E})$:

$$\mathbb{A} = \mathbb{A}_{[0]} + \mathbb{A}_{[1]} + \mathbb{A}_{[2]} + \dots$$

Proposition 1.39. *1. The operator $\mathbb{A}_{[1]}$ is a covariant derivative on the bundle $\mathscr{E}$ which preserves the sub-bundles $\mathscr{E}^+$ and $\mathscr{E}^-$.*

2. The operators $\mathbb{A}_{[i]}$ for $i\neq 1$ are given by the action of differential forms $\omega_{[i]}\in\mathscr{A}^i(M,\mathrm{End}(\mathscr{E}))$ on $\mathscr{A}(M,\mathscr{E})$, where $\omega_{[i]}$ lies in $\mathscr{A}^i(M,\mathrm{End}^-(\mathscr{E}))$ if i is even, and in $\mathscr{A}^i(M,\mathrm{End}^+(\mathscr{E}))$ if i is odd.

Proof. Let us decompose Leibniz's rule for $\mathbb{A}$ acting on $\Gamma(M,\mathscr{E})$ according to degree:

$$\mathbb{A}(fs) = \sum_{i=0}^{n}\mathbb{A}_{[i]}(fs) = df\otimes s + f\sum_{i=0}^{n}\mathbb{A}_{[i]}s.$$

It follows immediately from this that the operator $\mathbb{A}_{[1]}$ is a covariant derivative on the bundle $\mathscr{E}$; it preserves the $\mathbb{Z}_2$-grading of $\mathscr{E}$, since $\mathbb{A}_{[1]}$ has odd total degree, so that it is the direct sum of covariant derivatives on the bundles $\mathscr{E}^\pm$.

On the other hand, for $i\neq 1$, we see that $\mathbb{A}_{[i]}(fs) = f(\mathbb{A}_{[i]}s)$, in other words, that $\mathbb{A}_{[i]} : \Gamma(M,\mathscr{E}) \to \mathscr{A}^i(M,\mathscr{E})$ is given by the action of a differential form $\omega_{[i]}$ in $\mathscr{A}_{[i]}(M,\mathrm{End}(\mathscr{E}))$; from the fact that $\mathbb{A}$ is an odd operator it follows that $\omega_{[i]}$ is of odd total degree.

The actions of the operators $\mathbb{A}$ and $\omega_{[0]} + \mathbb{A}_{[1]} + \omega_{[2]} + \dots$ on $\mathscr{A}(M,\mathscr{E})$ must agree, since any two operators that agree when restricted to $\Gamma(M,\mathscr{E}) = \mathscr{A}^0(M,\mathscr{E})$ and satisfy Leibniz's rule are equal. □

Corollary 1.40. *The space of superconnections on $\mathscr{E}$ is an affine space modelled on the vector space $\mathscr{A}^-(M,\mathrm{End}(\mathscr{E}))$. Thus, if $\mathbb{A}_s$ is a smooth one-parameter family of superconnections, then $d\mathbb{A}_s/ds$ lies in $\mathscr{A}^-(M,\mathrm{End}(\mathscr{E}))$ for each s.*

We will call $\mathbb{A}_{[1]}$ the **covariant derivative component** of the superconnection $\mathbb{A}$.

As an example of a superconnection, let us take a trivial bundle on M with fibre a complex superspace E. If $L\in C^\infty(M,\mathrm{End}^-(E))$, we may form a superconnection by adding L to the trivial connection d. The curvature of the superconnection $d+L$ is then $F = dL + L^2$. Relative to the splitting $E = E^+\oplus E^-$, we may decompose L as

$$L = \begin{pmatrix} 0 & L^- \\ L^+ & 0 \end{pmatrix}$$

where $L^+ \in C^\infty(M, \mathrm{Hom}(E^+, E^-))$ and $L^- \in C^\infty(M, \mathrm{Hom}(E^-, E^+))$. In terms of this decomposition, the curvature of $d+L$ may be written

$$F = \begin{pmatrix} L^-L^+ & dL^- \\ dL^+ & L^+L^- \end{pmatrix}.$$

If $\mathscr{E}$ and $\mathscr{F}$ are superbundles, then $\mathscr{E} \otimes \mathscr{F}$ is the superbundle graded by

$$(\mathscr{E} \otimes \mathscr{F})^+ = \mathscr{E}^+ \otimes \mathscr{F}^+ \oplus \mathscr{E}^- \otimes \mathscr{F}^-,$$
$$(\mathscr{E} \otimes \mathscr{F})^- = \mathscr{E}^+ \otimes \mathscr{F}^- \oplus \mathscr{E}^- \otimes \mathscr{F}^+.$$

If $\mathbb{A}$ and $\mathbb{B}$ are superconnections respectively on $\mathscr{E}$ and $\mathscr{F}$, there is a tensor product superconnection $\mathbb{A} \otimes 1 + 1 \otimes \mathbb{B}$ on $\mathscr{E} \otimes \mathscr{F}$ defined by the formula

$$(\mathbb{A} \otimes 1 + 1 \otimes \mathbb{B})(\alpha \wedge \beta) = \mathbb{A}\alpha \wedge \beta + (-1)^{|\alpha|} \alpha . \mathbb{B}\beta,$$

for $\alpha \in \mathscr{A}(M, \mathscr{E})$ and $\beta \in \mathscr{A}(M, \mathscr{F})$.

If $\phi : M \to N$ is a smooth map and $\mathscr{E}$ is a superbundle on N with superconnnection $\mathbb{A}$, there is a unique superconnection $\phi^*\mathbb{A}$ on $\phi^*\mathscr{E}$ such that $\phi^*\mathbb{A}(\phi^*\alpha) = \phi^*(\mathbb{A}\alpha)$ for $\alpha \in \mathscr{A}(N, \mathscr{E})$, called the **pull-back** of $\mathbb{A}$ by ϕ.

1.5 Characteristic Classes

Let $\mathscr{E}$ be a superbundle, either real or complex, with superconnection $\mathbb{A}$, on a manifold M. Following Quillen, we will associate to this superconnection certain closed differential forms, called the characteristic forms of $\mathbb{A}$. These give invariants of the superbundle $\mathscr{E}$ in the de Rham cohomology $H^\bullet(M)$ of M; the cohomology classes thus obtained are called the characteristic classes of $\mathscr{E}$. We will only dealt with the special case of characteristic forms associated to invariant polynomials of the type $\mathrm{Str}(A^k)$. In Chapter 7, we will give a generalization of the theory of characteristic classes to the equivariant context, where a Lie group acts on the data.

In order to define the characteristic forms, we need the supertrace map on the space of differential forms $\mathscr{A}(M, \mathrm{End}(\mathscr{E}))$. This is defined by means of the supertrace $\mathrm{Str}_x : \mathrm{End}(\mathscr{E})_x \to \mathbb{C}$, defined in each fibre of $\mathrm{End}(\mathscr{E})$ by

$$\mathrm{Str} \begin{pmatrix} a & b \\ c & d \end{pmatrix} = \mathrm{Tr}(a) - \mathrm{Tr}(d).$$

Applying the construction of Definition 1.32 with auxiliary supercommutative algebra ΛT_x^*M, we obtain a bundle map

$$\mathrm{Str} : \Lambda T^*M \otimes \mathrm{End}(\mathscr{E}) \to \Lambda T^*M$$

which, applied to the sections of these bundles, gives the desired supertrace

$$\mathrm{Str} : \mathscr{A}(M, \mathrm{End}(\mathscr{E})) \to \mathscr{A}(M),$$

which is characterized by the formula $\operatorname{Str}(\alpha \otimes A) = \alpha \operatorname{Str}(A)$ for $\alpha \in \mathscr{A}(M)$ and $A \in \Gamma(M, \operatorname{End}(\mathscr{E}))$. The algebra $\mathscr{A}(M)$ is supercommutative, so by (1.25), this extension behaves like a supertrace from a superalgebra to $\mathbb{C}$, in that it vanishes on supercommutators. It is clear that the supertrace map preserves the $\mathbb{Z}_2$-gradings of these spaces, that is, it maps $\mathscr{A}^{\pm}(M, \operatorname{End}(\mathscr{E}))$ to $\mathscr{A}^{\pm}(M)$.

Suppose $f(z)$ is a polynomial in z. If $\mathbb{A}^2 \in \mathscr{A}^+(M, \operatorname{End}(\mathscr{E}))$ is the curvature of the superconnection $\mathbb{A}$ on $\mathscr{E}$, then $f(\mathbb{A}^2)$ is the element of $\mathscr{A}^+(M, \operatorname{End}(\mathscr{E}))$, defined by

$$f(\mathbb{A}^2) = \sum \frac{f^{(k)}(0)}{k!} (\mathbb{A}^2)^k.$$

If we apply the supertrace map

$$\operatorname{Str} : \mathscr{A}^+(M, \operatorname{End}(\mathscr{E})) \to \mathscr{A}^+(M)$$

to $f(\mathbb{A}^2)$, we obtain a differential form on M denoted by $\operatorname{Str} f(\mathbb{A}^2)$. We will call this differential form the **characteristic form**, or Chern-Weil form, of $\mathbb{A}$ corresponding to the polynomial $f(z)$.

Proposition 1.41. *1. The characteristic form* $\operatorname{Str}(f(\mathbb{A}^2))$ *is a closed differential form of even degree.*

2. (Transgression formula) If $\mathbb{A}_t$ is a differentiable one-parameter family of superconnections on $\mathscr{E}$, then

$$\frac{d}{dt} \operatorname{Str}(f(\mathbb{A}_t^2)) = d \operatorname{Str}\Big(\frac{d\mathbb{A}_t}{dt} f'(\mathbb{A}_t^2)\Big),$$

where $d\mathbb{A}_t/dt \in \mathscr{A}(M, \operatorname{End}(\mathscr{E}))$ is the derivative of $\mathbb{A}_t$ with respect to t.

3. If $\mathbb{A}_0$ and $\mathbb{A}_1$ are two superconnections on $\mathscr{E}$, the differential forms $\operatorname{Str}(f(\mathbb{A}_0^2))$ *and* $\operatorname{Str}(f(\mathbb{A}_1^2))$ *lie in the same de Rham cohomology class.*

Proof. The proof makes use of the following lemma.

Lemma 1.42. *For any $\alpha \in \mathscr{A}(M, \operatorname{End}(\mathscr{E}))$, $d(\operatorname{Str} \alpha) = \operatorname{Str}([\mathbb{A}, \alpha])$.*

Proof. In local coordinates on a subset $U \subset M$ and with respect to a trivialization of $\mathscr{E}$ over U, $\mathbb{A} = d + \omega$, where $\omega \in \mathscr{A}(U, \operatorname{End}(\mathscr{E}))$ is the superconnection form of $\mathbb{A}$ in this coordinate system, and hence $\operatorname{Str}([\mathbb{A}, \alpha]) = \operatorname{Str}([d, \alpha]) + \operatorname{Str}([\omega, \alpha])$. The second term $\operatorname{Str}([\omega, \alpha])$ vanishes, while the first equals $\operatorname{Str}(d\alpha)$. □

Using this lemma, we see that

$$d \operatorname{Str}(f(\mathbb{A}^2)) = \operatorname{Str}([\mathbb{A}, f(\mathbb{A}^2)]) = 0.$$

This proves that $\operatorname{Str}(f(\mathbb{A}^2))$ is closed; the fact that it has even degree is clear, since $f(\mathbb{A}^2)$ is even, and Str preserves degree.

Let $\mathbb{A}_t$ be a differentiable one-parameter family of superconnections on $\mathscr{E}$, with curvatures $\mathbb{A}_t^2$. We see that

$$\frac{d\mathbb{A}_t^2}{dt} = \Big[\mathbb{A}_t, \frac{d\mathbb{A}_t}{dt}\Big],$$

(where the bracket is, of course, an anticommutator, since both $\mathbb{A}_t$ and $d\mathbb{A}_t/dt$ are odd operators). We now use the following formula: if $\alpha_t : [0,1] \to \mathscr{A}^+(M, \mathrm{End}(\mathscr{E}))$ is a family of differential forms on M with values in $\mathrm{End}(\mathscr{E})$ of even total degree, then

$$\frac{d}{dt}\operatorname{Str}(f(\alpha_t)) = \operatorname{Str}(\dot{\alpha}_t f'(\alpha_t)),$$

where $\dot{\alpha}_t$ is the derivative of α_t with respect to t. To prove this, it suffices to consider the case $f(z) = z^n$:

$$\begin{aligned}\frac{d}{dt}\operatorname{Str}(\alpha_t^n) &= \operatorname{Str}\Bigl(\sum_{i=0}^{n-1} \alpha_t^i \dot{\alpha}_t \alpha_t^{n-i-1}\Bigr)\\ &= n \operatorname{Str}(\dot{\alpha}_t \alpha_t^{n-1})\end{aligned}$$

where the last step uses the fact that $\operatorname{Str}(\cdot)$ is a supertrace in order to replace $\operatorname{Str}(\alpha_t^i \dot{\alpha}_t \alpha_t^{n-i-1})$ by $\operatorname{Str}(\dot{\alpha}_t \alpha_t^{n-1})$. Applying this with $\alpha_t = \mathbb{A}_t^2$, we obtain

$$\begin{aligned}\frac{d}{dt}\operatorname{Str}(f(\mathbb{A}_t^2)) &= \operatorname{Str}\Bigl(\frac{d\mathbb{A}_t^2}{dt} f'(\mathbb{A}_t^2)\Bigr)\\ &= \operatorname{Str}\Bigl(\Bigl[\mathbb{A}_t, \frac{d\mathbb{A}_t}{dt} f'(\mathbb{A}_t^2)\Bigr]\Bigr) \quad \text{since } [\mathbb{A}_t, \mathbb{A}_t^2] = 0\\ &= d \operatorname{Str}\Bigl(\frac{d\mathbb{A}_t}{dt} f'(\mathbb{A}_t^2)\Bigr).\end{aligned}$$

If $\mathbb{A}_0$ and $\mathbb{A}_1$ are two superconnections and $\omega = \mathbb{A}_1 - \mathbb{A}_0$, then we may integrate the transgression formula applied to the family

$$\mathbb{A}_t = (1-t)\mathbb{A}_0 + t\mathbb{A}_1 = \mathbb{A}_0 + t\omega.$$

We see that

$$\operatorname{Str}(f(\mathbb{A}_1^2)) - \operatorname{Str}(f(\mathbb{A}_0^2)) = d\int_0^1 \operatorname{Str}(\omega f'(\mathbb{A}_t^2))\, dt$$

which shows explicitly that $\operatorname{Str}(f(\mathbb{A}_0^2))$ and $\operatorname{Str}(f(\mathbb{A}_1^2))$ differ by an exact form. □

There is another way to obtain the transgression formula, which is similar to Proposition 1.24. Introduce the space $M \times \mathbb{R}$, with projection $q : M \times \mathbb{R} \to M$. Using the family $\mathbb{A}_t$ of superconnections, we may define a superconnection $\tilde{\mathbb{A}}$ on the superbundle $q^*\mathscr{E}$, by the formula

$$(\tilde{\mathbb{A}}\beta)(x,s) = (\mathbb{A}_s\beta(\cdot,s))(x) + ds \wedge \frac{\partial\beta(x,s)}{\partial s}.$$

The curvature $\tilde{\mathscr{F}}$ of $\tilde{\mathbb{A}}$ is

$$\tilde{\mathscr{F}} = \mathscr{F}_s - \frac{d\mathbb{A}_s}{ds} \wedge ds,$$

where $\mathscr{F}_s = \mathbb{A}_s^2$ is the curvature of $\mathbb{A}_s$; thus,

$$f(\widetilde{\mathscr{F}}) = f\left(\mathscr{F}_s - \frac{d\mathbb{A}_s}{ds} \wedge ds\right).$$

Since $ds \wedge ds = 0$, we see as in the preceding proof that underneath the supertrace,

$$\operatorname{Str}(f(\widetilde{\mathscr{F}})) = \operatorname{Str}(f(\mathscr{F}_s)) - \operatorname{Str}\left(\frac{d\mathbb{A}_s}{ds} f'(\mathscr{F}_s)\right) \wedge ds.$$

Since $\operatorname{Str}(f(\widetilde{\mathscr{F}}))$ is closed in $\mathscr{A}(M \times \mathbb{R})$, we infer that

$$\frac{d \operatorname{Str}(f(\mathscr{F}_s))}{ds} = d \operatorname{Str}\left(\frac{d\mathbb{A}_s}{ds} f'(\mathscr{F}_s)\right).$$

A special case of the above construction of characteristic forms is that in which the bundle $\mathscr{E}$ is ungraded and the superconnection $\mathbb{A}$ is a covariant derivative ∇ with curvature $F \in \mathscr{A}^2(M, \operatorname{End}(\mathscr{E}))$; then $\operatorname{Str}(\nabla^2)^k$ is just the differential form $\operatorname{Tr}(F^k) \in \mathscr{A}^{2k}(M)$. In this case we can allow $f(z)$ to be a power series in z, since the curvature F is a nilpotent element in $\mathscr{A}(M, \operatorname{End}(\mathscr{E}))$. The differential form $\operatorname{Tr}(f(F))$ is usually called the Chern-Weil form associated to the invariant power series $A \mapsto \operatorname{Tr}(f(A))$ on $\mathfrak{gl}(N)$, (where N is the rank of $\mathscr{E}$). If $\mathbb{A}$ is a superconnection, we can allow f to be any power series with infinite radius of convergence.

It should be clear that any differential form which is a polynomial

$$P(\operatorname{Str}(f_1(\mathbb{A}^2)), \dots, \operatorname{Str}(f_k(\mathbb{A}^2)))$$

is a closed even differential form. Since the ring of invariant polynomials on $\mathfrak{gl}(N)$ is generated by the polynomials $\operatorname{Tr}(a^k)$, the ring of all forms thus obtained coincides with the classical ring of Chern-Weil forms, in the case of an ungraded bundle.

The reader is **warned** that topologists prefer to consider the characteristic forms defined by the formula $\operatorname{Str} f(-F/2\pi i)$, where F is the curvature of a covariant derivative ∇, because the Chern class, the cohomology class of $c(F) = \det(1 - F/2\pi i)$, defines an element of the integral cohomology of M only if these negative powers of $2\pi i$ are included. However, from the point of view of the local index theorem, this is not as natural, which is why we have preferred to give our less conventional definition.

Three cases of the construction given above are of special importance for us.

(1) If $\mathscr{E} = \mathscr{E}^+ \oplus \mathscr{E}^-$ is a complex $\mathbb{Z}_2$-graded vector bundle and $f(z) = \exp(-z)$, the characteristic form obtained is called the **Chern character** form of the superconnection,

$$\operatorname{ch}(\mathbb{A}) = \operatorname{Str}(e^{-\mathbb{A}^2}). \tag{1.30}$$

The Chern character form is additive with respect to taking the direct sum of superbundles with superconnections:

$$\operatorname{ch}(\mathbb{A} \oplus \mathbb{B}) = \operatorname{ch}(\mathbb{A}) + \operatorname{ch}(\mathbb{B}). \tag{1.31}$$

Furthermore, it is multiplicative with respect to taking the tensor product of superbundles with superconnections:

$$\mathrm{ch}(\mathbb{A}\otimes 1+1\otimes\mathbb{B})=\mathrm{ch}(\mathbb{A})\wedge\mathrm{ch}(\mathbb{B}). \tag{1.32}$$

These two formulas explain why the differential form $\mathrm{ch}(\mathbb{A})$ is called a character.

The transgression formula for the Chern character of a family of superconnections gives

$$\frac{d}{dt}\mathrm{ch}(\mathbb{A}_t)=-d\,\mathrm{Str}\left(\frac{d\mathbb{A}_t}{dt}e^{-\mathbb{A}_t^2}\right),$$

which implies that

$$\mathrm{ch}(\mathbb{A}_0)-\mathrm{ch}(\mathbb{A}_1)=d\int_0^1\mathrm{Str}\left(\frac{d\mathbb{A}_t}{dt}e^{-\mathbb{A}_t^2}\right)dt. \tag{1.33}$$

This proves the following proposition.

Proposition 1.43. *The cohomology class of* $\mathrm{ch}(\mathbb{A})$ *is independent of the superconnection* $\mathbb{A}$ *on* $\mathscr{E}$*; we will denote it by* $\mathrm{ch}(\mathscr{E})$.

In particular, we see that replacing $\mathbb{A}$ by its covariant derivative component $\nabla=\mathbb{A}_{[1]}$ does not change the cohomology class of the Chern character:

$$[\mathrm{ch}(\mathbb{A})]=[\mathrm{ch}(\nabla)].$$

However, the connection ∇ preserves the bundles $\mathscr{E}^{\pm}$, and if we denote by ∇^+ and ∇^- the restrictions of ∇ to $\mathscr{E}^+$ and $\mathscr{E}^-$, we have

$$\mathrm{ch}(\mathbb{A})=\mathrm{ch}(\nabla)+d\alpha=\mathrm{ch}(\nabla^+)-\mathrm{ch}(\nabla^-)+d\alpha$$

for some differential form α. In particular, if we take the corresponding cohomology classes, we obtain

$$\mathrm{ch}(\mathscr{E})=[\mathrm{ch}(\nabla^+)]-[\mathrm{ch}(\nabla^-)]=\mathrm{ch}(\mathscr{E}^+)-\mathrm{ch}(\mathscr{E}^-). \tag{1.34}$$

(2) If $\mathscr{E}$ is a real vector bundle with covariant derivative ∇ of curvature F, we associate to it its ***Â*-genus** form,

$$\hat{A}(\nabla)=\det{}^{1/2}\left(\frac{F/2}{\sinh F/2}\right)\in\mathscr{A}^{4\bullet}(M,\mathbb{R}). \tag{1.35}$$

Using the formula

$$\det(A)=\exp\mathrm{Tr}(\log A),$$

we see that $\hat{A}(\nabla)$ may be understood as the exponential of the characteristic form associated to the power series $\frac{1}{2}\log((z/2)/\sinh(z/2))$, that is,

$$\hat{A}(\nabla)=\exp\mathrm{Tr}\left(\frac{1}{2}\log\left(\frac{F/2}{\sinh(F/2)}\right)\right).$$

The $\hat{A}$-genus is multiplicative, in the sense that

$$\hat{A}(\nabla_1 \oplus \nabla_2) = \hat{A}(\nabla_1) \wedge \hat{A}(\nabla_2).$$

The reason that $\hat{A}(\nabla)$ only has non-vanishing terms in dimensions divisible by 4 is that the function $f(z)$ is an even function of z, so $f(\nabla^2)$ only involves even powers of the curvature ∇^2. In fact, the coefficients of the $\hat{A}$-genus of $\mathscr{E}$ are polynomials in what are called the Pontryagin classes of M, but we will not need to know how these are defined; the first few terms of the expansion of $\hat{A}(\nabla)$ in terms of the classes $\mathrm{Tr}(F^k)$ are as follows:

$$\hat{A}(\mathscr{E}, \nabla) = 1 - \frac{1}{2^2.12}\mathrm{Tr}(F^2) + \frac{1}{2^4.360}\mathrm{Tr}(F^4) + \frac{1}{2^4.288}\mathrm{Tr}(F^2)^2 + \dots \tag{1.36}$$

The cohomology class of $\hat{A}(\nabla)$ will be denoted $\hat{A}(\mathscr{E})$. In particular, the $\hat{A}$-genus $\hat{A}(M)$ of a manifold M is defined to be the $\hat{A}$-genus of its tangent bundle TM.

(3) If $\mathscr{E}$ is an ungraded complex vector bundle with a covariant derivative ∇ with curvature F, we define its **Todd genus** form by the formula

$$\mathrm{Td}(\nabla) = \exp \mathrm{Tr}\left(\log\left(\frac{F}{e^F - 1}\right)\right).$$

Thus

$$\mathrm{Td}(\nabla) = \det\left(\frac{F}{e^F - 1}\right) \in \mathscr{A}^{2\bullet}(M, \mathbb{C}).$$

Like $\hat{A}(\nabla)$, the Todd genus is multiplicative. The cohomology class of $\mathrm{Td}(\nabla)$ will be denoted $\mathrm{Td}(\mathscr{E})$. In particular, the Todd genus $\mathrm{Td}(M)$ of a complex manifold M is the Todd genus of its tangent bundle TM.

As we will see, the $\hat{A}$-genus and its cousin the Todd genus arise quite naturally in a number of situations in mathematics, for example, in the Atiyah-Singer index theorem and in the Weyl and Kirillov character formulas.

1.6 The Euler and Thom Classes

In this section, we will define a characteristic class, the Euler class, which differs from the characteristic classes defined in the last section, in that it is only defined for oriented bundles. However, we will see that it has a definition quite analogous to that of the Chern character. This section is a prerequisite only to reading Section 7.7.

We start by describing the Thom form of an oriented vector bundle. Let M be a compact oriented manifold and let $p : \mathscr{V} \to M$ be a real oriented vector bundle over M of rank N. Let $j : M \to \mathscr{V}$ be the inclusion of M into $\mathscr{V}$ as the zero section.

Definition 1.44. *A* **Thom form** *for the bundle $\mathscr{V}$ is a closed differential form $U \in \mathscr{A}_c^N(\mathscr{V})$ such that*

$$(2\pi)^{-N/2} \int_{\mathscr{V}/M} U = 1 \in \mathscr{A}^0(M).$$

We will show that a Thom form exists, following a method of Mathai-Quillen.

The following lemma is a fibred version of the homotopy used to prove the Poincaré lemma.

Lemma 1.45. *Let $\mathscr{R}$ be the vertical Euler vector field on $\mathscr{V}$, and let $h_t(v) = tv$. With $H : \mathscr{A}^\bullet(\mathscr{V}) \to \mathscr{A}^{\bullet-1}(\mathscr{V})$ defined by the formula*

$$H\beta = \int_0^1 h_t^*(\iota(\mathscr{R})\beta)t^{-1}dt.$$

we have the homotopy formula

$$\beta - p^*(j^*\beta) = (dH + Hd)\beta \quad \text{for all } \beta \in \mathscr{A}(\mathscr{V}).$$

Proof. First, note that the integral converges in the definition of H because $\mathscr{R}$ vanishes at 0. If $\beta \in \mathscr{A}(\mathscr{V})$, let $\beta_t = h_t^*\beta$, and observe that $\beta_0 = p^*j^*\beta$ and that $\beta_1 = \beta$. Differentiating by t and using integrating the Cartan homotopy formula

$$\mathscr{L}(\mathscr{R})\beta = d(\iota(\mathscr{R})\beta) + \iota(\mathscr{R})(d\beta),$$

we obtain the lemma. □

Corollary 1.46. *The maps*

$$\beta \longmapsto j^*\beta : H^\bullet(\mathscr{V}) \to H^\bullet(M) \quad \text{and}$$
$$\alpha \longmapsto p^*\alpha : H^\bullet(M) \to H^\bullet(\mathscr{V})$$

are inverses to each other.

We need another lemma, which is a consequence of Proposition 1.24.

Lemma 1.47. *Let $p : \mathscr{V} \to M$ be a real oriented vector bundle over a manifold M. There is a smooth bilinear map $m(\alpha, \beta)$ from $\mathscr{A}_c^i(\mathscr{V}) \times \mathscr{A}_c^j(\mathscr{V})$ to $\mathscr{A}_c^{i+j-\mathrm{rk}(\mathscr{V})-1}(\mathscr{V})$ such that*

$$(p^*p_*\alpha) \wedge \beta - \alpha \wedge (p^*p_*\beta) = dm(\alpha, \beta) - (m(d\alpha, \beta) + (-1)^{|\alpha|} m(\alpha, d\beta)).$$

Proof. Consider the one-parameter family k^t of submersions from the Whitney sum $\mathscr{V} \times_p \mathscr{V}$ to $\mathscr{V}$ given by the formula

$$(v, w) \in \mathscr{V}_x \times \mathscr{V}_x \longmapsto tv + (1-t)w \in \mathscr{V}_x.$$

If α and β are differential forms, we may form their external product $\alpha \boxtimes_p \beta$, which is a differential form on $\mathscr{V} \times_p \mathscr{V}$. We easily see that

$$k_*^0(\alpha \boxtimes_p \beta) = (p^*p_*\alpha) \wedge \beta,$$
$$k_*^1(\alpha \boxtimes_p \beta) = \alpha \wedge (p^*p_*\beta).$$

Thus, it suffices to take $m(\alpha, \beta)$ equal to

$$m(\alpha, \beta) = \int_0^1 \tilde{k}_*^t(\alpha \boxtimes_p \beta)\, dt.$$

□

The following result explains the importance of Thom classes.

Proposition 1.48. *Let M be a compact oriented manifold and let $p : \mathcal{V} \to M$ be a real oriented vector bundle over M of rank N, with Thom form $U \in \mathcal{A}_c^N(\mathcal{V})$.*

1. The maps

$$\beta \longmapsto (2\pi)^{-N/2} p_* \beta : H_c^\bullet(\mathcal{V}) \to H^{\bullet-N}(M) \quad \text{and}$$
$$\alpha \longmapsto U \wedge p^* \alpha : H^\bullet(M) \to H_c^{\bullet+N}(\mathcal{V})$$

are inverses to each other.

2. For every closed differential form β in $\mathcal{A}(\mathcal{V})$,

$$(2\pi)^{-N/2} \int_{\mathcal{V}} U \wedge \beta = \int_M j^* \beta.$$

Proof. By (1.16), $(2\pi)^{-N/2} p_*(U \wedge p^*\alpha) = \alpha$ for all $\alpha \in \mathcal{A}(M)$.

The other direction of (1) follows from the homotopy formula

$$\beta - (2\pi)^{-N/2} U \wedge (p^* p_* \beta) = (2\pi)^{-N/2} dm(U, \beta) - (-1)^N (2\pi)^{-N/2} m(U, d\beta),$$

which follows from Lemma 1.47 on setting $\alpha = (2\pi)^{-N/2} U$.

To prove (2), we use the homotopy H of Lemma 1.45. It follows that if $\beta \in \mathcal{A}(\mathcal{V})$,

$$(2\pi)^{-N/2} \int_{\mathcal{V}} U \wedge \beta = (2\pi)^{-N/2} \int_{\mathcal{V}} U \wedge p^*(j^*\beta) + (2\pi)^{-N/2} \int_{\mathcal{V}} U \wedge dH\beta + (2\pi)^{-N/2} \int_{\mathcal{V}} U \wedge H d\beta.$$

The first term on the right-hand side equals $\int_M j^*\beta$, the second term vanishes by Stokes's theorem, since U is closed, and the third term vanishes since β is closed. □

In this section, we will construct a Thom form using the Berezin integral. Let us start with a model problem, the construction of a volume form U on an oriented Euclidean vector space V with integral $\int_V U = (2\pi)^{N/2}$; this is a Thom class for V considered as a vector bundle over a point. If $\mathbf{x}^1, \dots, \mathbf{x}^N$ is an orthonormal coordinate system on V, we define U by

$$U(\mathbf{x}) = e^{-|\mathbf{x}|^2/2} d\mathbf{x}^1 \wedge \dots \wedge d\mathbf{x}^N.$$

We will show later how to remedy the fact that U does not have compact support.

Now, consider the identity map $V \to V$ to be an element of $\mathcal{A}^0(V, V)$, with exterior differential $d\mathbf{x} \in \mathcal{A}^1(V, V)$. The exponential $e^{-id\mathbf{x}}$ lies in $\mathcal{A}(V, \Lambda V)$. We extend the Berezin integral $T : \Lambda V \to \mathbb{R}$ to a map $T : \mathcal{A}(V, \Lambda V) \to \mathcal{A}(V)$ by

$$T(\alpha \otimes \xi) = \alpha T(\xi) \quad \text{for } \alpha \in \mathcal{A}(V) \text{ and } \xi \in \Lambda V.$$

Lemma 1.49. *1. The form U equals*

$$U(\mathbf{x}) = \varepsilon(N) T\left(e^{-|\mathbf{x}|^2/2 - id\mathbf{x}}\right),$$

where $\varepsilon(N) = 1$ if N is even and i if N is odd.
2. The integral of U over V equals $(2\pi)^{N/2}$.

Proof. Let e_k be an orthonormal basis of V dual to the basis $\mathbf{x}^k$ of V^*. We have

$$\begin{aligned} T\left(e^{-id\mathbf{x}}\right) &= T\Big(\prod_{k=1}^{N}(1 - id\mathbf{x}^k \otimes e_k)\Big) \\ &= (-i)^N T\big((d\mathbf{x}^1 \otimes e_1)\ldots(d\mathbf{x}^N \otimes e_N)\big) \\ &= (-i)^N(-1)^{N(N-1)/2} d\mathbf{x}_1 \wedge \ldots \wedge d\mathbf{x}_N, \end{aligned}$$

and (1) follows, since $(-i)^N(-1)^{N(N-1)/2}$ equals 1 if N is even and $-i$ if N is odd. The integral of U is calculated using the Gaussian integral

$$\int_{-\infty}^{\infty} e^{-x^2/2}\, dx = \sqrt{2\pi}.$$ □

We will define a Thom form on an oriented Euclidean vector bundle $p : \mathscr{V} \to M$ of rank N by modification of the formula of Lemma 1.49. Choosing a Euclidean connection $\nabla^{\mathscr{V}}$ on $\mathscr{V}$, we may replace $d\mathbf{x}$ by $\nabla^{\mathscr{V}}\mathbf{x} \in \mathscr{A}^1(\mathscr{V}, p^*\mathscr{V})$, where $\mathbf{x}$ is the tautological section of $p^*\mathscr{V}$. However, because the connection $\nabla^{\mathscr{V}}$ need not be flat, we will see that some modification to the definition of U is necessary.

Suppose $\mathscr{E}$ is an oriented Euclidean vector bundle of rank N over B. The Berezin integral T defines a map

$$T : \mathscr{A}(B, \Lambda\mathscr{E}) \to \mathscr{A}(B).$$

If ∇ is a covariant derivative on $\mathscr{E}$, we obtain a canonical covariant derivative on $\Lambda\mathscr{E}$.

Proposition 1.50. *If ∇ is a covariant derivative on $\mathscr{E}$ compatible with the metric on $\mathscr{E}$, then for any $\alpha \in \mathscr{A}(B, \Lambda\mathscr{E})$, we have*

$$dT(\alpha) = T(\nabla\alpha).$$

Proof. Let N be the dimension of the fibres of $\mathscr{E}$. The Berezin integral is given by pairing a section of $\Lambda^N\mathscr{E}$ with the section $T \in \Gamma(B, \Lambda^N\mathscr{E}^*)$ given with respect to a local orthonormal oriented frame e^i of $\mathscr{E}^*$ by the formula $e^1 \wedge \ldots \wedge e^N$. Since $\nabla T = 0$, the result follows by Leibniz's rule. □

We apply this proposition with $B = \mathscr{V}$ and $\mathscr{E} = p^*\mathscr{V}$, and with connection $\nabla^{p^*\mathscr{V}} = p^*\nabla^{\mathscr{V}}$. The space of sections $\mathscr{A} = \mathscr{A}(\mathscr{V}, \Lambda(p^*\mathscr{V}))$ is a bigraded algebra, since it may be decomposed as a direct sum of subspaces

$$\mathscr{A}^{i,j} = \mathscr{A}^i(\mathscr{V}, \Lambda^j p^*\mathscr{V}).$$

The covariant derivative $\nabla^{p^*\mathscr{V}}$ defines a map $\nabla : \mathscr{A}^{i,j} \to \mathscr{A}^{i+1,j}$. If $s \in \Gamma(\mathscr{V}, p^*\mathscr{V})$, the contraction $a(s)$ is defined by the formula

$$a(s)(\alpha \otimes (s_1 \wedge \ldots \wedge s_j)) = \sum_{k=1}^{j} (-1)^{|\alpha|+k-1} (s, s_k) \alpha \otimes (s_1 \wedge \ldots \wedge \widehat{s_k} \wedge \ldots \wedge s_j)$$

for $\alpha \in \mathscr{A}(\mathscr{V})$ homogeneous and $s_k \in \Gamma(\mathscr{V}, p^*\mathscr{V})$, $(1 \leq k \leq j)$. The contraction $a(s)$ defines a map $a(s) : \mathscr{A}^{i,j} \to \mathscr{A}^{i,j-1}$. We will identify $\mathfrak{so}(\mathscr{V})$ with $\Lambda^2\mathscr{V}$ by the map

$$A \in \mathfrak{so}(V) \longmapsto \sum_{i<j} (Ae_i, e_j) e_i \wedge e_j.$$

For any $s \in \mathscr{A}^{0,1} = \Gamma(\mathscr{V}, p^*\mathscr{V})$ and $\alpha \in \mathscr{A}$, $T(a(s)\alpha) = 0$, since $a(s)\alpha$ has no component in $\mathscr{A}^{\bullet,N}$. It follows that the Berezin integral $T : \mathscr{A}(\mathscr{V}, \Lambda p^*\mathscr{V}) \to \mathscr{A}(\mathscr{V})$ also satisfies the formula

$$dT(\alpha) = T((\nabla + a(s))\alpha)$$

Let us list some elements of the algebra $\mathscr{A}$:

1. the tautological section $\mathbf{x} \in \mathscr{A}^{0,1} = \Gamma(\mathscr{V}, p^*\mathscr{V})$;
2. the elements $|\mathbf{x}|^2 \in \mathscr{A}^{0,0}$ and $\nabla\mathbf{x} \in \mathscr{A}^{1,1}$, formed from $\mathbf{x}$;
3. identifying the curvature $F = (\nabla^{\mathscr{V}})^2 \in \mathscr{A}^2(M, \mathfrak{so}(\mathscr{V}))$ with an element of $\mathscr{A}^2(M, \Lambda^2\mathscr{V})$ and and pulling it back to $\mathscr{V}$ by the projection $p : \mathscr{V} \to M$, we obtain an element of $\mathscr{A}^{2,2} = \mathscr{A}^2(\mathscr{V}, \Lambda^2 p^*\mathscr{V})$; abusing notation, we will write this pull-back as F as well.

Lemma 1.51. *1. Let $\Omega = |\mathbf{x}|^2/2 + i\nabla\mathbf{x} + F \in \mathscr{A}$. Then*

$$(\nabla - ia(\mathbf{x}))\Omega = 0.$$

2. If $\rho \in C^\infty(\mathbb{R})$, define $\rho(\Omega) \in \mathscr{A}$ by the formula

$$\rho(\Omega) = \sum_{k=0}^{N} \frac{\rho^{(k)}(|\mathbf{x}|^2/2)}{k!} (i\nabla\mathbf{x} + F)^k,$$

Then $(\nabla - ia(\mathbf{x}))\rho(\Omega) = 0$.

Proof. By Leibniz's rule, we have $\nabla(|\mathbf{x}|^2) = -2a(\mathbf{x})\nabla\mathbf{x}$. By the definition of curvature, we have $\nabla(\nabla\mathbf{x}) = a(\mathbf{x})F$, while by the Bianchi identity, we have $\nabla F = 0$. Combining all of this with the obvious facts that $a(\mathbf{x})\mathbf{x} = |\mathbf{x}|^2$ and $a(\mathbf{x})|\mathbf{x}|^2 = 0$, we see that $(\nabla - ia(\mathbf{x}))\Omega = 0$.

The formula $(\nabla - ia(\mathbf{x}))\rho(\Omega) = 0$ now follows from the fact that $\nabla - ia(\mathbf{x})$ is a derivation of the algebra $\mathscr{A}$. □

Note that $\Omega \in \sum_{0\leq k\leq 2} \mathscr{A}^{k,k}$, so that $\rho(\Omega) \in \sum_{0\leq k\leq N} \mathscr{A}^{k,k}$. We see that $T(\rho(\Omega)) \in \mathscr{A}^N(\mathscr{V})$, and the above lemma shows that

$$dT(\rho(\Omega)) = T((\nabla - ia(\mathbf{x}))\rho(\Omega)) = 0,$$

so that $T(\rho(\Omega))$ is a closed N-form on $\mathcal{V}$.

We now define the form U on $\mathcal{V}$ by choosing ρ to be the function $\rho(x) = \varepsilon(N)e^{-x}$, where $\varepsilon(N) = 1$ if N is even and i if N is odd:

$$U = \varepsilon(N)T(e^{-\Omega}) = \varepsilon(N)T\left(e^{-(|\mathbf{x}|^2/2+i\nabla\mathbf{x}+F)}\right). \tag{1.37}$$

We will study in greater detail the analogies between this formula and that of the Chern character in Section 7.7. Observe that the Thom form depends on the orientation of $\mathcal{V}$ (through T), on its metric, and on its connection.

Proposition 1.52. *The form* $(2\pi)^{-N/2}U$ *has integral* 1 *over each fibre.*

Proof. To calculate $\int_{\mathcal{V}/M} U$, it suffices to consider the the case in which M is a point, so that $\mathcal{V}$ is a vector space; this is just Lemma 1.49. □

The Thom form U constructed above has rapid decay at infinity instead of being compactly supported. However, using the diffeomorphism from the interior of the unit ball bundle of $\mathcal{V}$ to $\mathcal{V}$ given by the formula

$$y \longmapsto \frac{y}{(1-|y|^2)^{1/2}},$$

we can pull back the Thom form U to obtain a Thom form with support in the unit ball bundle of $\mathcal{V}$.

Just as for the Chern character, there are transgression formulas for the Thom form which show how it changes when the metric is rescaled and when the connection changes. Let us first consider the effect of rescaling the metric: this is equivalent to replacing Ω by

$$\Omega_t = t^2|\mathbf{x}|^2/2 + it\nabla\mathbf{x} + F.$$

Let U_t denote the Thom form corresponding to Ω_t.

Proposition 1.53. *We have the transgression formula*

$$\frac{dU_t}{dt} = -i\varepsilon(N)\,dT\left(\mathbf{x}e^{-\Omega_t}\right).$$

Proof. Observe that

$$\frac{d\Omega_t}{dt} = t|\mathbf{x}|^2 + i\nabla\mathbf{x} = i(\nabla - ita(\mathbf{x}))\mathbf{x}.$$

Since $(\nabla - ita(\mathbf{x}))\Omega_t = 0$, we see that

$$\frac{d}{dt}e^{-\Omega_t} = -\frac{d\Omega_t}{dt}e^{-\Omega_t} = -i(\nabla - ita(\mathbf{x}))\left(\mathbf{x}e^{-\Omega_t}\right),$$

and hence that

$$\begin{aligned}\frac{dU_t}{dt} &= -i\varepsilon(N)T\left((\nabla - ita(\mathbf{x}))\left(\mathbf{x}e^{-\Omega_t}\right)\right)\\ &= -i\varepsilon(N)\,dT\left(\mathbf{x}e^{-\Omega_t}\right).\end{aligned}$$

□

By a very similar proof, we obtain a transgression formula for the Thom form under a change of connection. Let U_s be the Thom form corresponding to a smooth family of connections ∇_s on $\mathscr{V}$ compatible with its inner product.

The derivative of this family $d\nabla_s/ds$ is an element of $\mathscr{A}^1(M,\Lambda^2\mathscr{V})$; we also denote by $d\nabla_s/ds$ the element of $\mathscr{A}^{1,2} = \mathscr{A}^1(\mathscr{V},\Lambda^2\mathscr{V})$ obtained by pulling back $d\nabla_s/ds$ to $\mathscr{V}$, and let $\Omega_s = |\mathbf{x}|^2/2 + i\nabla_s\mathbf{x} + F_s$ be the element of $\mathscr{A}$ corresponding to the connection ∇_s.

Proposition 1.54. *We have the transgression formula*

$$\frac{dU_s}{ds} = -\varepsilon(N)\,d\,T\left(\frac{d\nabla_s}{ds}e^{-\Omega_s}\right).$$

Proof. It is easy to see that

$$\frac{d\Omega_s}{ds} = (\nabla_s - ia(\mathbf{x}))\frac{d\nabla_s}{ds}.$$

Using the formula $(\nabla_s - ia(\mathbf{x}))\Omega_s = 0$, we see that

$$\frac{d}{ds}e^{-\Omega_s} = -\frac{d\Omega_s}{ds}e^{-\Omega_s} = -(\nabla_s - ia(\mathbf{x}))\left(\frac{d\nabla_s}{ds}e^{-\Omega_s}\right),$$

and hence that

$$\begin{aligned}\frac{dU_s}{ds} &= -\varepsilon(N)\,T\left((\nabla_s - ia(\mathbf{x}))\left(\frac{d\nabla_s}{ds}e^{-\Omega_s}\right)\right)\\ &= -\varepsilon(N)\,d\,T\left(\frac{d\nabla_s}{ds}e^{-\Omega_s}\right).\end{aligned}$$

□

We now restrict attention to even-dimensional $\mathscr{V}$. Let $j : M \to \mathscr{V}$ be the inclusion of M in $\mathscr{V}$ as the zero-section. The pull-back j^*U is a closed differential form on M called the **Euler form** of $\mathscr{V}$, and denoted $\chi(\nabla^{\mathscr{V}})$. Thus, if F is the curvature of $\nabla^{\mathscr{V}}$, we have

$$j^*U = \chi(\nabla^{\mathscr{V}}) = \mathrm{Pf}(-F). \tag{1.38}$$

Proposition 1.55. *The Euler form $\chi(\nabla^{\mathscr{V}})$ is a closed differential form of degree N on M whose cohomology class depends only on the bundle $\mathscr{V}$ and its orientation and not on the metric and covariant derivative on $\mathscr{V}$.*

Proof. Choose a metric on $\mathscr{V}$. If ∇_s is a one-parameter family of Euclidean connections on $\mathscr{V}$ with corresponding Thom class U_s, the transgression formula (1.54) for U_s implies that

$$\chi(\nabla_0) - \chi(\nabla_1) = d\int_0^1 T\left(\frac{d\nabla_t}{dt}\wedge \exp_\Lambda(-F_t)\right)dt,$$

which is the analogue of the trangression formula (1.33) for the Chern character. This shows that the cohomology class of $\chi(\nabla)$ is independent of the connection used in its definition.

We can also show that the cohomology class of $\chi(\nabla)$ is independent of the metric used in its definition. If $(\cdot,\cdot)_0$ and $(\cdot,\cdot)_1$ are two metrics on $\mathscr{V}$, then we can find an invertible orientation preserving section of $\mathrm{End}(\mathscr{V})$ such that

$$(v,v)_1 = (gv,gv)_0.$$

If ∇_0 is a Euclidean covariant derivative for the metric $(\cdot,\cdot)_0$, then it is easy to see that $\nabla_1 = g^{-1}\cdot\nabla_0\cdot g$ is a Euclidean covariant derivative with respect to the metric $(\cdot,\cdot)_1$. If we now construct the Euler forms of the two covariant derivatives ∇_i with respect to the metrics $(\cdot,\cdot)_i$, we see that they are equal. □

A case of special importance is that in which the bundle $\mathscr{V}$ is the tangent bundle TM of an oriented even-dimensional Riemannian manifold. In this case, the differential form $\chi(\nabla)$, where ∇ is the Levi-Civita connection, is called the Euler form $\chi(M)$ of the manifold, and lies in $\mathscr{A}^n(M)$, where n is the dimension of M. For example, if M is a two-dimensional Riemannian manifold, and if $\{e_1,e_2\}$ is an orthonormal basis of T_xM for which $R(e_1,e_2) = \begin{pmatrix} 0 & \lambda \\ -\lambda & 0 \end{pmatrix}$, then $\chi(M)_x = \lambda e^1 \wedge e^2$.

Let $v \in \Gamma(M,\mathscr{V})$ be a section of the vector bundle $\mathscr{V}$. The pull-back v^*U is a closed differential form of degree N on M, given by the formula

$$v^*U = T\big(e^{-(|v|^2/2+i\nabla^{\mathscr{V}}v+F)}\big).$$

The transgression formula Proposition 1.53 implies that v^*U is cohomologous to the Euler class for any $v \in \Gamma(M,\mathscr{V})$:

$$\chi(\nabla^{\mathscr{V}}) - v^*U = -id\int_0^1 T\big(v\wedge e^{-(t^2|v|^2/2+it\nabla^{\mathscr{V}}v+F)}\big)\,dt.$$

The above results imply a form of the Gauss-Bonnet-Chern theorem, which is a formula for the Euler number of a manifold in terms of the integral of its Euler class. This is the simplest example of an index theorem.

Let $v \in \Gamma(M,TM)$ be a vector field on M. At a zero p of v, the Lie bracket

$$X \longmapsto [X,v]$$

induces an endomorphism $L_p(v)$ of T_pM; in a coordinate system in which $p=0$ and $v = \sum_{i=1}^n v^i\partial_i$, $L_p(v)$ is given by the formula

$$L_p(v)\partial_i = \sum_{j=1}^n \big(\partial_i v^j(p)\big)\partial_j.$$

A zero p of v is called **non-degenerate** if $L_p(v)$ is invertible. We denote by $\nu(p,v) \in \{\pm 1\}$ the sign of the determinant of $L_p(v)$.

If all of the zeros of v are non-degenerate, we say that v is non-degenerate. If M is compact, such a vector field has only a finite number of zeroes.

A non-degenerate vector field may be constructed with the help of a Morse function f, that is, a real smooth function on M such that at each critical point p of f, there is a coordinate system in which

$$f = f(p) + \sum_{i=1}^{n} \frac{\lambda_i \mathbf{x}_i^2}{2}, \text{where } \lambda_i \in \{\pm 1\}.$$

If we choose a Riemannian metric on M which is equal to $\sum_i d\mathbf{x}_i \otimes d\mathbf{x}_i$ in each of these coordinate neighbourhoods, we see that the vector field $\operatorname{grad} f$ equals $\sum_{i=1}^{n} \lambda_i \mathbf{x}^i \partial_i$ around the critical point p of f, and hence that $\nu(p, \operatorname{grad} f) = \prod_{i=1}^{n} \lambda_i$. The following theorem is Chern's generalization of the Gauss-Bonnet theorem to arbitrary orientable compact manifolds. The form in which we derive it here is not the usual statement of the Gauss-Bonnet Theorem, because it employs a definition of the Euler number in terms of the zeroes of a non-degenerate vector field on M; however, this agrees with the definition of the Euler number as alternating sum of the Betti numbers of M, as we will prove in Theorem 4.7.

Theorem 1.56 (Poincare-Hopf). *If v is a non-degenerate vector field on an orientable compact manifold M, then*

$$\sum_{\{p|v(p)=0\}} \nu(p,v) = (2\pi)^{-n/2} \int_M \chi(M).$$

In particular, the sum

$$\operatorname{Eul}(M) = \sum_{\{p|v(p)=0\}} \nu(p,v)$$

is independent of the non-degenerate vector field; it is called the **Euler number** *of the manifold.*

Proof. Choose a coordinate chart in a neighbourhood U^p of each zero p of v. In this coordinate chart, the tangent bundle is trivialized, $TU^p \cong U^p \times \mathbb{R}^n$. The vector field v defines a smooth map from U^p to $\mathbb{R}^n$ such that $v(p) = 0$, with invertible derivative $L_p(v)$ at p. It follows that v maps an open subset V^p of U^p diffeomorphically to a small ball $B \subset \mathbb{R}^n$. The orientations on V^p induced by M and by the diffeomorphism v to $\mathbb{R}^n$ differ by the sign $\nu(p,v)$. We may assume the open sets V^p to be disjoint. Choose a Riemannian metric on M which on V^p equals the metric induced by the diffeomorphism into $\mathbb{R}^n$. Since M is compact, we see that there is a positive constant $\varepsilon > 0$ such that $\|v\| \geq \varepsilon$ on the set $M \setminus \cup_p V^p$.

Let U be the Thom form of TM, with respect to the given Riemannian metric and its associated Levi-Civita connection. For each $t > 0$, we have the formula

$$\int_M v_t^* U = \int_M \chi(M).$$

Let $\psi : \mathbb{R}_+ \to [0,1]$ be any smooth function such that

$$\psi(s) = 1 \quad \text{if } s < \varepsilon^2/4$$
$$\psi(s) = 0 \quad \text{if } s > \varepsilon^2.$$

Then

$$\int_M v_t^* U = \int_M (1-\psi(\|v\|^2))v_t^* U + \int_M \psi(\|v\|^2) v_t^* U. \qquad (1.39)$$

Since $v_t^* U$ is of the form

$$v_t^* U = e^{-t^2\|v\|^2/2} \sum_{i=0}^n t^i \alpha_i,$$

where α_i are differential forms on M, it is rapidly decreasing as a function of t as $t \to \infty$ when $\|v\|^2 > \varepsilon^2/4$. We see that the first integral in (1.39) is a rapidly decreasing function of t. The second integral is a sum of integrals over the neighbourhoods V_p.

On V^p, the metric and connection are trivial and $v = \sum_{i=1}^n \mathbf{x}^i \partial_i$, so that

$$v_t^* U|_{V^p} = t^n \nu(p,v) e^{-t^2\|\mathbf{x}\|^2/2} d\mathbf{x}^1 \wedge \ldots \wedge d\mathbf{x}^n.$$

Thus

$$\int_{V_p} \psi(\|v\|^2) v_t^* U = t^n \nu(p,v) \int_{\mathbb{R}^n} \psi(\|\mathbf{x}\|^2) e^{-t^2\|\mathbf{x}\|^2/2} d\mathbf{x}^1 \wedge \ldots \wedge d\mathbf{x}^n.$$

Making the change of variables $\mathbf{x} \to t^{-1}\mathbf{x}$, we see that

$$\lim_{t\to\infty} t^n \int_{\mathbb{R}^n} \psi(\|\mathbf{x}\|^2) e^{-t^2\|\mathbf{x}\|^2/2} d\mathbf{x}^1 \wedge \ldots \wedge d\mathbf{x}^n = (2\pi)^{n/2}\psi(0) = (2\pi)^{n/2}.$$

Taking the sum over the zeroes p of v, the result follows. □

There is a generalization of this result, to the case where we only assume that the zeroes of v are isolated.

Definition 1.57. *The* **index** $\nu(p,v)$ *of a vector field v on a manifold M at an isolated zero $v(p) = 0$ is defined by choosing a coordinate system around p, in which case $\nu(p,v)$ is the degree of the map from the sphere $S_u = \{\mathbf{x} \in \mathbb{R}^n \mid \|\mathbf{x}\| = u\}$ to the unit sphere defined by*

$$\mathbf{x} \in S_u \longmapsto \|v(\mathbf{x})\|^{-1} v(\mathbf{x}).$$

The index $\nu(p,v)$ is an integer which is independent of the coordinate chart used in its definition, as well as the small parameter u.

We can again choose small balls V^p around the zeroes of v such that $\|v\| > \varepsilon$ on the complement of $\cup_p V^p$. The map sending $\mathbf{x} \in V^p$ to $v(\mathbf{x})$ is a covering map over the set of regular values, and using Sard's theorem, we see that

$$\int_{V^p} \psi(\|v\|^2) v_t^* U = t^n \nu(p,v) \int_{\mathbb{R}^n} \psi(\|v\|^2) e^{-t^2\|\mathbf{x}\|^2/2} d\mathbf{x}^1 \wedge \ldots \wedge d\mathbf{x}^n.$$

This is the outline of the proof of the following extension of Theorem 1.56.

Theorem 1.58 (Poincare-Hopf). *If v is a vector field with isolated zeroes on a compact oriented even dimensional manifold M, then*

$$\sum_{\{p \mid v(p)=0\}} \nu(p,v) = (2\pi)^{-n/2} \int_M \chi(M).$$

Bibliographic Notes

The book of Kobayashi-Nomizu [75] is perhaps the best reference for the theory of bundles and connections.

Section 2

For background on Riemannian manifolds, the reader may refer to the books of Milnor [84], Berger-Gauduchon-Mazet [19] and Kobayashi-Nomizu.

The proofs of Proposition 1.18 and 1.28 are taken from the second appendix of Atiyah-Bott-Patodi [8].

Sections 3 and 4

It was Berezin [27], motivated by the problem of understanding geometric quantization for fermions, who emphasized the strong analogy between calculus on superspaces and ordinary vector spaces.

The notion of a superconnection was developed by Quillen in order to provide a formalism for a local family index theorem for Dirac operators [94]. This hope was realised by Bismut, as we will see in Chapters 9 and 10. The notion of superconnection has turned out to be fruitful in other contexts, too.

Section 5

We have presented the theory of characteristic forms in the context of superconnections, following Quillen. The construction of characteristic forms for connections is due to Chern and Weil; see Kobayashi-Nomizu. Our interest is more in the characteristic forms themselves than in their cohomology classes. The reader interested in the topological point of view may consult Bott-Tu [45] and Milnor [86]. This section is also an introduction to Chapter 7, where we define equivariant characteristic forms; thus, the bibliographic notes of that chapter are relevant here.

Section 6

The results on the Thom form and the Euler form presented here are due to Mathai and Quillen [82], although we give more direct proofs which use the Berezin integral. The proof of Theorem 1.56 is close to Chern's proof of the Gauss-Bonnet-Chern theorem [49]; we have presented it in such a way as to show the analogy with the theorems of later chapters. For more on the proof of the Poincaré-Hopf Theorem 1.58, see Bott and Tu [45].

2

Asymptotic Expansion of the Heat Kernel

Given a Riemannian structure on a manifold M, one may construct the scalar Laplacian $\Delta = d^*d$, which is the operator on the space of functions on M corresponding to the quadratic form

$$(f, \Delta f) = \int_M (df, df)_x \, |dx|$$

This operator can be used to study the geometry of the manifold, and is also important because it arises frequently in physics problems, such as quantum mechanics and diffusion in curved spaces. All of the eigenfunctions of Δ are smooth functions and, if the manifold M is compact, Δ has a unique self-adjoint extension. The Laplacian has a heat kernel, which is the function $k_t(x,y)$ on $\mathbb{R}_+ \times M \times M$ which solves the equation

$$\partial_t k_t(x,y) + \Delta_x k_t(x,y) = 0,$$

subject to the initial condition that $\lim_{t \to 0} k_t(x,y)$ is the Dirac distribution along the diagonal $\delta(x,y)$. It turns out that the heat kernel $k_t(x,y)$ is a smooth function whose behaviour when t becomes small reflects the local geometry of the manifold M. In particular, there is an approximation to $k_t(x,y)$ near the diagonal, of the form

$$k_t(x,y) \sim (4\pi t)^{-n/2} e^{-d(x,y)^2/4t} \sum_{i=0}^{\infty} t^i f_i(x,y),$$

where $d(x,y)$ is the geodesic distance between x and y, the integer n is the dimension of M, and the functions $f_i(x,y)$ are given by explicit formulas when restricted to the diagonal.

In this chapter, we will construct the heat kernel of a more general class of operators which generalizes the scalar Laplacian to vector bundles. The construction is extremely intuitive geometrically, being based on the explicit formula involving a Gaussian for the heat kernel of the Laplacian on Euclidean space. Since the manifold on which we work is compact, we require no estimates more sophisticated than the summation of geometric series.

These generalized Laplacians are introduced in Section 1, while the construction of the heat kernel, when M is compact, occupies Sections 2–4. In Section 5, we give some applications, such as the smoothness of eigenfunctions of generalized Laplacians and the Weyl formula for the asymptotic number of eigenfunctions in terms of the volume of the manifold and its dimension. In Section 6, we discuss the dependence of the heat kernel on the Laplacian if the coefficients of the Laplacian are varied smoothly.

2.1 Differential Operators

Let M be a manifold, and let $\mathscr{E}$ be a vector bundle over M. The algebra of differential operators on $\mathscr{E}$, denoted by $\mathscr{D}(M,\mathscr{E})$, is the subalgebra of $\operatorname{End}(\Gamma(M,\mathscr{E}))$ generated by elements of $\Gamma(M,\operatorname{End}(\mathscr{E}))$ acting by multiplication on $\Gamma(M,\mathscr{E})$, and the covariant derivatives ∇_X, where ∇ is any covariant derivative on $\mathscr{E}$ and X ranges over all vector fields on M. The algebra $\mathscr{D}(M,\mathscr{E})$ has a natural filtration, defined by letting

$$\mathscr{D}_i(M,\mathscr{E}) = \Gamma(M,\operatorname{End}(\mathscr{E}))\cdot\operatorname{span}\{\nabla_{X_1}\dots\nabla_{X_j} \mid j \leq i\}.$$

We call an element of $\mathscr{D}_i(M,\mathscr{E})$ an i-th differential order operator. Note that if D is an i-th order operator and f is a smooth function, then $(\operatorname{ad} f)^i D$ is a zeroth order operator.

If A is any filtered algebra, that is, $A = \bigcup_{i=0}^{\infty} A_i$ with $A_i \subset A_{i+1}$ and $A_i \cdot A_j \subset A_{i+j}$, we may define the associated graded algebra

$$\operatorname{gr} A = \sum_{i=0}^{\infty} \operatorname{gr}_i A = \sum_{i=0}^{\infty} A_i/A_{i-1}.$$

where the component in degree i is $\operatorname{gr}_i A$. The projection from the graded algebra $\sum_{i=0}^{\infty} A_i$ to $\operatorname{gr} A$ is called the **symbol** map, and written σ; its component in degree i is the projection $A_i \to \operatorname{gr} A$ with kernel A_{i-1}.

We apply this formalism to the algebra of differential operators $\mathscr{D}(M,\mathscr{E})$. Let $S(TM)$ be the infinite dimensional graded vector bundle over M whose fibre at x is the symmetric algebra on T_xM.

Proposition 2.1. *The associated graded algebra*

$$\operatorname{gr}\mathscr{D}(M,\mathscr{E}) = \sum_{k=0}^{\infty} \mathscr{D}_k(M,\mathscr{E})/\mathscr{D}_{k-1}(M,\mathscr{E})$$

to the filtered algebra of differential operators is isomorphic to the space of sections of the bundle $S(TM)\otimes\operatorname{End}(\mathscr{E})$. The isomorphism

$$\sigma_k : \operatorname{gr}_k \mathscr{D}(M,\mathscr{E}) \to \Gamma(M, S^k(TM)\otimes\operatorname{End}(\mathscr{E}))$$

is given by the following formula: if $D \in \mathscr{D}_k(M,\mathscr{E})$, then for $x \in M$ and $\xi \in T_xM$,

$$\sigma_k(D)(x,\xi) = \lim_{t\to\infty} t^{-k}(e^{-itf}\cdot D\cdot e^{itf})(x) \in \operatorname{End}(\mathscr{E}_x), \tag{2.1}$$

where f is any smooth function on M such that $df(x) = \xi$.

Proof. Let D be the differential operator $a\nabla_{X_1}\dots\nabla_{X_k}$, where $X_1,\dots,X_k$ are vector fields on M and $a \in \Gamma(M,\mathrm{End}(\mathscr{E}))$. By Leibniz's rule, the operator $e^{-itf}\cdot D\cdot e^{itf}$ is a differential operator depending polynomially on t, of the form

$$(it)^k a(x)\langle X_1(x),\xi\rangle\dots\langle X_k(x),\xi\rangle + O(t^{k-1}).$$

It follows that the leading order is a zeroth-order operator, and that the limit $\lim_{t\to\infty} t^{-k}(e^{-itf}\cdot D\cdot e^{itf})(x)$ is independent of which function f with $df(x)=\xi$ we choose, and defines a linear isomorphism from $\mathrm{gr}_k\,\mathscr{D}(M,\mathrm{End}(\mathscr{E}))$ to $\Gamma(M,S^k(TM)\otimes\mathrm{End}(\mathscr{E}))$. □

By differentiation of (2.1), we see that

$$\sigma_k(D)(x,\xi) = \frac{(-i)^k}{k!}(\mathrm{ad}\,f)^k D.$$

If $D_i \in \mathscr{D}_{k_i}(M,\mathscr{E})$, $i=1,2$, then

$$\sigma_{k_1+k_2}(D_1D_2)(x,\xi) = \sigma_{k_1}(D_1)(x,\xi)\cdot\sigma_{k_2}(D_2)(x,\xi).$$

We will often write $\sigma(D)$ instead of $\sigma_k(D)$, if k is the degree of D.

We may identify $\Gamma(M,S(TM)\otimes\mathrm{End}(\mathscr{E}))$ with the subspace of sections in $\Gamma(T^*M,\pi^*\mathrm{End}(\mathscr{E}))$ which are polynomial along the fibres of T^*M. A differential operator D of order k is called **elliptic** if the section $\sigma_k(D)\in\Gamma(T^*M,\pi^*\mathrm{End}(\mathscr{E}))$ over the cotangent space is invertible over the open set $\{(x,\xi)\mid \xi\neq 0\}$.

Definition 2.2. *Let $\mathscr{E}$ be a vector bundle over a Riemannian manifold M. A* **generalized Laplacian** *on $\mathscr{E}$ is a second-order differential operator H such that*

$$\sigma_2(H)(x,\xi) = |\xi|^2.$$

Clearly, such an operator is elliptic. An equivalent way of stating this definition is that in any local coordinate system,

$$H = -\sum_{ij} g^{ij}(\mathbf{x})\partial_i\partial_j + \text{first-order part},$$

where $g^{ij}(\mathbf{x}) = (d\mathbf{x}^i, d\mathbf{x}^j)_{\mathbf{x}}$ is the metric on the cotangent bundle. By definition of the symbol, we can characterize a generalized Laplacian as follows:

Proposition 2.3. *An operator H is a generalized Laplacian if and only for any $f\in C^\infty(M)$,*

$$[[H,f],f] = -2|df|^2.$$

Let $\mathscr{E}$ be a vector bundle bundle on a Riemannian manifold M, with connection $\nabla^{\mathscr{E}}:\Gamma(M,\mathscr{E})\to\Gamma(M,T^*M\otimes\mathscr{E})$. Since M is Riemannian, it possesses a canonical connection, the Levi-Civita connection ∇, by means of which we define the tensor product connection on the bundle $T^*M\otimes\mathscr{E}$. In this way, we obtain an operator

$$\nabla^{T^*M\otimes\mathscr{E}}\nabla^{\mathscr{E}}:\Gamma(M,\mathscr{E})\to\Gamma(M,T^*M\otimes T^*M\otimes\mathscr{E})$$

as the composition of the two covariant derivatives.

Definition 2.4. *The Laplacian $\Delta^{\mathscr{E}}$ on $\Gamma(M,\mathscr{E})$ (which we denote by Δ if there is no ambiguity) is the second-order differential operator*

$$\Delta^{\mathscr{E}} s = -\operatorname{Tr}(\nabla^{T^*M\otimes\mathscr{E}}\nabla^{\mathscr{E}} s).$$

Here, we denote by $\operatorname{Tr}(S) \in \Gamma(M,\mathscr{E})$ *the contraction of an element $S \in \Gamma(M, T^*M \otimes T^*M \otimes \mathscr{E})$ with the metric $g \in \Gamma(M, TM \otimes TM)$.*

If $s \in \Gamma(M,\mathscr{E})$ and X and Y are two vector fields, then

$$(\nabla^{T^*M\otimes\mathscr{E}}\nabla^{\mathscr{E}} s)(X,Y) = (\nabla^{\mathscr{E}}_X\nabla^{\mathscr{E}}_Y - \nabla^{\mathscr{E}}_{\nabla_X Y})s.$$

If e_i is a local orthonormal frame of TM, we have the following explicit formula for $\Delta^{\mathscr{E}}$:

$$\Delta^{\mathscr{E}} = -\sum_i \left(\nabla^{\mathscr{E}}_{e_i}\nabla^{\mathscr{E}}_{e_i} - \nabla^{\mathscr{E}}_{\nabla_{e_i}e_i}\right).$$

On the other hand, with respect to the framing $\partial/\partial x_i$ defined by a coordinate system around a point in M, the operator $\Delta^{\mathscr{E}}$ equals

$$\Delta^{\mathscr{E}} = -\sum_{ij} g^{ij}(\mathbf{x})\left(\nabla^{\mathscr{E}}_{\partial_i}\nabla^{\mathscr{E}}_{\partial_j} - \sum_k \Gamma^k_{ij}\nabla^{\mathscr{E}}_{\partial_k}\right),$$

where Γ^k_{ij} are the components of the Levi-Civita connection, defined by $\nabla_i\partial_j = \sum_k \Gamma^k_{ij}\partial_k$. Thus, the Laplacian $\Delta^{\mathscr{E}}$ is a generalized Laplacian.

In particular the scalar Laplacian is given for $f \in C^\infty(M)$ by the formula

$$\Delta f = -\operatorname{Tr}(\nabla df) = -\sum_{ij} g^{ij}(\mathbf{x})\left(\partial_i\partial_j - \sum_k \Gamma^k_{ij}\partial_k\right) f.$$

If $\mathscr{E}_1$ and $\mathscr{E}_2$ are two bundles on M with connections $\nabla^{\mathscr{E}_1}$ and $\nabla^{\mathscr{E}_2}$, let $\Delta^{\mathscr{E}_1}$, $\Delta^{\mathscr{E}_2}$ and $\Delta^{\mathscr{E}_1\otimes\mathscr{E}_2}$ be the Laplacians defined with respect to the connections $\nabla^{\mathscr{E}_1}$, $\nabla^{\mathscr{E}_2}$ and $\nabla^{\mathscr{E}_1\otimes\mathscr{E}_2} = \nabla^{\mathscr{E}_1}\otimes 1 + 1\otimes\nabla^{\mathscr{E}_2}$. The following formula is clear: for $s_i \in \Gamma(M,\mathscr{E}_i)$,

$$\Delta^{\mathscr{E}_1\otimes\mathscr{E}_2}(s_1\otimes s_2) = (\Delta^{\mathscr{E}_1}s_1)\otimes s_2 - 2\operatorname{Tr}(\nabla^{\mathscr{E}_1}s_1\otimes\nabla^{\mathscr{E}_2}s_2) + s_1\otimes\Delta^{\mathscr{E}_2}s_2.$$

As we will now show, any generalized Laplacian is of the form $\Delta^{\mathscr{E}} + F$, where $\Delta^{\mathscr{E}}$ is the Laplacian associated to some connection $\nabla^{\mathscr{E}}$, and F is a section of the bundle $\operatorname{End}(\mathscr{E})$.

Proposition 2.5. *If H is a generalized Laplacian on a vector bundle $\mathscr{E}$, there exists a connection $\nabla^{\mathscr{E}}$ on $\mathscr{E}$ such that for any $f \in C^\infty(M)$, $[H,f]$ is the differential operator whose action on $\Gamma(M,\mathscr{E})$ is given by the formula*

$$[H,f] = -2\langle \operatorname{grad} f, \nabla^{\mathscr{E}}\rangle + \Delta f, \tag{2.2}$$

where Δ is the scalar Laplacian. Furthermore, $F = H - \Delta^{\mathscr{E}}$ is a zeroth-order operator.

Proof. Let us define an operator $\nabla^{\mathscr{E}}$ from $\Gamma(M,\mathscr{E})$ to $\mathscr{A}^1(M,\mathscr{E})$ by the formula

$$\langle f_0 \operatorname{grad} f_1, \nabla^{\mathscr{E}} s\rangle = \tfrac{1}{2} f_0(-H(f_1 s) + f_1(Hs) + (\Delta f_1)s),$$

for all f_0 and $f_1 \in C^\infty(M)$. The relation

$$\Delta(f_1 f_2) = (\Delta f_1) f_2 - 2(df_1, df_2) + f_1(\Delta f_2)$$

and the formula $[[H,f_1],f_2] = -2(df_1,df_2)$ implies that this definition is consistent, in the sense that

$$\langle f_0 \operatorname{grad}(f_1 f_2), \nabla^{\mathscr{E}} s\rangle = \langle f_0 f_1 \operatorname{grad} f_2 + f_0 f_2 \operatorname{grad} f_1, \nabla^{\mathscr{E}} s\rangle,$$

and that $\nabla^{\mathscr{E}}$ is a connection on the bundle $\mathscr{E}$:

$$[\langle \operatorname{grad} f, \nabla^{\mathscr{E}}\rangle, h] = [-\tfrac{1}{2}[H,f] + \tfrac{1}{2}\Delta f, h] = -\tfrac{1}{2}[[H,f],h] = \langle \operatorname{grad} f, dh\rangle.$$

It is now easy to show that $F = H - \Delta^{\mathscr{E}}$ is a zeroth-order operator: if f is any smooth function on M, then

$$[H - \Delta^{\mathscr{E}}, f] = 0,$$

since both $[H,f]$ and $[\Delta^{\mathscr{E}}, f]$ equal $-2\langle \operatorname{grad} f, \nabla^{\mathscr{E}}\rangle + \Delta f$. □

To summarize, a generalized Laplacian is constructed from the following three pieces of geometric data:

1. a metric g on M, which determines the second-order piece;
2. a connection $\nabla^{\mathscr{E}}$ on the vector bundle $\mathscr{E}$, which determines the first-order piece;
3. a section F of the bundle $\operatorname{End}(\mathscr{E})$, which determines the zeroth order piece.

If $\mathscr{E}$ is a vector bundle and $\mathscr{E}^*$ is its dual vector bundle, then the space $\Gamma_c(M,\mathscr{E})$ of compactly supported sections of $\mathscr{E}$ is naturally paired with the space of sections of $\mathscr{E}^* \otimes |\Lambda|$: if we take two sections $s_1 \in \Gamma_c(M,\mathscr{E})$ and $s_2 \in \Gamma(M, \mathscr{E}^* \otimes |\Lambda|)$, then their pointwise scalar product $\langle s_1, s_2\rangle_x$ is a section of the density bundle $|\Lambda|$, so can be integrated, giving a number $\int_M \langle s_1, s_2\rangle$, the pairing of the two sections. In particular, if $\mathscr{E}$ is a Hermitian bundle, then the space $\Gamma_c(M, \mathscr{E} \otimes |\Lambda|^{1/2})$ of half-densities has a natural inner-product.

Definition 2.6. *(1) If $\mathscr{E}_1$ and $\mathscr{E}_2$ are two vector bundles on M and $D : \Gamma(M,\mathscr{E}_1) \to \Gamma(M,\mathscr{E}_2)$ is a differential operator, the* **formal adjoint** *D^* of D is the differential operator*

$$D^* : \Gamma(M, \mathscr{E}_2^* \otimes |\Lambda|) \to \Gamma(M, \mathscr{E}_1^* \otimes |\Lambda|)$$

such that

$$\int_M \langle Ds, t\rangle = \int_M \langle s, D^* t\rangle$$

for $s \in \Gamma_c(M,\mathscr{E}_1)$ and $t \in \Gamma(M, \mathscr{E}_2^* \otimes |\Lambda|)$.

(2) If $\mathscr{E}$ is a Hermitian vector bundle and D a differential operator acting on $\Gamma(M, \mathscr{E} \otimes |\Lambda|^{1/2})$, we say that D is symmetric if it is equal to its formal adjoint, $D = D^*$.

It is easy to see that the symbol of the formal adjoint D^* is equal to $\sigma_i(D^*) = \sigma_i(D)^* \otimes 1_{|\Lambda|}$.

Although the bundle $|\Lambda|$ has no canonical trivialization, in most of this book, we will deal with Riemmanian manifolds, which have a canonical nowhere vanishing density $|dx| \in \Gamma(M,|\Lambda|)$. Using the identification of $\Gamma(M,\mathscr{E})$ with $\Gamma(M,\mathscr{E}\otimes|\Lambda|)$ which results from multiplication by $|dx|$, the formal adjoint of $D : \Gamma(M,\mathscr{E}_1) \to \Gamma(M,\mathscr{E}_2)$ becomes an operator $D^* : \Gamma(M,\mathscr{E}_2^*) \to \Gamma(M,\mathscr{E}_1^*)$. However, by taking into account the density bundles, we get the most geometrically invariant formulations of many theorems.

On a Riemannian manifold, the formal adjoint of the exterior differential $d : \mathscr{A}^\bullet(M) \to \mathscr{A}^{\bullet+1}(M)$ is an operator from the bundle $\Lambda^{\bullet+1}TM \otimes |\Lambda|$ to the bundle $\Lambda^\bullet TM \otimes |\Lambda|$. Using the identification of the bundles $\Lambda^\bullet T^*M$ and $\Lambda^\bullet TM \otimes |\Lambda|$ which comes from the Riemannian metric on M, we may identify it with a map $d^* : \mathscr{A}^{\bullet+1}(M) \to \mathscr{A}^\bullet(M)$. Thus, d^* is defined by the formula

$$\int_M (d\beta,\alpha)\,|dx| = \int_M (\beta,d^*\alpha)\,|dx|,$$

where α and β are compactly supported differential forms on M, and $|dx|$ is the Riemannian density.

If α is a one-form, then $d^*\alpha$ is a function called the **divergence** of α. Taking $\beta = 1$ in the preceding equation, we see that

$$\int_M d^*\alpha\,|dx| = 0 \tag{2.3}$$

if α is a compactly supported one-form.

If α is a one-form, the function $\operatorname{Tr}(\nabla\alpha)$ is given by the formula

$$\operatorname{Tr}(\nabla\alpha) = \sum_i e_i\alpha(e_i) - \alpha(\nabla_{e_i}e_i),$$

where e_i is a local orthonormal frame.

Proposition 2.7. *The divergence on one-forms is given by the formula*

$$d^*\alpha = -\operatorname{Tr}(\nabla\alpha),$$

and hence for any compactly supported C^1 one-form α, we have

$$\int_M \operatorname{Tr}(\nabla\alpha)\,|dx| = 0.$$

Proof. If X is a vector field on M, then

$$\nabla X \in \Gamma(M,T^*M\otimes TM) \cong \Gamma(M,\operatorname{End}(TM))$$

is the endomorphism of TM which acts on $Y \in \Gamma(M,TM)$ by the formula $(\nabla X)\cdot Y = \nabla_Y X$. Since ∇ is torsion free, we see that

$$\mathscr{L}(X)Y = [X,Y] = \nabla_X Y - (\nabla X)Y,$$

hence, using the fact that $|dx|$ is preserved by the Levi-Civita connection, that

$$\mathscr{L}(X)|dx| = \mathrm{Tr}(\nabla X)|dx|.$$

If $f \in C^\infty(M)$, it follows from the formula $\int_M \mathscr{L}(X)(f\,|dx|) = 0$ that

$$\int_M X(f)\,|dx| = -\int_M \mathrm{Tr}(\nabla X) f\,|dx|.$$

If α is the one-form on M corresponding to X under the Riemannian metric on M, so that $X(f) = (\alpha, df)$, we see that

$$\int_M (\alpha, df)\,|dx| = \int_M X(f)\,|dx| = -\int_M \mathrm{Tr}(\nabla\alpha) f\,|dx|. \qquad \square$$

The following proposition extends the above formula for the divergence, by giving a formula for $d^*\alpha$ in any degree.

Proposition 2.8. *Denote by $\iota : \Gamma(M, T^*M \otimes \Lambda T^*M) \to \Gamma(M, \Lambda T^*M)$ the contraction operator on differential forms defined by means of the Riemannian metric on M. Then*

$$d^* = -\iota \circ \nabla^{\Lambda T^*M}.$$

Proof. Since the Levi-Civita covariant derivative is torsion free, we see from Proposition 1.22 that $d = \varepsilon \circ \nabla$. Let $\beta_p \in \mathscr{A}^p(M)$, $\beta_{p+1} \in \mathscr{A}^{p+1}(M)$, and let α be the one-form given by the formula $\alpha(X) = (\beta_p, \iota(X)\beta_{p+1})$. Using Leibniz's rule and the fact that at each point $x \in M$, $\varepsilon^* = \iota$, we see that

$$(\varepsilon\nabla\beta_p, \beta_{p+1})_x = -(\beta_p, \iota\nabla\beta_{p+1})_x + \mathrm{Tr}(\nabla\alpha)_x.$$

Integration over M kills $\mathrm{Tr}(\nabla\alpha)$, and we obtain the proposition. $\square$

We will now calculate the formal adjoint of the Laplacian $\Delta^{\mathscr{E}}$. Recall that the bundle $|\Lambda|^s$ of s-densities on a Riemannian manifold has a natural connection, which preserves the canonical section $|dx|^s$. If $\Delta^{\mathscr{E}\otimes|\Lambda|^s}$ is the Laplacian associated to the connection $\nabla^{\mathscr{E}\otimes|\Lambda|^s}$ on the bundle $\mathscr{E} \otimes |\Lambda|^s$, we have

$$\Delta^{\mathscr{E}\otimes|\Lambda|^s}(\phi|dx|^s) = (\Delta^{\mathscr{E}}\phi)|dx|^s.$$

Because of the following result, it is often more natural to write formulas with half-densities. Let $\int_M \langle s_1, s_2\rangle$ denote the natural pairing between a compactly supported section s_1 of the bundle $\mathscr{E} \otimes |\Lambda|^{1/2}$, and a section s_2 of the bundle $\mathscr{E}^* \otimes |\Lambda|^{1/2}$. We consider on the bundle $\mathscr{E}^*$ the connection $\nabla^{\mathscr{E}^*}$ dual to the connection $\nabla^{\mathscr{E}}$ on $\mathscr{E}$.

Proposition 2.9. *If s_1 is a compactly supported C^2-section of $\mathscr{E} \otimes |\Lambda|^{1/2}$ and s_2 is a C^2-section of $\mathscr{E}^* \otimes |\Lambda|^{1/2}$, then*

$$\int_M \langle \Delta^{\mathscr{E}\otimes|\Lambda|^{1/2}} s_1, s_2 \rangle = \int_M \langle s_1, \Delta^{\mathscr{E}^*\otimes|\Lambda|^{1/2}} s_2 \rangle$$
$$= \int_M \operatorname{Tr}\langle \nabla^{\mathscr{E}\otimes|\Lambda|^{1/2}} s_1, \nabla^{\mathscr{E}^*\otimes|\Lambda|^{1/2}} s_2 \rangle,$$

where the last inner product is defined using the natural pairing between $T^*M \otimes \mathscr{E}$ and $T^*M \otimes \mathscr{E}^*$.

Proof. If we apply the Levi-Civita covariant derivative ∇ to the density $\langle \nabla^{\mathscr{E}\otimes|\Lambda|^{1/2}} s_1, s_2 \rangle_x$, we obtain the continuous section of $T^*M \otimes T^*M$

$$\nabla \langle \nabla^{\mathscr{E}\otimes|\Lambda|^{1/2}} s_1, s_2 \rangle_x =$$
$$\langle \nabla^{T^*M\otimes\mathscr{E}\otimes|\Lambda|^{1/2}} \nabla^{\mathscr{E}\otimes|\Lambda|^{1/2}} s_1, s_2 \rangle_x + \langle \nabla^{\mathscr{E}\otimes|\Lambda|^{1/2}} s_1, \nabla^{\mathscr{E}^*\otimes|\Lambda|^{1/2}} s_2 \rangle_x,$$

so that

$$\operatorname{Tr}\big(\nabla \langle \nabla^{\mathscr{E}\otimes|\Lambda|^{1/2}} s_1, s_2 \rangle\big)_x =$$
$$-\langle \Delta^{\mathscr{E}\otimes|\Lambda|^{1/2}} s_1, s_2 \rangle_x + \operatorname{Tr}\langle \nabla^{\mathscr{E}\otimes|\Lambda|^{1/2}} s_1, \nabla^{\mathscr{E}^*\otimes|\Lambda|^{1/2}} s_2 \rangle_x.$$

By Proposition 2.7, the term $\operatorname{Tr}\big(\nabla \langle \nabla^{\mathscr{E}\otimes|\Lambda|^{1/2}} s_1, s_2 \rangle\big)_x$ vanishes after integration over M. □

If $\mathscr{E}$ has a Hermitian metric, then the bundle $\mathscr{E}^*$ is naturally identified with $\mathscr{E}$, and there is a natural inner product on the space $\Gamma(M, \mathscr{E} \otimes |\Lambda|^{1/2})$. In this situation, the above proposition has the following consequence.

Corollary 2.10. *If the connection ∇ on $\mathscr{E}$ is compatible with the Hermitian metric on $\mathscr{E}$, then the Laplacian $\Delta^{\mathscr{E}\otimes|\Lambda|^{1/2}}$ on the bundle $\mathscr{E} \otimes |\Lambda|^{1/2}$ is symmetric with respect to the inner product on $\Gamma(M, \mathscr{E} \otimes |\Lambda|^{1/2})$.*

Applying Proposition 2.9 with $\mathscr{E} = M \times \mathbb{C}$ the trivial line bundle, we see in particular that the scalar Laplacian satisfies

$$\int_M (\Delta f, f)\,|dx| = \int_M (df, df)\,|dx|,$$

or, in other words, $\Delta f = d^* df$.

In the following lemma, we collect some formulas for the Laplacian on the bundle of half-densities in normal coordinates $\mathbf{x} \mapsto \exp_{x_0} \mathbf{x}$ around a point x_0. These formulas involve the function $j(\mathbf{x})$ on $T_{x_0}M$ defined in Chapter 1 by the property that the pull-back of the volume form dx on M by the map $\exp_{x_0}$ equals $j(\mathbf{x})\,d\mathbf{x}$, or, in other words,

$$j(\mathbf{x}) = |\det(d_{\mathbf{x}} \exp_{x_0})| = \det{}^{1/2}(g_{ij}(\mathbf{x})).$$

The reader should beware the distinction between the symbol x, denoting any coordinate system on the manifold M, and $\mathbf{x}$, denoting coordinates on the tangent space $T_{x_0}M$, or normal coordinates on M.

Let $\mathscr{R}$ be the radial vector field $\mathscr{R} = \sum \mathbf{x}^i \partial_{\mathbf{x}^i}$. Denote the Laplacians on functions and on half-densities by Δ.

Proposition 2.11. *In normal coordinates* $\mathbf{x}$ *around* x_0, *we have*

1. $\nabla(f|d\mathbf{x}|^{1/2}) = (df - \frac{1}{2} f \cdot d(\log j))|d\mathbf{x}|^{1/2}$, *for* $f \in C^\infty(M)$.
2. $\Delta(f|d\mathbf{x}|^{1/2}) = ((j^{1/2} \circ \Delta \circ j^{-1/2})f)|d\mathbf{x}|^{1/2}$.
3. $\Delta\|\mathbf{x}\|^2 = -2(n + \mathcal{R}(\log j))$.
4. *If* $q_t(\mathbf{x})$ *is the time-dependent half-density on a normal coordinate chart in* M *centred at* x_0 *given by the formula*
$$q_t(\mathbf{x}) = (4\pi t)^{-n/2} e^{-\|\mathbf{x}\|^2/4t} |d\mathbf{x}|^{1/2},$$
then
$$(\partial_t + \Delta - j^{1/2}(\Delta \cdot j^{-1/2}))q_t = 0.$$

Proof. Parts (1) and (2) follow from the formula
$$\nabla(j^{1/2}|d\mathbf{x}|^{1/2}) = \nabla|dx|^{1/2} = 0.$$

To prove (3), we use integration by parts: if $\phi \in C_c^\infty(T_{x_0}M)$, then
$$\int(\phi, \Delta\|\mathbf{x}\|^2)\, j(\mathbf{x})\, d\mathbf{x} = \int(d\phi, d\|\mathbf{x}\|^2)\, j(\mathbf{x})\, d\mathbf{x} = 2\int(\mathcal{R}\phi)\, j(\mathbf{x})\, d\mathbf{x};$$

here we have used that $(d\|\mathbf{x}\|^2, d\phi)/2 = \mathcal{R}\phi$, which is a consequence of Proposition 1.27 (3). Since the adjoint of the vector field $\mathcal{R}$ with respect to the density $|d\mathbf{x}|$ is $-n - \mathcal{R}$, we see that
$$\int(\phi, \Delta\|\mathbf{x}\|^2)\, j(\mathbf{x})\, d\mathbf{x} = -2\int(n + \mathcal{R}(\log j))\, \phi(\mathbf{x})\, j(\mathbf{x})\, d\mathbf{x}.$$

As for (4), Leibniz's rule shows that $\Delta(e^{-\|\mathbf{x}\|^2/4t}|d\mathbf{x}|^{1/2})$ equals
$$\Delta(e^{-\|\mathbf{x}\|^2/4t})|d\mathbf{x}|^{1/2} + t^{-1}e^{-\|\mathbf{x}\|^2/4t}\nabla_{\mathcal{R}}|d\mathbf{x}|^{1/2} + e^{-\|\mathbf{x}\|^2/4t}\Delta|d\mathbf{x}|^{1/2}$$
and that $\Delta(e^{-\|\mathbf{x}\|^2/4t})$ equals
$$\frac{1}{4t}e^{-\|\mathbf{x}\|^2/4t}(2(n + \mathcal{R}(\log j)) - t^{-1}\|\mathbf{x}\|^2).$$

From these formulas, it is easy to show that $(\partial_t + \Delta - j^{1/2}(\Delta \cdot j^{-1/2}))q_t$ vanishes. □

2.2 The Heat Kernel on Euclidean Space

Let $V = \mathbb{R}$ be the standard Euclidean vector space of dimension one, and let
$$\Delta = -\frac{d^2}{d\mathbf{x}^2}$$

be its Laplacian. If

$$q_t(\mathbf{x},\mathbf{y}) = (4\pi t)^{-1/2} e^{-(\mathbf{x}-\mathbf{y})^2/4t},$$

it is easy to check that

$$\left(\frac{\partial}{\partial t} + \Delta_{\mathbf{x}}\right) q_t(\mathbf{x},\mathbf{y}) = 0,$$

where $\Delta_{\mathbf{x}}$ denotes the Laplacian acting in the variable $\mathbf{x}$. This equation is a special case of Proposition 2.11 (4), since $j(\mathbf{x}) = 1$ on $\mathbb{R}$.

We will need the following lemma.

Lemma 2.12. *The Gaussian integrals*

$$a(k) = (4\pi)^{-1/2} \int_{\mathbb{R}} e^{-v^2/4} v^k \, dv$$

are given by the formula

$$a(k) = \begin{cases} \dfrac{k!}{(k/2)!}, & k \text{ even}, \\ 0, & k \text{ odd}. \end{cases}$$

Proof. Consider the generating function

$$A(t) = \sum_{k=0}^{\infty} \frac{t^k}{k!} a(k) = (4\pi)^{-1/2} \int_{\mathbb{R}} e^{-v^2/4+tv} \, dv.$$

By the change of variables $u = v - 2t$, we see that this integral is equal to

$$A(t) = (4\pi)^{-1/2} \int_{\mathbb{R}} e^{-u^2/4+t^2} \, du = e^{t^2},$$

from which the lemma easily follows. □

Define the norm on C^ℓ-functions by

$$\|\phi\|_\ell = \sup_{k\le\ell} \sup_{\mathbf{x}\in\mathbb{R}} \left| \frac{d^k}{d\mathbf{x}^k} \phi(\mathbf{x}) \right|.$$

Let Q_t be the operator on functions on $\mathbb{R}$ defined by

$$\begin{aligned}(Q_t\phi)(\mathbf{x}) &= \int_{\mathbb{R}} q_t(\mathbf{x},\mathbf{y}) \phi(\mathbf{y}) \, d\mathbf{y} \\ &= (4\pi t)^{-1/2} \int_{\mathbb{R}} e^{-(\mathbf{x}-\mathbf{y})^2/4t} \phi(\mathbf{y}) \, d\mathbf{y}.\end{aligned}$$

Proposition 2.13. *Let ℓ be an even number. If ϕ is a function on $\mathbb{R}$ with $\|\phi\|_{\ell+1}$ finite, then*

$$\left\| Q_t\phi - \sum_{k=0}^{\ell/2} \frac{(-t)^k}{k!} \Delta^k \phi \right\| \le O(t^{\ell/2+1}).$$

Proof. The change of variables $\mathbf{y} \mapsto \mathbf{x} + t^{1/2}v$ shows that

$$(Q_t\phi)(\mathbf{x}) = (4\pi)^{-1/2} \int_{\mathbf{R}} e^{-v^2/4} \phi(\mathbf{x} + t^{1/2}v)\, dv.$$

If $\|\phi\|_{\ell+1} < \infty$, Taylor's formula

$$\phi(\mathbf{x}+v) = \sum_{k=0}^{\ell} \frac{v^k}{k!}\phi^{(k)}(\mathbf{x}) + \frac{v^{\ell+1}}{\ell!} \int_0^1 (1-t)^\ell f^{(\ell+1)}(\mathbf{x}+tv)\, dt$$

shows that

$$\left\| \phi(\mathbf{x}+t^{1/2}v) - \sum_{k=0}^{\ell} \frac{t^{k/2}v^k}{k!}\phi^{(k)}(\mathbf{x}) \right\| \leq \frac{|v|^{\ell+1}t^{(\ell+1)/2}}{\ell!} \|\phi\|_{\ell+1}.$$

Thus,

$$\left\| Q_t\phi - \sum_{k=0}^{\ell} a(k) \frac{t^{k/2}}{k!}\phi^{(k)}(\mathbf{x}) \right\| \leq O(t^{(\ell+1)/2}).$$

The proposition follows immediately from this estimate and the formula of Lemma 2.12 for $a(k)$. □

In this chapter, we will show that there is an analogue of the function $q_t(\mathbf{x},\mathbf{y})$ on any compact Riemannian manifold, and also that the analogue of the above proposition holds.

In Section 5.2, we will use the fact that on any Euclidean vector space V, if $\phi \in C_c^\infty(V)$, there is an asymptotic expansion in powers of $t^{1/2}$,

$$\int_V e^{-\|a\|^2/4t} \phi(a)\, da \sim (4\pi t)^{\dim(V)/2} \sum_{k=0}^{\infty} (-t)^k (\Delta^k \phi)(0)$$

which we will denote by

$$\int_V^{\text{asymp}} e^{-\|a\|^2/4t} \phi(a) da.$$

This is proved by the same change of variables as was used in the proof of the Proposition 2.13.

2.3 Heat Kernels

In this section, we will define the heat kernel of a generalized Laplacian on a bundle $\mathcal{E} \otimes |\Lambda|^{1/2}$ over a compact Riemannian manifold M. As background to this definition, let us recall the kernel theorem of Schwartz. If $\mathcal{E}$ is a bundle on a compact manifold M, the space $\Gamma^{-\infty}(M,\mathcal{E})$ of **generalized sections** of a vector bundle $\mathcal{E}$ is defined as the topological dual space of the space of smooth sections $\Gamma(M,\mathcal{E}^* \otimes |\Lambda|)$. There is a natural embedding $\Gamma(M,\mathcal{E}) \subset \Gamma^{-\infty}(M,\mathcal{E})$. We will denote by $\|\cdot\|_\ell$ a C^ℓ-norm on the space $\Gamma^\ell(M,\mathcal{E})$ of C^ℓ-sections of a vector bundle $\mathcal{E}$ on M or on $M \times M$.

Let $\mathscr{E}_1$ and $\mathscr{E}_2$ be vector bundles on M, and let pr_1 and pr_2 be the projections from $M \times M$ onto the first and second factor M respectively. We denote by $\mathscr{E}_1 \boxtimes \mathscr{E}_2$ the vector bundle

$$\mathrm{pr}_1^* \mathscr{E}_1 \otimes \mathrm{pr}_2^* \mathscr{E}_2$$

over $M \times M$.

We will call a section $p \in \Gamma(M \times M, (\mathscr{F} \otimes |\Lambda|^{1/2}) \boxtimes (\mathscr{E}^* \otimes |\Lambda|^{1/2}))$ a **kernel**. Using p, we may define an operator

$$P : \Gamma^{-\infty}(M, \mathscr{E} \otimes |\Lambda|^{1/2}) \to \Gamma(M, \mathscr{F} \otimes |\Lambda|^{1/2})$$

by

$$(Ps)(x) = \int_{y \in M} p(x,y)\, s(y).$$

By this, we mean that $(Ps)(x)$ is the pairing of the distribution $s(\cdot)$ in $\Gamma^{-\infty}(M, \mathscr{E} \otimes |\Lambda|^{1/2})$ with the smooth function $p(x,\cdot) \in \Gamma(M, \mathscr{E}^* \otimes |\Lambda|^{1/2})$. An operator with a smooth kernel will be called a **smoothing operator**. Clearly, the composition of two operators P_1, P_2 with smooth kernels p_1, p_2 is the operator with smooth kernel

$$p(x,y) = \int_{z \in M} p_1(x,z)\, p_2(z,y);$$

$p_1(x,z)p_2(z,y)$ is a density in z, so can be integrated.

The following form of the **Schwartz kernel theorem** gives a characterization of operators with smooth kernels.

Proposition 2.14. *The map which assigns to the kernel $p(x,y)$ the corresponding operator $s \mapsto Ps$ is an isomorphism between the space of kernels $\Gamma(M \times M, (\mathscr{F} \otimes |\Lambda|^{1/2}) \boxtimes (\mathscr{E}^* \otimes |\Lambda|^{1/2}))$ and the space of bounded linear operators from $\Gamma^{-\infty}(M, \mathscr{E} \otimes |\Lambda|^{1/2})$ to $\Gamma(M, \mathscr{F} \otimes |\Lambda|^{1/2})$.*

We will sometimes use Dirac's notation for the kernel of an operator P: if x and y lie in M, then the kernel of an operator P at $(x,y) \in M \times M$ is written $\langle x \mid P \mid y \rangle$, so that

$$(P\phi)(x) = \int_{y \in M} \langle x \mid P \mid y \rangle \phi(y). \tag{2.4}$$

There are two quite different ways to investigate the heat kernel of H, that is, the kernel attached to the operator e^{-tH} $(t > 0)$ by the kernel theorem:

(1) If H is symmetric, if we can show that the operator H has a self-adjoint extension $\bar{H}$ that is bounded below, then by the spectral theorem, the operator $e^{-t\bar{H}}$, which we will simply denote by e^{-tH}, is well defined as a bounded operator on the Hilbert space of square integrable sections of $\mathscr{E} \otimes |\Lambda|^{1/2}$. If, in addition, we can show that $e^{-t\bar{H}}$ extends to an operator which maps distributional sections to smooth sections, then it follows by the kernel theorem that the kernel of $e^{-t\bar{H}}$ is smooth.

(2) The other method, which we will follow, starts by proving the existence of a smooth kernel $p_t(x,y)$ which possesses the properties that the kernel of the operator e^{-tH} should possess, without first constructing the operator $\bar{H}$. The advantage of

this method is that it gives more information about the operator e^{-tH} than the first method, is more closely related to the geometry of H, is technically more elementary, and does not require that H be symmetric.

The following definition summarizes the properties that the kernel of an operator e^{-tH} must have.

Definition 2.15. *Let H be a generalized Laplacian on $\mathscr{E}\otimes|\Lambda|^{1/2}$. A* **heat kernel** *for H is a continuous section $p_t(x,y)$ of the bundle $(\mathscr{E}\otimes|\Lambda|^{1/2})\boxtimes(\mathscr{E}^*\otimes|\Lambda|^{1/2})$ over $\mathbb{R}_+\times M\times M$, satisfying the following properties:*

1. *$p_t(x,y)$ is C^1 with respect to t, that is, $\partial p_t(x,y)/\partial t$ is continuous in (t,x,y);*
2. *$p_t(x,y)$ is C^2 with respect to $\mathbf{x}$, that is, the partial derivatives*
$$\frac{\partial^2 p_t(\mathbf{x},y)}{\partial\mathbf{x}^i\partial\mathbf{x}^j}$$
are continuous in $(t,\mathbf{x},y)$ for any coordinate system $\mathbf{x}$;
3. *$p_t(x,y)$ satisfies the* **heat equation**
$$(\partial_t+H_x)\,p_t(x,y)=0;$$
4. *$p_t(x,y)$ satisfies the boundary condition at $t=0$: if s is a smooth section of $\mathscr{E}\otimes|\Lambda|^{1/2}$, then*
$$\lim_{t\to 0}P_t s=s.$$
Here we denote by P_t the operator defined by the kernel p_t, and the limit is meant in the uniform norm $\|s\|_0=\sup_{x\in M}\|s(x)\|$, for any metric on $\mathscr{E}$.

Our first task is to prove that the heat kernel is unique. We consider the formal adjoint H^* of H, which is a generalized Laplacian on $\mathscr{E}^*\otimes|\Lambda|^{1/2}$.

Lemma 2.16. *Assume that the operator H^* has a heat kernel, which we denote by p_t^*. If $s(t,x)$ is a map from $\mathbb{R}_+$ to the space of sections of $\mathscr{E}\otimes|\Lambda|^{1/2}$ which is C^1 in t and C^2 in x, such that $\lim_{t\to 0}s_t=0$ and which satisfies the heat equation $(\partial_t+H)s_t=0$, then $s_t=0$.*

Proof. If $s_1(t,\cdot)$ and $s_2(t,\cdot)$ are C^2 sections of $\mathscr{E}\otimes|\Lambda|^{1/2}$ for each $t\geq 0$, we have
$$\int_M\langle Hs_1(t,\cdot),s_2(t,\cdot)\rangle=\int_M\langle s_1(t,\cdot),H^*s_2(t,\cdot)\rangle.$$
For any u in $\Gamma(M,\mathscr{E}^*\otimes|\Lambda|^{1/2})$, consider the function
$$f(\theta)=\int_{(x,y)\in M\times M}(s(\theta,x),p^*_{t-\theta}(x,y)u(y))$$
where $0<\theta<t$. Differentiating with respect to θ and using the heat equation Definition 2.15 (3), we see that $f(\theta)$ is constant. By considering the limits when θ tends to t and to 0, we obtain
$$\int_{x\in M}(s(t,x),u(x))=0$$
for every $u\in\Gamma(M,\mathscr{E}^*\otimes|\Lambda|^{1/2})$, so that $s_t=0$ for all $t>0$. □

Using this lemma, we obtain the following result.

Proposition 2.17. *1. Suppose there exists a heat kernel p_t^* for the operator H^*. Then there is at most one heat kernel p_t for H.*
2. Suppose there exist heat kernels p_t^* for H^*, and p_t for H. Then $p_t(x,y) = (p_t^*(y,x))^*$.
3. Suppose there exist heat kernels p_t^* for H^*, and p_t for H. Then the operators $P_t s(x) = \int_M p_t(x,y)s(y)\,dy$ form a semi-group.

Proof. Consider $f(\theta) = \int_M((P_\theta s)(x),(P^*_{t-\theta}u)(x))$ for $0 < \theta < t$, where s and u are smooth sections of $\Gamma(M,\mathscr{E}\otimes|\Lambda|^{1/2})$; as before, this is independent of θ and we obtain

$$(P_t s,u) = (s,P_t^* u)$$

by considering limits. This proves (1) and (2).

To show the semigroup property, we must show that if $s \in \Gamma(M,\mathscr{E}\otimes|\Lambda|^{1/2})$ and $\theta > 0$, then the unique solution s_t of the heat equation for H with boundary condition $\lim_{t\to 0} s_t = P_\theta s$ is given by $P_{t+\theta}s$; but this follows from the lemma. □

2.4 Construction of the Heat Kernel

On $\mathbb{R}^n$, the scalar heat kernel is given by the exact formula

$$q_t(x,y) = (4\pi t)^{-n/2}e^{-\|x-y\|^2/4t}.$$

For an arbitrary Riemannian manifold M and bundle $\mathscr{E}$, it is usually impossible to find an exact expression for the heat kernel $p_t(x,y)$. However, for many problems, the use of an approximate solution is sufficient. The true kernel p_t is then calculated from the approximate solution by a convergent iterative procedure.

In order to describe this technique, let us consider the following finite dimensional analogue of the construction of p_t. Let V be a finite dimensional vector space with a linear endomorphism H (we have in mind here though that $V = \Gamma(M,\mathscr{E}\otimes|\Lambda|^{1/2})$ and $H = \Delta^{\mathscr{E}\otimes|\Lambda|^{1/2}} + F$); we would like to construct the family of operators $P_t = e^{-tH}$.

Suppose that we are given a function $K_t : \mathbb{R}_+ \to \mathrm{End}(V)$ which is an approximate solution of the heat equation for small t in the sense that:

$$R_t = \frac{dK_t}{dt} + HK_t = O(t^\alpha), \quad \text{for some } \alpha \geq 0,$$

and such that $K_0 = 1$; such a function is also called a parametrix for the heat equation. The function R_t is called the remainder.

Definition 2.18. *The k-simplex Δ_k is the following subset of $\mathbb{R}^k$:*

$$\{(t_1,\dots,t_k) \mid 0 \leq t_1 \leq t_2 \dots \leq t_k \leq 1\}.$$

Often, we will parametrize Δ_k by the coordinates

$$\sigma_0 = t_1 \,,\ \sigma_i = t_{i+1} - t_i \,,\ \sigma_k = 1 - t_k,$$

which satisfy $\sigma_0 + \dots + \sigma_k = 1$ and $0 \le \sigma_i \le 1$. For $t > 0$, we will write $t\Delta_k$ for the rescaled simplex

$$\{(t_1,\dots,t_k) \mid 0 \le t_1 \le t_2 \dots \le t_k \le t\}.$$

Let v_k be the volume of Δ_k for the Lebesgue measure $dt_1 \dots dt_k$. Since $v_0 = 1$ and

$$v_k = \int_0^1 \operatorname{vol}(t_k\Delta_{k-1})\,dt_k = \int_0^1 t_k^{k-1} v_{k-1}\,dt_k = \frac{v_{k-1}}{k},$$

we see that $v_k = 1/k!$. The rapid decay of v_k will be important in the following proofs, when we must show that certain series converge.

Theorem 2.19. *Let $Q_t^k : \mathbb{R}_+ \to \operatorname{End}(V)$ be defined by the integral*

$$Q_t^k = \int_{t\Delta_k} K_{t-t_k} R_{t_k - t_{k-1}} \dots R_{t_2-t_1} R_{t_1}\, dt_1 \dots dt_k;$$

in particular, $Q_t^0 = K_t$. Then the sum of the convergent series $\sum_{k\ge 0} (-1)^k Q_t^k$ is equal to $P_t = e^{-tH}$, and $P_t = K_t + O(t^{1+\alpha})$.

Proof. The fundamental theorem of calculus shows that the derivative of the convolution $\int_0^t a(t-s)b(s)\,ds$ with respect to t equals

$$\int_0^t \frac{da}{dt}(t-s)b(s)\,ds + a(0)b(t).$$

If we apply this with $a(s) = K_s$ and $b(s) = R^{(k)}(s)$ defined by

$$R^{(k)}(s) = \int_{s\Delta_{k-1}} R_{s-t_{k-1}} \dots R_{t_2-t_1} R_{t_1}\, dt_1 \dots dt_{k-1},$$

then since $K_0 = 1$, we obtain

$$(d_t + H) Q_t^k = R^{(k+1)}(t) + R^{(k)}(t).$$

Thus, the sum with respect to k telescopes:

$$(d_t + H) \sum_{k=0}^{\infty} (-1)^k Q_t^k = 0.$$

To estimate $P_t - K_t$ for small t, we use the uniform bounds over $t\Delta_k$,

$$|K_{t-t_k}| \le C_0 \quad \text{and} \quad |R_{t_{i+1}-t_i}| \le C t^{\alpha},$$

which lead to the bound

$$|Q_t^k| \le C_0 C^k t^{k\alpha} \frac{t^k}{k!}, \quad \text{since } \operatorname{vol}(t\Delta_k) = \frac{t^k}{k!}.$$

It is clear from this that the series defining P_t converges, and that $P_t = K_t + O(t^{1+\alpha})$.

□

An important special case of this formula is the **Volterra series** for the exponential of a perturbed operator: if $H = H_0 + H_1 \in \mathrm{End}(V)$ and $K_t = e^{-tH_0}$, we see that $R_t = (d/dt + H)e^{-tH_0} = H_1 e^{-tH_0}$, and we obtain

$$e^{-t(H_0+H_1)} = e^{-tH_0} + \sum_{k=1}^{\infty}(-1)^k I_k,$$

where

$$I_k = \int_{t\Delta_k} e^{-(t-t_k)H_0} H_1 e^{-(t_k-t_{k-1})H_0} \dots H_1 e^{-t_1 H_0}\, dt_1 \dots dt_k.$$

This formula implies in particular that

$$\begin{aligned} e^{-t(H_0+H_1)} &= \sum_{k=0}^{\infty}(-t)^k \int_{\Delta_k} e^{-\sigma_0 t H_0} H_1 e^{-\sigma_1 t H_0} \dots H_1 e^{-\sigma_k t H_0}\, d\sigma_1 \dots d\sigma_k \\ &= e^{-tH_0} - t\int_0^1 e^{-\sigma t H_0} H_1 e^{-(1-\sigma)tH_0}\, d\sigma + \dots \end{aligned} \tag{2.5}$$

If H_z is a one-parameter smooth family in $\mathrm{End}(V)$, by setting $H_0 = H_z$ and $H_1 = \varepsilon dH_z/dz$ we obtain the following formula for the derivative of e^{-tH_z}:

$$\frac{d}{dz}(e^{-tH_z}) = -\int_0^t e^{-(t-s)H_z}\frac{dH_z}{dz} e^{-sH_z}\, ds. \tag{2.6}$$

We now turn to implementing this method in the case which interests us, namely $V = \Gamma(M, \mathcal{E} \otimes |\Lambda|^{1/2})$ and $H = \Delta^{\mathcal{E}\otimes|\Lambda|^{1/2}} + F$. In formulas which follow, we will omit the notation $dt_1 \dots dt_k$ for the Lebesgue measure on $t\Delta_k$, since this is the only measure which we will consider on this space. Imitating the above method in an infinite dimensional setting, the steps in the construction of the heat kernel are:

(1) Construction of an approximate solution $k_t(x,y)$, and study of the remainder $r_t(x,y) = (\partial_t + H_x)k_t(x,y)$;

(2) Proof of convergence of the series

$$\sum_{k=0}^{\infty}(-1)^k \int_{t\Delta_k}\int_{M^k} k_{t-t_k}(x,z_k) r_{t_k-t_{k-1}}(z_k,z_{k-1}) \dots r_{t_1}(z_1,y)$$

to the heat kernel $p_t(x,y)$.

The construction of k_t will be performed in the next section, by solving the heat equation in a formal power series in t. For the moment, we will only use the properties of the approximate solution $k_t(x,y)$ summarized in the following theorem, which will be proved in the next section.

Theorem 2.20. *For every positive integer N, there exists a smooth one-parameter family of smooth kernels $k_t^N(x,y)$ such that, for every integer ℓ*

1. *for every $T > 0$, the corresponding operators K_t form a uniformly bounded family of operators on the space $\Gamma^{\ell}(M, \mathcal{E} \otimes |\Lambda|^{1/2})$ for $0 < t < T$;*

2. *for every* $s \in \Gamma^\ell(M, \mathscr{E} \otimes |\Lambda|^{1/2})$, *we have* $\lim_{t\to 0} K_t s = s$ *with respect to the norm* $\|\cdot\|_\ell$;
3. *the kernel* $r_t(x,y) = (\partial_t + H_x)k_t^N(x,y)$ *satisfies the estimate*

$$\|r_t\|_\ell \le C(\ell)t^{(N-n/2)-\ell/2}.$$

We fix N and write k_t for k_t^N. For such a kernel k_t, we would like to consider the operator

$$Q_t^k = \int_{t\Delta_k} K_{t-t_k} R_{t_k - t_{k-1}} \dots R_{t_1}$$

defined by the kernel

$$q_t^k(x,y) = \int_{t\Delta_k}\int_{M^k} k_{t-t_k}(x,z_k) r_{t_k-t_{k-1}}(z_k,z_{k-1}) \dots r_{t_1}(z_1,y).$$

We will first prove that if N is chosen large enough, this integral is convergent and that the kernel q_t^k is differentiable to some order depending on N. To do this, we need some estimates.

Consider the kernel

$$r_t^{k+1}(x,y) = \int_{t\Delta_k}\int_{M^k} r_{t-t_k}(x,z_k) r_{t_k-t_{k-1}}(z_k,z_{k-1}) \dots r_{t_1}(z_1,y)$$

Lemma 2.21. *If N satisfies $N > (n+\ell)/2$, then r_t^{k+1} is of class C^ℓ with respect to x and y, and*

$$\|r_t^{k+1}\|_\ell \le C^{k+1} t^{(k+1)(N-n/2)-\ell/2} \operatorname{vol}(M)^k t^k/k!$$

Proof. If $N > (n+\ell)/2$, the section $(x,y) \to r_t(x,y)$ and its derivatives up to order ℓ extend continuously to $t=0$, so that the integral defining r_t^{k+1} is convergent. Now, the uniform estimates of the integrand over $t\Delta_k$ lead to the result, the factor $t^k/k!$ being equal to the volume of $t\Delta_k$. □

We now turn to the estimation of $q_t^k(x,y)$ and its derivatives.

Lemma 2.22. *Assume that $N > (n+\ell)/2$, and that $\ell \ge 1$.*

1. *The kernel $q_t^k(x,y)$ is of class C^ℓ with respect to x and y, and there exists a constant $\tilde{C}$ such that*

$$\|q_t^k\|_\ell < \tilde{C} C^k \operatorname{vol}(M)^{k-1} t^{k(N-n/2)-\ell/2} t^k/(k-1)!$$

for every $k \ge 1$.

2. *The kernel $q_t^k(x,y)$ is of class C^1 with respect to t, and*

$$(\partial_t + H_x)q_t^k(x,y) = r_t^{k+1}(x,y) + r_t^k(x,y).$$

Proof. We write the kernel q_t^k in the following form:

$$q_t^k(x,y) = \int_0^t \Big(\int_{z\in M} k_{t-s}(x,z) r_s^k(z,y) \Big) ds.$$

Since composition with the kernel k_s defines a uniformly bounded family of operators on $\Gamma^\ell(\mathscr{E} \otimes |\Lambda|^{1/2})$, the section

$$b(t,s,x,y) = \int_{z\in M} k_{t-s}(x,z) r_s^k(z,y)$$

depends continuously on $s \in [0,t]$, as do its derivatives up to order ℓ with respect to x and y. Furthermore, $b(t,t,x,y) = r_t^k(x,y)$; also

$$(\partial_t + H_x) b(t,s,x,y) = \int_{z\in M} r_{t-s}(x,z) r_s^k(z,y) = r^{k+1}(x,y),$$

so that $b(t,s,x,y)$ is differentiable with respect to t, and $\partial_t b(t,s,x,y)$ is a continuous function of $s \in [0,t]$. The proof of (2) now follows by differentiating $q_t^k(x,y) = \int_0^t b(t,s,x,y) ds$ with respect to t.

To prove (1), we use the uniform boundedness of K_t, which shows that for $0 \le s \le t$,

$$\|b(t,s)\|_\ell \le C' \|r_s^k\|_\ell.$$

Combining this with Lemma 2.21, we obtain

$$\|b(t,s)\|_\ell \le C' C^k t^{k(N-n/2)-\ell/2} \operatorname{vol}(M)^{k-1} t^{k-1}/(k-1)!$$

from which (1) follows. □

The following theorem shows that we can construct a heat kernel p_t starting from the approximate kernel k_t.

Theorem 2.23. *Assume that the kernel $k_t^N(x,y)$ satisfies the conditions of Theorem 2.20 with $N > n/2+1$.*

1. *For any ℓ such that $N > (n+\ell+1)/2$ the series*

$$p_t(x,y) = \sum_{k=0}^{\infty} (-1)^k q_t^k(x,y)$$

converges in the $\|\cdot\|_{\ell+1}$-norm (over $M \times M$) and defines a C^1-map from $\mathbb{R}_+$ to $\Gamma^\ell(M \times M, (\mathscr{E} \otimes |\Lambda|^{1/2}) \boxtimes (\mathscr{E}^* \otimes |\Lambda|^{1/2}))$ which satisfies the heat equation

$$(\partial_t + H_x) p_t(x,y) = 0.$$

2. *The kernel k_t^N approximates p_t in the sense that*

$$\|\partial_t^k (p_t - k_t^N)\|_\ell = O(t^{(N-n/2)-k-\ell/2+1}) \quad \textit{when } t \to 0.$$

3. *The kernel p_t is a heat kernel for the operator H.*

Proof. Our estimate on $\|q_t^k(x,y)\|_{\ell+1}$ proves the absolute convergence of the series defining $p_t(x,y)$ in the Banach space $\Gamma^{\ell+1}$, uniformly with respect to t. From the equation $\partial_t q_t^k = r_t^{k+1} + r_t^k - H_x q_t^k$, we obtain bounds for the derivative with respect to t in the C^ℓ norm. This shows the convergence of the series $\sum(-1)^k \partial_t q_t^k$ in Γ^ℓ, uniformly with respect to t, and the equation $(\partial_t + H_x)p_t(x,y) = 0$ follows from the same telescoping as in Theorem 2.19. To show (3), we must show that p_t satisfies the initial condition of a heat kernel. However, we know that k_t^N satisfies this initial condition, so that (3) follows from (2). □

2.5 The Formal Solution

Let $H = \Delta^{\mathcal{E}\otimes|\Lambda|^{1/2}} + F$ be a generalized Laplacian on the bundle $\mathcal{E}\otimes|\Lambda|^{1/2}$ over a compact manifold M. In this section, we will denote the connection $\nabla^{\mathcal{E}}$ on $\mathcal{E}$ simply by ∇. We denote by $|dx|^{1/2}$ the canonical half-density associated to the Riemannian metric on M. To construct an approximate solution to the heat equation for H, we start by finding a formal solution to the heat equation as a series of the form

$$k_t(x,y) = q_t(x,y)\sum_{i=0}^{\infty} t^i \Phi_i(x,y,H)\,|dy|^{1/2},$$

where the coefficients $(x,y)\mapsto \Phi_i(x,y,H)$ are smooth sections of the bundle $\mathcal{E}\boxtimes\mathcal{E}^*$ defined in a neighbourhood of the diagonal of $M\times M$ and q_t is modelled on the Euclidean heat kernel: in normal coordinates around y ($x = \exp_y \mathbf{x}$),

$$\begin{aligned} q_t(x,y) &= (4\pi t)^{-n/2} e^{-\|\mathbf{x}\|^2/4t}\,|d\mathbf{x}|^{1/2} \\ &= (4\pi t)^{-n/2} e^{-d(x,y)^2/4t}\, j(\mathbf{x})^{-1/2} |dx|^{1/2}. \end{aligned}$$

(The precise definition of a formal solution is given below). We fix $y\in M$ and write q_t for the section $x\mapsto q_t(x,y)$ of the bundle $|\Lambda|^{1/2}$. Abusing notation once more, we denote by $j^{1/2}$ the function on M around y given by $j^{1/2}(x) = j(\mathbf{x})^{1/2}$ if $x = \exp_y \mathbf{x}$.

To show the existence of a formal solution of the heat equation, we will make use of the following formula. Here, we write H for the operator $s\mapsto |dx|^{-1/2}\cdot H\cdot s\,|dx|^{1/2}$ on $\Gamma(M,\mathcal{E})$.

Proposition 2.24. *Let B be the operator acting on sections of $\mathcal{E}$ in a neighbourhood of y defined by the formula*

$$B = j^{1/2}\circ H\circ j^{-1/2}.$$

Then, for any time dependent section s_t of $\mathcal{E}$,

$$(\partial_t + H)s_t q_t = ((\partial_t + t^{-1}\nabla_{\mathcal{R}} + B)s_t)q_t.$$

Proof. Let $t \mapsto s_t$ be a smooth map from $\mathbb{R}_+$ to the space of smooth sections of $\mathscr{E}$ around y. By Leibniz's rule, we see that

$$(\partial_t + H)(s_t q_t) = ((\partial_t + H)s_t)q_t - 2(\nabla s_t, \nabla q_t) + s_t((\partial_t + \Delta)q_t),$$

where Δ is the scalar Laplacian. Proposition 2.11 gives explicit formulas for the last two terms:

$$-2\nabla q_t = (t^{-1}\mathscr{R} + d(\log j))q_t,$$
$$(\partial_t + \Delta)q_t = j^{1/2}(\Delta j^{-1/2})q_t.$$

From this, we see that

$$(\partial_t + H)(s_t q_t) = \left((\partial_t + H + t^{-1}\nabla_{\mathscr{R}} + \nabla_{d\log j} + j^{1/2}(\Delta j^{-1/2}))s_t\right) q_t.$$

By Leibniz's rule again, we obtain

$$j^{1/2} \circ H \circ j^{-1/2} = H + \nabla_{d\log j} + j^{1/2}(\Delta j^{-1/2}) \qquad \square$$

Proposition 2.24 makes clear what we mean by a formal solution of the heat equation.

Definition 2.25. *Let $\Phi_t(x,y)$ be a formal power series in t whose coefficients are smooth sections of $\mathscr{E} \boxtimes \mathscr{E}^*$ defined in a neighbourhood of the diagonal of $M \times M$. We will say that $q_t(x,y)\Phi_t(x,y)|dy|^{1/2}$ is a* **formal solution** *of the heat equation around y if $x \mapsto \Phi_t(x,y)$, considered as a section of the bundle $\mathscr{E} \otimes \mathscr{E}_y^*$ over a neighbourhood of y in M, satisfies the equation*

$$(\partial_t + t^{-1}\nabla_{\mathscr{R}} + B)\Phi_t(\cdot, y) = 0.$$

We can now prove the existence and uniqueness of the formal solution of the heat equation. Let x and y be sufficiently close points in M, and write $x = \exp_y \mathbf{x}$ for $\mathbf{x} \in T_yM$. Denote by $\tau(x,y) : \mathscr{E}_y \to \mathscr{E}_x$ the parallel transport along the geodesic curve $x_s = \exp_y s\mathbf{x} : [0,1] \to M$.

Theorem 2.26. *There exists a unique formal solution $k_t(x,y)$ of the heat equation*

$$(\partial_t + H_x)k_t(x,y) = 0$$

of the form

$$k_t(x,y) = q_t(x,y)\sum_{i=0}^{\infty} t^i \Phi_i(x,y,H)\,|dy|^{1/2},$$

such that $\Phi_0(y,y,H) = I_{\mathscr{E}_y} \in \operatorname{End}(\mathscr{E}_y)$. We have the following explicit formula for Φ_i:

$$\tau(x,y)^{-1}\Phi_i(x,y) = -\int_0^1 s^{i-1}\tau(x_s,y)^{-1}(B_x \cdot \Phi_{i-1})(x_s,y)ds.$$

In particular, $\Phi_0(x,y) = \tau(x,y)$.

Proof. In the left side of the equation

$$(\partial_t + t^{-1}\nabla_{\mathscr{R}} + B_x)\sum_{i=0}^{\infty} t^i \Phi_i(x,y,H) = 0,$$

we set the coefficients of each t^i equal to zero. This gives the system of equations:

$$\nabla_{\mathscr{R}}\Phi_0 = 0$$
$$(\nabla_{\mathscr{R}} + i)\Phi_i = -B_x\Phi_{i-1} \quad \text{if } i > 0.$$

Moreover, we know that $\Phi_0(y,y,H) = I_{\mathscr{E}_y} \in \operatorname{End}(\mathscr{E}_y)$, and that $\Phi_i(x,y,H)$ should be continuous at $x = y$.

The parallel transport $\tau(x,y) : \mathscr{E}_y \to \mathscr{E}_x$ along the geodesic $x_s = \exp_y s\mathbf{x} : [0,1] \to M$ satisfies the differential equation $\nabla_{\mathscr{R}}\tau = 0$ with boundary condition $\tau(y,y) = 1$, and we see that $\Phi_0(x,y,H) = \tau(x,y)$.

To obtain a formula for Φ_i in terms of Φ_{i-1}, we consider the function

$$\phi_i(s) = s^i \Phi_i(x_s, y, H).$$

where $x_s = \exp_y s\mathbf{x}$. Then $\phi_i(0) = 0$ and $\nabla_{d/ds}\phi_i = -s^{i-1}(B_x\Phi_{i-1})(x_s,y)$, and we obtain the required explicit formula for Φ_i. □

Example 2.27. We can use the explicit iterative formula for $\Phi_i(x,y)$ given above to calculate the subleading term $\Phi_1(x,x)$ in the asymptotic expansion of $k_t(x,y)$ on the diagonal. We have

$$\Phi_1(x,x) = -(B \cdot \Phi_0)(x,x).$$

However, since $\Phi_0(x,y) = \tau(x,y)$, we see that

$$\Phi_1(x,x) = -(j^{-1/2}(\Delta^{\mathscr{E}} + F)j^{1/2})(x,x).$$

Using Proposition 1.28, we see easily that this function equals $\frac{1}{3}r_M(x) - F(x)$, where $r_M(x)$ is the scalar curvature of M.

Consider the $\mathscr{E}_y$-valued function on T_yM defined by

$$U_i(\mathbf{x},y,H) = \tau(\exp_y \mathbf{x}, y)^{-1}\Phi_i(\exp_y \mathbf{x}, y, H) \quad \mathbf{x} \in T_yM.$$

Let L be the differential operator which acts on the space of $\mathscr{E}_y$-valued functions on M around y, defined by the formula

$$(L\phi)(x) = \tau(x,y)^{-1}B_x(\tau(x,y)\phi(x)).$$

Then we see that

$$U_i(\mathbf{x},y,H) = -\int_0^1 s^{i-1}(LU_{i-1})(x_s,y)ds,$$

where $x_s = \exp_y s\mathbf{x}$. It is clear from Propositions 1.18 and 1.28 that the coefficients $U_{i,\alpha}(y,H)$ in the Taylor expansion

$$\sum_{\alpha} U_{i,\alpha}(y,H)\mathbf{x}^{\alpha}$$

of the function $\mathbf{x} \mapsto U_i(\mathbf{x},y,H)$ are polynomials in the covariant derivatives at y of the Riemannian curvature, the curvature of the bundle $\mathscr{E}$, and the potential F, since this is true of the Taylor coefficients of the coefficients of the operator L at y.

Definition 2.28. *If $\psi : \mathbb{R}_+ \to [0,1]$ is a smooth function such that*

$$\psi(s) = 1, \quad \textit{if } s < \varepsilon^2/4,$$
$$\psi(s) = 0, \quad \textit{if } s > \varepsilon^2,$$

we will call ψ a **cut-off function**.

We will now use the formal solution to the heat equation to construct an approximate solution $k_t^N(x,y)$ satisfying the properties given in Theorem 2.20. To do this, we truncate the formal series and extend it from the diagonal to all of $M \times M$ by means of a cut-off function $\psi(d(x,y)^2)$ along the diagonal, where the small constant ε is chosen smaller than the injectivity radius of the manifold; thus, the exponential map is a diffeomorphism for $|\mathbf{x}| < \varepsilon$. We define the approxiamte solution by the formula

$$k_t^N(x,y) = \psi(d(x,y)^2)\, q_t(x,y) \sum_{i=0}^{N} t^i \Phi_i(x,y,H)|dy|^{1/2}. \tag{2.7}$$

In the next theorem, we will prove that k_t^N possesses the properties that were used to construct the true heat kernel in Section 3. Indeed, we will see that it satisfies somewhat stronger estimates than were stated in Theorem 2.20.

Theorem 2.29. *Let ℓ be an even positive integer.*

1. *For any $T > 0$ the kernels k_t^N, $0 \le t \le T$, define a uniformly bounded family of operators K_t^N on $\Gamma^\ell(M, \mathscr{E} \otimes |\Lambda|^{1/2})$, and*

$$\lim_{t \to 0} \left\| K_t^N s - s \right\|_\ell = 0.$$

2. *There exist differential operators D_k of order less than or equal to $2k$ such that D_0 is the identity and such that for any $s \in \Gamma^{\ell+1}(M, \mathscr{E} \otimes |\Lambda|^{1/2})$,*

$$\left\| K_t^N s - \sum_{k=0}^{\ell/2-j} t^k D_k s \right\|_{2j} = O(t^{(\ell+1)/2-j}).$$

3. *The kernel $r_t^N(x,y) = (\partial_t + H_x)k_t^N(x,y)$ satisfies the estimates*

$$\|\partial_t^k r_t^N\|_\ell < C t^{(N-n/2)-k-\ell/2},$$

for some constant C depending on ℓ and k.

Proof. We fix N and write K_t for the operator corresponding to the kernel k_t^N. In a neighbourhood of the diagonal, we write $y = \exp_x \mathbf{y}$, with $\mathbf{y} \in T_xM$, and identify $\mathcal{E}_x$ to $\mathcal{E}_y$ by parallel transport along the unique geodesic joining x and y. If s is a section of $\mathcal{E} \otimes |\Lambda|^{1/2}$, we denote by $s(x,\mathbf{y})$ the function of $\mathbf{y}$ such that $s(y) = s(x,\mathbf{y})|d\mathbf{y}|^{1/2}$. We see that

$$(K_t s)(x) = (4\pi t)^{-n/2} \int_{T_xM} e^{-\|\mathbf{y}\|^2/4t} \sum_{i=0}^{N} t^i \Psi_i(x,\mathbf{y}) s(x,\mathbf{y}) |d\mathbf{y}| \otimes |dx|^{1/2},$$

with smooth compactly supported coefficients

$$(x,\mathbf{y}) \longmapsto \Psi_i(x,\mathbf{y}) \in \mathrm{End}(\mathcal{E}_x)$$

Furthermore, $\Psi_0(x,0) = I_{\mathcal{E}_x}$.

The statement of the theorem is local in x, so we can assume that the vector spaces $\mathcal{E}_x$ and T_xM are fixed vector spaces E and V. The change of variables $\mathbf{y} = t^{1/2}v$ on V shows that $(K_t s)(x)$ equals

$$(4\pi)^{-n/2} \int_V e^{-\|v\|^2/4} \sum_{i=0}^{N} t^i \Psi_i(x,t^{1/2}v) s(x,t^{1/2}v) dv \otimes |dx|^{1/2},$$

from which (1) follows.

By Taylor's Theorem, there exist differential operators D_k of order less than $2k$ such that

$$\Big\| K_t^N s - \sum_{k=0}^{\ell/2} t^k D_k s \Big\|_0 = O(t^{\ell/2+1}).$$

A similar argument may be used to estimate the C^{2j}-norm of

$$K_t^N s - \sum_{k=0}^{\ell/2-j} t^k D_k s.$$

The bound on $r_t^N(x,y)$ follows from the formula for k_t^N by means of Proposition 2.24, which shows that

$$r_t^N(x,y) = [(\partial_t + t^{-1}\nabla_{\mathcal{R}} + B_x) \cdot (\psi(d(x,y)^2) \sum_{i=0}^{N} t^i \Phi_i(x,y,H))] q_t(x,y).$$

The right-hand side in this formula is made up of two kinds of terms, those which involve at least one differentiation of the factor $\psi(d(x,y)^2)$, and those which only involve time derivatives or derivatives of the functions Φ_i. The first type vanishes for $d(x,y) < \varepsilon/2$ and so is very easy to estimate, since for $d(x,y) > \varepsilon/2$

$$\partial_x^\alpha \psi(d(x,y)^2) e^{-d(x,y)^2/4t} = O(t^k),$$

for any k. From the system of differential equations defining the functions Φ_i, we see that the terms which do not involve a derivative of $\psi(d(x,y)^2)$ cancel, except for one remaining term equal to

$$t^N q_t(x,y)(B_x \Phi_N)(x,y),$$

which may be bounded by $t^{N-n/2}$. The argument used to bound the derivatives of $r_t(x,y)$ is similar, once one observes that

$$\partial_t e^{-x^2/t} = t^{-1}(x^2/t)e^{-x^2/t} = O(t^{-1}),$$
$$\partial_x e^{-x^2/t} = t^{-1/2}(-2x/t^{1/2})e^{-x^2/t} = O(t^{-1/2}). \qquad \square$$

As a consequence of Theorems 2.29 and 2.23, we have now proved the existence of a C^ℓ-heat kernel, for any operator $H = \Delta^{\mathscr{E}\otimes|\Lambda|^{1/2}} + F$ and any ℓ. From the unicity of the heat kernel (Proposition 2.17) it follows that the heat kernel $p_t(x,y,H)$ is smooth as a function of (x,y). The heat equation $(\partial_t + H_x)p_t(x,y) = 0$ implies of course that it is smooth as a function of (t,x,y).

Let us summarize the properties of the heat kernel that will be used in the proof of the local index theorem.

Theorem 2.30. *Let $p_t(x,y,H)$ be the heat kernel of the generalized Laplacian $H = \Delta^{\mathscr{E}\otimes|\Lambda|^{1/2}} + F$. There exist smooth sections $\Phi_i \in \Gamma(M \times M, \mathscr{E} \boxtimes \mathscr{E}^*)$ such that, for every $N > n = \dim(M)/2$, the kernel $k_t^N(x,y,H)$ defined by the formula*

$$(4\pi t)^{-n/2} e^{-d(x,y)^2/4t} \psi(d(x,y)^2) \sum_{i=0}^{N} t^i \Phi_i(x,y,H)\, j(x,y)^{-1/2} |dx|^{1/2} \otimes |dy|^{1/2}$$

is asymptotic to $p_t(x,y,H)$:

$$\|\partial_t^k (p_t(x,y,H) - k_t^N(x,y,H))\|_\ell = O(t^{N-n/2-\ell/2-k}).$$

The leading term $\Phi_0(x,y,H)$ is equal to the parallel transport $\tau(x,y) : \mathscr{E}_y \to \mathscr{E}_x$ with respect to the connection associated to H along the unique geodesic joining x and y.

If $s \in \Gamma(M, \mathscr{E} \otimes |\Lambda|^{1/2})$ is a smooth section of $\mathscr{E} \otimes |\Lambda|^{1/2}$, we see from Theorem 2.29 that $P_t s$, defined by

$$(P_t s)(x) = \int_{y \in M} p_t(x,y,H)s(y)$$

has an asymptotic expansion in $\Gamma(M, \mathscr{E} \otimes |\Lambda|^{1/2})$ with respect to t of the form $\sum t^k D_k s$. The heat equation $(\partial/\partial t + H)P_t s = 0$ implies that $D_k = (-H)^k/k!$. Thus, we obtain the following estimate:

$$\left\| P_t s - \sum_{i=0}^{k} \frac{(-tH)^i}{i!} s \right\|_j = O(t^{k+1}). \tag{2.8}$$

This asymptotic expansion justifies writing e^{-tH} for the operator P_t.

2.6 The Trace of the Heat Kernel

In this section, we will derive some of the consequences of the existence of a heat kernel for an operator of the form $H = \Delta^{\mathscr{E}\otimes|\Lambda|^{1/2}} + F$. In particular, we show that if H is symmetric, then it has a unique self-adjoint extension $\bar{H}$ in the space of square-integrable sections of the bundle $\mathscr{E}\otimes|\Lambda|^{1/2}$, and the kernel of the operator $e^{-t\bar{H}}$ is the heat kernel $p_t(x,y,H)$ of H.

We first recall some elementary facts on trace class operators on a Hilbert space. Let $\mathscr{H}$ be a Hilbert space with orthonormal basis e_i. An operator A is a **Hilbert-Schmidt** operator if

$$\|A\|_{\mathrm{HS}}^2 = \sum_i \|Ae_i\|^2 = \sum_{ij} |(Ae_i,e_j)|^2$$

is finite. The number $\|A\|_{\mathrm{HS}}$ is called the Hilbert-Schmidt norm of A. If A is a Hilbert-Schmidt operator, so is its adjoint A^* and $\|A\|_{\mathrm{HS}} = \|A^*\|_{\mathrm{HS}}$; also, if U is a bounded operator on $\mathscr{H}$ and A is an Hilbert-Schmidt operator, then UA and AU are Hilbert-Schmidt operators and $\|UA\|_{\mathrm{HS}} \leq \|U\|\,\|A\|_{\mathrm{HS}}$. If A and B are Hilbert-Schmidt operators, then for any orthonormal basis e_i of $\mathscr{H}$,

$$\sum_i |(ABe_i,e_i)| \leq \|A\|_{\mathrm{HS}}\|B\|_{\mathrm{HS}}.$$

Let M be a compact manifold and let $\mathscr{E}$ be a Hermitian vector bundle on M, and denote by $\Gamma_{L^2}(M,\mathscr{E}\otimes|\Lambda|^{1/2})$ the Hilbert space of square-integrable sections of $\mathscr{E}\otimes|\Lambda|^{1/2}$. We will also frequently consider the Hilbert space $\Gamma_{L^2}(M,\mathscr{E})$ where M is a compact manifold with canonical smooth positive density (for example, a Riemannian manifold with the Riemannian density).

Lemma 2.31. *The topological vector space underlying $\Gamma_{L^2}(M,\mathscr{E}\otimes|\Lambda|^{1/2})$ is independent of the metric on $\mathscr{E}$, although the metric on $\Gamma_{L^2}(M,\mathscr{E}\otimes|\Lambda|^{1/2})$ depends of course on the metric on $\mathscr{E}$. Similarly, the topological vector space underlying $\Gamma_{L^2}(M,\mathscr{E})$ is independent of the metric on $\mathscr{E}$ and the density on M.*

Proof. If $h_1(\cdot,\cdot)$ and $h_2(\cdot,\cdot)$ are two metrics on $\mathscr{E}$, then there is an invertible section $A \in \Gamma(M,\mathrm{End}(\mathscr{E}))$ such that $h_1(s,s) = h_2(As,As)$ for all $s \in \Gamma(M,\mathscr{E})$. Since A and A^{-1} induce bounded operators on $\Gamma_{L^2}(M,\mathscr{E}\otimes|\Lambda|^{1/2})$, the first part of the lemma follows.

If ω_1 and ω_2 are two smooth positive densities on M, then there is an everywhere positive function $\phi \in C^\infty(M)$ such that $\omega_1 = \phi\omega_2$. Since multiplication by $\phi^{1/2}$ and $\phi^{-1/2}$ induce bounded operators on $\Gamma_{L^2}(M,\mathscr{E})$, the independence of this space on the density on M follows. □

An operator K with square-integrable kernel

$$k(x,y) \in \Gamma_{L^2}(M\times M,(\mathscr{E}\otimes|\Lambda|^{1/2})\boxtimes(\mathscr{E}\otimes|\Lambda|^{1/2}))$$

is Hilbert-Schmidt, and

$$\|K\|_{\mathrm{HS}}^2 = \int_{(x,y)\in M\times M} \mathrm{Tr}(k(x,y)^*k(x,y)), \tag{2.9}$$

as follows from the definition of the Hilbert-Schmidt norm,

$$\|K\|_{\mathrm{HS}}^2 = \sum_{ij} |\langle Ke_i, e_j\rangle|^2.$$

In particular, we see that the heat kernel p_t of a generalized Laplacian H is the kernel of a Hilbert-Schmidt operator for all $t > 0$.

An operator K is said to be **trace-class** if it has the form AB, where A and B are Hilbert-Schmidt. For such an operator, the sum $\sum_i (Ke_i, e_i)$ is absolutely summable, and the number $\sum_i (Ke_i, e_i)$ is independent of the Hilbert basis; it is called the trace of K:

$$\mathrm{Tr}(K) = \sum_i (Ke_i, e_i).$$

Note that if A and B are Hilbert-Schmidt, then $\mathrm{Tr}(AB^*) = (A,B)_{\mathrm{HS}}$; also

$$\mathrm{Tr}(AB) = \mathrm{Tr}(BA) = \sum_{ij} \langle Ae_i, e_j\rangle \langle Be_j, e_i\rangle. \tag{2.10}$$

Observe that the restriction of the bundle $(\mathscr{E}\otimes|\Lambda|^{1/2}) \boxtimes (\mathscr{E}\otimes|\Lambda|^{1/2})$ on $M\times M$ to the diagonal is isomorphic to $\mathrm{End}(\mathscr{E})\otimes|\Lambda|$. If

$$a(x,y) \in \Gamma(M\times M, (\mathscr{E}\otimes|\Lambda|^{1/2}) \boxtimes (\mathscr{E}\otimes|\Lambda|^{1/2}))$$

is a smooth kernel on $M\times M$, denote by $\mathrm{Tr}(a(x,x))$ the section of $|\Lambda|$ obtained by taking the pointwise trace of $a(x,x) \in \Gamma(M, \mathrm{End}(\mathscr{E})\otimes|\Lambda|)$ acting on the bundle $\mathscr{E}$.

Proposition 2.32. *For all $t > 0$, the operator P_t with kernel p_t the heat kernel of a generalized Laplacian on a compact manifold M, is trace class, with trace given by the formula*

$$\mathrm{Tr}(P_t) = \int_{x\in M} \mathrm{Tr}(p_t(x,x)).$$

Proof. This follows from the fact that $P_t = (P_{t/2})^2$, and the fact that $P_{t/2}$ is Hilbert-Schmidt. The trace of P_t is given by the formula

$$\begin{aligned}\mathrm{Tr}(P_t) &= (P_{t/2}, P_{t/2}^*)_{\mathrm{HS}} \\ &= \int_{(x,y)\in M^2} \mathrm{Tr}\big(p_{t/2}(x,y)p_{t/2}(y,x)\big) \\ &= \int_{x\in M} \mathrm{Tr}(p_t(x,x)). \end{aligned}$$

□

Let $H = \Delta^{\mathscr{E}\otimes|\Lambda|^{1/2}} + F$ be a generalized Laplacian on $\mathscr{E}$, and consider the operator H as an unbounded operator on $\Gamma_{L^2}(M, \mathscr{E}\otimes|\Lambda|^{1/2})$ with domain equal to the space of smooth sections of $\mathscr{E}\otimes|\Lambda|^{1/2}$; its formal adjoint H^* will also be considered on the same domain.

Proposition 2.33. *The closure of H^* is equal to the adjoint of the operator H with domain $\Gamma(M, \mathscr{E} \otimes |\Lambda|^{1/2})$. In particular, if H is symmetric, then it is essentially self-adjoint; in other words, $\bar{H}$ is a self-adjoint extension of H, and is the only one.*

Proof. Consider the action of the formal adjoint H^* on the space of distributional sections of $\mathscr{E} \otimes |\Lambda|^{1/2}$. It is clear that the domain of the adjoint of H is

$$\mathscr{D} = \{s \in \Gamma_{L^2}(M, \mathscr{E} \otimes |\Lambda|^{1/2}) \mid H^*s \in \Gamma_{L^2}(M, \mathscr{E} \otimes |\Lambda|^{1/2})\},$$

where H^*s is defined in the distributional sense. Thus, we need to prove that for $s \in \mathscr{D}$, there is a sequence s_n of smooth sections of $\mathscr{E} \otimes |\Lambda|^{1/2}$ such that s_n converges to s and H^*s_n converges to H^*s in $\Gamma_{L^2}(M, \mathscr{E} \otimes |\Lambda|^{1/2})$. We will do this using the operators P_t with kernel $p_t(x,y,H)$; first, we need two lemmas.

Lemma 2.34. *If $s \in \Gamma_{L^2}(M, \mathscr{E} \otimes |\Lambda|^{1/2})$, then $\lim_{t \to 0} P_t s = s$ in the L^2-norm.*

Proof. If ϕ is a continuous section of $\mathscr{E} \otimes |\Lambda|^{1/2}$, we know that $P_t\phi$ converges to ϕ in the uniform norm, and hence in the L^2-norm. Writing $s = (s - \phi) + \phi$, we see that

$$\|P_t s - s\|_{L^2} \leq \|P_t(s - \phi)\|_{L^2} + \|P_t\phi - \phi\|_{L^2} + \|s - \phi\|_{L^2}.$$

Thus, it suffices to establish that for small t, the operators P_t are uniformly bounded on $\Gamma_{L^2}(M, \mathscr{E} \otimes |\Lambda|^{1/2})$

Theorem 2.30 shows that it is enough to prove the lemma for the operator Q_t associated to the kernel $q_t(x,y)|dx|^{1/2} \otimes |dy|^{1/2}$ with

$$q_t(x,y) = (4\pi t)^{-n/2} e^{-d(x,y)^2/4t} \psi(d(x,y)^2) |dx|^{1/2} \otimes |dy|^{1/2}.$$

There exists a constant C such that $\int_M |q_t(x,y)|\, dy \leq C$ for $t > 0$. By Schwarz's inequality, we see that if $f \in \Gamma_{L^2}(M, \mathscr{E} \otimes |\Lambda|^{1/2})$,

$$\begin{aligned}\int \left| \int q_t(x,y) f(y)\, dy \right|^2 dx &\leq \int \left(\int q_t(x,y)\, dy \right) \left(\int q_t(x,y) |f(y)|^2 dy \right) dx \\ &\leq C \int \int q_t(x,y) |f(y)|^2 dx\, dy \leq C^2 \|f\|^2. \end{aligned}$$

□

Lemma 2.35. *If $s \in \mathscr{D}$ and P_t^* is the operator with kernel $p_t(x,y,H^*) = p_t(y,x,H)^*$, then $H^* P_t^* s = P_t^* H^* s$.*

Proof. If $s \in \Gamma(M, \mathscr{E} \otimes |\Lambda|^{1/2})$, we have $HP_t s = P_t H s$, since both sides satisfy the heat equation and $P_t s - s$ converges uniformly to 0 as well as all its derivatives, when $t \to 0$. Taking adjoints, we see that $P_t^* H^* = H^* P_t^*$ on $\Gamma^{-\infty}(M, \mathscr{E} \otimes |\Lambda|^{1/2})$, and in particular on $\mathscr{D}$. □

Given s, we may take the sections s_n to be given by $P_{1/n}^* s$. As $n \to \infty$, the first lemma shows that $s_n \to s$ in Γ_{L^2}, while the second shows that $H^* s_n = P_{1/n}^* H^* s$, which again converges to $H^* s$ by the first lemma.

Applying this result with H replaced by its formal adjoint H^*, we see that the closure of H with domain $\Gamma(M, \mathscr{E} \otimes |\Lambda|^{1/2})$ is the adjoint of H^*. In particular, if H is symmetric, that is $H = H^*$, then $\bar{H}$ is self-adjoint, and we obtain Proposition 2.33.

□

From now on, we consider only the case in which H is symmetric. If we simultaneously diagonalize the trace-class self-adjoint operators P_t, we obtain a Hilbert basis of eigensections ϕ_i; the semigroup property for P_t now implies that for some $\lambda_i \in \mathbb{R}$,

$$P_t \phi_i = e^{-t\lambda_i} \phi_i.$$

Since $P_t \phi_i$ is a smooth section of $\mathscr{E} \otimes |\Lambda|^{1/2}$, we see that the eigensections ϕ_i are smooth. The eigenvalues λ_i are bounded below, since the operators P_t are bounded. The fact that the function $P_t \phi_i$ satisfies the heat equation implies that $H\phi_i = \lambda_i \phi_i$. Furthermore, since the operators P_t are trace-class, we see that $\sum_i e^{-t\lambda_i}$ is finite for each $t > 0$. This shows that the spectrum of $\bar{H}$ is discrete with finite multiplicity. Thus, we have proved the following result:

Proposition 2.36. *If H is a symmetric generalized Laplacian, then the operator P_t is equal to $e^{-t\bar{H}}$. The operator $\bar{H}$ has discrete spectrum, bounded below, and each eigenspace is finite dimensional, with eigenvectors given by smooth sections of $\mathscr{E} \otimes |\Lambda|^{1/2}$.*

We will not be very careful about distinguishing between the operator H and its closure $\bar{H}$, and will write e^{-tH} for the operator $P_t = e^{-t\bar{H}}$. From the spectral decomposition of H it follows that

$$p_t(x,y) = \sum_j e^{-t\lambda_j} \phi_j(x) \otimes \phi_j(y).$$

Proposition 2.37. *If H is a generalized Laplacian, and if $P_{(0,\infty)}$ is the projection onto the eigenfunctions of H with positive eigenvalue, then for each $\ell \in \mathbb{N}$, there exists a constant $C(\ell) > 0$ such that for t sufficiently large,*

$$\| \langle x \mid P_{(0,\infty)} e^{-tH} P_{(0,\infty)} \mid y \rangle \|_\ell \leq C(\ell) e^{-t\lambda_1/2},$$

where λ_1 is the smallest non-zero eigenvalue of H.

Proof. For t large, we may write

$$\begin{aligned} &\langle x \mid P_{(0,\infty)} e^{-tH} P_{(0,\infty)} \mid y \rangle \\ &\quad = \int_{(z_1,z_2)\in M^2} k_{1/2}(x,z_1) \langle z_1 \mid P_{(0,\infty)} e^{-(t-1)H} P_{(0,\infty)} \mid z_2 \rangle k_{1/2}(z_2,y). \end{aligned}$$

The kernel $k_{1/2}(x,y)$ is smooth, so we only have to show that

$$\int_{(x,y)\in M^2} |\langle x \mid P_{(0,\infty)} e^{-(t-1)H} P_{(0,\infty)} \mid y \rangle|^2 \leq C e^{-t\lambda_1}.$$

However, the left-hand side is nothing but the square of the Hilbert-Schmidt norm of $P_{(0,\infty)} e^{-(t-1)H} P_{(0,\infty)}$. Thus, if $0 < \lambda_1 \leq \lambda_2 \leq \ldots$ are the positive eigenvalues of H, repeated according to their multiplicities, we must show that

$$\sum_{j\geq 1} e^{-2(t-1)\lambda_j} \leq C e^{-t\lambda_1}.$$

Thus, we see for t large that

$$\begin{aligned}\sum_{j\geq 1} e^{-2(t-1)\lambda_j} &\leq \Big(\sum_{j\geq 1} e^{-(t-2)\lambda_j}\Big)\cdot \sup_{j\geq 1} e^{-t\lambda_j}\\ &\leq \Big(\sum_{j\geq 1} e^{-(t-2)\lambda_j}\Big)\cdot e^{-t\lambda_1} \leq \Big(\sum_j e^{-\lambda_j}\Big)\cdot e^{-t\lambda_1}\\ &= \Big(\int_{x\in M} \langle x \mid P_{(0,\infty)} e^{-H} P_{(0,\infty)} \mid x\rangle\Big)\cdot e^{-t\lambda_1}. \qquad \square\end{aligned}$$

In particular, if H is positive, that is, if it is non-negative and has vanishing kernel, then $P_{(0,\infty)}$ is the identity, and each $\ell \in \mathbb{N}$, there exists a constant $C(\ell) > 0$ such that for t sufficiently large,

$$\|p_t(x,y)\|_\ell \leq C(\ell) e^{-t\lambda_1/2},$$

where λ_1 is the smallest eigenvalue of H.

The **Green operator** G of a positive generalized Laplacian H is by definition the inverse of H on $\Gamma_{L^2}(M, \mathcal{E}\otimes|\Lambda|^{1/2})$. Clearly G is bounded on $\Gamma_{L^2}(M,\mathcal{E}\otimes|\Lambda|^{1/2})$. In the following theorem, we discuss the properties of G.

Theorem 2.38. *Assume that H is a positive generalized Laplacian. Let G be the inverse of H on $\Gamma_{L^2}(M,\mathcal{E}\otimes|\Lambda|^{1/2})$. Then we have the integral representation*

$$G^k = \frac{1}{(k-1)!}\int_0^\infty e^{-tH} t^{k-1}\, dt$$

for each natural number k. The operator G^k has a C^ℓ kernel for

$$k > 1 + (\dim(M) + \ell)/2.$$

Proof. If $k \geq 1$, define the operator G^k by the integral

$$G^k = \frac{1}{(k-1)!}\int_0^\infty e^{-tH} t^{k-1}\, dt.$$

If ϕ is an eigenfunction of H with eigenvalue λ, it is easy to see that the operator G^k applied to ϕ equals

$$\frac{1}{(k-1)!}\Big(\int_0^\infty e^{-t\lambda} t^{k-1}\, dt\Big)\phi = \lambda^{-k}\phi.$$

We will show that for k sufficiently large, the integral

$$g^k(x,y) = \frac{1}{(k-1)!}\int_0^\infty p_t(x,y) t^{k-1}\, dt$$

converges and defines a kernel which is in fact the kernel of the operator G^k. We start by splitting the integral into two pieces,

$$g^k(x,y) = \frac{1}{(k-1)!}\int_0^1 p_t(x,y) t^{k-1}\, dt + \frac{1}{(k-1)!}\int_1^\infty p_t(x,y) t^{k-1}\, dt.$$

We bound the first integral by means of the following lemma.

Lemma 2.39. *The family of kernels $t^{k-1}p_t(x,y)$, $(t \le 1)$, is uniformly bounded in the C^ℓ-norm when $k > 1 + (\ell + n)/2$.*

Proof. Using the asymptotic expansion for the heat kernel $p_t(x,y)$, we see that it suffices to prove the lemma for the kernel

$$q_t(x,y) = \psi(d(x,y)^2)t^{-\dim(M)/2}e^{-d(x,y)^2/4t},$$

where ψ is a smooth function on $\mathbb{R}_+$ which equals 1 near zero and 0 in $[\varepsilon^2,\infty)$. Using the fact that $|\partial^\alpha e^{-d(x,y)^2/4t}| \le Ct^{-|\alpha|/2}$, it is easily seen that for $k > 1 + (\ell + n)/2$, the family of kernels $t^{k-1}\partial^\alpha q_t(x,y)$ is uniformly continuous for $0 \le t \le 1$, $\alpha \le \ell$. □

We will complete the proof by showing that the second integral converges to a smooth kernel, using Proposition 2.37. Indeed, for any C^ℓ-norm,

$$\frac{1}{(k-1)!}\int_1^\infty \|p_t(x,y)\|_\ell t^{k-1}\,dt \le \frac{C(\ell)}{(k-1)!}\int_1^\infty e^{-t\lambda_1/2}t^{k-1}\,dt,$$

which is finite. □

In Section 9.6, we will extend this theorem to fractional powers of the Green operator.

It is straightforward to extend this theorem to an arbitrary self-adjoint generalized Laplacian H. We define the Green operator of H to be the inverse of H on $\ker(H)^\perp$, and to vanish on $\ker(H)$. The integral representation of G is similar to the above one, except that we must handle the non-positive eigenvalues separately: if $\lambda_{-m} \le \cdots \le \lambda_{-1} < 0$ are the strictly negative eigenvalues of H, and if P_{λ_i} is the projection onto the eigenspace of H with eigenvalue λ_i, then we see that

$$G^k = \frac{1}{(k-1)!}\int_0^\infty P_{(0,\infty)}e^{-tH}t^{k-1}\,dt + \sum_{i=1}^m (\lambda_{-i})^{-k}P_{-\lambda_i}, \tag{2.11}$$

Since the operators $P_{-\lambda_i}$ are smoothing operators, the estimates on the kernel of G^k may be carried out in exactly the same fashion as when H is positive.

The above theorem has the following important consequence, known as elliptic regularity.

Corollary 2.40. *Let H be a symmetric generalized Laplacian on $\mathcal{E} \otimes |\Lambda|^{1/2}$.*

1. *If $s \in \Gamma_{L^2}(M, \mathcal{E} \otimes |\Lambda|^{1/2})$ is such that $H^k s$, in the sense of generalized sections, is square-integrable for all k, then s is smooth.*
2. *If A is a bounded operator on $\Gamma_{L^2}(M, \mathcal{E} \otimes |\Lambda|^{1/2})$ which commutes with H, and if $s \in \Gamma(M, \mathcal{E} \otimes |\Lambda|^{1/2})$ is smooth, then $As \in \Gamma(M, \mathcal{E} \otimes |\Lambda|^{1/2})$ is smooth.*

Proof. We may assume that H is positive by adding a large positive constant to H. It follows that $1 = G^k \circ H^k = H^k \circ G^k$.

Since G^k commutes with H, we see that $s = G^k H^k s$. By hypothesis, $H^k s$ is square-integrable, hence by the above estimates, $G^k H^k s$ lies in C^ℓ for $\ell \le 2k - n - 1$. Letting k tend to ∞, we obtain the first part.

The second part is proved in an analogous way, using the formula

$$As = (G^k \circ H^k \circ A)s = (G^k \circ A \circ H^k)s.$$

If $s \in \Gamma(M, \mathscr{E} \otimes |\Lambda|^{1/2})$ is smooth, then so is $H^k s$; in particular, $H^k s$ is square-integrable. Since A is bounded on the Hilbert space $\Gamma_{L^2}(M, \mathscr{E} \otimes |\Lambda|^{1/2})$, we see that $AH^k s \in \Gamma_{L^2}(M, \mathscr{E} \otimes |\Lambda|^{1/2})$. It follows that $As = G^k(AH^k s)$ is C^ℓ for $\ell \leq 2k - \dim(M) - 1$. Taking k large, we obtain the result. □

It follows from Theorem 2.30 and the asymptotic expansion of $p_t(x,y)$ that the trace of the heat operator e^{-tH} possesses an asymptotic expansion in a Laurent series in $t^{1/2}$ for small t:

$$\operatorname{Tr}(e^{-tH}) = \int_M \operatorname{Tr}(p_t(x,x)) \sim (4\pi t)^{-n/2} \sum_{i=0}^{\infty} t^i \int_M \operatorname{Tr}(\Phi_i(x,x,H))\, dx.$$

In particular, using the fact that $\Phi_0(x,x,H)$ is equal to $I_{\mathscr{E}}$, we obtain Weyl's Theorem.

Theorem 2.41 (Weyl). *Let λ_i be the eigenvalues of $\bar{H}$, each counted with multiplicities. Then as $t \to 0$, we have the asymptotic formula for the Laplace transform of the spectral measure of $\bar{H}$:*

$$\sum_i e^{-t\lambda_i} = (4\pi t)^{-n/2} \operatorname{rk}(\mathscr{E}) \operatorname{vol}(M) + O(t^{-n/2+1}).$$

Note that, at least in principle, it is possible to compute all the coefficients $\Phi_i(x,x,H)$ in the asymptotic expansion of the heat kernel $p_t(x,y)$ explicitly, so that we can, if we desire, obtain a full asymptotic expansion for the Laplace transform of the spectral measure of H.

It is possible to derive from this theorem information about the asymptotic distribution of the eigenvalues of H at infinity, by means of Karamata's Theorem.

Theorem 2.42 (Karamata). *Let $d\mu(\lambda)$ be a positive measure on $\mathbb{R}_+$ such that the integral*

$$\int_0^\infty e^{-t\lambda}\, d\mu(\lambda)$$

converges for $t > 0$, and such that

$$\lim_{t \to 0} t^\alpha \int_0^\infty e^{-t\lambda}\, d\mu(\lambda) = C$$

for some positive constants α and C. If $f(x)$ is a continuous function on the unit interval $[0,1]$, then

$$\lim_{t \to 0} t^\alpha \int_0^\infty f(e^{-t\lambda}) e^{-t\lambda}\, d\mu(\lambda) = \frac{C}{\Gamma(\alpha)} \int_0^\infty f(e^{-t}) t^{\alpha-1} e^{-t}\, dt.$$

Proof. By Weierstrass's Theorem, we can approximate the function $f(x)$ arbitrarily closely by a polynomial p:

$$|f(x)-p(x)|\leq \varepsilon \quad \text{for } x\in[0,1].$$

It follows that it suffices to prove the limit formula for polynomials on the unit interval, and, in fact, for the monomial $f(x)=x^k$. But this is a straightforward calculation: on the one hand,

$$\lim_{t\to 0} t^\alpha \int_0^\infty e^{-(k+1)t\lambda}\,d\mu(\lambda)=C(k+1)^{-\alpha},$$

while on the other,

$$\frac{C}{\Gamma(\alpha)}\int_0^\infty e^{-kt}t^{\alpha-1}e^{-t}\,dt$$

also equals $C(k+1)^{-\alpha}$. □

Using Karamata's Theorem we obtain the following restatement of Weyl's theorem.

Corollary 2.43. *If $N(\lambda)$ is the number of eigenvalues of H that are less than λ, then*

$$N(\lambda)\sim\frac{\operatorname{rk}(\mathscr{E})\operatorname{vol}(M)}{(4\pi)^{n/2}\Gamma(n/2+1)}\lambda^{n/2}.$$

Proof. Taking in Karamata's Theorem a decreasing sequence of continuous functions converging to the function equal to x^{-1} on the interval $[1/e,1]$ and zero on the interval $[0,1/e)$, we see that

$$\lim_{t\to 0} t^\alpha \int_0^{t^{-1}} d\mu(\lambda)=\frac{C}{\Gamma(\alpha)}\int_0^1 t^{\alpha-1}\,dt=\frac{C}{\Gamma(\alpha+1)}.$$

The corollary follows by considering a constant a such that $H+a$ is positive, and applying the above result to the spectral measure $\mu=\sum_i \delta_{\lambda_i}$ of $H+a$, for which $\alpha=n/2$ and $C=(4\pi)^{-n/2}\operatorname{rk}(\mathscr{E})\operatorname{vol}(M)$. □

It is interesting to see how this works in the simplest possible case, where M is equal to the circle of circumference 2π, and H is the scalar Laplacian Δ. The eigenvalues of the Laplacian are $\lambda=0$, with multiplicity 1, and the squares $\lambda=n^2$ ($n\geq 1$), with multiplicity 2. It follows that $N(\lambda)=[2\sqrt{\lambda}+1]$. On the other hand, Weyl's Theorem tells us that

$$N(\lambda)\sim\frac{\operatorname{vol}(S^1)}{\sqrt{4\pi}\,\Gamma(3/2)}\sqrt{\lambda}$$

and it is easily checked that the coefficient of $\sqrt{\lambda}$ equals 2 (since $\operatorname{vol}(S^1)=2\pi$ and $\Gamma(3/2)=\sqrt{\pi}/2$).

In the next proposition, we will extend the formula for the trace of the heat kernel to any operator with smooth kernel. The proof is an application of the Green operator.

Proposition 2.44. *If A is an operator with smooth kernel*

$$a(x,y) \in \Gamma(M \times M, (\mathcal{E} \otimes |\Lambda|^{1/2}) \boxtimes (\mathcal{E} \otimes |\Lambda|^{1/2})),$$

then A is trace class, with trace given by the formula

$$\mathrm{Tr}(A) = \int_{x \in M} \mathrm{Tr}(a(x,x)).$$

Proof. Let H be a positive generalized Laplacian on the bundle $\mathcal{E}$, with Green operator G. By Theorem 2.38, the operator G^k has continuous kernel for $k > 1 + (\dim(M)/2)$, and hence is Hilbert-Schmidt.

The operator A_k associated to the smooth kernel $H_x^k a(x,y)$ is Hilbert-Schmidt, as is any operator with smooth kernel, and $A = G^k A_k$ is the product of two Hilbert-Schmidt operators and hence is trace class. The trace of A is now easy to calculate:

$$\begin{aligned}\mathrm{Tr}(A) &= \mathrm{Tr}(G^k A_k) \\ &= \int_{(x,y) \in M^2} \mathrm{Tr}(g^k(x,y) H_y^k a(y,x)) \\ &= \int_{x \in M} \mathrm{Tr}(a(x,x)).\end{aligned}$$

□

Let D be a differential operator acting on $\Gamma(M, \mathcal{E} \otimes |\Lambda|^{1/2})$. If K is an operator with smooth kernel, then so are DK and KD, and hence they are trace class. More precisely, the operator DK has smooth kernel $D_x k(x,y)$ while integration by parts shows that KD has smooth kernel $D_y^* k(x,y)$.

Proposition 2.45. *Let D be a differential operator acting on the bundle $\mathcal{E}$. If K is an operator with smooth kernel, then* $\mathrm{Tr}(DK) = \mathrm{Tr}(KD)$.

Proof. Choose a positive generalized Laplacian on the bundle $\mathcal{E}$, and let G be its Green operator. For sufficiently large N, the operator DG^N has a continuous kernel, and hence is a Hilbert-Schmidt operator. Thus,

$$\begin{aligned}\mathrm{Tr}(DK) &= \mathrm{Tr}((DG^N)(H^N K)) = \mathrm{Tr}((H^N K)(DG^N)) \quad \text{by (2.10)} \\ &= \mathrm{Tr}((H^N KD)G^N) = \mathrm{Tr}(G^N(H^N KD)) = \mathrm{Tr}(KD).\end{aligned}$$

□

Let us construct an asymptotic expansion for the kernel of the operator DP_t, where D is a differential operator.

Proposition 2.46. *If P_t is the heat kernel of a generalized Laplacian H and D is a differential operator of degree m on the bundle $\mathcal{E}$, then there exist smooth kernels $\Psi_i(x,y,H,D)$ such that the kernel $\langle x \mid DP_t \mid y \rangle$ satisfies*

$$\Big\| \langle x \mid DP_t \mid y \rangle - h_t(x,y) \sum_{i=-m}^{N} t^i \Psi_i(x,y,H,D) \Big\| = O(t^{N-n/2}),$$

where $h_t(x,y) = (4\pi t)^{-n/2} e^{-d(x,y)^2/4t} \psi(d(x,y)^2) |dx|^{1/2} \otimes |dy|^{1/2}$, and ψ is a cut-off function.

Proof. If we apply the differential operator $\partial/\partial \mathbf{x}_i$ to the Euclidean heat kernel

$$q_t(\mathbf{x},\mathbf{y}) = (4\pi t)^{-n/2} e^{-|\mathbf{x}-\mathbf{y}|^2/4t},$$

we obtain

$$-q_t(\mathbf{x},\mathbf{y})\frac{\mathbf{x}_i - \mathbf{y}_i}{2t}.$$

By induction, we see that $\partial_{\mathbf{x}}^{\alpha} q_t$ has the form

$$\partial_{\mathbf{x}}^{\alpha} q_t(\mathbf{x},\mathbf{y}) = q_t(\mathbf{x},\mathbf{y}) \sum_{|\beta|+2j\leq|\alpha|} c_{\beta,j} \frac{(\mathbf{x}-\mathbf{y})^{\beta}}{t^{|\beta|+j}}.$$

The proof now follows by combining this with Theorem 2.30. □

From the proof of Proposition 2.46, we see that the restriction of the kernel of the operator DP_t to the diagonal has the form

$$(4\pi t)^{-n/2} \sum_{i=-[m/2]}^{\infty} t^i \Psi_i(x,x,H,D)\,|dx|,$$

since the restriction of the coefficients $\Psi_i(x,y,H,D)$ to the diagonal vanishes for $i < -m/2$. It follows that the trace $\operatorname{Tr}(DP_t)$ has an asymptotic expansion of the form

$$\operatorname{Tr}(DP_t) \sim (4\pi t)^{-n/2} \sum_{i=-[m/2]}^{\infty} t^i a_i,$$

where

$$a_i = \int_M \operatorname{Tr}(\Psi_i(x,x,H,D))\,|dx|.$$

It is also possible to give an asymptotic expansion of the trace of the operator KP_t, if K is a smoothing operator on $\Gamma(M,\mathscr{E}\otimes|\Lambda|^{1/2})$.

Proposition 2.47. *There is an asymptotic expansion of the form*

$$\operatorname{Tr}(KP_t) \sim \operatorname{Tr}(K) + \sum_{i=1}^{\infty} t^i a_i.$$

Proof. Let $k(x,y)\,|dx|^{1/2}\otimes|dy|^{1/2}$ be the kernel of K, and let $k_y \in \Gamma(M,\mathscr{E})$ be the section $x \mapsto k(x,y)$. Since $\operatorname{Tr}(KP_t) = \operatorname{Tr}(P_t K)$, we see that

$$\operatorname{Tr}(KP_t) = \int_M \operatorname{Tr}((P_t k_x)(x))\,|dx|.$$

Applying the asymptotic expansion (2.8), we see that

$$P_t k_y \sim \sum_{i=0}^{\infty} \frac{(-t)^i H^i}{i!} k_y,$$

from which the proposition follows. □

2.7 Heat Kernels Depending on a Parameter

Let M be a compact manifold and let $\mathscr{E}$ be a vector bundle over M. Let $(H^z \mid z \in \mathbb{R})$, be a smooth family of generalized Laplacians; in other words, we are given the following data, from which we construct the generalized Laplacians H^z in the canonical way:

1. a family g_z of Riemannian metrics on M depending smoothly on $z \in \mathbb{R}$;
2. a family of connections $\nabla^z = \nabla^{\mathscr{E}} + \omega^z$ on the bundle $\mathscr{E}$, where ω^z is a family of one-forms in $\mathscr{A}^1(M, \mathrm{End}(\mathscr{E}))$ depending smoothly on $z \in \mathbb{R}$;
3. a smooth family of potentials $F^z \in \Gamma(M, \mathrm{End}(\mathscr{E}))$.

Our goal in this section is to prove the following result.

Theorem 2.48. *If H^z is a smooth family of generalized Laplacians, then for each $t > 0$, the corresponding family of heat kernels $p_t(x,y,z)$ defines a smooth family of smoothing operators on the bundle $\mathscr{E}$. Furthermore, the derivative of $p_t(x,y,z)$ with respect to z is given by* **Duhamel's formula**

$$\frac{\partial}{\partial z} p_t(x,y,z) = -\int_0^t \Big(\int_{y_1 \in M} p_{t-s}(x,y_1,z)(\partial_z H^z)_{y_1} p_s(y_1,y,z) \Big) ds;$$

alternatively, writing this in operator form, we have

$$\frac{\partial}{\partial z} e^{-tH^z} = -\int_0^t e^{-(t-s)H^z} (\partial_z H^z) e^{-sH^z} \, ds.$$

As a first step in the proof, we will prove a result about the dependence of the formal heat kernel of the family H^z on the parameter z. Since we are only interested in results local in z, we may assume without loss of generality that the injectivity radius is everywhere bounded below by $\varepsilon > 0$. Let $\psi : \mathbb{R}_+ \to [0,1]$ be a cut-off function.

Lemma 2.49. *For each $t > 0$, the formal solution $k_t^N(x,y,z) = k_t^N(x,y,H^z)$ depends smoothly on z. Given $T > 0$ and z lying in a compact subset of $\mathbb{R}$, the family of kernels $\partial_z^j k_t^N(x,y,z)$, $0 \le t \le T$, form a uniformly bounded collection of operators on $\Gamma^\ell(M, \mathscr{E} \otimes |\Lambda|^{1/2})$ for each $\ell \ge 0$.*

Proof. Consider the explicit formula for $k_t^N(x,y,z)$:

$$(4\pi t)^{-n/2} e^{-d_z(x,y)^2/4t} \psi(d_z(x,y)^2) \sum_{i=0}^{N} t^i \Phi_i(x,y,z) j_z^{-1/2}(x,y) \, |d_z x|^{1/2} \, |d_z y|^{1/2}.$$

In this formula, $d_z(x,y)^2$ is the distance between x and $y \in M$, $j_z(x,y)$ is the Jacobian of the exponential map, and $|d_z x|$ is the Riemannian volume, all with respect to the Riemannian metric g_z. All of these objects depend smoothly on z, as does the parallel transport $\tau^z(x,y)$ in $\mathscr{E}$ with respect to the connection ∇^z.

We can show, by induction on i, that the terms $\Phi_i(x,y,z)$ in the asymptotic expansion depend smoothly on z near the diagonal in $M \times M$. Indeed, we have the following formula for $\Phi_i(x,y,z)$ in terms of $\Phi_{i-1}(x,y,z)$:

$$\tau^z(x,y)^{-1}\Phi_i(x,y,z) = -\int_0^1 s^{i-1}\tau^z(x_s,y)^{-1}(B_x^z \cdot \Phi_{i-1})(x_s,y,z)\,ds.$$

It remains to observe that the operator B^z depends smoothly on z, as is clear from the formula

$$B^z = j_z^{1/2} \circ H^z \circ j_z^{-1/2}.$$

The proof of uniform boundedness of the kernels $k_t^N(x,y,z)$ for z in a compact subset of $\mathbb{R}$ is the same as in the proof of Theorem 2.29. □

We now turn to the proof of Theorem 2.48.

Proof. Consider the formula for $p_t(x,y,z)$,

$$\sum_{k=0}^{\infty}(-t)^k \int_{\Delta_k}\int_{(y_1,\dots,y_k)\in M^k} k^N_{\sigma_0 t}(x,y_1,z) r^N_{\sigma_1 t}(y_1,y_2,z)\dots r^N_{\sigma_k t}(y_k,y,z),$$

where $r_t^N(x,y,z) = (\partial_t + H_x^z)k_t^N(x,y,z)$. By Lemma 2.49, we know that $r_t^N(x,y,z)$ depends smoothly on z. From now on, restrict z to lie in a bounded interval. It is easy to see that r_t^N satisfies the uniform estimate

$$\left\|\frac{\partial^m r_t^N}{\partial z^m}\right\|_\ell \le C(\ell)t^{N-n/2-\ell/2-m},$$

where the C^ℓ-norm is that of $\Gamma^\ell(M\times M, (\mathscr{E}\otimes|\Lambda|^{1/2})\boxtimes(\mathscr{E}\otimes|\Lambda|^{1/2}))$.

Consider the space of all families of C^ℓ-sections of $\mathscr{E}\otimes|\Lambda|^{1/2}$ which are C^k in the parameter z, with norm

$$\sum_{m\le k}\left\|\frac{\partial^m s}{\partial z^m}\right\|_\ell.$$

Application of the operator $K_t^N(z)$ with kernel $k_t^N(x,y,z)$, $0\le t\le 1$, gives a uniformly bounded family of linear operators for this norm; this follows from the fact that the operator $\partial_z^m K_t^N(z)$ equals

$$\sum_{j=0}^{m}\binom{m}{j}(\partial_z^j K_t^N(z))\partial_z^{m-j},$$

combined with the fact that the family $\partial_z^j K_t^N(z)$ is uniformly bounded on $\Gamma^\ell(M,\mathscr{E}\otimes|\Lambda|^{1/2})$ for $t\in[0,1]$. Thus, we see that the series for $\partial_z^m p_t(x,y,z)$ converges uniformly in the C^ℓ-norm as long as $N > m + (\ell+n)/2$. Since N is arbitrary, this proves the smoothness of $p_t(x,y,z)$ in z.

Let us now prove Duhamel's formula for $\partial p_t(x,y,z)/\partial z$. Let ϕ be a smooth section of $\mathscr{E}\otimes|\Lambda|^{1/2}$. If we apply the operator $\partial_t + H^z$ to the section $\partial_z P_t^z\phi \in \Gamma(M,\mathscr{E}\otimes|\Lambda|^{1/2})$, we obtain

$$\begin{aligned}(\partial_t + H^z)\partial_z P_t^z\phi &= \partial_z(\partial_t + H^z)P_t^z\phi - (\partial_z H^z)P_t^z\phi\\ &= -(\partial_z H^z)P_t^z\phi.\end{aligned}$$

From this, it is easy to see that

$$A_t^z(x) = \frac{\partial}{\partial z}(P_t^z\phi)(x) + \int_0^t \int_{(y_1,y_2)} p_{t-s}(x,y_1,z)(\partial_z H^z)_{y_1} p_s(y_1,y_2,z)\,\phi(y_2)\,ds$$

satisfies the heat equation $(\partial_t + H^z)A_t^z = 0$. We will show that $A_t^z(x)$ converges to zero as $t \to 0$; by the uniqueness of solutions to the heat equation, it follows that it vanishes. Since ϕ is an arbitrary element of $\Gamma(M, \mathscr{E} \otimes |\Lambda|^{1/2})$, this proves Duhamel's formula.

We have the following asymptotic expansion for $P_t^z\phi$ in $\Gamma(M, \mathscr{E} \otimes |\Lambda|^{1/2})$:

$$P_t^z\phi \sim \sum_{k=0}^{\infty} \frac{(-tH^z)^k}{k!}\phi.$$

Furthermore, we may take derivatives of both sides of this asymptotic expansion for $\partial_z^m P_t^z\phi$. In particular,

$$\lim_{t\to 0}\Big(\int_{y\in M} \frac{\partial}{\partial z} p_t(x,y,z)\phi(y)\Big) = 0.$$

On the other hand, the section

$$\int_{(y_1,y_2)\in M^2} p_{t-s}(\cdot,y_1,z)(\partial_z H^z)_{y_1} p_s(y_1,y_2,z)\,\phi(y_2)$$

is a smooth function of (s,t,x,z), and hence

$$\lim_{t\to 0}\int_0^t \Big(\int_{(y_1,y_2)\in M^2} p_{t-s}(\cdot,y_1,z)(\partial_z H^z)_{y_1} p_s(y_1,y_2,z)\,\phi(y_2)\Big)\,ds = 0. \qquad \square$$

The following corollary will be used repeatedly in the rest of the book.

Corollary 2.50. *If H^z is a one-parameter smooth family of generalized Laplacians on a vector bundle $\mathscr{E}$ over a compact manifold, then*

$$\frac{\partial}{\partial z}\operatorname{Str}\big(e^{-tH^z}\big) = -t\operatorname{Str}\Big(\frac{\partial H^z}{\partial z}e^{-tH^z}\Big).$$

Proof. If $p_t(x,y,z)$ is the kernel of the operator e^{-tH^z}, then Duhamel's formula shows that

$$\frac{\partial}{\partial z}p_t(x,y,z) = -\int_0^t \int_{y_1\in M} p_{t-s}(x,y_1,z)(\partial_z H^z)_{y_1} p_s(y_1,y,z)\,ds.$$

Taking the supertrace of both sides gives

$$\begin{aligned}\frac{\partial}{\partial z}\operatorname{Str}(e^{-tH^z}) &= -\int_0^t \operatorname{Str}\Big(e^{(s-t)H^z}\frac{\partial H^z}{\partial z}e^{-sH^z}\Big)\,ds\\ &= -\int_0^t \operatorname{Str}\Big(\frac{\partial H^z}{\partial z}e^{-tH^z}\Big)\,ds,\end{aligned}$$

from which the result is clear, since the integrand is independent of s. $\square$

Bibliographic Notes

The pioneers of the approach to the analysis of the heat kernel that we present in this chapter were Minakshisundaram and Pleijel [87]. A more recent work which presents some applications to Riemannian geometry is the book by Berger-Gauduchon-Mazet [19]; we have followed this book closely in Sections 2–4.

3

Clifford Modules and Dirac Operators

In this chapter, we begin the study of Dirac operators, which are a generalization of the operator discovered by Dirac in his study of the quantum mechanics of the electron.

The problem that Dirac asked was the following (he was actually working in four-dimensional Minkowski space and not in a Riemannian manifold): what are the first-order differential operators whose square is the Laplacian? Generalizing the question, one is motivated to ask: if M is a manifold and $\mathscr{E}$ is a vector bundle over M, what are the first-order differential operators D whose square is a generalized Laplacian on $\Gamma(M,\mathscr{E})$?

At this level of generality, the answer is quite easy to give. The operator D, if written in local coordinates, takes the form

$$\mathsf{D} = \sum_k a^k(x)\partial_k + b(x),$$

where $a^k(x)$ and $b(x)$ are sections of $\operatorname{End}(\mathscr{E})$, and it is easily seen that $\sum_k a^k(x)\partial_k$ transforms as a section of the bundle $\operatorname{Hom}(T^*M, \operatorname{End}(\mathscr{E}))$; this section is $-i$ times the symbol of D introduced in Chapter 2,

$$\sigma(\mathsf{D})\Big(x, \sum_k \xi_k\, dx^k\Big) = i\sum_k a^k(x)\xi_k.$$

The square of D is

$$\mathsf{D}^2 = \tfrac{1}{2}\sum_{ij}(a^i(x)a^j(x) + a^j(x)a^i(x))\partial_i\partial_j + \text{first order operator},$$

so that D^2 is a generalized Laplacian if and only if for any ξ and $\eta \in T_x^*M$, we have

$$\langle a(x),\xi\rangle\langle a(x),\eta\rangle + \langle a(x),\eta\rangle\langle a(x),\xi\rangle = -2(\xi,\eta)_x,$$

where $(\cdot,\cdot)_x$ is the metric on T_x^*M. This is the defining relation for the Clifford algebra of T_x^*M, which we introduce in Section 1, and study the representations of in Section 2.

In Sections 3–5, we will define and study the notion of a generalized Dirac operator, and in particular, introduce the index, which is our main object of interest in this book. Our general application of the term "Dirac operator" should be carefully distinguished from its more usual use, which reserves this name for the operators introduced by Lichnerowicz and Atiyah-Singer, which act on a twisted spinor bundle. We will discuss this special case in Section 6. Also in Section 6, we show how some of other the classical operators of differential geometry are generalized Dirac operators, such as $d+d^*$ on a Riemannian manifold, and $\bar{\partial}+\bar{\partial}^*$ on a Kähler manifold.

3.1 The Clifford Algebra

The first step in defining the Dirac operator is to understand the linear algebra underlying it: the Clifford algebra.

Definition 3.1. *Let V be a real vector space with quadratic form Q (which need not be non-degenerate). The* **Clifford algebra** *of (V,Q), denoted by $C(V,Q)$, is the algebra over $\mathbb{R}$ generated by V with the relations*

$$v\cdot w+w\cdot v=-2Q(v,w) \quad \text{for } v \text{ and } w \text{ in } V,$$

in other words, $v^2=-Q(v)$ for all $v\in V$.

When Q is fixed, we may write $C(V)$ for $C(V,Q)$, and abbreviate $Q(a,b)$ to (a,b). The Clifford algebra of (V,Q) is the solution to the following universal problem.

Proposition 3.2. *If A is an algebra and $c: V\to A$ is a linear map satisfying*

$$c(w)c(v)+c(v)c(w)=-2Q(v,w),$$

for all $v,w\in V$, then there is a unique algebra homomorphism from $C(V,Q)$ to A extending the given map from V to A.

The Clifford algebra may be realized as the quotient of the tensor algebra $T(V)$ by the ideal generated by the set

$$\mathscr{I}_Q=\{v\otimes w+w\otimes v+2Q(v,w)\mid v,w\in V\}.$$

Observe that the algebra $T(V)$ is a $\mathbb{Z}_2$-graded algebra, or superalgebra, with $\mathbb{Z}_2$-grading obtained from the natural $\mathbb{N}$-grading after reduction mod 2. Since the generating set of the above ideal is contained in the evenly graded subalgebra of $T(V)$, it follows that $C(V)$ itself is a superalgebra, $C(V)=C^+(V)\oplus C^-(V)$, with $V\subset C^-(V)$. We will mainly consider modules over $\mathbb{R}$ or $\mathbb{C}$ for the Clifford algebra which are $\mathbb{Z}_2$-graded, by which we mean that the module E is a superspace $E=E^+\oplus E^-$, and the Clifford action is even with respect to this grading:

$$C^+(V)\cdot E^\pm\subset E^\pm,$$
$$C^-(V)\cdot E^\pm\subset E^\mp.$$

Since the algebra $T(V)$ carries a natural action of the group $\mathrm{O}(V,Q)$ of linear maps on V preserving the quadratic form Q and the above ideal is invariant under this action, it follows that the Clifford algebra carries a natural action of $\mathrm{O}(V,Q)$ as well.

Let $a \mapsto a^*$ be the anti-automorphism of $T(V)$ such that $v \in V$ is sent to $-v$. Since the ideal $\mathcal{J}_Q$ is stable under this map, we obtain an anti-automorphism $a \mapsto a^*$ of $C(V)$.

Definition 3.3. *If Q is a positive-definite quadratic form, we say that a Clifford module E of $C(V)$ with an inner-product is* **self-adjoint** *if $c(a^*) = c(a)^*$. This is equivalent to the operators $c(v)$, $v \in V$, being skew-adjoint.*

We will denote by $c(v)$ the action of an element of V on a Clifford module of $C(V)$, and in particular, on the Clifford algebra $C(V)$ itself. If E is a $\mathbb{Z}_2$-graded Clifford module, we denote by $\mathrm{End}_{C(V)}(E)$ the algebra of endomorphisms of E supercommuting with the action of $C(V)$.

Our first example of a Clifford module is the exterior algebra of V. To define the Clifford module action of $C(V)$ on ΛV, by the above lemma, we need only specify how V acts on ΛV. To do this, we introduce the notations $\varepsilon(v)\alpha$ for the exterior product of v with α, and $\iota(v)$ for the contraction with the covector $Q(v,\cdot) \in V^*$. We now define the Clifford action by the formula

$$c(v)\alpha = \varepsilon(v)\alpha - \iota(v)\alpha. \tag{3.1}$$

To check that this defines a Clifford module action on ΛV, we use the formula

$$\varepsilon(v)\iota(w) + \iota(w)\varepsilon(v) = Q(v,w). \tag{3.2}$$

If Q is positive-definite, the operator $\iota(v)$ is the adjoint of $\varepsilon(v)$, so that the Clifford module ΛV is self-adjoint.

Definition 3.4. *The* **symbol map** $\sigma : C(V) \to \Lambda V$ *is defined in terms of the Clifford module structure on ΛV by*

$$\sigma(a) = c(a)1 \in \Lambda V,$$

where $1 \in \Lambda^0 V$ is the identity in the exterior algebra ΛV. Note that $\sigma(1) = 1$.

Let e_i be an orthogonal basis of V, and denote by c_i the element of $C(V)$ corresponding to e_i.

Proposition 3.5. *The symbol map σ has an inverse, denoted $\mathbf{c} : \Lambda V \to C(V)$, which is given by the formula*

$$\mathbf{c}(e_{i_1} \wedge \ldots \wedge e_{i_j}) = c_{i_1} \ldots c_{i_j}.$$

We call $\mathbf{c}$ the **quantization** map. It follows that $C(V)$ has the same dimension as ΛV, namely $2^{\dim(V)}$. In fact, σ is an isomorphism of $\mathbb{Z}_2$-graded $\mathrm{O}(V)$-modules.

The Clifford algebra has a natural increasing filtration $C(V) = \bigcup_i C_i(V)$, defined as the smallest filtration such that

$$C_0(V) = \mathbb{R},$$
$$C_1(V) = V \oplus \mathbb{R}.$$

It follows that $a \in C(V)$ lies in $C_i(V)$ if and only if it lies in the span of elements of the form $v_1 \dots v_k$, where $v_j \in V$ and $k \leq i$. The following result is analogous to the identification of $\operatorname{gr} \mathscr{D}(M, \mathscr{E})$ with sections of $S(TM) \otimes \operatorname{End}(\mathscr{E})$ in the theory of differential operators; in fact, these two results are unified in the theory of supermanifolds.

Proposition 3.6. *The associated graded algebra* $\operatorname{gr} C(V)$ *is naturally isomorphic to the exterior algebra* ΛV*, the isomorphism being given by sending* $v_1 \wedge \dots \wedge v_i \in \Lambda^i V$ *to* $\sigma_i(v_1 \cdots v_i) \in \operatorname{gr}_i C(V)$.

The symbol map σ *extends the symbol map* $\sigma_i : C_i(V) \to \operatorname{gr}_i C(V) \cong \Lambda^i V$*, in the sense that if* $a \in C_i(V)$*, then* $\sigma(a)_{[i]} = \sigma_i(a)$*. The filtration* $C_i(V)$ *may be written*

$$C_i(V) = \sum_{k=0}^{i} C^k(V), \quad \text{where } C^k(V) = \mathbf{c}(\Lambda^k V).$$

Using σ, the Clifford algebra $C(V)$ may be identified with the exterior algebra ΛV with a twisted, or quantized, multiplication $\alpha \cdot_Q \beta$. We will need the following formula later: if $v \in V$ and $a \in C(V)$, then

$$\sigma([v, a]) = -2\iota(v)\,\sigma(a) \tag{3.3}$$

We will now show that if V is a real vector space with positive-definite scalar product, then there is a natural embedding of the Lie algebra $\mathfrak{so}(V)$ into $C(V)$.

Proposition 3.7. *The space* $C^2(V) = \mathbf{c}(\Lambda^2 V)$ *is a Lie subalgebra of* $C(V)$*, with bracket the commutator in* $C(V)$*. It is isomorphic to the Lie algebra* $\mathfrak{so}(V)$*, under the map* $\tau : C^2(V) \to \mathfrak{so}(V)$ *obtained by letting* $a \in C^2(V)$ *act on* $C^1(V) \cong V$ *by the adjoint action:*

$$\tau(a) \cdot v = [a, v].$$

Proof. It is easy to check that the map $\tau(a)$ really does preserve $C^1(V)$, so defines a Lie algebra homomorphism from $C^2(V)$ to $\mathfrak{gl}(V)$. To see that it maps into $\mathfrak{so}(V)$, observe that

$$Q(\tau(a) \cdot v, w) + Q(v, \tau(a) \cdot w) = -\tfrac{1}{2}[[a, v], w] - \tfrac{1}{2}[v, [a, w]].$$

Since $[[v, w], a] = 0$, Jacobi's identity shows that this vanishes.

The map τ must be an isomorphism, since it is injective and since the dimensions of $C^2(V)$ and $\mathfrak{so}(V)$ are the same, namely $n(n-1)/2$. □

We will make constant use in this book of the fact that a matrix $A \in \mathfrak{so}(V)$ corresponds to the Clifford element

$$\tau^{-1}(A) = \tfrac{1}{2} \sum_{i<j} (Ae_i, e_j) c_i c_j. \tag{3.4}$$

Note the following confusing fact: if we identify A with an element of $\Lambda^2 V$ by the standard isomorphism

$$A \in \mathfrak{so}(V) \longmapsto \sum_{i<j} (Ae_i, e_j) e_i \wedge e_j,$$

then $\mathbf{c}(A)$ equals

$$\mathbf{c}(A) = \sum_{i<j} (Ae_i, e_j) c_i c_j,$$

which differs from $\tau^{-1}(A)$ by a factor of two.

If $a \in C(V)$, then we may form its exponential in $C(V)$, which we will denote by $\exp_C(a)$; on the other hand, if $\alpha \in \Lambda V$, we can exponentiate it in ΛV, to form $\exp_\Lambda(\alpha)$.

Let $a \in \Lambda^2 V$ and consider the exponential $\exp_\Lambda a$ in the algebra ΛV. If T is the Berezin integral on ΛV introduced in (1.28), then by Proposition 1.36, we have the following result.

Lemma 3.8. *Assume that V is even-dimensional and oriented. If $a \in C^2(V)$, so that $\sigma(a) \in \Lambda^2 V$, then*

$$T(\exp_\Lambda \sigma(a)) = 2^{-\dim(V)/2} \det{}^{1/2}(\tau(a)).$$

We will often use the following formula for the exponential of an element of $C^2(V)$ inside $C(V)$. If v and w are vectors in V such that $(v,v) = (w,w) = 1$ and $(v,w) = 0$, then

$$\begin{aligned} \exp_C t(v \cdot w) &= \sum_{k=0}^{\infty} \frac{t^k}{k!} (v \cdot w)^k \\ &= \cos t + (\sin t) v \cdot w. \end{aligned} \tag{3.5}$$

If we are given a Lie subalgebra of the Lie algebra underlying a finite dimensional algebra, then we can exponentiate it inside this algebra to obtain an associated Lie group.

Definition 3.9. *The group* Spin(V) *is the group obtained by exponentiating the Lie algebra $C^2(V)$ inside the Clifford algebra $C(V)$.*

The adjoint action τ of $C^2(V)$ on V exponentiates to an orthogonal action still denoted by τ of Spin(V) on V. For $g \in \mathrm{Spin}(V)$ and $v \in V$, we have the fundamental relation

$$gvg^{-1} = \tau(g) \cdot v. \tag{3.6}$$

Indeed, writing $g = \exp_C(a)$ with $a \in \Lambda^2 V$, we see from the fact that $[a,v] = \tau(a)v$ that $\exp_C(a) v \exp_C(a)^{-1} = \exp(\tau(a)) \cdot v$.

Proposition 3.10. *If* $\dim(V) > 1$, *the homomorphism*

$$\tau : \mathrm{Spin}(V) \to \mathrm{SO}(V)$$

is a double covering.

Proof. The map τ is clearly surjective, since the exponential map is surjective on $\mathrm{SO}(V)$. Let $g \in \mathrm{Spin}(V)$ be such that $\tau(g) = 1$. Then $[g,v] = 0$ for all $v \in V$. Formula (3.3) implies that $\iota(v)\sigma(g) = 0$ for all $v \in V$, so that g is a scalar.

If $\dim(V) > 1$, then $-1 \in \mathrm{Spin}(V)$, as follows from (3.5) with $t = \pi$. Consider the canonical anti-automorphism $a \mapsto a^*$ of $C(V)$. On $C^2(V)$, this anti-automorphism equals multiplication by -1, since if v and w are orthogonal vectors,

$$(v\cdot w)^* = (-w)\cdot(-v) = -v\cdot w.$$

Since

$$\exp_C(a)^* = \exp_C(a^*) = \exp_C(-a)$$

for $a \in C^2(V)$, we see that $g \cdot g^* = 1$ for all $g \in \mathrm{Spin}(V)$. For g in the kernel of τ, and hence scalar, the relation $g \cdot g^* = 1$ implies that $g = \pm 1$. Since $-1 \in \mathrm{Spin}(V)$, we see that τ is a double cover. □

Any Clifford module restricts to a representation of the group $\mathrm{Spin}(V) \subset C^+(V)$; this gives a nice way to construct those representations of $\mathrm{Spin}(V)$ which do not descend to representations of $\mathrm{SO}(V)$. These are known as spinor representations, and will be discussed in the next section.

Proposition 3.11. *If the Clifford module E is self-adjoint, then the representation of* $\mathrm{Spin}(V)$ *on E is unitary.*

Proof. The representation of $C^2(V)$ on E is skew-adjoint, since $c(a)^* = c(a^*) = -c(a)$ for $a \in C^2(V)$. It follows that the Clifford action of $C^2(V)$ exponentiates to a unitary representation of $\mathrm{Spin}(V)$. □

Definition 3.12. *Let V be a vector space. If* $X \in \mathrm{End}(V)$, *let*

$$J_V(X) = \frac{\sinh(X/2)}{X/2} = \frac{e^{X/2} - e^{-X/2}}{X},$$

and let

$$j_V(X) = \det(J_V(X)) = \det\left(\frac{\sinh(X/2)}{X/2}\right).$$

Let us compare the two functions $\sigma(\exp_C(a))$ and $\exp_\Lambda \sigma(a)$ from $C(V)$ to ΛV. If $X \in \mathrm{End}(V)$, let

$$H_V(X) = \frac{X/2}{\tanh(X/2)}$$

be the analytic map from $\mathrm{End}(V)$ to itself which is defined on the set of X whose spectral radius is less than 2π. Since $H_V(0) = 1$, there is a well defined square root $H_V^{1/2}(X) \in \mathrm{GL}(V)$ if X lies in a neighbourhood of 0. For $a \in C^2(V)$, we will denote $H_V(\tau(a))$ by $H_V(a)$, and $j_V(\tau(a))$ by $j_V(a)$. We denote by $\alpha \mapsto g\cdot\alpha$ the automorphism of the algebra ΛV corresponding to an element $g \in \mathrm{GL}(V)$.

Proposition 3.13. *If $a \in C^2(V) = \mathrm{c}(\Lambda^2 V)$ is sufficiently small, we have*

$$\sigma(\exp_C(a)) = j_V^{1/2}(a) \det(H_V^{1/2}(a)) \left(H_V^{-1/2}(a) \cdot \exp_\Lambda(\sigma(a))\right).$$

Proof. We can choose an orthonormal basis $(e_i)_{i=1}^n$ of V such that

$$a = \theta_1 e_1 e_2 + \theta_2 e_3 e_4 + \dots.$$

In this way we reduce the proof to the case in which V is two-dimensional and $a = \theta e_1 e_2$. In this case, $\exp_C a = \cos\theta + \sin\theta e_1 e_2$, while $j_V^{1/2}(a) = \theta^{-1}\sin\theta$, and the matrix $H_V(a)$ equals $\theta\cot\theta$. □

The preceding proof shows that on the Cartan subalgebra of $\mathfrak{so}(V)$ of matrices of the form

$$Xe_{2i-1} = 2\theta_i e_{2i}, \quad 1 \le i \le \dim(V)/2,$$
$$Xe_{2i} = -2\theta_i e_{2i-1},$$

we have the formula

$$j_V(X) = \prod_{i=1}^{\ell} \left(\frac{\sin\theta_i}{\theta_i}\right)^2.$$

It follows that on this subalgebra, $j_V(X)$ has an analytic square root

$$j_V^{1/2}(X) = \prod_{i=1}^{\ell} \frac{\sin\theta_i}{\theta_i}.$$

We may now apply the following theorem of Chevalley, which is proved in Chapter 7 (see Theorem 7.28).

Proposition 3.14. *Let G be a connected compact Lie group with Lie algebra $\mathfrak{g}$,a nd let T be a maximal torus of G, with Lie algebra $\mathfrak{t}$ and Weyl group $W(G,T)$. If ϕ is an analytic function on $\mathfrak{t}$ which is invariant under $W(G,T)$, then it extends to an analytic function on all of $\mathfrak{g}$.*

Corollary 3.15. *The function $j_V^{1/2}(X)$ has an analytic extension to $\mathfrak{so}(V)$.*

Let T be the Berezin integral on ΛV. Since $T(g\cdot\alpha) = \det(g)T(\alpha)$, we obtain the following corollary of Proposition 3.13.

Proposition 3.16. *If $a \in C^2(V)$, then*

$$T(\sigma(\exp_C(a))) = j_V^{1/2}(a)\, T(\exp_\Lambda \sigma(a)).$$

3.2 Spinors

In this section, we will construct the spinor representation of the Clifford algebra of a Euclidean vector space V. We will suppose that V has even dimension n, and, in addition, that an orientation on V has been chosen, to remove an ambiguity which otherwise exists in defining the $\mathbb{Z}_2$-grading of the spinor space.

In the following lemma, we do not assume that V is even-dimensional.

Lemma 3.17. *Let e_j, $1 \le j \le n$, be an oriented, orthonormal basis of V, and define the* **chirality operator**

$$\Gamma = i^p e_1 \dots e_n,$$

where $p = n/2$ if n is even, and $p = (n+1)/2$ if n is odd. Then $\Gamma \in C(V) \otimes \mathbb{C}$ does not depend on the basis of V used in its definition. It satisfies $\Gamma^2 = 1$, and super-anticommutes with v for $v \in V$, in other words, $\Gamma v = -v\Gamma$ if n is even, while $\Gamma v = v\Gamma$ if n is odd.

It follows from this lemma that Γ belongs to the centre of $C(V) \otimes \mathbb{C}$ if n is odd, while if n is even and E is a complex Clifford module, we can define a $\mathbb{Z}_2$-grading on E by

$$E^{\pm} = \{v \in E \mid \Gamma v = \pm v\}.$$

Furthermore, when the dimension of V is divisible by four, Γ belongs to the real Clifford algebra $C(V)$, so that in that case, real Clifford modules are also $\mathbb{Z}_2$-graded.

Definition 3.18. *A* **polarization** *of the complex space $V \otimes \mathbb{C}$ is a subspace P which is isotropic, that is, $Q(w,w) = 0$ for all $w \in P$, and such that $V \otimes \mathbb{C} = P \oplus \bar{P}$; here, we extend the bilinear form Q on $V \otimes \mathbb{C}$ by complex linearity. The polarization is called oriented if there is an oriented orthonormal basis e_i of V, such that P is spanned by the vectors $\{e_{2j-1} - ie_{2j} \mid 1 \le j \le n/2\}$.*

Proposition 3.19. *If V is an even-dimensional oriented Euclidean vector space, then there is a unique $\mathbb{Z}_2$-graded Clifford module $S = S^+ \oplus S^-$, called the* **spinor module**, *such that*

$$C(V) \otimes \mathbb{C} \cong \mathrm{End}(S).$$

In particular, $\dim(S) = 2^{n/2}$ *and* $\dim(S^+) = \dim(S^-) = 2^{(n/2)-1}$.

Proof. Given a polarization of V, finding a realization of S is very simple. The bilinear form Q on $V \otimes \mathbb{C}$ places P and $\bar{P}$ in duality. The spinor space S is now defined to be equal to the exterior algebra ΛP of P. An element v of V is represented on S by splitting it into its components in P and $\bar{P}$; the first of these acts on ΛP by exterior product, while the second, which may be considered to be an element of P^* by the above duality, acts by contraction (there are some constants inserted to make the relations of the Clifford algebra come out precisely): if $s \in S = \Lambda P$, we have

$$c(w) \cdot s = 2^{1/2} \varepsilon(w) s, \quad \text{if } w \in P,$$
$$c(\bar{w}) \cdot s = -2^{1/2} \iota(\bar{w}) s, \quad \text{if } \bar{w} \in \bar{P} \cong P^*.$$

To see that $C(V)\otimes\mathbb{C}$ is in fact isomorphic to $\mathrm{End}(S)$, we only need to observe that $\dim(C(V)) = \dim(S)^2$, and that no element of $C(V)$ acts as the zero map on S. The uniqueness of the spinor representation is now a consequence of the fact that the algebra of matrices is simple, that is, it has a unique irreducible module.

We will now show that if the polarization P of V is oriented, the operator $c(\Gamma)$ on $S = \Lambda P$ is equal to $(-1)^k$ on $\Lambda^k P$, so that

$$S^{\pm} = \Lambda^{\pm} P. \tag{3.7}$$

In terms of the vectors $w_j = 2^{-1/2}(e_{2j-1} - ie_{2j})$, we may rewrite Γ as

$$\Gamma = 2^{-n/2}(w_1\bar{w}_1 - \bar{w}_1 w_1)\dots(w_{n/2}\bar{w}_{n/2} - \bar{w}_{n/2}w_{n/2}).$$

Thus, Γ acts on ΛP by

$$(-1)^{n/2}(\varepsilon(w_1)\iota(\bar{w}_1) - \iota(\bar{w}_1)\varepsilon(w_1))\dots(\varepsilon(w_{n/2})\iota(\bar{w}_{n/2}) - \iota(\bar{w}_{n/2})\varepsilon(w_{n/2}));$$

this is easily seen to equal $(-1)^k$ on $\Lambda^k P$. □

Since $\mathrm{Spin}(V) \subset C^+(V)$, it follows that both S^+ and S^- are representations of $\mathrm{Spin}(V)$. They are called the half-spinor representations, and have a great importance in differential geometry.

Definition 3.20. *We denote by ρ the spinor representation of* $\mathrm{Spin}(V)$ *in* $S = S^+ \oplus S^-$.

There is a natural supertrace on $C(V)$, defined by

$$\mathrm{Str}(a) = \mathrm{Tr}_S(\Gamma a) = \begin{cases} \mathrm{Tr}_{S^+}(a) - \mathrm{Tr}_{S^-}(a), & \text{if } a \in C^+(V), \\ 0, & \text{if } a \in C^-(V). \end{cases} \tag{3.8}$$

We will derive an explicit formula for this supertrace; before doing so, we will prove a result about abstract supertraces on $C(V)$.

Proposition 3.21. *If the quadratic form Q is non-degenerate, then there is, up to a constant factor, a unique supertrace on $C(V)$, equal to $T\circ\sigma$. The supertrace $\mathrm{Str}(a)$ defined in (3.8) equals*

$$\mathrm{Str}(a) = (-2i)^{n/2}\, T\circ\sigma(a).$$

Proof. We will show that $[C(V),C(V)] = C_{n-1}(V)$. It follows that any supertrace on $C(V)$ must vanish on $C_{n-1}(V)$, and hence be proportional to $T\circ\sigma$.

Let $(e_i)_{i=1}^n$ be an orthonormal basis of V. For any multi-index $I \subset \{1,\dots,n\}$, we form the Clifford product $c_I = \prod_{i\in I} c_i$; the set

$$\{c_I \mid I \subset \{1,\dots,n\}\}$$

is a basis for $C(V)$. If $|I| < n$, then there is at least one j such that $j \notin I$, and we obtain

$$c(e_I) = -\tfrac{1}{2}[c_j, c_j c_I].$$

On the other hand, the space of supertraces on $C(V)$ is at least one-dimensional, since we have constructed a non-vanishing supertrace in (3.8). To calculate the normalization of this particular supertrace, it suffices to calculate its value on the single element Γ, which by (3.7) equals $\dim(\Lambda P) = 2^{n/2}$. In this way, we obtain the formula $\mathrm{Str}(a) = (-2i)^{n/2} T \circ \sigma(a)$. □

In particular, for $a \in C^2(V)$, we see by Corollary 3.16 that

$$\mathrm{Str}(\exp_C a) = (-2i)^{n/2} j_V^{1/2}(a) T(\exp_\Lambda \sigma(a)). \tag{3.9}$$

Lemma 3.22. *If* $g \in \mathrm{Spin}(V)$, *then* $T(\sigma(g))^2 = \det((1-\tau(g))/2)$.

Proof. The calculation is easily reduced to the case in which $\dim(V) = 2$ and $g = \cos\theta + \sin\theta e_1 e_2$, so that $\det((1-\tau(g))/2) = \frac{1}{4}(1-e^{2i\theta})(1-e^{-2i\theta}) = \sin^2\theta$. □

It follows from Lemma 3.22 that if $g \in \mathrm{Spin}(V)$ is such that $\det(1-\tau(g))$ is invertible, then $T(\sigma(g)) \in \Lambda^n V$ is non-zero. Define $\varepsilon(g)$ to equal $+1$ (respectively -1) if $T(\sigma(g))$ is a positive (negative) element of $\Lambda^n V$. With this notation, we obtain the following formula for the supertrace of the Spin-representation. This result expresses in a quantitative way the notion that the spinors are a "square-root" representation.

Proposition 3.23. *If* $g \in \mathrm{Spin}(V)$, *then*

$$\mathrm{Str}(\rho(g)) = i^{-n/2} \varepsilon(g) |\det(1-\tau(g))|^{1/2}.$$

When $g \in \mathrm{SO}(V)$, the eigenvalues of g occur either in conjugate complex pairs, or are ± 1, and hence $\det(1-g) \geq 0$. The following result does not mention the spinor representation, but we state it here because the proof makes use of this representation.

Proposition 3.24. *If* $g \in \mathrm{SO}(V)$, *then the function* $\det(1 - g\exp A)$ *of* $A \in \mathfrak{so}(V)$ *has an analytic square root.*

Proof. The function $i^{n/2}\,\mathrm{Str}(\rho(\tilde{g}\exp A))$, where $\tilde{g}$ is any element of $\mathrm{Spin}(V)$ which covers $g \in \mathrm{SO}(V)$, is an analytic square root of $\det(1 - g\exp A)$. □

Definition 3.25. *If* $g \in \mathrm{SO}(V)$, *then* $D_g(A)$, $A \in \mathfrak{so}(V)$, *is the analytic square root of* $\det(1 - g\exp A)$ *such that* $D_g(0) = |\det(1-g)^{1/2}|$.

We will use the more suggestive notation $\det^{1/2}(1 - g\exp A)$ for $D_g(A)$.
Next, we show that the spinor representation is self-adjoint.

Proposition 3.26. *There is, up to a constant, a unique scalar product on the spinor space* S *for which it becomes a self-adjoint Clifford module. With respect to this metric, the spaces* S^+ *and* S^- *are orthogonal to each other.*

Proof. Uniqueness is clear, since it is implied by the irreducibility of S as a $C(V)$-module. To show the existence of a scalar product, we construct one explicitly on the realization ΛP that we obtained above. Indeed, it is sufficient to have a scalar product on the space P, which can be extended in the usual way to all of ΛP. We choose the scalar product furnished by $(w, \bar{w})$, where $\bar{w}$ is the complex conjugate of w and is thus an element of $\bar{P}$. In particular, it is immediate that $S^+ = \Lambda^+ P$ is orthogonal under this scalar product to $S^- = \Lambda^- P$.

To show that this metric makes the Clifford module S self-adjoint, it suffices to show that the operator $c(v)$ is skew-adjoint for $v \in V$. But an element of V, being real, has the form $w + \bar{w}$; thus it acts on ΛP by $\sqrt{2}(\varepsilon(w) - \iota(\bar{w}))$, which is clearly skew-adjoint. □

The next result shows that the spinor space S is the building block for complex representations of $C(V)$.

Proposition 3.27. *If V is an even-dimensional real Euclidean vector space, then every finite-dimensional $\mathbb{Z}_2$-graded complex module E of the Clifford algebra $C(V)$ is isomorphic to $W \otimes S$, for the $\mathbb{Z}_2$-graded complex vector space $W = \operatorname{Hom}_{C(V)}(S, E)$ which carries a trivial $C(V)$-action, and* $\operatorname{End}(W)$ *is isomorphic to* $\operatorname{End}_{C(V)}(E)$.

Proof. This follows from the fact that any finite dimensional module for the matrix algebra $\operatorname{End}(S)$ is of the form $W \otimes S$. The isomorphism between $\operatorname{Hom}_{C(V)}(S, E) \otimes S$ and E is given by $w \otimes s \mapsto w(s)$. □

The space W will be called the **twisting space** for the Clifford module E.

Since $\operatorname{Str}_S(\Gamma) = 2^{n/2}$, the supertrace over W of an element $F \in \operatorname{End}(W) \cong \operatorname{End}_{C(V)}(E)$ is given by the formula

$$\operatorname{Str}_W(F) = 2^{-n/2} \operatorname{Str}_E(\Gamma F). \tag{3.10}$$

Motivated by this, we make the following definition.

Definition 3.28. *The* **relative supertrace** $\operatorname{Str}_{E/S} : \operatorname{End}_{C(V)}(E) \to \mathbb{C}$ *equals*

$$\operatorname{Str}_{E/S}(F) = 2^{-n/2} \operatorname{Str}_E(\Gamma F).$$

If W is ungraded, the relative supertrace equals the W-trace, and is proportional to the E-trace:

$$\operatorname{Tr}_W(F) = \operatorname{Str}_{E/S}(F) = 2^{-n/2} \operatorname{Tr}_E(F).$$

We end this section with some formulas for the special case in which V is a complex Hermitian vector space; these formulas will be of use in studying the $\bar{\partial}$-operator on a Kähler manifold. In this case, the group $\mathrm{U}(V)$ of unitary transformations of V is a subgroup of $\mathrm{SO}(V)$. Let us consider the decomposition

$$V \otimes \mathbb{C} = V^{1,0} \oplus V^{0,1},$$

where $V^{1,0}$ and $V^{0,1}$ are the spaces on which the complex structure J of V acts by $+i$ and $-i$ respectively. Then $V^{1,0}$ is a polarization of $V \otimes \mathbb{C}$ and we can take $S = \Lambda V^{1,0}$

as the spinor space. On the other hand a complex transformation of V leaves $V^{1,0}$ stable, so there is a natural representation λ of $\mathrm{U}(V)$ on $\Lambda V^{1,0}$. The representation λ and the restriction of the spinor representation ρ of $\mathrm{Spin}(V)$ to $\tau^{-1}\mathrm{U}(V)$ differ by the character $\det_{\mathbb{C}}^{1/2}$ of the double cover $\tau^{-1}(\mathrm{U}(V))$ of $\mathrm{U}(V)$. More precisely, the infinitesimal representations of the Lie algebras on S, which we denote also by λ and ρ, are related by the following result.

Lemma 3.29. *Let $a \in C^2(V)$. If $\tau(a)$ is complex linear on V, then*

$$\rho(a) = \lambda(\tau(a)) - \tfrac{1}{2}\operatorname{Tr}_{V^{1,0}}\tau(a).$$

Proof. Choose a complex basis w_j of $V^{1,0}$ such that $\tau(a)w_j = \lambda_j w_j$ for every j, and $(w_j, \bar{w}_i) = \delta_{ij}$. Then $a = \frac{1}{4}\sum_{1\le i\le n/2}\lambda_i(\bar{w}_i w_i - w_i\bar{w}_i)$, so that

$$\lambda(\tau(a)) = \sum \lambda_i \varepsilon(w_i)\iota(\bar{w}_i),$$
$$\rho(a) = \tfrac{1}{2}\sum \lambda_i(\varepsilon(w_i)\iota(\bar{w}_i) - \iota(\bar{w}_i)\varepsilon(w_i)) = \lambda(\tau(a)) - \tfrac{1}{2}\sum \lambda_i.$$

From these formulas, the result is evident. □

3.3 Dirac Operators

Definition 3.30. *If M is a Riemannian manifold, the* **Clifford bundle** *$C(M)$ is the bundle of Clifford algebras over M whose fibre at $x \in M$ is the Clifford algebra $C(T_x^*M)$ of the Euclidean spaces T_x^*M.*

The Clifford bundle $C(M)$ is an associated bundle to the orthonormal frame bundle,

$$C(M) = \mathrm{O}(M) \times_{\mathrm{O}(n)} C(\mathbb{R}^n).$$

From this representation of $C(M)$, we see that it inherits a connection, the Levi-Civita connection, which is compatible with the Clifford product, in the sense that

$$\nabla(ab) = (\nabla a)b + a(\nabla b) \quad \text{for all } a \text{ and } b \in \Gamma(M, C(M)).$$

Definition 3.31. *The symbol map σ*

$$\sigma : C(M) \to \Lambda T^*M,$$

which identifies the Clifford bundle $C(M)$ with the exterior bundle ΛT^*M, is defined by means of the symbol map $\sigma_x : C(T_x^*M) \to \Lambda T_x^*M$.

Definition 3.32. *1. A* **Clifford module** *$\mathcal{E}$ on an even-dimensional Riemannian manifold M is a $\mathbb{Z}_2$-graded bundle $\mathcal{E}$ on M with a graded action of the bundle of algebras $C(M)$ on it, which we write as follows:*

$$(a, s) \longmapsto c(a)s, \quad \text{where } a \in \Gamma(M, C(M)) \text{ and } s \in \Gamma(M, \mathcal{E}).$$

2. *If $\mathscr{E}$ is a Clifford module with metric h, for which $\mathscr{E}^+$ and $\mathscr{E}^-$ are orthogonal, we say that the Clifford module $\mathscr{E}$ is* **self-adjoint** *if the Clifford action is self-adjoint at each point of M, in other words, if the operators $c(a)$ with $a \in T^*M$ are skew-adjoint.*
3. *If $\mathscr{W}$ is a vector bundle, then the* **twisted** *Clifford module obtained from $\mathscr{E}$ by twisting with $\mathscr{W}$ is the bundle $\mathscr{W} \otimes \mathscr{E}$, with Clifford action $1 \otimes c(a)$.*

The most basic example of a Clifford module is the bundle of spinors. In order to define this, we need a spin-structure on the manifold.

Definition 3.33. *A* **spin-structure** *on a manifold M is a* Spin(n)*-principal bundle* Spin(M) *on M such that the cotangent bundle of M is isomorphic to the associated bundle* $\mathrm{Spin}(M) \times_{\mathrm{Spin}(n)} \mathbb{R}^n$.

Note that a manifold with a spin-structure is, in particular, an oriented Riemannian manifold, with frame bundle

$$\mathrm{SO}(M) = \mathrm{Spin}(M) \times_{\mathrm{Spin}(n)} \mathrm{SO}(n).$$

Thus the principal bundle Spin(M) is a double cover of SO(M) and inherits from SO(M) the Levi-Civita connection.

There may be topological obstructions to the existence of a spin-structure on M, while on the other hand, there may be more than one. This question is understood using the classification of isomorphism classes of principal G-bundles on a manifold G by the cohomology group $H^1(M,G)$, combined with the Bockstein long exact sequence

$$0 \to H^1(M,\mathbb{Z}_2) \to H^1(M,\mathrm{Spin}(n)) \to H^1(M,\mathrm{SO}(n)) \to H^2(M,\mathbb{Z}_2)$$

associated to the short exact sequence of groups

$$1 \to \mathbb{Z}_2 \to \mathrm{Spin}(n) \to \mathrm{SO}(n) \to 1.$$

The frame bundle SO(M) defines an element $\alpha \in H^1(M,\mathrm{SO}(n))$ whose image in $H^2(M,\mathbb{Z}_2)$ may be identified with the second Stieffel-Whitney class of M, written $w_2(M)$. This at least motivates the following result.

Proposition 3.34. *An orientable manifold M has a spin-structure if and only if its second Stieffel-Whitney class $w_2(M) \in H^2(M,\mathbb{Z}_2)$ vanishes. If this is the case, then the different spin-structures are parametrized by elements of $H^1(M,\mathbb{Z}_2) \cong \mathrm{Hom}(\pi_1(M),\mathbb{Z}_2)$.*

If M is an even-dimensional spin manifold, the **spinor bundle** $\mathscr{S}$ is defined to be the associated bundle

$$\mathscr{S} = \mathrm{Spin}(M) \times_{\mathrm{Spin}(n)} S.$$

It is clear that $\mathscr{S}$ is a Clifford module, since the action of $C(\mathbb{R}^n)$ on S leads to an action of the associated bundle $C(M) = \mathrm{Spin}(M) \times_{\mathrm{Spin}(n)} C(\mathbb{R}^n)$ on $\mathscr{S}$. The following is the analogue for manifolds of Proposition 3.27.

Proposition 3.35. *If M is an even-dimensional oriented manifold with spin-structure, every Clifford module $\mathscr{E}$ is a twisted bundle $\mathscr{E} = \mathscr{W} \otimes \mathscr{S}$.*

Proof. We may define a complex vector bundle $\mathscr{W}$ such that $\mathscr{E} \cong \mathscr{W} \otimes \mathscr{S}$ as Clifford modules by the formula $\mathscr{W} = \operatorname{End}_{C(M)}(\mathscr{S}, \mathscr{E})$. The isomorphism between $\operatorname{Hom}_{C(M)}(\mathscr{S}, \mathscr{E}) \otimes \mathscr{S}$ and $\mathscr{E}$ is given by $w \otimes s \mapsto w(s)$. □

Locally, we may trivialize the bundle $C(M)$ by choosing an orthonormal frame of the cotangent bundle. Over such an open set U, the Clifford bundle $C(M) \cong U \times C(\mathbb{R}^n)$. Hence, we see that a spinor bundle $\mathscr{S}$ always exists locally, and we can always locally decompose a Clifford module $\mathscr{E}$ as $\operatorname{Hom}_{C(M)}(\mathscr{S}, \mathscr{E}) \otimes \mathscr{S}$.

If $a \in \mathscr{A}(M, C(M))$ is a Clifford algebra-valued differential form on M, we can define an operator $c(a)$ which acts on the space $\mathscr{A}(M, \mathscr{E})$ by means of the following formula: if α and β are differential forms on M, a is a Clifford algebra section, and s is a section of $\mathscr{E}$, all homogeneous with respect to the $\mathbb{Z}_2$-grading, then

$$(c(\alpha \otimes a))(\beta \otimes s) = (-1)^{|a| \cdot |\beta|} (\alpha \wedge \beta) \otimes (c(a)s) \tag{3.11}$$

The interest of Clifford modules is that they are naturally associated to a class of first-order differential operators, known as Dirac operators. We will use a more general definition of Dirac operators than is usual, since it fits in better with the notion of a superconnection, and does not make any of the proofs more difficult.

Definition 3.36. *A* **Dirac operator** D *on a $\mathbb{Z}_2$-graded vector bundle $\mathscr{E}$ is a first-order differential operator of odd parity on $\mathscr{E}$,*

$$\mathsf{D} : \Gamma(M, \mathscr{E}^{\pm}) \to \Gamma(M, \mathscr{E}^{\mp}),$$

such that D^2 is a generalized Laplacian.

Applying the results of Chapter 2, we obtain the following result. From now on, when discussing the heat kernel of a generalized Laplacian H on a Riemannian manifold M and vector bundle $\mathscr{E}$, we will use the Riemannian density to identify the heat kernel $\langle x \mid e^{-tH} \mid y \rangle$ of H with a section of $\mathscr{E} \boxtimes \mathscr{E}^*$ over $M \times M$.

Proposition 3.37. *1. If D is a Dirac operator on a compact manifold M, then D^2, acting on $\Gamma(M, \mathscr{E})$, has a smooth heat kernel*

$$\langle x \mid e^{-t\mathsf{D}^2} \mid y \rangle \in \Gamma(M \times M, \mathscr{E} \boxtimes \mathscr{E}^*).$$

2. If the manifold M is compact, the Dirac operator D has finite dimensional kernel. In particular, a symmetric Dirac operator D is essentially self-adjoint.

Proof. The operator D^2 has a finite dimensional kernel, and hence so does the Dirac operator D. The proof that D is essentially self-adjoint uses the heat kernel $\langle x \mid e^{-t\mathsf{D}^2} \mid y \rangle$ in the same way as the proof of the corresponding result for D^2 given in Proposition 2.33 did. □

The purpose of this section is to work out what are the geometric data needed to specify a Dirac operator: this is analogous to the one-to-one correspondence between generalized Laplacians and the three pieces of data consisting of a metric on M, a connection on $\mathscr{E}$ and a potential $F \in \Gamma(M, \mathrm{End}(\mathscr{E}))$.

If we are given a Dirac operator on a vector-bundle $\mathscr{E}$, then $\mathscr{E}$ inherits a natural Clifford module structure.

Proposition 3.38. *The action of T^*M on $\mathscr{E}$ defined by*

$$[\mathsf{D}, f] = c(df), \quad \textit{where } f \textit{ is a smooth function on } M,$$

is a Clifford action, which is self-adjoint with respect to a metric on $\mathscr{E}$ if the operator D is symmetric. Conversely, any differential operator D such that $[\mathsf{D}, f] = c(df)$ for all $f \in C^\infty(M)$ is a Dirac operator.

Proof. First of all, this definition is unambiguous, since if f and h are both smooth functions on M, we have

$$c(d(fh)) = [\mathsf{D}, fh] = fc(dh) + hc(df).$$

To check the relations for $c(df)$ to be a Clifford action, we need only observe that

$$c(df)^2 = \tfrac{1}{2}[[\mathsf{D}, f], [\mathsf{D}, f]] = \tfrac{1}{2}[[\mathsf{D}^2, f], f] = -|df|^2,$$

since D^2 is by assumption a generalized Laplacian; we have used the fact that $[[\mathsf{D}, f], f] = 0$ and that $[\mathsf{D}, [\mathsf{D}, A]] = [\mathsf{D}^2, A]$, which is true for any odd operator D and even operator A. This equation also shows that any operator satisfying $[\mathsf{D}, f] = c(df)$ for all $f \in C^\infty(M)$ is in fact a Dirac operator. Finally, if D is self-adjoint with respect to some metric on $\mathscr{E}$, then for all real functions f on M,

$$c(df)^* = [\mathsf{D}, f]^* = [f^*, \mathsf{D}^*] = -c(df). \qquad \square$$

It is clear that there is always at least one Dirac operator on a Clifford module; we can construct one explicitly by taking the operators $\sum_i c(dx^i)\partial_i$ on local coordinate patches and combining them by means of a partition of unity.

If D is a Dirac operator on a bundle $\mathscr{E}$ and A is an odd section of the bundle $\mathrm{End}(\mathscr{E})$, the operator $\mathsf{D} + A$ will be another Dirac operator on $\mathscr{E}$ corresponding to the same Clifford action on $\mathscr{E}$. Conversely, any two Dirac operators on $\mathscr{E}$ corresponding to the same Clifford action will differ by such a section. Indeed, if D_0 and D_1 are two Dirac operators on $\mathscr{E}$,

$$[\mathsf{D}_1 - \mathsf{D}_0, f] = c(df) - c(df) = 0$$

for all $f \in C^\infty(M)$, and hence $\mathsf{D}_1 - \mathsf{D}_0$ may be represented by the action of a section of $\mathrm{End}^-(\mathscr{E})$. Thus, the collection of all Dirac operators on a Clifford module is an affine space modelled on $\Gamma(M, \mathrm{End}^-(\mathscr{E}))$. In order to sharpen this identification, we will introduce the notion of a Clifford superconnection on a Clifford module.

Definition 3.39. *1. If $\nabla^{\mathscr{E}}$ is a connection on a Clifford module $\mathscr{E}$, we say that $\nabla^{\mathscr{E}}$ is a* **Clifford connection** *if for any $a \in \Gamma(M, C(M))$ and $X \in \Gamma(M, TM)$,*

$$[\nabla^{\mathscr{E}}_X, c(a)] = c(\nabla_X a).$$

In this formula, ∇_X is the Levi-Civita covariant derivative extended to the bundle $C(M)$.

2. If $\mathbb{A}$ is a superconnection on a Clifford module $\mathscr{E}$, we say that $\mathbb{A}$ is a **Clifford superconnection** *if for any $a \in \Gamma(M, C(M))$,*

$$[\mathbb{A}, c(a)] = c(\nabla a). \tag{3.12}$$

In this formula, ∇a is the Levi-Civita covariant derivative of a, which is an element of $\mathscr{A}^1(M, C(M))$.

It follows from this definition that if $\mathbb{A}$ is a Clifford superconnection, then the formula (3.12) holds for all $a \in \mathscr{A}(M, C(M))$.

The collection of all Clifford superconnections on $\mathscr{E}$ is an affine space based on $\mathscr{A}^-(M, \mathrm{End}_{C(M)}(\mathscr{E}))$, the space of sections of $\Lambda T^*M \otimes \mathrm{End}_{C(M)}(\mathscr{E})$ of odd total degree; this follows from the fact that if $\mathbb{A}$ is a Clifford superconnection on $\mathscr{E}$ and $a \in \mathscr{A}^-(M, \mathrm{End}_{C(M)}(\mathscr{E}))$, then $\mathbb{A} + a$ is again a Clifford superconnection.

The Levi-Civita connection $\nabla^{\mathscr{S}}$ on the spinor bundle $\mathscr{S}$ of a spin manifold is a Clifford connection. If ω_{ij}^k are the coefficients of the Levi-Civita connection with respect to an orthonormal frame e_i of the tangent bundle, defined by $\nabla_{\partial_i} e_j = \omega_{ij}^k e_k$, then (3.4) shows that $\nabla^{\mathscr{S}}$ is given in local coordinates by the formula

$$\nabla^{\mathscr{S}}_{\partial_i} = \partial_i + \tfrac{1}{4} \sum_{jk} \omega_{ij}^k c^j c^k. \tag{3.13}$$

Proposition 3.40. *Let M be a spin manifold and let $\mathscr{E} = \mathscr{W} \otimes \mathscr{S}$ be a twisted spinor bundle.*

1. If $\mathbb{A}^{\mathscr{W}}$ is a superconnection on the bundle $\mathscr{W}$, then the tensor product superconnection $\mathbb{A} = \mathbb{A}^{\mathscr{W}} \otimes 1 + 1 \otimes \nabla^{\mathscr{S}}$ on the twisted Clifford module $\mathscr{W} \otimes \mathscr{S}$ is a Clifford superconnection. We call $\mathbb{A}^{\mathscr{W}}$ the **twisting superconnection** *associated to the Clifford superconnection $\mathbb{A}$, and denote it by $\mathbb{A}^{\mathscr{E}/S}$.*

2. There is a one-to-one correspondence between Clifford superconnections on the twisted Clifford module $\mathscr{W} \otimes \mathscr{S}$ and superconnections on the twisting bundle $\mathscr{W}$, induced by the correspondence $\mathbb{A}^{\mathscr{W}} \mapsto (\mathbb{A}^{\mathscr{W}} \otimes 1 + 1 \otimes \nabla^{\mathscr{S}})$.

Proof. Part (1) follows from the formula for the Clifford action on $\mathscr{W} \otimes \mathscr{S}$, which sends $a \in C(M)$ to $1 \otimes c(a) \in \Gamma(M, \mathrm{End}(\mathscr{W} \otimes \mathscr{S}))$. Thus, we see that

$$[\mathbb{A}^{\mathscr{W}} \otimes 1 + 1 \otimes \nabla^{\mathscr{S}}, 1 \otimes c(a)] = 1 \otimes [\nabla^{\mathscr{S}}, c(a)] = 1 \otimes c(\nabla a).$$

To see that the assignment of a twisted Clifford superconnection on $\mathscr{W} \otimes \mathscr{S}$ to a superconnection on $\mathscr{W}$ is a bijection, we simply observe that the space of Clifford superconnections on $\mathcal{W} \otimes \mathcal{S}$ is an affine space modelled on the space of sections $\mathcal{A}^-(M, \mathrm{End}_{C(M)}(\mathcal{W} \otimes \mathcal{E})) = \mathcal{A}^-(M, \mathrm{End}(\mathcal{W}))$, as is the space of superconnections on $\mathcal{W}$.

If $\mathbb{A}$ is a Clifford superconnection on the twisted Clifford module $\mathcal{W} \otimes \mathcal{S}$, we define an operator $\mathbb{A}^{\mathcal{W}}$ acting on sections of $\mathcal{W}$ by the formula

$$(\mathbb{A}^{\mathcal{W}} u) \otimes s = \mathbb{A}(u \otimes s) - u \otimes (\nabla^{\mathcal{S}} s),$$

where s is a non-vanishing section of $\mathcal{S}$. It is easy to see that this formula is independent of the section s used, since any two sections s_1 and s_2 of $\mathcal{S}$ are related locally by the formula $s_2 = c(a)s_1$ for some section $a \in \Gamma(M, C(M))$ of the Clifford bundle $C(M)$. We then see that

$$\begin{aligned}\mathbb{A}(u \otimes s_2) - u \otimes (\nabla^{\mathcal{S}} s_2) &= \mathbb{A}(u \otimes (c(a)s_1) - u \otimes (\nabla^{\mathcal{S}} c(a)s_1) \\ &= c(a)\big(\mathbb{A}(u \otimes s_1) - u \otimes (\nabla^{\mathcal{S}} s_1)\big). \qquad \square\end{aligned}$$

Corollary 3.41. *There exists a Clifford superconnection on any Clifford module $\mathcal{E}$.*

Proof. To construct a Clifford connection on the Clifford module $\mathcal{E}$, it is sufficient to do this locally and then patch the local Clifford connections together by means of a partition of unity. But locally, we may decompose $\mathcal{E}$ as $\mathcal{W} \otimes \mathcal{S}$. It suffices to take any connection $\nabla^{\mathcal{W}}$ and form the Clifford connection $\nabla^{\mathcal{W}} \otimes 1 + 1 \otimes \nabla^{\mathcal{S}}$. □

We wil now study the relationship between Clifford superconnections and Dirac operators. Let us start with the case of a Clifford connection $\nabla^{\mathcal{E}}$. We may define a Dirac operator on $\mathcal{E}$ by means of the following composition:

$$\Gamma(M, \mathcal{E}) \xrightarrow{\nabla^{\mathcal{E}}} \Gamma(M, T^*M \otimes \mathcal{E}) \xrightarrow{c} \Gamma(M, \mathcal{E})$$

In local coordinates, this operator may be written

$$\mathsf{D} = \sum_i c(dx^i) \nabla^{\mathcal{E}}_{\partial_i},$$

from which it is clear that it is a Dirac operator, that is, that $[\mathsf{D}, f] = c(df)$. Often in the results of this book, we will have to restrict attention to operators in the subclass of Dirac operators constructed from a Clifford connection. In terms of an orthonormal frame e_i of the tangent bundle TM over an open set U with dual frame e^i of T^*M, we may also write

$$\mathsf{D} = \sum_i c(e^i) \nabla^{\mathcal{E}}_{e_i}.$$

If M is a spin manifold, the Dirac operator on $\mathcal{S}$ associated to the Levi-Civita connection $\nabla^{\mathcal{S}}$ is the operator often referred to as the Dirac operator associated to the spin-structure, with no further qualification. Other authors reserve the term Dirac operator for twisted versions of this operator; we have chosen a far more liberal definition of Dirac operators because many results of index theory deserve to be stated in greater generality.

Let $\mathbb{A}$ be a Clifford superconnection on the Clifford module $\mathscr{E}$. We can define a first-order differential operator on $\Gamma(M,\mathscr{E})$, which we will write as $\mathsf{D}_{\mathbb{A}}$, by composing the superconnection with the Clifford multiplication map:

$$\Gamma(M,\mathscr{E}) \xrightarrow{\mathbb{A}} \mathscr{A}(M,\mathscr{E}) \xrightarrow{\mathbf{c}} \Gamma(M,C(M)\otimes\mathscr{E}) \xrightarrow{c} \Gamma(M,\mathscr{E})$$

Here, we have made use of the isomorphism

$$\mathscr{A}(M,\mathscr{E}) \cong \Gamma(M,C(M)\otimes\mathscr{E})$$

given by the quantization map $\mathbf{c} : \Lambda T^*M \to C(M)$ of Proposition 3.5. With respect to a local coordinate system and orthonormal frame e_i of the tangent bundle in which

$$\mathbb{A} = \sum_{i=1}^{n} dx^i \otimes \partial_i + \sum_{I\subset\{1,\dots,n\}} e^I \otimes A_I,$$

A_I being sections of $\operatorname{End}(\mathscr{E})$, we have

$$\mathsf{D}_{\mathbb{A}} = \sum_{i=1}^{n} c(dx^i)\partial_i + \sum_{I\subset\{1,\dots,n\}} c(e^I)A_I.$$

Proposition 3.42. *The map sending a Clifford superconnection $\mathbb{A}$ to the operator $\mathsf{D}_{\mathbb{A}}$ is a one-to-one correspondence between Clifford superconnections and Dirac operators compatible with the given Clifford action on $\mathscr{E}$.*

Proof. First, we must check that the operator $\mathsf{D}_{\mathbb{A}}$ is in fact a Dirac operator compatible with the given Clifford action. This follows from Leibniz's formula

$$\mathbb{A}(fs) = df\otimes s + f\cdot\mathbb{A}s \in \mathscr{A}(M,\mathscr{E}), \quad \text{where } f\in C^\infty(M) \text{ and } s\in\Gamma(M,\mathscr{E}).$$

If we apply the Clifford map

$$\mathscr{A}(M,\mathscr{E}) \to \Gamma(M,C(M)\otimes\mathscr{E}) \to \Gamma(M,\mathscr{E})$$

to both sides, we obtain the desired equation, namely

$$\mathsf{D}_{\mathbb{A}}(fs) = c(df)s + f\mathsf{D}_{\mathbb{A}}s.$$

It remains to show that any Dirac operator on $\mathscr{E}$ may be realised in a unique way in the form $\mathsf{D}_{\mathbb{A}}$ for some Clifford superconnection $\mathbb{A}$. The uniqueness is immediately deduced from the fact that any two Clifford superconnections differ by an element of $\mathscr{A}^-(M,\operatorname{End}_{C(M)}(\mathscr{E}))$. If D_0 and D_1 are two Dirac operators on $\mathscr{E}$, $\mathsf{D}_1 - \mathsf{D}_0$ may be represented by the action of a section of the bundle $\operatorname{End}^-(\mathscr{E})$. However, applying the symbol map

$$\Gamma(M,\operatorname{End}^-(\mathscr{E})) \cong \Gamma(M,C(M)\otimes\operatorname{End}_{C(M)}(\mathscr{E}))^- \to \mathscr{A}^-(M,\operatorname{End}(\mathscr{E}))$$

to this section, we obtain a differential form $\omega\in\mathscr{A}^-(M,\operatorname{End}_{C(M)}(\mathscr{E}))$. It follows from this that $\mathsf{D}_1 = \mathsf{D}_0 + \mathbf{c}(\omega)$. If D_0 is the Dirac operator associated to the Clifford superconnection $\mathbb{A}$, we see that D_1 is the Dirac operator associated to the Clifford superconnection $\mathbb{A}+\omega$. □

Proposition 3.43. *The curvature* $\mathbb{A}^2 \in \mathscr{A}(M, \operatorname{End}(\mathscr{E}))$ *of a Clifford superconnection* $\mathbb{A}$ *on* $\mathscr{E}$ *decomposes under the isomorphism* $\operatorname{End}(\mathscr{E}) \cong C(M) \otimes \operatorname{End}_{C(M)}(\mathscr{E})$ *as follows:*

$$\mathbb{A}^2 = R^{\mathscr{E}} + F^{\mathscr{E}/S}.$$

In this formula, $R^{\mathscr{E}} \in \mathscr{A}^2(M, C(M)) \subset \mathscr{A}^2(M, \operatorname{End}(\mathscr{E}))$ *is the action of the Riemannian curvature* R *of* M*, on the bundle* $\mathscr{E}$*, given by the formula*

$$R^{\mathscr{E}}(e_i, e_j) = \tfrac{1}{4} \sum_{kl} (R(e_i, e_j)e_k, e_l) c^k c^l,$$

and $F^{\mathscr{E}/S} \in \mathscr{A}(M, \operatorname{End}_{C(M)}(\mathscr{E}))$ *is an invariant of* $\mathbb{A}$*, called the* **twisting curvature** *of the Clifford module* $\mathscr{E}$*.*

Proof. Define $F^{\mathscr{E}/S}$ to equal $\mathbb{A}^2 - R^{\mathscr{E}}$. If we can show that the operation $\varepsilon(F^{\mathscr{E}/S})$ of exterior multiplication by $F^{\mathscr{E}/S}$, acting on $\mathscr{A}(M, \mathscr{E})$, commutes with the operators $c(a)$, where $a \in \Gamma(M, T^*M) \subset \Gamma(M, C(M))$, then it will follow immediately that $F^{\mathscr{E}/S}$ is actually a differential form with values in $\operatorname{End}_{C(M)}(\mathscr{E})$. But this commutation follows immediately from the the fact that $\mathbb{A}$ is compatible with the Clifford action, as expressed in (3.12):

$$[\mathbb{A}^2, c(a)] = [\mathbb{A}, [\mathbb{A}, c(a)]] = [\mathbb{A}, c(\nabla a)] = c(\nabla^2 a).$$

To finish the proof, we use the fact that $\nabla^2 a = Ra$,where R is the Riemannian curvature of M and that $[R^{\mathscr{E}}, c(a)] = c(Ra)$, so that

$$[\mathbb{A}^2, c(a)] = c(Ra) = [R^{\mathscr{E}}, c(a)]. \qquad \square$$

Let M be a spin manifold and $\mathscr{E} = \mathscr{W} \otimes \mathscr{S}$ be a twisted spinor bundle. If $\mathbb{A}^{\mathscr{W}}$ is a superconnection on $\mathscr{W}$, with curvature $F^{\mathscr{W}}$, and if $\mathbb{A} = \mathbb{A}^{\mathscr{W}} \otimes 1 + 1 \otimes \nabla^{\mathscr{S}}$ is the corresponding twisted superconnection on $\mathscr{E} = \mathscr{W} \otimes \mathscr{S}$, then the curvature $\mathbb{A}^2$ equals

$$\mathbb{A}^2 = R^{\mathscr{E}} + F^{\mathscr{W}}.$$

Thus, the twisting curvature $F^{\mathscr{E}/S}$ of $\mathbb{A}$ equals the curvature $F^{\mathscr{W}}$ of the twisting superconnection $\mathbb{A}^{\mathscr{W}}$; this explains why we call it the twisting curvature.

In the next proposition we compute the formal adjoint of a Dirac operator. Recall that the adjoint ∇^* of an ordinary connection ∇ on a Hermitian vector bundle is defined by the equality of one-forms

$$d(s_1, s_2) = (\nabla s_1, s_2) + (s_1, \nabla^* s_2) \quad \text{for all } s_i \in \Gamma(M, \mathscr{E}).$$

Likewise, if $\mathbb{A} = \nabla + \omega$ is a superconnection, where $\omega \in \mathscr{A}^-(M, \operatorname{End}(\mathscr{E}))$, then $\mathbb{A}^*$ is defined to equal $\nabla^* + \omega^*$, where by ω^* we mean the differential form $\sum_{i=0}^n (-1)^{i(i+1)/2} \omega_{[i]}^*$; the reason for the sign $(-1)^{i(i+1)/2}$ is that the differential form $df_1 \dots df_i = (-1)^{i(i-1)/2} df_i \dots df_1$ and taking adjoints will introduce an additional $(-1)^i$ sign.

Proposition 3.44. *The adjoint of the Dirac operator* $\mathsf{D}_{\mathbb{A}}$ *is the Dirac operator* $\mathsf{D}_{\mathbb{A}^*}$*.*

Proof. It is easy to see that it suffices to prove this when $\mathbb{A}$ is a connection ∇. Let s_1 and s_2 be two sections of $\mathcal{E}$ and let X be the vector field on M given by $\alpha(X) = (s_1, c(\alpha)s_2)$ for $\alpha \in \mathcal{A}^1(M)$. Thus ∇X belongs to $\Gamma(M, TM \otimes T^*M)$. It is easy to see that

$$(\mathsf{D}_{\mathbb{A}} s_1, s_2)_x = (s_1, \mathsf{D}_{\mathbb{A}^*} s_2)_x - \operatorname{Tr}(\nabla X)_x.$$

The integral of the last term vanishes by Proposition 2.7, proving the proposition. □

Dirac operators are characterized by the formula $[\mathsf{D}, f] = c(df)$ for all $f \in C^\infty(M)$. In the next proposition, we calculate $[\mathsf{D}, c(\theta)]$ when θ is a one-form on M and D is a Dirac operator. Let $\nabla^{\mathcal{E}}$ be the connection associated to the generalized Laplacian D^2 on the bundle $\mathcal{E}$.

Proposition 3.45. *If $\theta \in \mathcal{A}^1(M)$ is dual to the vector field X, then*

$$[\mathsf{D}, c(\theta)] = -2\nabla^{\mathcal{E}}_X + \mathbf{c}(d\theta) + d^*\theta,$$

where we recall that

$$\mathbf{c}(d\theta) = \sum_{i<j} (d\theta)(e_i, e_j) c(e^i) c(e^j).$$

Proof. It suffices to prove the theorem for differential forms of the type $f dg$. Let $X = \operatorname{grad} g$ be the vector field dual to dg. Observe that

$$[\mathsf{D}, c(f dg)] = f[\mathsf{D}, c(dg)] + c(df) c(dg).$$

Since $c(df)c(dg) = \mathbf{c}(df \wedge dg) - (df, dg)$, we see that

$$[\mathsf{D}, c(f dg)] = f[\mathsf{D}, c(dg)] + \mathbf{c}(d(f dg)) - (df, dg).$$

Now observe that

$$\begin{aligned}[\mathsf{D}, c(dg)] &= [\mathsf{D}, [\mathsf{D}, g]] = [\mathsf{D}^2, g] \\ &= -2\nabla^{\mathcal{E}}_X + \Delta g,\end{aligned}$$

since D^2 is a generalized Laplacian associated to the connection $\nabla^{\mathcal{E}}$. We see that

$$[\mathsf{D}, c(f dg)] = -2f\nabla^{\mathcal{E}}_X + \mathbf{c}(d(f dg)) + f(d^* dg) - (df, dg).$$

Using the fact that $d^*(f dg) = f\Delta g - (df, dg)$, the result follows. □

3.4 Index of Dirac Operators

Let $\mathcal{E}$ be a $\mathbb{Z}_2$-graded vector bundle on a compact Riemannian manifold M, and let $\mathsf{D} : \Gamma(M, \mathcal{E}) \to \Gamma(M, \mathcal{E})$ be a self-adjoint Dirac operator. We denote by $\mathsf{D}^\pm$ the restrictions of D to $\Gamma(M, \mathcal{E}^\pm)$, so that

$$\mathsf{D} = \begin{pmatrix} 0 & \mathsf{D}^- \\ \mathsf{D}^+ & 0 \end{pmatrix},$$

where $\mathsf{D}^- = (\mathsf{D}^+)^*$.

An important invariant that can be associated to the operator D is its index. If $E = E^+ \oplus E^-$ is a finite-dimensional superspace, define its dimension to be

$$\dim(E) = \dim(E^+) - \dim(E^-).$$

The superspace $\ker(\mathsf{D})$ is finite dimensional.

Definition 3.46. *The* **index space** *of the self-adjoint Dirac operator* D *is its kernel*

$$\ker(\mathsf{D}) = \ker(\mathsf{D}^+) \oplus \ker(\mathsf{D}^-).$$

The **index** *of* D *is the dimension of the superspace* $\ker(\mathsf{D})$*,*

$$\operatorname{ind}(\mathsf{D}) = \dim(\ker(\mathsf{D}^+)) - \dim(\ker(\mathsf{D}^-)).$$

Lemma 3.47. *If* $u = \begin{pmatrix} 0 & u^- \\ u^+ & 0 \end{pmatrix}$ *is an odd self-adjoint endomorphism of a superspace* $E = E^+ \oplus E^-$*, then* $\operatorname{coker}(u^+) \cong \ker(u^-)^*$*, where* $\operatorname{coker}(u^+) = E^- / \operatorname{im}(u^+)$.

Proof. Since u^- is the adjoint of u^+, we have for all $e \in E^+$ and $f \in E^-$ that

$$\langle u^+ e, f \rangle = \langle e, u^- f \rangle,$$

so that $f \in \ker(u^-)$ if and only if $\langle u^+ e, f \rangle = 0$ for all $e \in E^+$. □

We will now prove an analogous result for self-adjoint Dirac operators. Thus let $\mathscr{E}$ be a $\mathbb{Z}_2$-graded Clifford module on a compact Riemannian manifold M of even dimension, and let D be a self-adjoint Dirac operator on $\Gamma(M, \mathscr{E})$. As above, we denote by $\mathsf{D}^{\pm}$ the restrictions of D to $\Gamma(M, \mathscr{E}^{\pm})$.

Proposition 3.48. *If* D *is a self-adjoint Dirac operator on a Clifford module* $\mathscr{E}$ *over a compact manifold* M*, then*

$$\Gamma(M, \mathscr{E}^{\pm}) = \ker(\mathsf{D}^{\pm}) \oplus \operatorname{im}(\mathsf{D}^{\mp}),$$

and hence

$$\operatorname{ind}(\mathsf{D}) = \dim(\ker(\mathsf{D}^+)) - \dim(\operatorname{coker}(\mathsf{D}^+)).$$

Proof. Since D is self-adjoint, $(\mathsf{D}^2 s, s) = (\mathsf{D}s, \mathsf{D}s)$ and $\ker(\mathsf{D}) = \ker(\mathsf{D}^2)$.

Let G be the Green operator for the square of D; this is the operator which inverts D^2 on the orthogonal complement of $\ker(\mathsf{D})$. Since D^2 is a generalized Laplacian, G preserves $\Gamma(M, \mathscr{E})$ by Proposition 2.38. Let P_0 be the projection operator onto $\ker(\mathsf{D})$. If $s \in \Gamma(M, \mathscr{E})$, then $s = \mathsf{D}(\mathsf{D}Gs) + P_0 s$ is a decomposition of s into a section $\mathsf{D}(\mathsf{D}Gs)$ in the image of the D and a section in the kernel of D. This shows that

$$\Gamma(M, \mathscr{E}^{\pm}) = \operatorname{im}(\mathsf{D}^{\mp}) \oplus \ker(\mathsf{D}^{\pm}).$$ □

From this proposition, we see that the index $\operatorname{ind}(\mathsf{D})$ may be identified with the dimension of the superspace $\ker(\mathsf{D}^+) \oplus \operatorname{coker}(\mathsf{D}^+)$ if D is self-adjoint. This is the usual definition of the index of D^+. However, we prefer to phrase the definition of the index in a more symmetric form, involving both D^+ and D^-, so we adopt our first definition of the index. It is also for this reason that we write $\operatorname{ind}(\mathsf{D})$ and not the more usual notation $\operatorname{ind}(\mathsf{D}^+)$.

The index of D is a topological invariant of the manifold M and the Clifford bundle $\mathscr{E}$: that is, if D^z is a one-parameter family of operators on $\Gamma(M,\mathscr{E})$ which are Dirac operators with respect to a family of metrics g^z on M and Clifford actions c^z of $C(M,g^z)$ on $\mathscr{E}$, then the index of D^z is independent of z. We will prove this by means of a beautiful formula for $\operatorname{ind}(\mathsf{D})$ in terms of the heat kernel $e^{-t\mathsf{D}^2}$ called the McKean-Singer formula.

If A is a trace-class operator on a $\mathbb{Z}_2$-graded Hilbert space $\mathscr{H}^\pm$, then its supertrace is defined as in the finite dimensional case by the formula

$$\operatorname{Str}(A) = \begin{cases} \operatorname{Tr}_{\mathscr{H}^+}(A) - \operatorname{Tr}_{\mathscr{H}^-}(A) & \text{if } A \text{ is even,} \\ 0 & \text{if } A \text{ is odd.} \end{cases}$$

Consider the Hilbert spaces $\mathscr{H}^\pm$ of L^2-sections of $\mathscr{E}^\pm$. The following lemma is the analogue for the supertrace of Proposition 2.45.

Lemma 3.49. *If D is a differential operator on $\mathscr{E}$ and if K has a smooth kernel, then* $\operatorname{Str}[D,K] = 0$.

Proof. If K and D are both even, this follows immediately from Lemma 2.45. On the other hand, if D and K are both odd, that is, $D = \begin{pmatrix} 0 & D^- \\ D^+ & 0 \end{pmatrix}$ and $K = \begin{pmatrix} 0 & K^- \\ K^+ & 0 \end{pmatrix}$, then $\operatorname{Str}(DK+KD) = \operatorname{Tr}[K^-,D^+] + \operatorname{Tr}[D^-,K^+] = 0$, again by Proposition 2.45. □

The supertrace of the projection onto the kernel of D is clearly equal to the index of D:

$$\operatorname{Str}(P_0) = \dim(\ker(\mathsf{D}^+)) - \dim(\ker(\mathsf{D}^-)) = \operatorname{ind}(\mathsf{D}).$$

The McKean-Singer formula generalizes this formula, and will allow us to use the kernel of $e^{-t\mathsf{D}^2}$ to calculate the index of D.

Theorem 3.50 (McKean-Singer). *Let $\langle x \mid e^{-t\mathsf{D}^2} \mid y\rangle$ be the heat kernel of the operator D^2. Then for any $t > 0$*

$$\operatorname{ind}(\mathsf{D}) = \operatorname{Str}(e^{-t\mathsf{D}^2}) = \int_M \operatorname{Str}(\langle x \mid e^{-t\mathsf{D}^2} \mid x\rangle)\, dx.$$

Proof. We give two proofs of this formula. The first one uses the spectral theorem for the essentially self-adjoint operator D^2. If λ is a real number, let $n_\lambda^\pm$ be the dimension of the λ-eigenspace $\mathscr{H}_\lambda^\pm$ of the generalized Laplacian D^2, acting on $\Gamma(M,\mathscr{E}^\pm)$. We have the following formula for the supertrace of $e^{-t\mathsf{D}^2}$:

$$\operatorname{Str}\left(e^{-t\mathsf{D}^2}\right) = \sum_{\lambda \geq 0} (n_\lambda^+ - n_\lambda^-) e^{-t\lambda}.$$

Since the Dirac operator D commutes with D^2, it interchanges $\mathscr{H}_\lambda^+$ and $\mathscr{H}_\lambda^-$, and moreover induces an isomorphism between them if $\lambda \neq 0$, since in that case, the composition

$$\mathscr{H}_\lambda^+ \xrightarrow{\mathsf{D}^+} \mathscr{H}_\lambda^- \xrightarrow{\lambda^{-1}\mathsf{D}^-} \mathscr{H}_\lambda^+$$

is equal to the identity. Thus, $n_\lambda^+ - n_\lambda^- = 0$ for $\lambda > 0$, and we are left with $n_0^+ - n_0^-$, which is of course nothing but the index of D.

The second proof is more algebraic; we give it because it is a model for the proof of more general versions of the McKean-Singer formula. Let $a(t)$ be the function $\operatorname{Str}(e^{-t\mathsf{D}^2})$. If $P_1 = 1 - P_0$ is the projection onto the orthogonal complement of $\ker(\mathsf{D})$, then by Lemma 2.37, for t large,

$$\left|\operatorname{Str}\big(e^{-t\mathsf{D}^2} - P_0\big)\right| = \left|\int_M \operatorname{Str}(\langle x \mid P_1 e^{-t\mathsf{D}^2} P_1 \mid x\rangle)\, dx\right| \leq C \operatorname{vol}(M) e^{-t\lambda_1/2},$$

where λ_1 is the smallest non-zero eigenvalue of D^2. This shows that

$$a(\infty) = \lim_{t\to\infty} \operatorname{Str}(e^{-t\mathsf{D}^2}) = \operatorname{ind}(\mathsf{D}).$$

The proof is completed by showing that the function $a(t)$ is independent of t, so that $a(t) = a(\infty) = \operatorname{ind}(\mathsf{D})$. We do this by differentiating with respect to t, which is legitimate because the heat equation tells us that the operator $d\big(e^{-t\mathsf{D}^2}\big)/dt$ has a smooth kernel equal to $-\mathsf{D}^2 e^{-t\mathsf{D}^2}$, and is thus trace-class for $t > 0$. We see that

$$\begin{aligned}\frac{d}{dt} a(t) &= -\operatorname{Str}\big(\mathsf{D}^2 e^{-t\mathsf{D}^2}\big)\\ &= -\tfrac{1}{2} \operatorname{Str}\big([\mathsf{D}, \mathsf{D} e^{-t\mathsf{D}^2}]\big),\end{aligned}$$

since D is odd, and this vanishes by Lemma 3.49. □

With the McKean-Singer formula in hand, it is easy to see that the index of a smooth one-parameter family of Dirac operators $(\mathsf{D}^z \mid z \in \mathbb{R})$ is a smooth function of z, since the heat kernel of $(\mathsf{D}^z)^2$ is. However, the index is an integer, so it follows that it is independent of s. Even without using the fact that the index is an integer, we can show explicitly that the variation of the supertrace of $e^{-t\mathsf{D}^2}$ vanishes, by means of the following result.

Theorem 3.51. *The index is an invariant of the manifold M and Clifford module* $\mathscr{E}$.

Proof. Given a one-parameter family of Dirac operators D^z on $\mathscr{E}$, we see that $(\mathsf{D}^z)^2$ is a one-parameter family of Laplacians. Applying Lemma 2.50 to the McKean-Singer formula, it follows that

$$\begin{aligned}\frac{d}{dz} \operatorname{ind}(\mathsf{D}^z) &= \frac{d}{dz} \operatorname{Str}\big(e^{-t(\mathsf{D}^z)^2}\big)\\ &= -t \operatorname{Str}\left(\left[\frac{d\mathsf{D}^z}{dz}, \mathsf{D}^z e^{-t(\mathsf{D}^z)^2}\right]\right).\end{aligned}$$

Once again, this last quantity vanishes by Lemma 3.49. □

3.5 The Lichnerowicz Formula

If $\mathbb{A}$ is a Clifford connection, there is an explicit formula for $\mathsf{D}_{\mathbb{A}}^2$ which is of great importance in differential geometry, as we will see when we describe examples of Clifford modules in the next section.

Theorem 3.52 (Lichnerowicz formula). *Let $\mathbb{A}$ be a Clifford connection on the Clifford module $\mathscr{E}$. Denote the Laplacian with respect to the connection $\mathbb{A}$ by $\Delta^{\mathbb{A}}$. If r_M is the scalar curvature of M, then*

$$\mathsf{D}_{\mathbb{A}}^2 = \Delta^{\mathbb{A}} + \mathbf{c}(F^{\mathscr{E}/S}) + \frac{r_M}{4},$$

where $F^{\mathscr{E}/S} \in \mathscr{A}^2(M, \mathrm{End}_{C(M)}(\mathscr{E}))$ is the twisting curvature of the Clifford connection $\mathbb{A}$ defined in Proposition 3.43, and

$$\mathbf{c}(F^{\mathscr{E}/S}) = \sum_{i<j} F^{\mathscr{E}/S}(e_i, e_j) c(e^i) c(e^j).$$

Proof. Choose a covering of the manifold by coordinate patches. In a coordinate system, we may write the Dirac operator as $\mathsf{D}_{\mathbb{A}} = \sum c(dx^i)\mathbb{A}_i$, where $\mathbb{A}_i$ is the covariant differentiation in the direction $\partial/\partial x_i$. Squaring this formula gives

$$\begin{aligned}\mathsf{D}^2 &= \tfrac{1}{2}\sum_{ij}[c(dx^i)\mathbb{A}_i, c(dx^j)\mathbb{A}_j] \\ &= \tfrac{1}{2}\sum_{ij}[c(dx^i), c(dx^j)]\mathbb{A}_i\mathbb{A}_j + \sum_{ij} c(dx^i)[\mathbb{A}_i, c(dx^j)]\mathbb{A}_j \\ &\quad + \tfrac{1}{2}\sum_{ij} c(dx^i)c(dx^j)[\mathbb{A}_i, \mathbb{A}_j] \\ &= -\sum_{ij} g^{ij}(\mathbb{A}_i\mathbb{A}_j + \sum_k \Gamma_{ij}^k \mathbb{A}_k) + \sum_{i<j} c(dx^i)c(dx^j)[\mathbb{A}_i, \mathbb{A}_j].\end{aligned}$$

In this calculation, ∇ is the Levi-Civita connection on M, and Γ_{ij}^k are its coefficients with respect to the frame ∂_i, which satisfy $\nabla_i dx^j = -\Gamma_{ik}^j dx^k$. Here, we have used the compatibility of the connection $\mathbb{A}$ with the Clifford action, so that

$$[\mathbb{A}_i, c(dx^j)] = c(\nabla_i dx^j) = -\Gamma_{ik}^j c(dx^k),$$

and the symmetry $\Gamma_{ik}^j = \Gamma_{ki}^j$ which follows from $\nabla_i \partial_k = \nabla_k \partial_i$, so that

$$\sum_{ik} \Gamma_{ik}^j c(dx^i)c(dx^k) = \tfrac{1}{2}\sum_{ik} \Gamma_{ik}^j [c(dx^i), c(dx^k)].$$

Let $\mathbb{A}^2$ be the curvature tensor of the connection $\mathbb{A}$, and let e_i, $(1 \le i \le n)$, be an orthonormal frame of TM. The above calculation gives the formula

$$\begin{aligned}\mathsf{D}_{\mathbb{A}}^2 &= \Delta^{\mathbb{A}} + \sum_{ij} c(dx^i)c(dx^j)[\mathbb{A}_i, \mathbb{A}_j] \\ &= \Delta^{\mathbb{A}} + \sum_{i<j} c(e^i)c(e^j)\mathbb{A}^2(e_i, e_j).\end{aligned} \tag{3.14}$$

By Proposition 3.43, we see that

$$\sum_{i<j} c(e^i)c(e^j)\mathbb{A}^2(e_i,e_j) = -\tfrac{1}{8}\sum_{ijkl} R_{klij}c(e^i)c(e^j)c(e^k)c(e^l) + \mathbf{c}(F^{\mathscr{E}/S}),$$

where R_{ijkl} are the components of the Riemannian curvature of M.

Observe that $c(e^i)c(e^j)c(e^k)$ equals

$$\tfrac{1}{6}\sum_{\sigma\in S_3} \operatorname{sgn}(\sigma)c(e^{\sigma(i)})c(e^{\sigma(j)})c(e^{\sigma(k)}) - \delta^{ij}c(e^k) - \delta^{jk}c(e^i) + \delta^{ki}c(e^j).$$

Using the fact that the antisymmetrization of R_{klij} over ijk vanishes, we obtain

$$\sum_{ijkl} R_{klij}c(e^i)c(e^j)c(e^k)c(e^l) = -\sum_{ijl} c(e^i)c(e^l)R_{jlij} + \sum_{ijl} c(e^j)c(e^l)R_{ilij}.$$

We see that

$$\sum_{ijkl} R_{klij}c(e^i)c(e^j)c(e^k)c(e^l) = 2\sum_{ij} c(e^j)c(e^i)\sum_k R_{ikjk}.$$

But $\sum_{ij} c(e^j)c(e^i)R_{ikjk} = -\sum_i R_{ikik}$, since $R_{ikjk} = R_{jkik}$. Hence, the right-hand side equals $-2r_M$, and the Lichnerowicz formula follows. □

3.6 Some Examples of Clifford Modules

In this section, we will show that the classical linear first-order differential operators of differential geometry are Dirac operators with respect to very natural Clifford modules and Clifford connections. By **classical**, we mean an operator considered by Atiyah and Singer [13]. Our exposition follows closely that of Atiyah and Bott [6].

The operators that we describe here, rather briefly, are those for which the index theorem of the next chapter might be considered to be classical; both the Gauss-Bonnet-Chern theorem and the Riemann-Roch-Hirzebruch theorem turn out to be special cases, in which we take as our Clifford module respectively the operator $d+d^*$ on $\mathscr{A}(M)$, and the operator $\bar\partial+\bar\partial^*$ on $\mathscr{A}^{0,\bullet}(M,\mathscr{E})$, where M is a Kähler manifold and $\mathscr{E}$ is a holomorphic Hermitian bundle.

The De Rham Operator

Our first example is related to the de Rham operator

$$0 \to \mathscr{A}^0(M) \xrightarrow{d_0} \mathscr{A}^1(M) \xrightarrow{d_1} \mathscr{A}^2(M) \xrightarrow{d_2}$$

On a Riemannian manifold, the bundle ΛT^*M is a Clifford module by the action of (3.1): if $\alpha\in\Gamma(M,T^*M)$ and $\beta\in\mathscr{A}(M)$, then $c(\alpha)\beta = \varepsilon(\alpha)\beta - \iota(\alpha)\beta$. Clearly, the Levi-Civita connection on the bundle ΛT^*M is compatible with this Clifford action.

Proposition 3.53. *The Dirac operator associated to the Clifford module ΛT^*M and its Levi-Civita connection is the operator $d+d^*$, where*

$$d^* : \mathscr{A}^\bullet(M) \to \mathscr{A}^{\bullet-1}(M)$$

is the adjoint of the exterior differential d.

Proof. Since ∇ is torsion-free, Proposition 1.22 shows that d equals $\varepsilon \circ \nabla$, where ∇ is the Levi-Civita connection. On the other hand, Proposition 2.8 shows that $d^* = -\iota \circ \nabla$. It follows that $d+d^* = (\varepsilon - \iota) \circ \nabla = c \circ \nabla$. □

The square $dd^* + d^*d$ of the operator $d+d^*$ is called the Laplace-Beltrami operator. Recall the statement of de Rham's theorem: the cohomology

$$H^i(\mathscr{A}(M), d) = \ker(d_i)/\operatorname{im}(d_{i-1})$$

of the de Rham complex is isomorphic to the singular cohomology $H^i(M,\mathbb{R})$. A differential form in the kernel of $dd^* + d^*d$ is called a **harmonic form.**

It is useful to consider a more general situation than the above, in which $\mathscr{E}^\bullet \to M$ is a $\mathbb{Z}$-graded vector bundle (with $\mathscr{E}^i$ non-zero for only a finite number of values of i) with first-order differential operator $d : \Gamma(M, \mathscr{E}^\bullet) \to \Gamma(M, \mathscr{E}^{\bullet+1})$, such that $d^2 = 0$ and $\Delta = dd^* + d^*d$ is a generalized Laplacian. We will write $H^i(d)$ for the cohomology of this complex,

$$H^i(d) = \ker(d_i)/\operatorname{im}(d_{i-1}).$$

In this situation, the operator $d+d^*$ is a Dirac operator, since $(d+d^*)^2 = dd^* + d^*d$ is a generalized Laplacian; the $\mathbb{Z}_2$-grading on the bundle $\sum_i \mathscr{E}^i$ is obtained by taking the $\mathbb{Z}$-degree modulo two. It is clear from the following equation that $\ker(\Delta) = \ker(d+d^*)^2 = \ker(d) \cap \ker(d^*)$:

$$\int_M (\alpha, (dd^* + d^*d)\alpha) = \int_M (d\alpha, d\alpha) + (d^*\alpha, d^*\alpha).$$

Theorem 3.54 (Hodge). *If $(\mathscr{E}, d)$ is as above, then the kernel $\ker(\Delta_i)$ of the operator $\Delta_i = d_{i+1}^* d_i + d_{i-1} d_i^*$ is finite-dimensional in each dimension, and naturally isomorphic to $H^i(d)$ by the decomposition $\ker(d_i) = \operatorname{im}(d_{i-1}) \oplus \ker(\Delta_i)$.*

The Dirac operator $d+d^*$ on the bundle $\sum_i \mathscr{E}^i$ has index equal to the Euler number of the complex $(\mathscr{E}, d)$, defined by $\sum_i (-1)^i \dim(H^i(d))$.

Proof. Since the Laplacian Δ has a smooth heat kernel, we see that it is essentially self-adjoint, has a discrete spectrum with finite multiplicities, and smooth eigenfunctions. Thus, the space $\ker(\Delta_i)$ coincides with the zero-eigenspace of the closure of $d_{i+1}^* d_i + d_{i-1} d_i^*$ in the Hilbert space $\Gamma_{L^2}(M, \mathscr{E}^i)$ of square-integrable sections of $\mathscr{E}^i$.

On the orthogonal complement to $\ker(\Delta)$, we can invert the generalized Laplacian Δ to obtain its Green operator G, which we extend by zero on $\ker(\Delta)$; moreover, if $\alpha \in \Gamma(M, \mathscr{E}^i)$ is smooth, then $G\alpha \in \Gamma(M, \mathscr{E}^i)$ is also smooth (Corollary 2.38).

Observe that Δ commutes with d and d^* as does G. The operator d^*G on $\Gamma(M,\mathscr{E})$ satisfies

$$[d,d^*G] = (dd^* + d^*d)G = 1 - P_0,$$

where P_0 is the orthogonal projection onto $\ker(\Delta)$. Thus, d^*G is what is known as a homotopy between the identity operator on $\Gamma_{L^2}(M,\mathscr{E}^\bullet)$ and P_0, and gives us an explicit decomposition $\ker(d_i) = \operatorname{im}(d_{i-1}) \oplus \ker(\Delta_i)$.

Thus, we may identify the kernel of the Dirac operator $d+d^*$ with $H^\bullet(d)$. The dimension of the kernel of $d+d^*$ on the space of even-degree sections $\sum_i \Gamma(M,\mathscr{E}^{2i})$ is equal to $\sum_i \dim(H^{2i}(d))$, while the dimension of its kernel on $\sum_i \Gamma(M,\mathscr{E}^{2i+1})$ is equal to $\sum_i \dim(H^{2i+1}(d))$. Thus, the index of $d+d^*$ equals

$$\sum_i \big(\dim(H^{2i}(d)) - \dim(H^{2i+1}(d))\big),$$

which equals the Euler number of $(\mathscr{E},d)$ by definition. □

Corollary 3.55. *The kernel of the Laplace-Beltrami operator $dd^* + d^*d$ on $\mathscr{A}^i(M)$ is naturally isomorphic to the de Rham cohomology space $H^i(M,\mathbb{R})$. The index of the Dirac operator $d+d^*$ on $\mathscr{A}(M)$ is equal to the* **Euler number**

$$\operatorname{Eul}(M) = \sum_{i=1}^{n} (-1)^i \dim(H^i(M,\mathbb{R}))$$

of the manifold M.

The Euler number defined using the de Rham cohomology is equal to the Euler number defined in Theorem 1.56 by means of a vector field with non-degenerate zeroes; this may be shown using Morse theory, but would take us too far afield to discuss here.

Corollary 3.56 (Künneth). *If M and N are two compact manifolds,*

$$H^\bullet(M\times N) \cong H^\bullet(M) \otimes H^\bullet(N).$$

Proof. Choose Riemannian metrics on M and N, with Laplace-Beltrami operators Δ_M and Δ_N. The Laplace-Beltrami $\Delta_{M\times N}$ on $M\times N$ with the product metric is $\Delta_M \otimes 1 + 1 \otimes \Delta_N$. Using the spectral decomposition of the positive operators Δ_M and Δ_N, we see that the zero eigenspace of $\Delta_{M\times N}$ is the tensor product of the zero eigenspaces of Δ_M and Δ_N. □

The Lichnerowicz formula for the square of a Dirac operator, when applied to the operator $d+d^*$, gives a formula for the Laplace-Beltrami operator which is known as Weitzenböck's formula. Let us denote by $\Delta^{\Lambda T^*M}$ the Laplacian of the exterior bundle ΛT^*M. The bundle ΛT^*M is the bundle associated to the principal bundle $\mathrm{GL}(M)$ by the representation $\lambda : \mathrm{GL}(n) \to \Lambda(\mathbb{R}^n)^*$ of (1.26), so that the curvature of the Levi-Civita connection on the bundle ΛT^*M is given by the formula

$$\lambda(R) = \sum_{k<l} e^k \wedge e^l \sum_{ij} R_{ijkl} \varepsilon(e^i) \iota(e^j). \tag{3.15}$$

Hence by (3.14),

$$(d+d^*)^2 = \Delta^{\Lambda T^*M} + \tfrac{1}{2} \sum_{ijkl} R_{ijkl} (\varepsilon^k - \iota^k)(\varepsilon^l - \iota^l) \varepsilon^i \iota^j.$$

Using the fact that the antisymmetrization of R_{ijkl} over three indices vanishes, we see that

$$\sum \varepsilon^i \varepsilon^j \varepsilon^k \iota^l R_{ijkl} = 0,$$
$$\sum \iota^i \iota^j \varepsilon^k \iota^l R_{ijkl} = 0.$$

Retaining only the non-zero terms, we obtain **Weitzenböck's formula**: the decomposition of the Laplace-Beltami operator as a generalized Laplacian is

$$(d+d^*)^2 = \Delta^{\Lambda T^*M} - \sum_{ijkl} R_{ijkl} \varepsilon^k \iota^l \varepsilon^i \iota^j. \tag{3.16}$$

Observe that on the space of one-forms, the Weitzenböck formula becomes

$$(d+d^*)^2 = \Delta^{\Lambda T^*M} + \mathrm{Ric},$$

where the Ricci curvature is identified with a section of the bundle $\mathrm{End}(T^*M)$ by means of the Riemannian metric on M. This follows from the following formula,

$$\sum R_{ijkl} \varepsilon^i \iota^j \varepsilon^k \iota^l = -\sum R_{ijkl} \varepsilon^i \varepsilon^k \iota^j \iota^l - \sum \mathrm{Ric}_{ij} \varepsilon^i \iota^j,$$

where the first term on the right-hand side vanishes on one-forms. In particular, if the manifold M has Ricci curvature uniformly bounded below by a positive constant c, then the Laplace-Beltrami operator is bounded below by c on one-forms, since the operator $\Delta^{\Lambda T^*M}$ itself positive. By Hodge's Theorem, this implies that the first de Rham cohomology space of the manifold vanishes; the Hurewicz theorem then tells us that the manifold has finite fundamental group. This result is a simple example of the so-called **vanishing theorems** that follow from lower bounds for the square of the Dirac operator, and which are proved by combining assumptions on curvature and the Lichnerowicz theorem.

There is a generalization of the de Rham complex, in which we twist the bundle of differential forms by a flat bundle $\mathscr{W}$, with covariant derivative ∇:

$$0 \to \mathscr{A}^0(M,\mathscr{W}) \xrightarrow{\nabla_0} \mathscr{A}^1(M,\mathscr{W}) \xrightarrow{\nabla_1} \mathscr{A}^2(M,\mathscr{W}) \xrightarrow{\nabla_2}$$

This sequence is a complex precisely when $\nabla^2 = 0$, that is, when the bundle $\mathscr{W}$ is flat. There is a simple way to characterize flat bundles, as an associated bundle $\widetilde{M} \times_{\pi_1(M)} W$, where W is representation of $\pi_1(M)$. Here, $\widetilde{M}$ is the covering space of M, which is a principal bundle with structure group $\pi_1(M)$ the fundamental group of M, acting on M by means of the so-called deck transformations.

The Signature Operator

There is a variant of the first example, in which the Dirac operator is the same but the definition of the $\mathbb{Z}_2$-grading on the Clifford module ΛT^*M is changed. In order to define this new grading, we use the Hodge star operator, which is defined on an oriented Riemannian manifold.

Definition 3.57. *If V is an oriented Euclidean vector space with complexification $V_{\mathbb{C}}$, the* **Hodge star** *operator $\star$ on $\Lambda V_{\mathbb{C}}$ equals the action of the chirality element $\Gamma \in C(V)\otimes\mathbb{C}$, defined in Proposition 3.17, on the Clifford module $\Lambda V_{\mathbb{C}}$.*

This definition differs from the usual one by a power of i in order that $\star^2 = 1$.

Applying this operator to each fibre of the complexified exterior bundle $\Lambda_{\mathbb{C}}T^*M = \Lambda T^*M \otimes_{\mathbb{R}} \mathbb{C}$ of an oriented n-dimensional Riemannian manifold M, we obtain the Hodge star operator

$$\star : \Lambda^k_{\mathbb{C}}T^*M \to \Lambda^{n-k}_{\mathbb{C}}T^*M.$$

Proposition 3.58. *1. If $\alpha \in \mathscr{A}^1(M,\mathbb{C})$, then $\star^{-1}\varepsilon(\alpha)\star = (-1)^n\iota(\alpha)$.*
2. If α and β are k-forms on M, then

$$\int_M \alpha\wedge\star\beta = (-1)^{kn+k(k-1)/2}\, i^p \int_M (\alpha,\beta)\,dx,$$

where dx is the Riemannian volume form on M corresponding to the chosen orientation, and $p = n/2$ for even and $(n+1)/2$ for n odd.
3. The operator d^* equals $(-1)^{n+1}\star^{-1}d\star$ on the space of k-forms.
In these formulas, we can replace $\star^{-1}$ by $\star$, since $\star^2 = 1$.

Proof. (1) If e_i is an orthonormal frame of V, we may assume that $\alpha = e_n$, and then we see that

$$\varepsilon(e_n)c(e_1)\dots c(e_n) = (-1)^{n-1}c(e_1)\dots c(e_{n-1})\varepsilon(e_n)c(e_n).$$

But for any $v \in V$, we have $\varepsilon(v)c(v) = \varepsilon(v)(\varepsilon(v) - \iota(v)) = -(\varepsilon(v) - \iota(v))\iota(v)$, from which the formula follows.

(2) Let $\alpha,\beta \in \Lambda^k V$. We will prove by induction on k that

$$\alpha\wedge\star\beta = (-1)^{kn+k(k-1)/2} i^p (\alpha,\beta) e_1\wedge\dots\wedge e_n.$$

For $k = 0$, $\beta \in \Lambda^0 V = \mathbb{C}$, we have by definition

$$\star\beta = i^p \beta e_1\wedge\dots\wedge e_n.$$

If $\alpha \in \Lambda^k V$, $\beta \in \Lambda^{k+1}V$ and $e \in V$, we have

$$\begin{aligned}
\alpha\wedge e\wedge(\star\beta) &= (-1)^n \alpha\wedge\star\iota(e)\beta \\
&= (-1)^{n+kn+k(k-1)/2} i^p (\alpha,\iota(e)\beta) e_1\wedge\dots\wedge e_n \\
&= (-1)^{n+kn+k(k-1)/2} i^p (\varepsilon(e)\alpha,\beta) e_1\wedge\dots\wedge e_n \\
&= (-1)^{n+kn+k(k-1)/2} (-1)^k i^p (\alpha\wedge e,\beta),
\end{aligned}$$

which is the desired formula for $k+1$.

(3) To prove this, we simply insert (2) in the formula which defines d^*, and apply Stokes' formula: if $\alpha \in \mathscr{A}^k(M)$ and $\beta \in \mathscr{A}^{k+1}(M)$, then

$$\begin{aligned}\int_M (\alpha, d^*\beta)\,dx = \int_M (d\alpha,\beta)\,dx &= (-1)^{(k+1)n+k(k+1)/2} i^{-p} \int_M d\alpha \wedge \star\beta \\ &= -(-1)^{k+(k+1)n+k(k+1)/2} i^{-p} \int_M \alpha \wedge d\star\beta \\ &= -(-1)^n \int_M (\alpha, \star d \star \beta)\,dx. \qquad \square\end{aligned}$$

Corollary 3.59 (Poincare duality). *Let M be an oriented compact n-dimensional manifold. The bilinear form $\int_M \alpha\wedge\beta$ induces a non-degenerate pairing*

$$H^k(M) \times H^{n-k}(M) \to \mathbb{R}.$$

Proof. Let $\alpha \in \mathscr{A}^k(M)$ be a harmonic form representing an element of $H^k(M)$; thus, $\star\alpha$ is closed. If $\int_M \alpha\wedge\beta = 0$ for all $\beta \in H^{n-k}(M)$, then in particular $\int_M \alpha \wedge (\star\alpha) = 0$. It follows that $\int_M |\alpha|^2\,dx = 0$, and hence that $\alpha = 0$. □

Suppose that $n = \dim(M)$ is even. Then the operator $\star$ anticommutes with the Dirac operator $d + d^* = d - \star d \star$, hence with the Laplace-Beltrami operator. Since $\star^2 = 1$ and $\star$ anticommutes with the action of $c(v) = \varepsilon(v) - \iota(v)$, we can define another $\mathbb{Z}_2$-grading on the exterior bundle $\Lambda T^*M \otimes \mathbb{C}$ using the operator $\star$; the differential forms satisfying $\star\alpha = +\alpha$ (respectively $\star\alpha = -\alpha$) are called **self-dual** (respectively anti-self-dual). If the dimension of M is divisible by four, Γ is actually an element of the real Clifford algebra, so $\star$ operates on the space of real valued forms.

Note as a special case of Proposition 3.58 that if $\dim(M) = n$ is even and α and $\beta \in \mathscr{A}^{n/2}(M)$, then

$$\int_M \alpha\wedge\beta = \varepsilon_p \int_M (\alpha, \star\beta)\,dx,$$

where $\varepsilon_p = 1$ if $n = 4l$ and $\varepsilon_p = i$ if $n = 4l+2$

Let us recall the definition of the signature of a quadratic form.

Definition 3.60. *Let Q be a quadratic form on a real vector space V. Choose a basis in which*

$$Q(x) = x_1^2 + \ldots + x_p^2 - x_{p+1}^2 - \cdots - x_{p+q}^2.$$

The **signature** *$\sigma(Q)$ is the number $p - q$, which depends only on the quadratic form Q.*

If M is a manifold of even dimension n, the bilinear form on $H^{n/2}(M,\mathbb{R})$ defined by $(\alpha,\beta) \mapsto \int_M \alpha\wedge\beta$ satisfies

$$\int_M \alpha\wedge\beta = (-1)^{n/2} \int_M \beta\wedge\alpha$$

Thus, if the dimension of M is divisible by four, this quadratic form is symmetric. The signature of M is by definition the signature of the restriction of this quadratic form to the space $H^{n/2}(M,\mathbb{R})$, and we denote it by $\sigma(M)$.

If M is an oriented compact Riemannian manifold of dimension divisible by four, the **signature operator** of M is the Dirac operator $d+d^*$ for the Clifford module ΛT^*M with $\mathbb{Z}_2$-grading coming from the Hodge star $\star$.

Proposition 3.61. *The index of the signature operator $d+d^*$ equals the signature $\sigma(M)$, so by the McKean-Singer formula,*

$$\sigma(M) = \operatorname{Tr}(\star e^{-t\Delta}),$$

where Δ is the Laplace-Beltrami operator.

Proof. Let $\alpha_i \in \mathscr{A}^k(M)$ be a basis for the harmonic forms $\mathscr{H}^k$, where $k < n/2$. Note that the differential forms $\alpha_i^\pm = \alpha_i \pm \star\alpha_i$ form a basis for the direct sum of the spaces of harmonic forms $\mathscr{H}^k \oplus \mathscr{H}^{n-k}$ since $\star$ maps the space of harmonic forms into itself. Now, α_i^+ is self-dual and α_i^- is anti-self-dual, thus the index of the Dirac operator $d+d^*$ restricted to $\mathscr{A}^k(M) \oplus \mathscr{A}^{n-k}(M)$ is zero.

Now consider the harmonic forms of degree $n/2$. We may find a basis α_i such that $\star\alpha_i = \alpha_i$ for $i \le m$ and $\star\alpha_i = -\alpha_i$ for $i > m$; it is clear that the index of $d+d^*$ with respect to the $\mathbb{Z}_2$-grading defined by $\star$ is $2m-n$. Since n is divisible by four, the signature form may be rewritten as follows: if $\alpha \in \mathscr{H}^{n/2}$ satisfies $\star\alpha = \pm\alpha$, then

$$\int_M \alpha \wedge \alpha = \int_M (\alpha, \star\alpha)\, dx = \pm \int_M |\alpha|^2\, dx;$$

from this we see that the signature of the manifold is equal to $2m-n$ as well. □

If $\mathscr{W}$ is a $\mathbb{Z}_2$-graded vector bundle on M with connection, we may twist the Dirac operator $d+d^*$ by $\mathscr{W}$ to obtain a twisted signature operator on the tensor product $\Lambda T^*M \otimes \mathscr{W}$, whose index is denoted $\sigma(\mathscr{W})$. It is clear from the homotopy invariance of the index that the signature $\sigma(\mathscr{W})$ is an invariant which depends solely on the manifold M, its orientation, and the vector bundle $\mathscr{W}$.

The Dirac Operator on a Spin Manifold

Let $\mathscr{S}$ be the spinor bundle over an even-dimensional spin manifold M. The most basic example of a Dirac operator is the Dirac operator on $\mathscr{S}$ associated to the Levi-Civita connection $\nabla^{\mathscr{S}}$ on $\mathscr{S}$; this operator is often referred to as **the** Dirac operator on M. A section of the spinor bundle $\mathscr{S}$ lying in the kernel of D is known as a **harmonic spinor**. More generally, we may consider the Dirac operator $\mathsf{D}_{\mathscr{W}\otimes\mathscr{S}}$ on a twisted spinor bundle $\mathscr{W} \otimes \mathscr{S}$ with respect to a Clifford connection of the form $\nabla^{\mathscr{W}\otimes\mathscr{S}} = \nabla^{\mathscr{W}} \otimes 1 + 1 \otimes \nabla^{\mathscr{S}}$. If $F^{\mathscr{W}}$ is the curvature of $\nabla^{\mathscr{W}}$, then the twisting curvature $F^{\mathscr{E}/S}$ of $\nabla^{\mathscr{W}\otimes\mathscr{S}}$ equals $F^{\mathscr{W}}$, and the Lichnerowicz formula becomes

$$\mathsf{D}^2_{\mathscr{W}\otimes\mathscr{S}} = \Delta^{\mathscr{W}\otimes\mathscr{S}} + \sum_{i<j} F^{\mathscr{W}}(e_i, e_j) c(e^i) c(e^j) + \frac{r_M}{4}.$$

In particular, for the spinor bundle $\mathscr{S}$ itself, we have

$$\mathsf{D}^2 = \Delta^{\mathscr{S}} + \frac{r_M}{4}.$$

Lichnerowicz used this formula to prove the following result.

Proposition 3.62. *If M is a compact spin manifold whose scalar curvature is non-negative, and strictly positive at at least one point, then the kernel of the Dirac operator on the spinor bundle $\mathcal{S}$ vanishes; in particular, its index is 0.*

Proof. If $s \in \Gamma(M, \mathcal{S})$, then

$$\int_M (\mathsf{D}^2 s, s)\, dx = \int_M \|\nabla s\|^2\, dx + \frac{1}{4} \int_M r_M \|s\|^2\, dx.$$

Thus, if $\mathsf{D}s = 0$, we obtain $\nabla s = 0$ and $r_M s = 0$. Then (s,s) is constant on M, hence $s = 0$. □

On a spin manifold M, the signature operator of the last section may be rewritten as a twisted Dirac operator in an explicit way. Indeed, from the isomorphism of Clifford modules

$$\Lambda T^*M \otimes \mathbb{C} \cong C(M) \otimes \mathbb{C} \cong \operatorname{End}(\mathcal{S}) \cong \mathcal{S}^* \otimes \mathcal{S},$$

we see that the Dirac operator $d + d^*$ on ΛT^*M is the Dirac operator obtained by twisting the Dirac operator on the spinor bundle $\mathcal{S}$ by the bundle $\mathcal{S}^*$. The two different $\mathbb{Z}_2$-gradings on ΛT^*M correspond to the two different ways of grading the twisting bundle $\mathcal{S}^*$: if we treat it as a $\mathbb{Z}_2$-graded bundle, we recover the grading of ΛT^*M according to the parity of the degree, while if we treat $\mathcal{S}^*$ as an ungraded bundle, we recover the grading coming from the Hodge star $\star$.

The $\bar\partial$-Operator on a Kähler Manifold

Our last example of a Dirac operator is important in algebraic geometry.

Definition 3.63. *An* **almost-complex manifold** *M is a manifold of real dimension $2n$ with a reduction of its structure group to the group $\mathrm{GL}(n,\mathbb{C})$; that is, there is given a $\mathrm{GL}(n,\mathbb{C})$-principal bundle P and an isomorphism between the vector bundles $P \times_{\mathrm{GL}(n,\mathbb{C})} \mathbb{C}^n$ and TM.*

An equivalent way of defining an almost-complex structure is to give a section J of $\operatorname{End}(TM)$ satisfying $J^2 = -1$.

On an almost-complex manifold, the complexification of the tangent bundle splits into two pieces, called the holomorphic and anti-holomorphic tangent spaces,

$$TM \otimes_{\mathbb{R}} \mathbb{C} = T^{1,0}M \oplus T^{0,1}M,$$

on which J acts by i and $-i$, respectively. In terms of a local frame X_i, Y_i, $(1 \le i \le n)$, such that $JX_i = Y_i$ and $JY_i = -X_i$, we see that $T^{1,0}M$ is spanned by $W_j = (X_j - iY_j)$, and $T^{0,1}M$ is spanned by $\bar W_j = (X_j + iY_j)$.

It follows from this splitting that if $\mathcal{W}$ is a complex vector bundle on M, the space of differential forms is bigraded,

$$\mathcal{A}(M,\mathcal{W}) = \sum_{0 \le p,q \le n} \mathcal{A}^{p,q}(M,\mathcal{W}) = \Gamma(M, \Lambda^p(T^{1,0}M)^* \otimes \Lambda^q(T^{0,1}M)^* \otimes \mathcal{W}).$$

A **Hermitian structure** on an almost-complex manifold M is a Hermitian metric on the tangent space TM. Underlying the Hermitian structure is a Riemannian structure, and its associated Levi-Civita connection. However, in general the Levi-Civita connection does not preserve $T^{0,1}M$ and $T^{1,0}M$. If M is a Hermitian almost-complex manifold, and $\mathscr{W}$ a vector bundle on M, then the vector bundle of anti-holomorphic differential forms $\Lambda(T^{0,1}M)^* \otimes \mathscr{W}$ is a Clifford module, with the following action: if $f \in \mathscr{A}^1(M)$ decomposes as $f = f^{1,0} + f^{0,1}$ with $f^{1,0} \in \mathscr{A}^{1,0}(M)$ and $f^{0,1} \in \mathscr{A}^{0,1}(M)$, then its Clifford action on $v \in \Gamma(M, \Lambda(T^{0,1}M)^* \otimes \mathscr{W})$ equals

$$c(f)v = \sqrt{2}(\varepsilon(f^{1,0}) - \iota(f^{0,1}))v.$$

Note that if $\mathscr{W}$ is Hermitian, $\varepsilon(f^{1,0}) - \iota(f^{0,1})$ is skew-adjoint, since $f^{0,1} = \overline{f^{1,0}}$, and the Clifford action is self-adjoint.

From now on, we will assume that M is a **complex manifold**, in other words, there is a collection of local charts for M parametrized by an open ball in $\mathbb{C}^n$ for which the transition functions are holomorphic. Let z^i be a set of local holomorphic coordinates. Then $T^{1,0}M$ is spanned by the vectors ∂_{z^i}, while $T^{0,1}M$ is spanned by the vectors $\partial_{\bar{z}^i}$. The elements in $\mathscr{A}^{p,q}(M)$ have the form

$$\sum f_{i_1,\dots,i_p,j_1,\dots,j_q} dz^{i_1} \wedge \cdots \wedge dz^{i_p} \wedge d\bar{z}^{j_1} \wedge \cdots \wedge d\bar{z}^{j_q}$$

On a complex manifold, the exterior differential d splits into two pieces, $d = \partial + \bar{\partial}$, where ∂ increases the degree of a form (p,q) by $(1,0)$ and $\bar{\partial}$ increases it by $(0,1)$, and $\partial^2 = \partial\bar{\partial} + \bar{\partial}\partial = \bar{\partial}^2 = 0$. Both of these operators satisfies Leibniz's rule:

$$\partial(\alpha \wedge \beta) = \partial\alpha \wedge \beta + (-1)^{|\alpha|}\alpha \wedge \partial\beta,$$
$$\bar{\partial}(\alpha \wedge \beta) = \bar{\partial}\alpha \wedge \beta + (-1)^{|\alpha|}\alpha \wedge \bar{\partial}\beta.$$

Now let $\mathscr{W}$ be a complex vector bundle over M with covariant derivative ∇. With respect to the decomposition $T^*M \otimes_{\mathbb{R}} \mathbb{C} = (T^{1,0}M)^* \oplus (T^{0,1}M)^*$, ∇ splits into two pieces $\nabla = \nabla^{1,0} + \nabla^{0,1}$, which satisfy the following Leibniz formulas: if $\alpha \in \mathscr{A}(M)$ and $v \in \mathscr{A}(M, \mathscr{W})$, then

$$\nabla^{1,0}(\alpha \wedge v) = \partial\alpha \wedge v + (-1)^{|\alpha|}\alpha \wedge \nabla^{1,0}v,$$
$$\nabla^{0,1}(\alpha \wedge v) = \bar{\partial}\alpha \wedge v + (-1)^{|\alpha|}\alpha \wedge \nabla^{0,1}v.$$

Definition 3.64. *A* **holomorphic** *vector bundle on a complex manifold M is a complex vector bundle $\mathscr{W}$ which has a chart, that is, a covering of M by open sets U_i and framings $\phi_i : \mathscr{W}_i|U_i \to U_i \times \mathbb{C}^N$, in which the transition functions $\phi_i \circ \phi_j^{-1} : U_i \cap U_j \to$ $\mathrm{GL}(n, \mathbb{C})$ are complex analytic.*

If $\mathscr{W}$ is a holomorphic vector bundle over M, there is a unique operator $\bar{\partial}$: $\mathscr{A}^{p,q}(M, \mathscr{W}) \to \mathscr{A}^{p,q+1}(M, \mathscr{W})$ which in each coordinate patch of a chart with complex analytic transition functions equals

$$\sum_{i=1}^{n} \varepsilon(d\bar{z}^i)\frac{\partial}{\partial\bar{z}^i}.$$

The sheaf $\mathcal{O}(\mathcal{W})$ of holomorphic sections of a holomorphic bundle $\mathcal{W}$ is the kernel of $\bar\partial$ acting on the sheaf $\Gamma(M,\mathcal{W}) = \mathcal{A}^{0,0}(M,\mathcal{W})$ of smooth sections of $\mathcal{W}$. We say that a covariant derivative ∇ on a holomorphic vector bundle is holomorphic if $\nabla^{0,1} = \bar\partial$.

Proposition 3.65. *If $\mathcal{W}$ is a holomorphic vector bundle with a Hermitian metric $h(s,t)$, there exists a unique holomorphic covariant derivative on $\mathcal{W}$ preserving the metric.*

Proof. If ∇ is holomorphic and preserves the Hermitian metric h, then taking the component of $dh(s,t)$ lying in $\mathcal{A}^{0,1}(M)$, we see that

$$\bar\partial h(s,t) = h(\nabla^{0,1}s,t) + h(s,\nabla^{1,0}t).$$

If moreover $\nabla^{0,1} = \bar\partial$, this equation determines $\nabla^{1,0}$. It is then easy to check that $\nabla = \nabla^{1,0} + \nabla^{0,1}$ is a covariant derivative. □

Let ∇ be a holomorphic covariant derivative on $\mathcal{W}$ which preserves a Hermitian metric h. Since the curvature is skew-adjoint as a section of $\Lambda^2 T^*M \otimes \mathrm{End}(\mathcal{W})$, we see that $(\nabla^{1,0})^2 = 0$, so that the curvature of ∇ is the section of $\mathcal{A}^{1,1}(M,\mathrm{End}(\mathcal{W}))$ equal to $[\nabla^{1,0},\nabla^{0,1}]$.

It is interesting to see how the Hermitian metric and holomorphic covariant derivative corresponding to it are related in a chart. If the Hermitian metric on $\mathcal{W}$ is given at a point $z \in M$ by a Hermitian matrix $h(z)$, then the corresponding covariant derivative $\nabla^{\mathcal{W}}$ equals

$$\nabla^{\mathcal{W}} = d + h^{-1}\partial h.$$

Its curvature is the $(1,1)$-form

$$F_h = h^{-1}\bar\partial\partial h - (h^{-1}\bar\partial h) \wedge (h^{-1}\partial h).$$

If $\mathcal{W}$ is a line bundle, h is a positive real function and the curvature is thus

$$F_h = \bar\partial\partial \log(h).$$

If s is a non-vanishing holomorphic section of $\mathcal{W}$ over the chart, we see that

$$F_h = \bar\partial\partial \log(\|s\|^2).$$

If M is a complex manifold with a Hermitian metric h, the real part of h restricted to TM is a Riemannian metric on M, while the imaginary part Ω of h restricted to TM is a two-form on M: if X and Y are vector fields on M, then

$$\Omega(X,Y) = \mathrm{Im}\, h(X,Y) = \mathrm{Im}\, \overline{h(Y,X)} = -\mathrm{Im}\, h(Y,X) = -\Omega(Y,X).$$

If X and Y are vector fields on M, we see that $g(X,Y) = \Omega(JX,Y)$, where J is the almost-complex structure on M.

We call a complex manifold M **Kähler** if Ω is a closed two-form. The following result gives another characterization of Kähler manifolds.

Proposition 3.66. *A complex Hermitian manifold M is Kähler if and only if the bundles $T^{1,0}M$ and $T^{0,1}M$ are preserved by the Levi-Civita connection ∇ of the underlying Riemannian structure, or in other words, if $\nabla J = 0$, where J is the almost-complex structure of M.*

Proof. Using the fact that $(\cdot,\cdot) = \Omega(J\cdot,\cdot)$, we see that

$$\begin{aligned}((\nabla_X J)Y,Z) &= (\nabla_X(JY),Z) + \Omega(\nabla_X Y,Z)\\ &= -X\Omega(Y,Z) + \Omega(\nabla_X Y,Z) + \Omega(Y,\nabla_X Z)\\ &= -(\nabla_X \Omega)(Y,Z).\end{aligned}$$

Since the Levi-Civita connection is torsion free, the right hand side of this equation when antisymmetrized over $\{X,Y,Z\}$ is nothing but $d\Omega(X,Y,Z)$. Thus, if ∇J vanishes, we see that $d\Omega = 0$.

Using the explicit formula for the Levi-Civita connection (1.18), we compute that in a holomorphic coordinate system z^i

$$\begin{aligned}&2(\nabla_{\partial_{z^j}}\partial_{z^k},\partial_{z^l}) = 0,\\ &2(\nabla_{\partial_{\bar z^j}}\partial_{z^k},\partial_{z^l}) = \partial_{z^k}(\partial_{z^l},\partial_{\bar z^j}) - \partial_{z^l}(\partial_{\bar z^j},\partial_{z^k}) = i\,d\Omega(\partial_{z_k},\partial_{z^l},\partial_{\bar z^j}),\\ &2(\nabla_{\partial_{\bar z^j}}\partial_{z^k},\partial_{\bar z^l}) = \partial_{\bar z^j}(\partial_{z^k},\partial_{\bar z^l}) - \partial_{\bar z^l}(\partial_{\bar z^j},\partial_{z^k}) = i\,d\Omega(\partial_{\bar z_j},\partial_{z^k},\partial_{\bar z^l}).\end{aligned}$$

Thus we see that if $d\Omega = 0$, the Levi-Civita covariant derivative preserves sections of $T^{1,0}M$ and induces on the complex vector bundle $T^{1,0}M$ the canonical holomorphic covariant derivative. □

This result has the following beautiful consequence.

Proposition 3.67. *Let $\mathcal{W}$ be a holomorphic vector bundle with Hermitian metric on a Kähler manifold. The tensor product of the Levi-Civita connection with the canonical connection of $\mathcal{W}$ is a Clifford connection on the Clifford module $\Lambda(T^{0,1}M)^* \otimes \mathcal{W}$, with associated Dirac operator $\sqrt{2}(\bar\partial + \bar\partial^*)$.*

Proof. It is clear that the Levi-Civita connection is a Clifford connection; a quick look at the formula for the Clifford action of T^*M on $\Lambda(T^{0,1}M)^*$ shows that what is needed is that the splitting $TM \otimes_{\mathbb{R}} \mathbb{C} = T^{1,0}M \oplus T^{0,1}M$ is preserved by the Levi-Civita connection, in other words, that $\nabla J = 0$.

If Z_i is a local frame of $T^{1,0}M$, with dual frame $Z^i \in (T^{1,0}M)^*$, then

$$d = \sum_i \varepsilon(Z^i)\nabla_{Z_i} + \varepsilon(\bar Z^i)\nabla_{\bar Z_i},$$

from which we see that

$$\bar\partial = \sum_i \varepsilon(\bar Z^i)\nabla_{\bar Z_i}.$$

If we choose a holomorphic trivialisation of $\mathcal{W}$ over an open set $U \subset M$, so that $\mathcal{W}|_U \cong U \times \mathbb{C}^N$, the operator $\bar\partial_{\mathcal{W}}$ may be identified over U with the operator $\bar\partial$ on

$\mathscr{A}(U,\mathbb{C}^N)$. On the other hand, if ∇ is the tensor product of the Levi-Civita connection of $\Lambda(T^{1,0}M)^*$ with the holomorphic connection of $\mathscr{W}$, then the operator

$$\sum_i \varepsilon(\bar{Z}^i)\nabla_{\bar{Z}_i},$$

also coincide with $\bar{\partial}$ in the holomorphic trivialisation of $\mathscr{W}$ over U, and thus must be equal to $\bar{\partial}_{\mathscr{W}}$.

It remains to show that

$$\bar{\partial}^* = -\sum_i \iota(Z^i)\nabla_{Z^i}.$$

Let α be the one-form on M such that for any $\beta_q \in \mathscr{A}^{p,q}(M,\mathscr{W})$ and $\beta_{q+1} \in \mathscr{A}^{p,q+1}(M,\mathscr{W})$,

$$\alpha(X) = (\beta_q, \iota(X^{0,1})\beta_{q+1}).$$

Since ∇ preserves the splitting $TM \otimes_{\mathbb{R}} \mathbb{C} = T^{1,0}M \oplus T^{0,1}M$, the contraction $\operatorname{Tr}(\nabla\alpha)$ of the two-form $\nabla\alpha$ with the Hermitian structure on $TM \otimes_{\mathbb{R}} \mathbb{C}$ equals

$$\operatorname{Tr}(\nabla\alpha) = \sum_i \bar{Z}_i\alpha(\bar{Z}_i) - \alpha(\nabla_{\bar{Z}_i}\bar{Z}_i),$$

where Z_i is a local orthonormal frame of $T^{1,0}M$ with respect to the Hermitian structure on $T^{1,0}M$. Thus

$$\Big((\sum_i \varepsilon(\bar{Z}^i)\nabla_{\bar{Z}_i})\beta_q, \beta_{q+1}\Big)_x = -\Big(\beta_q, (\sum_i \iota(Z^i)\nabla_{Z^i})\beta_{q+1}\Big)_x + \operatorname{Tr}(\nabla\alpha)_x.$$

The last term vanishes after integration over M and we obtain the desired formula for $\bar{\partial}^*$, from which it is clear that $\sqrt{2}(\bar{\partial} + \bar{\partial}^*) = c \circ \nabla$. □

The cohomology of the complex

$$0 \to \mathscr{A}^{0,0}(M,\mathscr{W}) \xrightarrow{\bar{\partial}} \mathscr{A}^{0,1}(M,\mathscr{W}) \xrightarrow{\bar{\partial}} \mathscr{A}^{0,2}(M,\mathscr{W}) \xrightarrow{\bar{\partial}}$$

is called the **Dolbeault cohomology** of the holomorphic bundle $\mathscr{W}$. Dolbeault's theorem shows that this is isomorphic to the sheaf cohomology space $H^\bullet(M,\mathscr{O}(\mathscr{W}))$. Hodge's theorem gives an explicit realization of this cohomology as the kernel of the Laplace-Beltrami operator on $\mathscr{A}^{0,i}(M,\mathscr{W})$.

Theorem 3.68 (Hodge). *The kernel of the Dirac operator $\sqrt{2}(\bar{\partial} + \bar{\partial}^*)$ on the Clifford module $\Lambda(T^{0,1}M)^* \otimes \mathscr{W}$ is naturally isomorphic to the sheaf cohomology space $H^\bullet(M,\mathscr{O}(\mathscr{W}))$.*

Corollary 3.69. *The index of $\bar{\partial} + \bar{\partial}^*$ on the Clifford module $\Lambda(T^{0,1}M)^* \otimes \mathscr{W}$ is equal to the Euler number of the holomorphic vector bundle $\mathscr{W}$:*

$$\operatorname{ind}(\bar{\partial} + \bar{\partial}^*) = \operatorname{Eul}(\mathscr{W}) = \sum (-1)^i \dim(H^i(M,\mathscr{O}(\mathscr{W}))).$$

The Lichnerowicz formula for the square of the operator $\bar{\partial}+\bar{\partial}^*$ on a Kähler manifold is called the Bochner-Kodaira formula. Define the generalized Laplacian $\Delta^{0,\bullet}$ on $\mathscr{A}^{0,\bullet}(M,\mathscr{W})$ by the formula

$$\int_M (\Delta^{0,\bullet}s,s)\,dx = \int_M (\nabla^{0,1}s,\nabla^{0,1}s)\,dx.$$

Lemma 3.70. *If Z_i is a local orthonormal frame of $T^{1,0}M$ for the Hermitian metric on M, we have the formula*

$$\Delta^{0,\bullet} = -\sum_i (\nabla_{Z_i}\nabla_{\bar{Z}_i} - \nabla_{\nabla_{Z_i}\bar{Z}_i}).$$

Proof. Introduce the (1,0)-form $\alpha(X) = (\nabla_{X^{1,0}}s,s)$. Then

$$\operatorname{Tr}(\nabla\alpha) = \sum_i \bar{Z}_i\alpha(Z_i) - \alpha(\nabla_{\bar{Z}_i}Z_i),$$

since $T^{1,0}M$ is preserved by the Levi-Civita connection. Thus

$$\Big(-\sum_i (\nabla_{Z_i}\nabla_{\bar{Z}_i} - \nabla_{\nabla_{Z_i}\bar{Z}_i})s,s\Big)_x = (\nabla^{0,1}s,\nabla^{0,1}s)_x - \operatorname{Tr}(\nabla\alpha)_x.$$

The lemma follows by integrating over M, using Proposition 2.7. □

The **canonical line bundle** of a Kähler manifold is the holomorphic line bundle $K = \Lambda^n(T^{1,0}M)^*$ on M. The curvature of $K^* = \Lambda^n T^{1,0}M$ with respect to the Levi-Civita connection is the (1,1)-form $\sum(RZ_i,\bar{Z}_i)$, where R is the Riemannian curvature of M.

Proposition 3.71 (Bochner-Kodaira). *Let $\mathscr{W}$ be an Hermitian holomorphic vector bundle over the Kähler manifold M. In a local holomorphic coordinate system,*

$$\bar{\partial}\bar{\partial}^* + \bar{\partial}^*\bar{\partial} = \Delta^{0,\bullet} + \sum_{ij} \varepsilon(d\bar{z}^i)\iota(dz^j)F^{\mathscr{W}\otimes K^*}(\partial_{z^j},\partial_{\bar{z}^i}).$$

Proof. The proof is similar to the proof of Lichnerowicz's formula. Writing $\bar{\partial} = \sum\varepsilon(d\bar{z}^i)\nabla_{\partial/\partial\bar{z}^i}$ and $\bar{\partial}^* = -\sum\iota(dz^i)\nabla_{\partial/\partial z^i}$, we see that

$$\bar{\partial}\bar{\partial}^* + \bar{\partial}^*\bar{\partial} = \Delta^{0,\bullet} + \sum_{ij} \varepsilon(d\bar{z}^i)\iota(dz^j)\big(R^+(\partial_{z^j},\partial_{\bar{z}^i}) + F^{\mathscr{W}}(\partial_{z^j},\partial_{\bar{z}^i})\big)$$

where R^+ is the curvature of $\Lambda(T^{0,1}M)^*$. Thus it only remains to show that

$$\sum_{ij} \varepsilon(d\bar{z}^i)\iota(dz^j)R^+(\partial_{z^j},\partial_{\bar{z}^i}) = \sum_{ij} \varepsilon(d\bar{z}^i)\iota(dz^j)F^{K^*}(\partial_{z^j},\partial_{\bar{z}^i}).$$

The left side is equal to

$$\sum_{ijkl} \varepsilon(\bar{Z}_i^*)\iota(Z_j^*)\varepsilon(\bar{Z}_l^*)\iota(Z_k^*)(R(Z_j,\bar{Z}_i)Z_k,\bar{Z}_l) = \sum_{ijk} \varepsilon(\bar{Z}_i^*)\iota(Z_k^*)(R(Z_j,\bar{Z}_i)Z_k,\bar{Z}_j).$$

Using the fact that the Levi-Civita connection is torsion-free, we see that

$$R(Z_j,\bar{Z}_i)Z_k + R(Z_k,Z_j)\bar{Z}_i + R(\bar{Z}_i,Z_k)Z_j = 0,$$

and the fact that $R(Z_k,Z_j) = 0$, we see that

$$R(Z_j,\bar{Z}_i)Z_k = R(Z_k,\bar{Z}_i)Z_j.$$

It follows that

$$\sum_{ijk} \varepsilon(\bar{Z}_i^*)\iota(Z_k^*)(R(Z_j,\bar{Z}_i)Z_k,\bar{Z}_j) = \sum_{ik} \varepsilon(\bar{Z}_i^*)\iota(Z_k^*)F^{K^*}(Z_k,\bar{Z}_i),$$

completing the proof of the theorem. □

This formula implies some important vanishing results. If $\mathscr{L}$ is a Hermitian holomorphic line bundle with curvature $F = \sum F_{ij} dz^i \wedge d\bar{z}^j$, we say that $\mathscr{L}$ is positive if the Hermitian form $v \mapsto F(v,\bar{v})$ on $T^{1,0}M$ is positive. This positivity condition is related to the existence of global holomorphic sections of $\mathscr{L}$. For example, if s is a global holomorphic section of $\mathscr{W}$, and x is a point in M at which $|s|^2$ attains a strict maximum, then it is easy to see that $F_x = \bar{\partial}\partial \log |s|^2$ is positive on $T_x^{1,0}M$.

Proposition 3.72 (Kodaira).

1. *If $\mathscr{L}$ is a Hermitian holomorphic line bundle on a compact Kähler manifold such that $\mathscr{L} \otimes K^*$ is positive, then*
$$H^i(M,\mathscr{O}(\mathscr{L})) = 0 \quad \text{for } i > 0.$$
2. *If $\mathscr{L}$ is a positive Hermitian holomorphic line bundle and $\mathscr{W}$ is a Hermitian holomorphic vector bundle on M, then for m sufficiently large*
$$H^i(M,\mathscr{O}(\mathscr{L}^m \otimes \mathscr{W})) = 0 \quad \text{for } i > 0.$$

Proof. (1) Denote by $F^{\mathscr{L}\otimes K^*} = \sum F_{ij}^{\mathscr{L}\otimes K^*} dz^i \wedge d\bar{z}^j$ the curvature of $\mathscr{L} \otimes K^*$ and by $\lambda(F^{\mathscr{L}\otimes K^*})$ the endomorphism $\sum_{ij} \varepsilon(d\bar{z}^i)\iota(dz^j)F_{ij}^{\mathscr{L}\otimes K^*}$ of $\Lambda^i(T^{0,1}M)^*$. Then the Bochner-Kodaira formula shows that

$$\begin{aligned}\int_M ((\bar{\partial}\bar{\partial}^* + \bar{\partial}^*\bar{\partial})\alpha,\alpha) &= \int_M (\nabla^{0,1}\alpha,\nabla^{0,1}\alpha) + \int_M (\lambda(F^{\mathscr{L}\otimes K^*})\alpha,\alpha) \\ &\geq \int_M (\lambda(F^{\mathscr{L}\otimes K^*})\alpha,\alpha).\end{aligned}$$

The right-hand side is strictly positive if $i > 0$ and $\alpha \neq 0$; thus the conditions $\bar{\partial}\alpha = 0$ and $\bar{\partial}^*\alpha = 0$ imply that $\alpha = 0$.

(2) For the bundle $\mathscr{L}^m \otimes \mathscr{W}$, the Bochner-Kodaira formula gives

$$\bar{\partial}\bar{\partial}^* + \bar{\partial}^*\bar{\partial} = \Delta_m^{0,\bullet} + m\lambda(F^{\mathscr{L}}) + \text{other curvature terms independent of } m.$$

For m sufficiently large the term $m\lambda(F^{\mathscr{L}})$ dominates the other curvature terms, and we obtain (2). □

The reason for the importance of this result is that it shows that for m large, $\dim(H^0(M,\mathcal{O}(\mathcal{L}^m\otimes\mathcal{W}))) = \mathrm{Eul}(\mathcal{L}^m\otimes\mathcal{W})$. Riemann-Roch-Hirzebruch theorem that we prove in the next chapter gives an explicit formula for $\mathrm{Eul}(\mathcal{L}^m\otimes\mathcal{W})$ and hence for $\dim(H^0(M,\mathcal{O}(\mathcal{L}^m\otimes\mathcal{W})))$. This combination of an index theorem with a vanishing theorem is a very powerful technique in differential geometry.

Bibliographic Notes

The book of Lawson and Michelson [80] is a valuable reference for many of the topics in this chapter.

Section 1

The theory of Clifford algebras and their representations is classical; we mention as references Chevalley [51], Atiyah-Bott-Shapiro [9], and Karoubi [72].

Sections 2 and 3

The characterization of spin manifolds in terms of $w_2(M)$ of Proposition 3.34 is proved by Borel and Hirzebruch [39] (see also Milnor [85] and Lawson and Michelson [80]). The Dirac operator on spin manifolds appears in Atiyah-Singer [13]; they introduced the idea of twisting a Dirac operator. Lichnerowicz also studied this operator in [81], proving the fundamental formula for its square; he also proves the vanishing theorem for harmonic spinors on manifolds of positive scalar curvature.

Sections 4 and 5

The supertrace formula for the index of a Dirac operator is formulated by Atiyah-Bott [5] in terms of the zeta-function of D^2; the introduction of the heat kernel in this context is due to McKean-Singer [83].

Section 6

We do not attempt here to give references to the original articles on the geometric Dirac operators discussed in this section. However, there are some good textbooks on this subject, among which we mention Gilkey [67], Lawson and Michelson [80] for Dirac operators, and Wells [105] for the $\bar{\partial}$-operator. Many examples are contained in the article of Atiyah-Bott [6].

4

Index Density of Dirac Operators

Let M be a compact oriented Riemannian manifold of even dimension n. Let D be a Dirac operator on a Clifford module $\mathscr{E}$ on M associated to a Clifford connection $\nabla^{\mathscr{E}}$. Since D is an elliptic operator, the Atiyah-Singer Index Theorem gives a formula for the index of D in terms of the Chern character of $\mathscr{E}$ and the $\hat{A}$-genus of M. In this chapter, we will give a proof of this formula which uses the asymptotic expansion of the heat kernel of D^2 for t small. This approach was suggested by Atiyah-Bott [5] and McKean-Singer [83], and first carried out by Patodi [90] and Gilkey [65]. However, our method differs from theirs, in that the $\hat{A}$-genus and Chern character emerge in a natural way from purely local calculations, while previous proofs all required that the index density be determined by calculating the index of a range of examples.

We will prove a refined form of the local index theorem mentioned above. We define a filtration on sections of the bundle $\operatorname{End}(\mathscr{E})$ over $M \times \mathbb{R}_+$. We take the heat kernel of the operator D^2 restricted to the diagonal, and calculate it explicitly modulo lower order terms with respect to this filtration. In this way, we obtain a generalization of the local index theorem which extracts a differential form on M from the heat kernel for small times; this differential form is nothing other than the product of the $\hat{A}$-genus $\hat{A}(M)$ with respect to the Riemannian curvature and $e^{-F^{\mathscr{E}/S}} \in \mathscr{A}(M, \operatorname{End}_{C(M)} \mathscr{E})$, the exponential of the twisting curvature of $\mathscr{E}$.

4.1 The Local Index Theorem

Let $k_t(x,x) = \langle x \mid e^{-t\mathsf{D}^2} \mid x \rangle$ be the restriction of the heat kernel of D^2 to the diagonal. In this chapter, we will calculate the leading term in the asymptotic expansion of $\operatorname{Str}(k_t(x,x))$ as $t \to 0$. In particular, when combined with the McKean-Singer formula, this will imply a formula for the index of a Dirac operator.

The heat kernel $k_t(x,x)$ is a section of the bundle of filtered algebras $\operatorname{End}(\mathscr{E}) \cong C(M) \otimes \operatorname{End}_{C(M)}(\mathscr{E})$, where the filtration is induced by the filtration of $C(T^*M)$ and elements of $\operatorname{End}_{C(M)}(\mathscr{E})$ are given degree zero. We denote by $C_i(M)$ the sub-bundle of $C(M)$ of Clifford elements of degree less than or equal to i. The associated bundle of graded algebras is the bundle $\Lambda T^*M \otimes \operatorname{End}_{C(M)}(\mathscr{E})$. Let

$$\hat{A}(M) = \det{}^{1/2}\left(\frac{R/2}{\sinh(R/2)}\right)$$

be the $\hat{A}$-genus form of the manifold M with respect to the Riemannian curvature R.

Let $F^{\mathscr{E}/S}$ be the twisting curvature of the connection $\nabla^{\mathscr{E}}$, which was defined in (3.43). The aim of this chapter is to prove the following theorem.

Theorem 4.1. *Consider the asymptotic expansion of $k_t(x,x)$,*

$$k_t(x,x) \sim (4\pi t)^{-n/2} \sum_{i=0}^{\infty} t^i k_i(x)$$

with coefficients $k_i \in \Gamma(M, C(M) \otimes \mathrm{End}_{C(M)}(\mathscr{E}))$.

1. The coefficient k_i is a section of the bundle $C_{2i}(M) \otimes \mathrm{End}_{C(M)}(\mathscr{E})$.

2. Let $\sigma(k) = \sum_{i=0}^{i=n/2} \sigma_{2i}(k_i) \in \mathscr{A}(M, \mathrm{End}_{C(M)}(\mathscr{E}))$. Then

$$\sigma(k) = \det{}^{1/2}\left(\frac{R/2}{\sinh(R/2)}\right) \exp(-F^{\mathscr{E}/S}).$$

This result need not hold for a Dirac operator associated to a general Clifford superconnection; for a discussion of the modifications which must be made in the general case, see the Additional Remarks at the end of this chapter.

The $i = 0$ piece of Theorem 4.1 is precisely Weyl's formula

$$\lim_{t\to 0} (4\pi t)^{n/2} k_t(x,x) = \mathrm{Id}_{\mathscr{E}}.$$

Thus, Theorem 4.1 is an extension of Weyl's theorem in the case of a generalized Laplacian which is the square of a Dirac operator associated to a Clifford connection.

The idea of the proof is to work in normal coordinates $\mathbf{x} \in U$ around a point $x_0 \in M$, where U is an open neighbourhood of zero in $V = T_{x_0}M$. Using the parallel transport map for the connection $\nabla^{\mathscr{E}}$, we may trivialize the bundle $\mathscr{E}$ over the set U, obtaining an identification of $\Gamma(U, \mathrm{End}(\mathscr{E}))$ with $C^\infty(U, \mathrm{End}(E))$, where $E = \mathscr{E}_{x_0}$. There is an isomorphism σ of the vector space $\mathrm{End}(E) = C(V^*) \otimes \mathrm{End}_{C(V^*)}(E)$ with $\Lambda V^* \otimes \mathrm{End}_{C(V^*)}(E)$. We introduce a rescaling on the space of functions on $\mathbb{R}_+ \times U$ with values in $\Lambda V^* \otimes \mathrm{End}_{C(V^*)}(E)$ by the formula

$$(\delta_u a)(t,\mathbf{x}) = \sum_{i=0}^{n} u^{-i/2} a(ut, u^{1/2}\mathbf{x})_{[i]}.$$

Using this rescaling, we may restate Theorem 4.1 as follows: if

$$k(t,\mathbf{x}) = \sigma(\langle \exp_{x_0} \mathbf{x} \mid e^{-t\mathrm{D}^2} \mid x_0 \rangle),$$

then

$$\lim_{u\to 0} (u^{n/2} \delta_u k)|_{(t,\mathbf{x})=(1,0)} = (4\pi)^{-n/2} \hat{A}(M) \exp(-F^{\mathscr{E}/S}).$$

The rescaling operator δ_u induces a filtration on the algebra of differential operators acting on $C^\infty(\mathbb{R}_+ \times U, \Lambda V^* \otimes \mathrm{End}_{C(V^*)}(E))$; an operator D has filtration degree m if

$$\lim_{u\to 0} u^{m/2} \delta_u D \delta_u^{-1}$$

exists. Thus, a polynomial $P(\mathbf{x})$ has degree $-\deg(P)$, a polynomial $P(t)$ has degree $-2\deg(P)$, a derivative $\partial/\partial \mathbf{x}^i$ has degree one, a derivative $\partial/\partial t$ has degree two, an exterior multiplication operator ε^i has degree one, and an interior multiplication operator ι^i has degree -1. It will be shown, using Lichnerowicz's formula, that the operator D^2 has degree two; this is why Theorem 4.1 holds. Indeed, up to operators of lower order, we will see that the operator D^2 may be identified with a harmonic oscillator with differential form coefficients.

Given r and $f \in \mathbb{R}$, let H be the harmonic oscillator on the real line

$$H = -\frac{d^2}{dx^2} + \frac{r^2x^2}{16} + f.$$

In Section 3, we will prove Mehler's formula, which states that the heat kernel $p_t(x,y;H)$ of the harmonic oscillator H is equal to

$$\left(\frac{tr/2}{4\pi t \sinh(tr/2)}\right)^{1/2} \exp\left(-\frac{r}{8t}(\coth(tr/2)(x^2+y^2) - 2\operatorname{cosech}(tr/2)xy) - tf\right).$$

The above filtration argument reduces the calculation of $(\delta_u k)(t,\mathbf{x})$ to the calculation of the heat kernel of a harmonic oscillator, with differential form coefficients, in which r is replaced by the Riemannian curvature R of M and f by the twisting curvature $F^{\mathscr{E}/S}$ of $\mathscr{E}$. The $\hat{A}$-genus of the manifold M and the twisting curvature $F^{\mathscr{E}/S}$ appear as consequences of the above formula for $p_t(0,0;H)$.

The main consequence of Theorem 4.1 is the local index theorem for Dirac operators, which is a strengthening of Atiyah and Singer's result. By the McKean-Singer formula, for every $t > 0$,

$$\operatorname{ind}(\mathsf{D}) = \int_M \operatorname{Str}(k_t(x,x))\,dx.$$

Theorem 4.1 implies that the integrand itself has a limit when t converges to zero, if D is associated to a connection. Indeed, as the supertrace of an element $a \in C(M)$ vanishes on all elements of Clifford filtration strictly less than $n = \dim(M)$, the first part of Theorem 4.1 implies that

$$\operatorname{Str}(k_t(x,x)) \sim (4\pi t)^{-n/2} \sum_{i \geq n/2} t^i \operatorname{Str}(k_i(x)),$$

thus there are no poles in the asymptotic expansion of $\operatorname{Str}(k_t(x,x))$. Furthermore, as the left-hand side $\operatorname{ind}(\mathsf{D})$ of the McKean-Singer formula is independent of t, we necessarily have

$$\operatorname{ind}(\mathsf{D}) = (4\pi)^{-n/2} \int_M \operatorname{Str}(k_{n/2}(x))\,dx,$$

while the integrals of all other terms $\int_M \operatorname{Str}(k_j(x))\,dx$, $j \neq n/2$, vanish.

To identify the term $\operatorname{Str}(k_{n/2}(x))$ as a characteristic form on M, we must introduce some notation. If we write a Clifford module $\mathscr{E}$ locally as $\mathscr{W} \otimes \mathscr{S}$, where $\mathscr{S}$ is a spinor bundle and $\mathscr{W}$ is an auxiliary bundle, then by (3.28), for $a \in \Gamma(M, \operatorname{End}(\mathscr{W})) \cong \Gamma(M, \operatorname{End}_{C(M)}(\mathscr{E}))$,

$$\operatorname{Str}_{\mathscr{W}}(a) = 2^{-n/2} \operatorname{Str}_{\mathscr{E}}(\Gamma a),$$

where $\Gamma \in \Gamma(M, C(M))$ is the chirality operator. We will avoid making use of this decomposition, which is often only possible locally. Instead, we use the relative supertrace of Definition 3.28; the relative supertrace of $a \in \Gamma(M, \operatorname{End}_{C(M)}(\mathscr{E}))$ is

$$\operatorname{Str}_{\mathscr{E}/S}(a) = 2^{-n/2} \operatorname{Str}_{\mathscr{E}}(\Gamma a).$$

We extend the relative supertrace to a linear map

$$\operatorname{Str}_{\mathscr{E}/S} : \mathscr{A}(M, \operatorname{End}_{C(M)}(\mathscr{E})) \to \mathscr{A}(M),$$

and define, for a Clifford superconnection $\nabla^{\mathscr{E}}$ on $\mathscr{E}$, the **relative Chern character** form of the bundle $\mathscr{E}$ by the formula

$$\operatorname{ch}(\mathscr{E}/S) = \operatorname{Str}_{\mathscr{E}/S}(\exp(-F^{\mathscr{E}/S})).$$

The relative Chern character $\operatorname{ch}(\mathscr{E}/S)$ is a closed differential form on M. The cohomology class of this differential form is independent of the choice of Clifford superconnection on $\mathscr{E}$, and will also be denoted by $\operatorname{ch}(\mathscr{E}/S)$. In the special case where the Clifford module $\mathscr{E} = \mathscr{W} \otimes \mathscr{S}$ and $\nabla^{\mathscr{E}}$ is the tensor product of the Levi-Civita connection on $\mathscr{S}$ and a connection on $\mathscr{W}$ with curvature $F^{\mathscr{W}}$, we see that $\operatorname{ch}(\mathscr{E}/S) = \operatorname{Str}_{\mathscr{W}}(\exp(-F^{\mathscr{W}}))$ is the Chern character form of the twisting bundle $\mathscr{W}$.

If $a \in \Gamma(M, C(M))$ and $b \in \Gamma(M, \operatorname{End}_{C(M)}(\mathscr{E}))$, then the point-wise supertrace of the section $a \otimes b \in \Gamma(M, C(M) \otimes \operatorname{End}_{C(M)}(\mathscr{E})) \cong \Gamma(M, \operatorname{End}(\mathscr{E}))$ was shown in (3.21) to equal the Berezin integral

$$\operatorname{Str}_{\mathscr{E}}(a(x) \otimes b(x)) = (-2i)^{n/2} \sigma_n(a(x)) \operatorname{Str}_{\mathscr{E}/S}(b(x)),$$

so that

$$\operatorname{Str}_{\mathscr{E}}(k_{n/2})(x) = (-2i)^{n/2} \operatorname{Str}_{\mathscr{E}/S}(\sigma_n(k_{n/2}(x))).$$

Thus Theorem 4.1 implies the following theorem for the index of a Dirac operator associated to a Clifford connection.

Theorem 4.2 (Patodi, Gilkey). *Let M be a compact oriented Riemannian manifold of even dimension n, with Clifford module $\mathscr{E}$ and Clifford connection $\nabla^{\mathscr{E}}$, and let D be the associated Dirac operator. If $k_t(x,x)\,|dx| \in \Gamma(M, \operatorname{End}(\mathscr{E}) \otimes |\Lambda|)$ is the restriction of the heat kernel of the operator D^2 to the diagonal, then $\lim_{t\to 0} \operatorname{Str}(k_t(x,x))\,|dx|$ exists and is the volume-form on M obtained by taking the n-form piece of*

$$(2\pi i)^{-n/2} \det{}^{1/2}\left(\frac{R/2}{\sinh(R/2)}\right) \operatorname{Str}_{\mathscr{E}/S}(\exp(-F^{\mathscr{E}/S})).$$

It is important to note that there is no *a priori* reason to suppose that the limit $\lim_{t\to 0} \operatorname{Str} k_t(x,x)$ exists at each point, and indeed, this is not necessarily true for Dirac operators which are not associated to a Clifford connection. However, since the index of a Dirac operator is independent of the Clifford superconnection used to define it, we obtain the Atiyah-Singer formula for the index of an arbitrary Dirac operator.

Theorem 4.3 (Atiyah-Singer). *The index of a Dirac operator on a Clifford module $\mathcal{E}$ over a compact oriented even-dimensional manifold is given by the cohomological formula*

$$\operatorname{ind}(\mathsf{D}) = (2\pi i)^{-n/2} \int_M \hat{A}(M) \operatorname{ch}(\mathcal{E}/S).$$

The factor $(2\pi i)^{-n/2}$ is absent from the usual statement of this theorem, since it is part of the topologist's definition of the characteristic classes, the purpose of which is to ensure that the characteristic Chern classes are integral. However, in this book, we prefer the more natural (to the geometer) definitions of the $\hat{A}$-genus form

$$\hat{A}(M) = \det{}^{1/2}\left(\frac{R/2}{\sinh(R/2)}\right)$$

and of the Chern character form $\operatorname{ch}(\mathcal{E}) = \operatorname{Str}(\exp(-\nabla^2))$.

There is an extension of Theorem 4.1 to a Dirac operator associated to an arbitrary Clifford superconnection. Let $\mathcal{E}$ be a Clifford module on the manifold M, and let $\mathbb{A}$ be a Clifford superconnection on $\mathcal{E}$. Introduce a family of rescaled superconnections

$$\mathbb{A}_t = \sum_{i=0}^{n} t^{(1-i)/2} \mathbb{A}_{[i]}.$$

Let $F^{\mathcal{E}/S}$ be the twisting curvature of the superconnection $\mathbb{A}$.

Theorem 4.4. *Let D_t be the Dirac operator associated to the Clifford superconnection $\mathbb{A}_{t^{-1}}$, and let $k_u(t,x,y) = \langle x \mid e^{-u\mathsf{D}_t^2} \mid y\rangle$ be the heat kernel of the generalized Laplacian $t\mathsf{D}_t^2$.*

1. *There exist sections $\Phi_i \in C^\infty(\mathbb{R}_+, C_i(M) \otimes \operatorname{End}_{C(M)}(\mathcal{E}))$ of the Clifford bundle of M such that as $t \to 0$,*

$$k_u(t,x,x) \sim (4\pi u t)^{-n/2} \sum_{i=0}^{n} t^{i/2} \Phi_i(u,x).$$

2. $$\sum_{i=0}^{n} \sigma_i(\Phi_i(u,x)) = \det{}^{1/2}\left(\frac{uR/2}{\sinh(uR/2)}\right) \exp(-uF^{\mathcal{E}/S})$$

See the bibliographic notes for a discussion of proofs, and of yet further generalizations of this result.

In the rest of this section, we will derive the usual formulas for the index density $(2\pi i)^{-n/2} \hat{A}(M) \operatorname{ch}(\mathcal{E}/S)$ for each of the four classical Dirac operators that were described in the last chapter. The resulting formulas are special cases of the Atiyah-Singer Index Theorem for elliptic operators.

The Gauss-Bonnet-Chern Theorem

We consider the Clifford bundle $\mathscr{E} = \Lambda T^*M$ with its Levi-Civita connection $\nabla^{\mathscr{E}}$. In order to analyse what the above index theorem says for the case of the de Rham operator, we must calculate $F^{\mathscr{E}/S}$ in this case. We will do this in a local orthonormal frame e^i.

We start with some general results about the Clifford module $E = \Lambda V$ of the Clifford algebra $C(V)$, for V an even-dimensional space. From the isomorphisms

$$S^* \otimes S \cong \operatorname{End}(S) \cong C(V) \otimes_{\mathbb{R}} \mathbb{C} \xrightarrow{\sigma} \Lambda V \otimes_{\mathbb{R}} \mathbb{C},$$

we see that in this case, we may identify the auxiliary bundle W with S^*. Thus, there are two natural $\mathbb{Z}_2$-gradings on S^*: the grading $S^* = (S^*)^+ \oplus (S^*)^-$ induced by the grading on ΛV, and the grading defined by the chirality operator, for which S^* is purely even. We denote the first corresponding supertrace on $\operatorname{End}_{C(V)}(\Lambda V)$ by $\operatorname{Str}_{\Lambda V/S}$; the second one, the relative trace, we denote by

$$\operatorname{Tr}_{\Lambda V/S} = 2^{-n/2} \operatorname{Tr}_{\Lambda V}.$$

Let e^i be a positive orthonormal basis of V. Consider the following elements of $\operatorname{End}(E)$:

$$c^i = \varepsilon(e^i) - \iota(e^i), \quad b^i = \varepsilon(e^i) + \iota(e^i).$$

We have the relations $[c^i, c^j] = -2\delta^{ij}$, $[b^i, b^j] = 2\delta^{ij}$, and $[c^i, b^j] = 0$, so that the algebra $\operatorname{End}_{C(V)}(E)$ is generated by the elements b^i. We freely identify $C^2(V)$ and $\Lambda^2 V$ by the map σ.

Lemma 4.5. *If $a = \sum_{i<j} a_{ij} e^i e^j \in \Lambda^2 V$, define $b(a) = \sum_{i<j} a_{ij} b^i b^j \in \operatorname{End}_{C(V)}(E)$. We have the following formulas:*

$$\operatorname{Str}_{\Lambda V/S}(e^{b(a)}) = (-2i)^{n/2} j_V(a)^{1/2} \operatorname{Pf}_\Lambda(a),$$
$$\operatorname{Tr}_{\Lambda V/S}(e^{b(a)}) = 2^{n/2} \det(\cosh(\tau(a)/2))^{1/2}.$$

Proof. By definition, $\operatorname{Str}_{\Lambda V/S}(e^{b(a)}) = (-2i)^{-n/2} \operatorname{Str}_{\Lambda V}(c^1 \cdots c^n e^{b(a)})$. It suffices to prove the lemma when $\dim(V) = 2$ and $a = \theta e^1 e^2$, so that $b(a) = \theta b^1 b^2$ and $\exp(\theta b^1 b^2) = \cos\theta + (\sin\theta) b^1 b^2$.

It is easily checked that $\operatorname{Str}_{\Lambda V}(c^1 c^2) = 0$ and that $\operatorname{Str}_{\Lambda V}(c^1 c^2 b^1 b^2) = -4$, so that

$$\begin{aligned} \operatorname{Str}_{\Lambda V/S}(e^{b(a)}) &= (-2i)^{-1} \operatorname{Str}_{\Lambda V}(c^1 c^2 \cos\theta + (\sin\theta) c^1 c^2 b^1 b^2) \\ &= -2i \sin\theta. \end{aligned}$$

On the other hand, $j_V(\theta e^1 e^2)^{1/2} = \theta^{-1} \sin\theta$, while $\operatorname{Pf}_\Lambda(\theta e^1 e^2) = \theta$. Putting this all together proves the first formula.

On the other hand, $\det(\cosh_V(\tau(\theta e^1 e^2)/2)^{1/2} = \cos\theta$, while

$$\operatorname{Tr}_{\Lambda V}(e^{\theta b^1 b^2}) = \operatorname{Tr}_{\Lambda V}(\cos\theta) + \operatorname{Tr}_{\Lambda V}((\sin\theta) b^1 b^2) = 4\cos\theta,$$

since $\operatorname{Tr}_{\Lambda V}(b^1 b^2) = 0$. This proves the second formula. □

The curvature tensor $(\nabla^{\mathscr{E}})^2$ is equal to $\sum_{ij,k<l} R_{ijkl}\varepsilon^i\iota^j e^k \wedge e^l$, which when rewritten in terms of the operators c^i and b^i equals

$$-\tfrac{1}{4}\sum_{ij,k<l} R_{ijkl}(c^i+b^i)(c^j-b^j)e^k\wedge e^l = -\tfrac{1}{4}\sum_{ij,k<l} R_{ijkl}(c^ic^j-b^ib^j)e^k\wedge e^l,$$

where we have used the antisymmetry of R_{ijkl} in i and j to show that

$$\sum_{ij} R_{ijkl}(c^ib^j-b^ic^j) = \sum_{ij} R_{ijkl}(c^ib^j+c^jb^i) = 0.$$

From this formula and (1.20), it is easy to see that $F^{\mathscr{E}/S}$ is given by the formula

$$F^{\mathscr{E}/S} = -\tfrac{1}{4}\sum_{ij}(Re_i,e_j)b^ib^j.$$

Let us now calculate the term $\hat{A}(M)\operatorname{Str}_{\mathscr{E}/S}(\exp(-F^{\mathscr{E}/S}))$ which enters in the formula for the index of $d+d^*$. Let $\chi(TM)\in\mathscr{A}^n(M)$ be the Euler form of the manifold M associated to the Levi-Civita connection (1.38).

Proposition 4.6. *We have* $\hat{A}(M)\operatorname{Str}_{\mathscr{E}/S}(\exp(-F^{\mathscr{E}/S})) = i^{n/2}\chi(TM)$.

Proof. When $a=\sum_{i<j}a_{ij}e_i\wedge e_j\in\Lambda^2V$, we have seen that

$$j_V(a)^{-1/2}\operatorname{Str}_{\mathscr{E}/S}\Big(\exp\sum_{ij}a_{ij}b^ib^j\Big) = (-2i)^{n/2}\operatorname{Pf}_\Lambda(-a).$$

By analytic continuation of this identity, we see that the same formula remains true when $a=\frac{1}{2}\sum_{i<j}(Re_i,e_j)e^i\wedge e^j$ is an element of Λ^2V with two-form coefficients, and we obtain the proposition. □

As a result, we see that the index formula for

$$\operatorname{Eul}(M) = \operatorname{ind}(d+d^*) = \sum_{i=0}^{n}(-1)^i\dim(H^i(M)),$$

the Euler number of the manifold, simplifies greatly, and we obtain the following consequence of Theorem 4.1.

Theorem 4.7 (Gauss-Bonnet-Chern). *The Euler number of an even-dimensional oriented manifold M is given by the formula*

$$\operatorname{Eul}(M) = (2\pi)^{-n/2}\int_M \operatorname{Pf}(-R).$$

For example, if the manifold is two-dimensional, we recover the classical Gauss-Bonnet theorem

$$\operatorname{Eul}(M) = \frac{1}{2\pi}\int_M r_M\,dx.$$

Of course, this is possibly the most difficult proof of this basic theorem of differential geometry, with the exception of proofs using quantum field theory. Indeed, the usual proof of the Gauss-Bonnet-Chern Theorem was given in Section 1.6, although we did not give there the topological argument, involving Morse theory, which is needed to show that the Euler number calculated from a Morse function equals the Euler number calculated from the de Rham cohomology. However, the above proof does provide a test for the index theorem.

The Hirzebruch Signature Theorem

To derive Hirzebruch's formula for the index of the signature operator, we have only to combine the formula that we have already obtained for $F^{\mathcal{E}/S}$ with the general index theorem. Recall that the grading of ΛT^*M for which the supertrace is computed for the signature operator is the grading induced by the Hodge star-operator $\star$.

Define the characteristic form $L(M) \in \mathcal{A}^{4\bullet}(M)$ by

$$L(M) = \det{}^{1/2}\left(\frac{R/2}{\tanh(R/2)}\right);$$

it is known as the ***L*-genus**. By Lemma 4.5, $\hat{A}(M)\operatorname{Tr}_{\mathcal{E}/S}(\exp(-F^{\mathcal{E}/S})) = 2^{n/2}L(M)$, and we obtain the following result.

Theorem 4.8 (Hirzebruch). *Let M be an oriented Riemannian manifold of dimension divisible by four; then the signature $\sigma(M)$ is given by the formula*

$$\sigma(M) = (\pi i)^{-n/2}\int_M L(M).$$

It follows by the same calculation that the twisted signature with respect to an auxiliary vector bundle $\mathcal{W}$ is given by the formula

$$\sigma(\mathcal{W}) = (\pi i)^{-n/2}\int_M L(M)\operatorname{ch}(\mathcal{W}).$$

This formula is due to Atiyah-Singer [13], and is one of the main steps in their calculation of the index of an arbitrary elliptic operator on a compact manifold.

The first non-trivial term in the Taylor expansion of the L-genus may be easily calculated:

$$L(M) = 1 + \frac{1}{24}\operatorname{Tr}(R^2) + \dots.$$

From this, we see that for a four-manifold,

$$\sigma(M) = -\int_M \frac{\operatorname{Tr}(R^2)}{24\pi^2}.$$

The Index Theorem for the Dirac Operator

In the case of the Dirac operator on a spinor bundle over a spin manifold, the curvature $F^{\mathscr{E}/S} = 0$, so that

$$\operatorname{ind}(\mathsf{D}) = (2\pi i)^{-n/2} \int_M \hat{A}(M).$$

This formula has a number of immediate corollaries:

1. The index of the Dirac operator is independent of the spin-structure.
2. If the right-hand side (which is always a rational number) is not an integer, then the manifold M does not have a spin-structure.
3. If the manifold M has a metric with respect to which the scalar curvature r_M is positive and non-zero, then the integral of the $\hat{A}$-genus is zero, since by Lichnerowicz's vanishing theorem, the index of the Dirac operator vanishes.

Using the formula for the first few terms in the expansion of $\hat{A}(R)$ that we gave in (1.36), we see that if $\dim(M) = 4$, then $\sigma(M) = -8\operatorname{ind}(\mathsf{D})$.

Consider the twisted Clifford module $\mathscr{E} = \mathscr{W} \otimes \mathscr{S}$, where $\mathscr{W}$ is an auxiliary Hermitian vector bundle with a Hermitian connection of curvature $F^{\mathscr{W}}$. In the index formula for the twisted Dirac operator $\mathsf{D}^{\pm}_{\mathscr{W}}$ associated to the Clifford connection obtained by twisting the Levi-Civita connection on $\mathscr{S}$ by the connection of $\mathscr{W}$, the term $F^{\mathscr{E}/S}$ is nothing but the curvature $F^{\mathscr{W}}$ of the twisting bundle, so we obtain the following case of the index theorem, which is in some ways the most fundamental.

Theorem 4.9 (Atiyah-Singer). *Let M be an oriented spin manifold of even dimension. Let $\mathscr{S}$ be the spin bundle and let $\mathscr{W}$ be a vector bundle over M. If $\mathsf{D}_{\mathscr{W}}$ is the twisted Dirac operator on $\Gamma(M, \mathscr{W} \otimes \mathscr{S})$, then*

$$\operatorname{ind}(\mathsf{D}_{\mathscr{W}}) = (2\pi i)^{-n/2} \int_M \hat{A}(M)\operatorname{ch}(\mathscr{W}).$$

The Riemann-Roch-Hirzebruch Theorem

Let M be a Kähler manifold, and consider the Clifford module $\mathscr{E} = \Lambda(T^{0,1}M)^*$. Let $\nabla^{\mathscr{E}}$ be the Clifford connection on this Clifford module induced from the Levi-Civita connection. Let us calculate the curvature $F^{\mathscr{E}/S}$ of this Clifford module.

Considering the Riemannian curvature R to be a matrix with two-form coefficients, the curvature operator $(\nabla^{\mathscr{E}})^2$ is

$$\sum_{kl} (Rw_k, \bar{w}_l)\varepsilon(\bar{w}^l)\iota(w^k)$$

where w_k is a basis of $T^{1,0}(M)$ and w^k is the dual basis, while the $\operatorname{End}(\mathscr{E})$-valued two-form $R_{\mathscr{E}}$ of Proposition 3.43 equals

$$\tfrac{1}{4}\sum_{kl}(Rw_k, \bar{w}_l)c(w^k)c(\bar{w}^l) + \tfrac{1}{4}\sum_{kl}(R\bar{w}_l, w_k)c(\bar{w}^l)c(w^k).$$

Thus, if the basis w_k is orthonormal, we have

$$(\nabla^{\mathscr{E}})^2 = R^{\mathscr{E}} + \tfrac{1}{2}\sum_k (Rw_k, \bar{w}_k).$$

If $\mathscr{W}$ is a holomorphic Hermitian vector bundle and $\nabla^{\mathscr{E}}$ is the Clifford connection on $\Lambda^{0,\bullet}T^*M \otimes \mathscr{W}$ obtained by twisting the Levi-Civita connection with the connection of $\mathscr{W}$ compatible with the metric, we obtain

$$F^{\mathscr{E}/S} = \tfrac{1}{2}\operatorname{Tr}_{T^{1,0}M}(R^+) + F^{\mathscr{W}},$$

where R^+ is the curvature of the bundle $T^{1,0}M$ and $F^{\mathscr{W}}$ the curvature of the bundle $\mathscr{W}$.

Since $TM \otimes_{\mathbb{R}} \mathbb{C} = T^{1,0}M \oplus T^{0,1}M$, we see that

$$\hat{A}(M) = \det\Big(\frac{R^+}{e^{R^+/2} - e^{-R^+/2}}\Big),$$

so that

$$\hat{A}(M)\operatorname{Tr}_{\mathscr{E}/S}(\exp(-F^{\mathscr{E}/S})) = \operatorname{Td}(M)\operatorname{Tr}(\exp(-F^{\mathscr{W}})),$$

where $\operatorname{Td}(M)$ is the Todd genus of the complex manifold M:

$$\operatorname{Td}(M) = \det\Big(\frac{R^+}{e^{R^+} - 1}\Big) = \det\Big(\frac{R^+}{e^{R^+/2} - e^{-R^+/2}}\Big)\exp(-\operatorname{Tr}(R^+/2)).$$

We obtain the following theorem, which is due to Hirzebruch in the case of projective manifolds, and to Atiyah and Singer in general.

Theorem 4.10 (Riemann-Roch-Hirzebruch). *The Euler number of the holomorphic bundle $\mathscr{W}$ is given by the formula*

$$\operatorname{Eul}(\mathscr{W}) = (2\pi i)^{-n/2}\int_M \operatorname{Td}(M)\operatorname{ch}(\mathscr{W}).$$

For example, if M is a Riemann surface (has complex dimension one) and curvature $R^+ \in \mathscr{A}^2(M,\mathbb{C})$, and if $\mathscr{L}$ is a line-bundle with curvature $F \in \mathscr{A}^2(M,\mathbb{C})$, then $\operatorname{Td}(M) = 1 - R^+/2$ and $\operatorname{ch}(\mathscr{L}) = 1 - F$, and we obtain the classical Riemann-Roch Theorem,

$$\operatorname{ind}(\bar{\partial}_{\mathscr{L}}) = \dim H^0(M,\mathscr{L}) - \dim H^1(M,\mathscr{L}) = \frac{-1}{4\pi i}\int_M (R^+ + 2F).$$

If we take $\mathscr{L} = \mathbb{C}$ to be the trivial line-bundle in this formula, we see that

$$\dim H^0(M) - \dim H^{0,1}(M) = 1 - g = \frac{-1}{4\pi i}\int_M R^+,$$

where $g = \dim H^{1,0}(M) = \dim H^{0,1}(M) = \frac{1}{2}\dim H^1(M)$ is the genus of M. If we define the degree $\deg(\mathscr{L}) \in \mathbb{Z}$ of the line bundle $\mathscr{L}$ by the formula

$$\deg(\mathscr{L}) = \frac{1}{-2\pi i}\int_M F,$$

we may restate the Riemann-Roch Theorem in its classical form,

$$\mathrm{Eul}(\mathscr{L}) = 1 - g + \deg(\mathscr{L}).$$

If M is a compact complex manifold which is not Kähler, then for any holomorphic vector bundle $\mathscr{W}$ we may still form the Clifford module $\Lambda(T^{0,1}M)^* \otimes \mathscr{W}$, and the operator $\sqrt{2}(\bar\partial + \bar\partial^*)$ is a generalized Dirac operator. The index of the operator $\bar\partial + \bar\partial^*$ is equal to the Euler number of the bundle $\mathscr{W}$ by Hodge's Theorem, as for a Kähler manifold,

$$\mathrm{ind}(\bar\partial + \bar\partial^*) = \mathrm{Eul}(\mathscr{W}) = \sum_{i=0}^{n/2}(-1)^i \dim H^{0,i}(M,\mathscr{W}).$$

However, although the operator $\sqrt{2}(\bar\partial + \bar\partial^*)$ is a Dirac operator associated to a superconnection, it is no longer the Dirac operator associated to a connection, so we cannot expect the local index theorem to hold. Nevertheless, there is a natural Dirac operator D on this Clifford module which does satisfy the local index theorem, which is formed simply by taking the Dirac operator associated to the Levi-Civita connection of the bundle $\Lambda(T^{0,1}M)^* \otimes \mathscr{W}$. It is not hard to extend our above treatment to this operator, and it is easy to see that its index is given by the same formula as in the Riemann-Roch-Hirzebruch Theorem:

$$\mathrm{ind}(\mathsf{D}) = (2\pi i)^{-n/2}\int_M \mathrm{Td}(M)\,\mathrm{ch}(\mathscr{W}).$$

Since the Dirac operator D and $\sqrt{2}(\bar\partial + \bar\partial^*)$ are both Dirac operators for the same Clifford module, they must have the same index. In this way, we see that the Riemann-Roch-Hirzebruch Theorem is valid on any compact complex manifold, but not the local formula.

4.2 Mehler's Formula

In this section, we derive Mehler's formula for the heat kernel of the harmonic oscillator, in the form in which it is needed for the proof of Theorem 4.1. In order to give an idea of the result in its most basic form, let us consider the harmonic oscillator on the real line:

$$H = -\frac{d^2}{dx^2} + x^2.$$

We would like to find a function $p_t(x,y)$ satisfying the following requirements: for each $t \geq 0$, the map

$$\phi \longmapsto \phi_t(x) = \int_{-\infty}^{\infty} p_t(x,y)\,\phi(y)\,dy$$

is bounded on the space of test functions $\mathscr{S}(\mathbb{R})$, satisfies the heat equation

$$(\partial_t + H_x)\phi_t(x) = 0,$$

and the initial condition

$$\lim_{t\to 0}\int_{-\infty}^{\infty} p_t(x,y)\,\phi(y)\,dy = \phi(x).$$

To find a solution to this problem, one is guided by the fact that the operator H is quadratic in differentiation and multiplication to seek a solution which is a Gaussian function of x and y; in addition, the solution must clearly be symmetric in x and y, since the harmonic oscillator is self-adjoint:

$$p_t(x,y) = \exp(a_t x^2/2 + b_t xy + a_t y^2/2 + c_t).$$

Applying the heat operator to this ansatz, and dividing by $p_t(x,y)$, we find

$$\dot{p}_t(x,y) + Hp_t(x,y) = \\ (\dot{a}_t x^2/2 + \dot{b}_t xy + \dot{a}_t y^2/2 + \dot{c}_t - (a_t x + b_t y)^2 - a_t + x^2)\,p_t(x,y).$$

In particular, we obtain the following ordinary differential equations for the coefficients:

$$\dot{a}_t/2 = a_t^2 - 1 = b_t^2 \quad\text{and}\quad \dot{c}_t = a_t.$$

These equations have the solutions $a_t = -\coth(2t+C)$, $b_t = \operatorname{cosech}(2t+C)$ and $c_t = -1/2\log\sinh(2t+C) + D$, and in order to determine the values of the integration constants C and D, we simply have to substitute in the initial conditions for $p_t(x,y)$, which show that $C = 0$ and $D = \log(2\pi)^{-1/2}$. Thus, we have obtained the following solution to the heat equation for H:

$$p_t(x,y) = (2\pi\sinh 2t)^{-1/2}\exp\Big(-\tfrac{1}{2}\big((\coth 2t)(x^2+y^2) - 2(\operatorname{cosech} 2t)xy\big)\Big).$$

This formula is known as **Mehler's formula.**

We will mainly use this formula when $y = 0$. By change of variables (or by direct calculation), for any real numbers r, f, it follows that the function $p_t(x,r,f)$ defined by

$$(4\pi t)^{-1/2}\left(\frac{tr/2}{\sinh(tr/2)}\right)^{1/2}\exp(-(tr/2)\coth(tr/2)x^2/4t - tf) \tag{4.1}$$

is a solution of the equation

$$\left(\partial_t - \frac{d^2}{dx^2} + \frac{r^2x^2}{16} + f\right)p_t(x) = 0.$$

We will now generalize the above argument so that it applies to the formal heat equation that we use to prove Theorem 4.1, as well as in the proof of the index theorem for families in Chapter 10.

Let V be an n-dimensional Euclidean vector space with orthonormal basis e^i. Let $\mathscr{A}$ be a finite dimensional commutative algebra over $\mathbb{C}$ with identity (in the applications $\mathscr{A}$ will be the even part of an exterior algebra).

If R is an $n \times n$ antisymmetric matrix with coefficients in $\mathscr{A}$, we associate to it the $\mathscr{A}$-valued one-form ω on V given by

$$\omega = \frac{1}{4}\sum_{ij} R_{ij} x_j dx_i.$$

The operator $\nabla_i = \partial_i + \omega(\partial_i) = \partial_i + \frac{1}{4}\sum_j R_{ij} x_j$ acts on the space of $\mathscr{A}$-valued functions on V.

Definition 4.11. *Let R be an $n \times n$ antisymmetric matrix and let F be an $N \times N$ matrix, both with coefficients in the commutative algebra $\mathscr{A}$. The generalised harmonic operator H associated to R and F is the differential operator acting on $\mathscr{A} \otimes \operatorname{End}(\mathbb{C}^N)$-valued functions on V defined by the formula*

$$H = -\Big(\sum_i \nabla_i^2\Big) + F = -\sum_i \Big(\partial_i + \frac{1}{4}\sum_j R_{ij} x_j\Big)^2 + F.$$

Consider the following $\mathscr{A}$-valued function defined on the space of $n \times n$ antisymmetric matrices with values in $\mathscr{A}$:

$$j_V(R) = \det\left(\frac{e^{R/2} - e^{-R/2}}{R/2}\right).$$

Since $j_V(0) = 1$, $j_V^{-1/2}(tR)$ is defined for t sufficiently near 0, and is an analytic function of t with a Taylor series of the form

$$j_V^{-1/2}(tR) = 1 + \sum_{k=1}^{\infty} t^k f_k(R), \tag{4.2}$$

where $f_k(R)$ is a homogeneous polynomial function of degree $k \geq 1$ with respect to the coefficients R_{ij} of the matrix R. Similarly, $(tR/2)\coth(tR/2)$ is an $n \times n$-symmetric matrix with coefficients in $\mathscr{A}$, defined for t small, and with Taylor expansion $1 + \sum_{k=1}^{\infty} t^{2k} c_k R^{2k}$. Thus the $\mathscr{A}$-valued quadratic form

$$\Big\langle x \Big| \frac{tR}{2} \coth\Big(\frac{tR}{2}\Big) \Big| x \Big\rangle$$

is defined for t small, and has a Taylor series of the form

$$\|x\|^2 + \sum_{k=1}^{\infty} c_k t^{2k} \langle x | R^{2k} | x \rangle. \tag{4.3}$$

Proposition 4.12. *The kernel $p_t(x, R, F)$ taking values in $\mathscr{A} \otimes \operatorname{End}(\mathbb{C}^N)$, defined for t small by the formula*

$$p_t(x,R,F) = (4\pi t)^{-n/2} j_V^{-1/2}(tR) \exp\Big(-\frac{1}{4t}\Big\langle x \Big| \frac{tR}{2}\coth\Big(\frac{tR}{2}\Big)\Big| x\Big\rangle\Big) \exp(-tF)$$

is a solution of the heat equation

$$(\partial_t + H_x) p_t(x) = 0.$$

Proof. We have to prove that $dp_t/dt = -Hp_t$. Both sides of this formula are analytic with respect to the coefficients R_{ij} of the matrix R. Thus it is sufficient to prove the result when R has real coefficients. Choosing an appropriate orthonormal basis of V, it suffices to verify the above formula when V is two dimensional and $Re^1 = re^2$, $Re^2 = -re^1$, so that

$$H = -(\partial_1^2 + \partial_2^2) - \tfrac{1}{2}r(x_2\partial_1 - x_1\partial_2) - (\tfrac{1}{4}r)^2(x_1^2 + x_2^2) + F$$

and

$$p_t(x_1, x_2) = (4\pi t)^{-1}\frac{tr/2}{\sin(tr/2)}\exp(-(tr/2)\cot(tr/2)\|x\|^2/4t)\exp(-tF).$$

Since the function $\|x\|^2$ is annihilated by the infinitesimal rotation $x_2\partial_1 - x_1\partial_2$, the fact that p_t satisfies the heat equation follows from (4.1) which gives the heat kernel for the harmonic oscillator on the real line, except that r must be replaced by ir. □

In the applications we will make of the above formula, the coefficients of the matrices R and F will be nilpotent, so that the kernel will be defined for all t. Let $q_t(x) = (4\pi t)^{-n/2}e^{-\|x\|^2/4t}$ be the solution of the heat equation for the Euclidean Laplacian on V. The kernel $p_t(x, R, F)$ has an asymptotic expansion of the form

$$p_t(x, R, F) = q_t(x)\sum_{k=0}^{\infty} t^k \Phi_k(x),$$

for small t, with $\Phi_0 = 1$.

We will prove again the unicity result for formal solutions of the heat equation for the harmonic oscillator, even though a theorem of greater generality has already been proved in Theorem 2.26. Observe that H is the Laplacian on the Euclidean vector space V associated to the non-trivial connection $\nabla = d + \omega$. Let $\mathcal{R} = \sum_i x_i\partial_i$ be the radial vector field on V. If s_t is a $\mathcal{A} \otimes \mathrm{End}(\mathbb{C}^N)$-valued smooth function on V, then

$$(\partial_t + H)q_t s_t = q_t(\partial_t + t^{-1}\mathcal{R} + H)s_t;$$

this follows from Proposition 2.24, if we bear in mind that the antisymmetry of R implies that $\omega(\mathcal{R}) = 0$, so that $\nabla_{\mathcal{R}} = \mathcal{R}$.

Let $\Phi_t(x)$ be a formal power series in t, whose coefficients are smooth $\mathcal{A} \otimes \mathrm{End}(\mathbb{C}^N)$-valued functions defined in a neighbourhood of 0 in V. We will say that $q_t(x)\Phi_t(x)$ is a formal solution of the heat equation $(\partial_t + H)p_t = 0$, if $\Phi_t(x)$ satisfies the equation

$$(\partial_t + t^{-1}\mathcal{R} + H)\Phi_t = 0.$$

Theorem 4.13. *For any $a_0 \in \mathcal{A} \otimes \mathrm{End}(\mathbb{C}^N)$, there exists a unique formal solution $p_t(x, R, F, a_0)$ of the heat equation*

$$(\partial_t + H_x)p_t(x) = 0$$

of the form

$$p_t(x) = q_t(x) \sum_{k=0}^{\infty} t^k \Phi_k(x)$$

and such that $\Phi_0(0) = a_0$. *The function* $p_t(x,R,F,a_0)$ *is given by the formula*

$$(4\pi t)^{-n/2} j_V^{-1/2}(tR) \exp\Big(-\frac{1}{4t}\Big\langle x \Big| \frac{tR}{2} \coth\Big(\frac{tR}{2}\Big)\Big| x\Big\rangle\Big) \exp(-tF) a_0.$$

Proof. Although this follows from the results of Chapter 2, it is worth giving the direct proof in this special case, since it is made so easy by our formula for the formal solution. On the left side of the equation

$$(\partial_t + t^{-1}\mathscr{R} + H_x) \sum_{k=0}^{\infty} t^k \Phi_k(x) = 0,$$

we set the coefficients of each t^k to zero. This gives the system of equations

$$\mathscr{R}\Phi_0 = 0$$
$$(\mathscr{R}+k)\Phi_k = -H_x \Phi_{k-1} \quad \text{if } k > 0.$$

Thus we see that Φ_0 is the constant function equal to a_0; uniqueness is then clear, since if $a_0 = 0$, then $\Phi_k = 0$ for all k.

The function $p_t(x,R,F)$ that we found above is analytic in t for small t and is a solution of the heat equation, thus the corresponding formal expansion $p_t(x,R,F) = q_t(x) \sum_{k=0}^{\infty} t^k \Phi_k(x)$, is a formal solution. □ □

4.3 Calculation of the Index Density

Let $\mathscr{E}$ be a graded Clifford module on an even-dimensional Riemannian manifold M, with Clifford connection $\nabla^{\mathscr{E}}$ and associated Dirac operator D. Recall Lichnerowicz's formula, Theorem 3.52, for D^2

$$\mathsf{D}^2 = \Delta + \sum_{i<j} F^{\mathscr{E}/S}(e_i,e_j) c^i c^j + \tfrac{1}{4} r_M,$$

where Δ is the Laplacian of the bundle $\mathscr{E}$ with respect to the connection $\nabla^{\mathscr{E}}$, e_i a local orthonormal frame of TM with dual frame e^i of T^*M, c^i equals $c(e^i)$, and $F^{\mathscr{E}/S}(e_i,e_j) \in \operatorname{End}_{C(M)}(\mathscr{E})$ are the coefficients of the twisting curvature of the Clifford connection $\nabla^{\mathscr{E}}$.

Fix $x_0 \in M$ and trivialize the vector bundle $\mathscr{E}$ in a neighbourhood of x_0 by parallel transport along geodesics. More precisely, let $V = T_{x_0}M$, $E = \mathscr{E}_{x_0}$ and $U = \{\mathbf{x} \in V \mid \|\mathbf{x}\| < \varepsilon\}$, where ε is smaller than the injectivity radius of the manifold M at x_0. We identify U by means of the exponential map $\mathbf{x} \mapsto \exp_{x_0} \mathbf{x}$ with a neighbourhood of x_0 in M. For $x = \exp_{x_0} \mathbf{x}$, the fibre $\mathscr{E}_x$ and E are identified by the parallel transport map $\tau(x_0,x) : \mathscr{E}_x \to E$ along the geodesic $x_s = \exp_{x_0} s\mathbf{x}$. Thus the space $C^\infty(U,\mathscr{E})$ of sections of $\mathscr{E}$ over U may be identified with the space of E-valued C^∞-functions on

V, defined in the neighbourhood U. A differential operator D is written in this local trivialization as

$$D = \sum_{\alpha} a_\alpha(\mathbf{x})\partial_{\mathbf{x}}^{\alpha},$$

where $a_\alpha(\mathbf{x}) \in \mathrm{End}(E)$.

Let us compute D^2 in the above trivialization. For this, we need to compute the Clifford action and the connection. Choose an orthonormal basis ∂_i of $V = T_{x_0}M$, with dual basis $d\mathbf{x}^i$ of $T^*_{x_0}M$, and let $c^i = c(d\mathbf{x}^i) \in \mathrm{End}(E)$. Let S be the spinor space of V^* and let $W = \mathrm{Hom}_{C(V^*)}(S, E)$ be the auxiliary vector space such that $E = S \otimes W$, so that $\mathrm{End}(E) \cong \mathrm{End}(S) \otimes \mathrm{End}(W) \cong C(V^*) \otimes \mathrm{End}(W)$. Let e_i be the local orthonormal frame obtained by parallel transport along geodesics from the orthonormal basis ∂_i of $T_{x_0}M$, and let e^i be the dual frame of T^*M.

Lemma 4.14. *In the trivialization of $\mathscr{E}$ by parallel transport along geodesics, the* $\mathrm{End}(E)$*-valued function $c(e^i)_{\mathbf{x}}$ is the constant endomorphism c^i.*

Proof. Let $\mathscr{R}$ be the radial vector field on V. In the trivialization of $\mathscr{E}$ given by parallel transport along geodesics, the covariant derivative $\nabla^{\mathscr{E}}_{\mathscr{R}}$ is, by definition, differentiation by $\mathscr{R}$. By construction, $\nabla_{\mathscr{R}} e^i = 0$, and since $\nabla^{\mathscr{E}}$ is a Clifford connection, $[\nabla^{\mathscr{E}}, c(e^i)] = c(\nabla e^i)$. Thus $\mathscr{R} \cdot c(e^i)_{\mathbf{x}} = 0$, so that the Clifford action of the cotangent vector $e^i_{\mathbf{x}}$ is constant and equal to c^i. □

It follows from the above lemma that in this trivialization the bundle $\mathrm{End}_{C(M)}(\mathscr{E})$, restricted to U, is the trivial bundle with fibre $\mathrm{End}_{C(V)}(E) = \mathrm{End}(W)$.

Lemma 4.15. *The action of the covariant derivative $\nabla^{\mathscr{E}}_{\partial_i}$ on $C^\infty(U, E)$ is given by the formula*

$$\nabla^{\mathscr{E}}_{\partial_i} = \partial_i + \frac{1}{4} \sum_{j;k<l} R_{klij}\mathbf{x}^j c^k c^l + \sum_{k<l} f_{ikl}(\mathbf{x}) c^k c^l + g_i(\mathbf{x}),$$

where $R_{klij} = (R(\partial_i, \partial_j)\partial_l, \partial_k)_{x_0}$ is the Riemannian curvature at x_0, and

$$f_{ikl}(\mathbf{x}) = O(|\mathbf{x}|^2) \in C^\infty(U),$$
$$g_i(\mathbf{x}) = O(|\mathbf{x}|) \in C^\infty(U, \mathrm{End}_{C(V^*)}(E)) = \Gamma(U, \mathrm{End}(W)),$$

are error terms.

Proof. If $F^{\mathscr{E}}$ is the curvature of the connection $\nabla^{\mathscr{E}}$, and R is the Riemannian curvature of M, Proposition 3.43 shows that

$$F^{\mathscr{E}} = \tfrac{1}{2} \sum_{i<j;k<l} (R(\partial_i, \partial_j)e_k, e_l) c^k c^l \, d\mathbf{x}^i \wedge d\mathbf{x}^j + F^{\mathscr{E}/S}(\partial_i, \partial_j) d\mathbf{x}^i \wedge d\mathbf{x}^j,$$

where

$$F^{\mathscr{E}/S}(\partial_i, \partial_j) \in \Gamma(U, \mathrm{End}_{C(V^*)}(E)) \cong \Gamma(U, \mathrm{End}(W)), \quad \text{and}$$
$$(R(\partial_i, \partial_j)e_k, e_l) \in C^\infty(U).$$

Observe that $(R(\partial_i, \partial_j)e_k, e_l)_{x_0} = -R_{klij}$, since at the point x_0, $e_i = \partial_i$. If we apply the formula $\mathscr{L}(\mathscr{R})\omega = \iota(\mathscr{R})F^{\mathscr{E}}$ (see 1.12), we obtain the lemma. □

Let $p_t(x,x_0)$ be the heat kernel of the operator D^2. We transfer this kernel to the neighbourhood U of $0 \in V$, thinking of it as taking values in $\mathrm{End}(E)$, by writing

$$k(t,\mathbf{x}) = \tau(x_0,x)p_t(x,x_0),$$

where $x = \exp_{x_0}\mathbf{x}$. Since we may identify $\mathrm{End}(E) = C(V^*) \otimes \mathrm{End}(W)$ with $\Lambda V^* \otimes \mathrm{End}(W)$ by means of the full symbol map σ, $k(t,\mathbf{x})$ is identified with a $\Lambda V^* \otimes \mathrm{End}(W)$-valued function on U. Consider the space $\Lambda V^* \otimes \mathrm{End}(W)$ as a $C(V^*) \otimes \mathrm{End}(W)$ module, where the action of $C(V^*)$ on ΛV^* is the usual one $c(\alpha) = \varepsilon(\alpha) - \iota(\alpha)$. The next lemma shows that $k(t,\mathbf{x})$ is a solution to a heat equation on the open manifold U.

Lemma 4.16. *Let L be the differential operator on $U \subset V$, with coefficients in $C(V^*) \otimes \mathrm{End}(W)$, defined by the formula*

$$L = -\sum_i \Big((\nabla^{\mathcal{E}}_{e_i})^2 - \nabla^{\mathcal{E}}_{\nabla_{e_i} e_i} \Big) + \tfrac{1}{4} r_M + \sum_{i<j} F^{\mathcal{E}/S}(e_i,e_j) c^i c^j.$$

The $\Lambda V^* \otimes \mathrm{End}(W)$-valued function $k(t,\mathbf{x})$ satisfies the differential equation

$$(\partial_t + L)k(t,\mathbf{x}) = 0.$$

Proof. Since the kernel $p_t(x,x_0) \in \mathcal{E}_x \otimes \mathcal{E}^*_{x_0}$ satisfies the differential equation

$$(\partial_t + D_x^2)p_t(x,x_0) = 0,$$

it follows that $\tau(x_0,x)p_t(x,x_0) \in C^\infty(U,\mathrm{End}(E))$ satisfies the differential equation $(\partial_t + L)\tau(x_0,x)p_t(x,x_0) = 0$, where $C(V^*) \otimes \mathrm{End}(W) \cong \mathrm{End}(E)$ acts on the left on E. But under the symbol isomorphism $\mathrm{End}(E) \cong C(V^*) \otimes \mathrm{End}(W) \cong \Lambda V^* \otimes \mathrm{End}(W)$, the action of $C(V^*) \otimes \mathrm{End}(W)$ on E intertwines with the usual action of $C(V^*) \otimes \mathrm{End}(W)$ on $\Lambda V^* \otimes \mathrm{End}(W)$. □

To explain the method which we will use to prove the local index theorem, let us explain it in a simplified setting. Let H be a generalized Laplacian on a vector bundle $\mathcal{E}$. If $0 < u \leq 1$ is a small parameter, let us rescale space-time by sending t to ut and $\mathbf{x}$ to $u^{1/2}\mathbf{x}$; t rescales by the same power of u as $\|\mathbf{x}\|^2$ because the heat equation satisfied by $k(t,\mathbf{x})$ has two space derivatives but only one time derivative. Consider the rescaled heat kernel $k(u,t,\mathbf{x})$, defined by

$$k(u,t,\mathbf{x}) = u^{n/2}k(ut,u^{1/2}\mathbf{x}). \tag{4.4}$$

The factor of $u^{n/2}$ in this definition is included because the heat kernel is a density in the variable $\mathbf{x}$; to put this in a more elementary way, it is needed in order that the rescaled kernel continues to satisfy the initial condition $\lim_{t\to 0} k(u,t,\mathbf{x}) = \delta(\mathbf{x})$ for all $0 < u \leq 1$. Note that the Euclidean heat kernel $q_t(\mathbf{x}) = (4\pi t)^{-n/2}e^{-\|\mathbf{x}\|^2/4t}$ is invariant under this rescaling.

If $\Psi(t,\mathbf{x}) = \sum_{i=0}^\infty t^i \Psi_i(\mathbf{x})$ is a formal power series in t with polynomial coefficients $\Psi_i(\mathbf{x})$, we define the rescaled series by

$$\Psi(u,t,\mathbf{x}) = \sum_{i=0}^{\infty}(ut)^i\Psi_i(u^{1/2}\mathbf{x}).$$

This is a formal series in $u^{1/2}$ with coefficients polynomial functions of $(t,\mathbf{x})$.

Proposition 4.17. *There exist* $\mathrm{End}(E)$*-valued polynomials* $\Psi_i(\mathbf{x})$ *on* V *with* $\Psi_0(0) = 1$*, such that the function* $u \mapsto k(u,t,\mathbf{x})$ *has an asymptotic expansion in* $u^{1/2}$ *when* $u \to 0$ *of the form*

$$k(u,t,\mathbf{x}) \sim q_t(\mathbf{x})\sum_{i=0}^{\infty}(ut)^i\Psi_i(u^{1/2}\mathbf{x}).$$

This expansion is uniform for $(t,\mathbf{x})$ *lying in compact subsets of* $(0,1)\times U$ *and asymptotic expansions for the derivatives* $u \mapsto \partial_t^k\partial_{\mathbf{x}}^{\alpha}(k(u,t,\mathbf{x}))$ *may be obtained by differentiation.*

Proof. By Theorem 2.30, there exist functions $\phi_i \in C^\infty(U,\mathrm{End}(E))$, with $\phi_0(0) = 1$, such that for any $\mathbf{x} \in U$,

$$\left\| k(t,\mathbf{x}) - q_t(\mathbf{x})\sum_{i=0}^{N} t^i\phi_i(\mathbf{x})\right\| \le C(N)t^{N-n/2}.$$

Using the bounds $|x^k e^{-x^2/4t}| \le C_k t^{k/2}$, we can replace $\phi_i(\mathbf{x})$ by its Taylor expansion $\psi_i(\mathbf{x})$ of order $2(N-i)$:

$$\left\| k(t,\mathbf{x}) - q_t(\mathbf{x})\sum_{i=0}^{N} t^i\psi_i(\mathbf{x})\right\| \le C'(N)t^{N-n/2}.$$

Thus we obtain

$$\left\| k(u,t,\mathbf{x}) - q_t(\mathbf{x})\sum_{i=0}^{N} (ut)^i\psi_i(u^{1/2}\mathbf{x})\right\| \le C'(N)u^N,$$

for $(t,\mathbf{x})$ lying in compact subsets of $(0,1)\times U$ and $0 < u \le 1$, which gives the desired asymptotic expansion of the function $u \mapsto k(u,t,\mathbf{x})$. The argument for the derivatives of $k(u,t,\mathbf{x})$ is similar. □

The main idea of our proof of the index theorem is to modify the rescaling of space-time of (4.4) by inclusion of a "rescaling" of ΛV^*. Given $0 < u \le 1$ and $\alpha \in C^\infty(\mathbb{R}_+ \times U, \Lambda V^* \otimes \mathrm{End}(W))$, let $\delta_u\alpha$ equal

$$(\delta_u\alpha)(t,\mathbf{x}) = \sum_{i=0}^{n} u^{-i/2}\alpha(ut,u^{1/2}\mathbf{x})_{[i]}.$$

This rescaling has the following effect on operators on $C^\infty(\mathbb{R}_+ \times U, \Lambda V^* \otimes \mathrm{End}(W))$:

$$\begin{aligned}
\delta_u \phi(\mathbf{x}) \delta_u^{-1} &= \phi(u^{1/2}\mathbf{x}) \quad \text{for } \phi \in C^\infty(U), \\
\delta_u \partial_t \delta_u^{-1} &= u^{-1} \partial_t, \\
\delta_u \partial_i \delta_u^{-1} &= u^{-1/2} \partial_i, \\
\delta_u \varepsilon(\alpha) \delta_u^{-1} &= u^{-1/2} \varepsilon(\alpha) \quad \text{for } \alpha \in V^*, \\
\delta_u \iota(\alpha) \delta_u^{-1} &= u^{1/2} \iota(\alpha).
\end{aligned}$$

Definition 4.18. *The rescaled heat kernel $r(u,t,\mathbf{x})$ is defined by*

$$r(u,t,\mathbf{x}) = u^{n/2}(\delta_u k)(t,\mathbf{x}).$$

The above choice is motivated by the fact that

$$\lim_{u\to 0} r(u,t=1,\mathbf{x}=0) = \lim_{u\to 0} \sum_{i=0}^{n} u^{(n-i)/2} k_u(x_0,x_0)_{[i]}.$$

The right-hand side of this equation is precisely what we would like to prove to be equal to

$$(4\pi)^{-n/2} \hat{A}(M) \exp(-F^{\mathscr{E}/S}).$$

It is clear that $r(u,t,\mathbf{x})$ satisfies the differential equation

$$(\partial_t + u\delta_u L \delta_u^{-1}) r(u,t,\mathbf{x}) = 0.$$

The most important step in the proof of Theorem 4.1 is the calculation of the leading term of the asymptotic expansion of $L(u) = u\delta_u L \delta_u^{-1}$ as a function of u; we show that

$$L(u) = K + O(u^{1/2}),$$

where K is a harmonic oscillator of the type that we discussed in the last section. To show that this implies that $r(u,t,\mathbf{x})$ has a limit as $u \to 0$, we use the following lemma, which is a consequence of Proposition 4.17.

Lemma 4.19. *There exist $\Lambda V^* \otimes \mathrm{End}(W)$-valued polynomials $\gamma_i(t,\mathbf{x})$ on $\mathbb{R} \times V$, such that for every integer N, the function $r^N(u,t,\mathbf{x})$ defined by*

$$r^N(u,t,\mathbf{x}) = q_t(\mathbf{x}) \sum_{i=-n}^{2N} u^{i/2} \gamma_i(t,\mathbf{x})$$

approximates $r(u,t,\mathbf{x})$ in the following sense: for $N > j + |\alpha|/2$, there is a constant $C(N,j,\alpha)$ such that

$$\|\partial_t^j \partial_{\mathbf{x}}^\alpha (r(u,t,\mathbf{x}) - r^N(u,t,\mathbf{x}))\| \le C(N,j,\alpha) u^N,$$

for $0 < u \le 1$ and $(t,\mathbf{x}) \in (0,1) \times U$. Furthermore, $\gamma_i(0,0) = 0$ if $i \neq 0$, while $\gamma_0(0,0) = 1$.

Proof. By Proposition 4.17, there exist $\Lambda V^* \otimes \mathrm{End}(W)$-valued polynomial functions $\Psi_i(\mathbf{x})$ with $\Psi_0(0) = 1$ such that

$$\left\| k(t,\mathbf{x}) - q_t(\mathbf{x}) \sum_{i=0}^{N} t^i \Psi_i(\mathbf{x}) \right\| \le C t^{N-n/2}.$$

If we project out the p-form component of this equation, we see that

$$\left\| k(t,\mathbf{x})_{[p]} - q_t(\mathbf{x}) \sum_{i=0}^{N} t^i \Psi_i(\mathbf{x})_{[p]} \right\| \le C t^{N-n/2}.$$

Rescaling by $u \in (0,1)$, we obtain

$$\left\| r(u,t,\mathbf{x})_{[p]} - u^{-p/2} q_t(\mathbf{x}) \sum_{i=0}^{N} (ut)^i \Psi_i(u^{1/2}\mathbf{x})_{[p]} \right\| \le C u^{N-p/2} t^{N-n/2}.$$

It is clear that we must define $\gamma_j(t,\mathbf{x})_{[p]}$ to be the coefficient of $u^{j/2}$ in the sum

$$u^{-p/2} \sum_{i=0}^{(j+p)/2} (ut)^i \Psi_i(u^{1/2}\mathbf{x})_{[p]},$$

which is a polynomial on $\mathbb{R}_+ \times V$ with values in $\Lambda^p V^* \otimes \mathrm{End}(W)$. It is clear that the sum $\gamma_j(t,\mathbf{x}) = \sum_{p=0}^{n} \gamma_j(t,\mathbf{x})_{[p]}$ satisfies $\delta_u \gamma_j = u^{j/2} \gamma_j$. Furthermore, $\gamma_j(t,\mathbf{x})_{[p]} = 0$ for $j < -p$, and hence $\gamma_j(t,\mathbf{x}) = 0$ for $j < -n$.

We see that for $(t,\mathbf{x}) \in (0,1) \times U$ and $0 < u \le 1$,

$$\left\| r(u,t,\mathbf{x}) - q_t(\mathbf{x}) \sum_{i=-n/2}^{2N} u^{i/2} \gamma_i(t,\mathbf{x}) \right\| \le C u^N.$$

The argument for the derivatives is similar. The values of $\gamma_i(0,0)$ are determined by

$$\sum_{i=-n/2}^{\infty} u^{i/2} \gamma_i(0,0) = (\delta_u \Psi)(0,0) = 1. \qquad \square$$

Using the fact that $L(u) = K + O(u^{1/2})$, which will be proved later, we can now show that there are no poles in the Laurent series expansion in $u^{1/2}$ of $r(u,t,\mathbf{x})$,

$$r(u,t,\mathbf{x}) \sim q_t(\mathbf{x}) \sum_{i=-n}^{\infty} u^{i/2} \gamma_i(t,\mathbf{x}).$$

Expanding the equation $(\partial_t + L(u)) r(u,t,\mathbf{x}) = 0$ in a Laurent series in $u^{1/2}$, we see that the leading term $u^{-\ell/2} q_t(\mathbf{x}) \gamma_{-\ell}(t,\mathbf{x})$ of the asymptotic expansion of $r(u,t,\mathbf{x})$ satisfies the heat equation $(\partial_t + K_{\mathbf{x}})(q_t(\mathbf{x}) \gamma_{-\ell}(t,\mathbf{x})) = 0$. Since formal solutions of the heat equation for the harmonic oscillator are uniquely determined by $\gamma_{-\ell}(0,0)$, by the results of the last section, and since $\gamma_{-\ell}(0,0) = 0$ for $\ell > 0$, we see that $\gamma_{-\ell}$ is identically equal to zero unless $\ell = 0$. In particular, we see that there are no poles in the Laurent expansion of $r(u,t,\mathbf{x})$ in powers of $u^{1/2}$.

The other thing that we learn from this argument is that the leading term of the expansion of $r(u,t,\mathbf{x})$, namely $r(0,t,\mathbf{x}) = q_t(\mathbf{x})\gamma_0(t,\mathbf{x})$, satisfies the heat equation for the operator $L(0) = K$, with initial condition

$$\gamma_0(0,0) = 1.$$

Thus, to calculate $r(0,t,\mathbf{x})$, we have only to obtain a formula for K, and to apply the results of the last section.

In order to state the formula for K, let us denote by R the matrix with nilpotent entries equal to

$$\mathsf{R}_{ij} = (R_{x_0}\partial_i, \partial_j) \in \Lambda^2 V^*, \tag{4.5}$$

where R_{x_0} is the Riemannian curvature of M at x_0, operating on the algebra ΛV^* by exterior multiplication. Similarly, let F be the element of $\Lambda^2 V^* \otimes \mathrm{End}(W)$ obtained by evaluating the twisting curvature $F^{\mathcal{E}/S}$ at the point x_0; again, it acts on $\Lambda V^* \otimes \mathrm{End}(W)$ by exterior multiplication.

Proposition 4.20. *The family of differential operators $L(u) = u\delta_u L \delta_u^{-1}$ acting on $C^\infty(U, \Lambda V^* \otimes \mathrm{End}(W))$ has a limit K when u tends to 0, given by the formula*

$$K = -\sum_i \Big(\partial_i - \tfrac{1}{4}\sum_j \mathsf{R}_{ij}\mathbf{x}_j\Big)^2 + \mathsf{F}.$$

Proof. By Lemma 4.15, the differential operator $\nabla^{\mathcal{E},u}_{\partial_i} = u^{1/2}\delta_u \nabla^{\mathcal{E}}_{\partial_i}\delta_u^{-1}$ is equal to

$$\partial_i + \tfrac{1}{4}\sum_{j;k<l} R_{klij}\mathbf{x}^j(\varepsilon^k - u\iota^k)(\varepsilon^l - u\iota^l)$$
$$+ u^{-1/2}\sum_{k<l} f_{ikl}(u^{1/2}\mathbf{x})(\varepsilon^k - u\iota^k)(\varepsilon^l - u\iota^l) + u^{1/2} g_i(u^{1/2}\mathbf{x}).$$

Since $f_{ikl}(\mathbf{x}) = O(|\mathbf{x}|^2)$, we see that $\nabla^{\mathcal{E},u}_{\partial_i}$ has a limit as $u \to 0$, equal to

$$\partial_i + \frac{1}{4}\sum_{j;k<l} (R_{x_0})_{klij}\mathbf{x}^j \varepsilon^k \varepsilon^l.$$

Using the fundamental symmetry of the curvature $R_{ijkl} = R_{klij}$ and the definition (4.5) of $\mathsf{R}_{ij} = \sum_{k<l}(R_{x_0})_{jikl}\varepsilon^k\varepsilon^l$, the above limit is seen to equal

$$\nabla^{\mathcal{E},0}_{\partial_i} = \partial_i - \frac{1}{4}\sum_j \mathsf{R}_{ij}\mathbf{x}^j.$$

The operator $L(u)$ equals $L_1(u) + L_2(u)$, where

$$L_1(u) = -\sum_i (\nabla^{\mathcal{E},u}_{e_i})^2 + \sum_{i<j} F^{\mathcal{E}/S}(e_i,e_j)(u^{1/2}\mathbf{x})(\varepsilon^i - u\iota^i)(\varepsilon^j - u\iota^j)$$

$$L_2(u) = \tfrac{1}{4} u r_M(u^{1/2}\mathbf{x}) + u^{1/2}\sum_i \nabla^{\mathcal{E},u}_{\nabla_{e_i}e_i}.$$

Clearly, $\lim_{u\to 0} L_2(u) = 0$, since its leading term is the sum of $ur_M(x_0)/4$ and $u^{1/2}$ times $\lim_{u\to 0}\sum_i \nabla^{\mathcal{E},u}_{\nabla_{e_i}e_i}$, which we have just shown to be nonsingular. The first term of $L_1(u)$ has limit

$$-\sum_i (\nabla^{\mathcal{E},0}_{\partial_i})^2 = -\sum_i (\partial_i - \tfrac{1}{4}\sum_j \mathsf{R}_{ij}\mathbf{x}^j)^2,$$

while its second term converges to

$$\lim_{u\to 0}\sum_{i<j} F^{\mathcal{E}}(e_i,e_j)(u^{1/2}\mathbf{x})(\varepsilon^i - u\iota^i)(\varepsilon^j - u\iota^j) = \sum_{i<j}\mathsf{F}(\partial_i,\partial_j)\varepsilon^i\varepsilon^j = \mathsf{F},$$

since the vector fields e_i coincide with ∂_i when $\mathbf{x} = 0$. □

The operator $L(0) = K$ is a generalized harmonic oscillator, in the sense of Definition 4.11, associated to the $n\times n$ antisymmetric curvature matrix $-\mathsf{R}_{ij}$ and the $N\times N$-matrix F (where $N = \dim(W)$), with coefficients in the commutative algebra $\Lambda^+ V$. Applying Theorem 4.13, we obtain the following result.

Theorem 4.21. *The limit* $\lim_{u\to 0} r(u,t,\mathbf{x})$ *exists, and is given by the formula*

$$(4\pi t)^{-n/2}\det{}^{1/2}\left(\frac{t\mathsf{R}/2}{\sinh t\mathsf{R}/2}\right)\exp\left(-\frac{1}{4t}\left\langle \mathbf{x}\left|\frac{t\mathsf{R}}{2}\coth\left(\frac{t\mathsf{R}}{2}\right)\right|\mathbf{x}\right\rangle - t\mathsf{F}\right).$$

Setting $t = 1$ and $\mathbf{x} = 0$, we obtain Theorem 4.1.

Bibliographic Notes

Theorem 4.1 is proved in Getzler [61], using a symbol calculus for pseudodifferential operators which combines the usual symbol for pseudodifferential operators with the Clifford symbol σ. It is also in this article that the approximation of D^2 by a harmonic oscillator is introduced. Another exposition of this proof may be found in Roe's book [97].

Earlier proofs which were special cases of the proof of [61] were those of the Gauss-Bonnet-Chern theorem by Patodi [89], and the Riemann-Roch theorem for Riemann surfaces by Kotake [78]. The previous proofs of the local index theorem, by Patodi [90], Gilkey [65], and Atiyah-Bott-Patodi [8], use power counting and symmetries of the Riemannian curvature to show that the top-order piece $\sigma_n(k_t(x,x))$ must be a characteristic form given by a universal formula for all n-dimensional manifolds, and then calculate sufficient special cases of the global index theorem to characterize this characteristic form completely. Thus, this route to calculating the local index density is less direct. This approach to the local index theorem is explained in Gilkey's book [67].

There have been a number of other proofs of the local index theorem similar to ours, among which we mention Bismut [28], Friedan-Windey [60] and Yu [108].

Also, the work of the physicist Alvarez-Gaumé [1] was influential in the development of this point of view.

Getzler [62] gives a proof of Theorem 4.1 based on the Feynman-Kac formula in stochastic differential geometry. The same proof is used in Getzler [63] to prove Theorem 4.4. This theorem is also proved in Bismut-Cheeger [34] by stochastic estimates. Both of these papers obtain global estimates for the error

$$k_u(t,x,x) - (4\pi ut)^{-n/2} \det^{1/2}\Big(\frac{uR/2}{\sinh(uR/2)}\Big) \exp(-uF^{\mathcal{E}/S})$$

as a function of u and x.

There is a generalization of Theorem 4.4 to the infinite dimensional situation of Chapter 9. This generalization may be used to give an alternative approach to the family index theorem, and is also an important tool in further generalizations of this theorem; see Bismut-Cheeger [34] for a typical example of its use.

The Gauss-Bonnet-Chern theorem is due to Chern [49]; we gave an adaptation of his original proof in Section 1.6. The local Gauss-Bonnet-Chern theorem is due to Patodi [89].

The signature theorem is proved by Hirzebruch in [70], using cobordism theory; in [71], Hirzebruch tells the story of his discovery of this theorem. The twisted signatures were first calculated by Atiyah-Singer [13], also using cobordism theory. In the same article, they introduce the twisted Dirac operator and calculate its index. Their work was motivated by one of the central problems of analysis: to calculate the index of an arbitrary elliptic differential operator on a compact manifold. Atiyah and Singer used Bott periodicity to show that the index of any elliptic pseudodifferential operator may be written in terms of the numbers $\sigma(\mathcal{W})$. References for their general index theorem are two seminars directed by Cartan [48] and Palais [88], and of course Atiyah's Collected Works [2]; an excellent account of Atiyah and Singer's discovery of their theorem may be found in the introduction to [2].

The local index theorem is proved for the $\bar{\partial}$-operator on Kähler manifolds by Patodi [89], and for the twisted signature operator by Gilkey [65]. Atiyah-Bott-Patodi [8] give a unified treatment of the local index theorem for the classical Dirac operators by Gilkey's method.

The index of the $\bar{\partial}$-operator was first calculated by Hirzebruch [70], in the special case of smooth projective varieties; another proof, which proves a much more general result, and led to the introduction of K-theory into the study of index problems, is due to Grothendieck [40]. Atiyah and Singer gave the first proof valid for an arbitrary compact complex manifold, as a corollary of their general index theorem. The local index theorem for these operators was proved by Kotake [78] for Riemann surfaces, and by Patodi [89] and Gilkey [65] in general.

5

The Exponential Map and the Index Density

In this chapter, we will give another proof of Theorem 4.1 for a Dirac operator D, independent of the one in Chapter 4. This proof generalizes easily to obtain the fixed point formula for the equivariant index of a group of isometries, as we will see in the next chapter. A feature of the proof is that it gives an explanation for the striking similarity between the $\hat{A}$-genus and the Jacobian of the exponential map on a Lie group, both of which involve the j-**function**

$$j(X) = \frac{\sinh(X/2)}{X/2}.$$

Indeed, the $\hat{A}$-genus will appear naturally through the Jacobian of the exponential map on the principal bundle $SO(M)$ over M: this Jacobian can be computed along the fibres in terms of the curvature of M, and this is where the j-function makes its appearance, as we will see in Section 1.

Let us explain the idea of this proof. Let $P \to M$ be a principal bundle endowed with a connection, with compact structure group G, and let $\mathscr{E}$ be a vector bundle associated to P, with Laplacian $\Delta^{\mathscr{E}}$. It is easy to see that the kernel k_t of the heat semi-group $e^{-t\Delta^{\mathscr{E}}}$ is obtained from the scalar heat kernel h_t on the manifold P by averaging over the fibres $P_x \cong G$. This gives integral formulas for the coefficients of the asymptotic expansion of $k_t(x,x)$, as we will see in Section 2.

Let M be a spin manifold of dimension n, and let $\mathrm{Spin}(M)$ be the corresponding principal bundle with structure group $\mathrm{Spin}(n)$; $\mathrm{Spin}(M)$ is a double cover of $SO(M)$. Let D be the Dirac operator on the spin bundle $\mathscr{S}$ of M. Since by Lichnerowicz's formula, the square of the Dirac operator is a generalized Laplacian, we obtain a formula for the restriction $k_t(x,x) = \langle x \mid e^{-t\mathsf{D}^2} \mid x\rangle$ of the heat kernel to the diagonal in terms of the scalar heat kernel h_t of the Riemannian manifold $\mathrm{Spin}(M)$: apart from minor factors, k_t is given by the integral

$$k_t(x,x) = \int_{\mathrm{Spin}(n)} h_t(p, pg)\rho(g)^{-1}\, dg,$$

where $\rho : \mathrm{Spin}(n) \to \mathrm{End}(S)$ is the spin representation and p is an element of $\mathrm{Spin}(M)$ lying above $x \in M$. In exponential coordinates, this is a Gaussian integral over $\Lambda^2\mathbb{R}^n$, which we identify with the Lie algebra of $\mathrm{Spin}(n)$. Using this formula, it is easy to compute $\sigma(k)$, and the answer only involves the first term of the asymptotic expansion of h_t. The elimination of the singular part of $\mathrm{Str}(k_t(x,x))$ is reduced to the following simple lemma: if ξ is a nilpotent element of an algebra $\mathscr{A}$ (which for us will be an exterior algebra), and if f is a polynomial in one variable, then

$$\lim_{t\to 0}(4\pi t)^{-1/2}\int_{\mathbb{R}} e^{-(x-\xi)^2/4t} f(x)\,dx = f(\xi).$$

As we saw in Chapter 2, the first term in the asymptotic expansion of h_t can be computed explicitly, for any Riemannian manifold, in terms of the Jacobian of the Riemannian exponential map, so we obtain the equality $\sigma(k) = \hat{A}(M)$. This argument is extended to the case of a twisted Dirac operator, in which case h_t must be replaced by the heat kernel of a generalized Laplacian associated to the twisting bundle $\mathscr{W}$, and we obtain $\sigma(k) = \hat{A}(M)\exp(-F^{\mathscr{E}/S})$; the twisting curvature $F^{\mathscr{E}/S}$ is brought into this formula by the parallel transport factor in the first term of the asymptotic expansion of h_t.

5.1 Jacobian of the Exponential Map on Principal Bundles

Let G be a compact Lie group with Lie algebra $\mathfrak{g}$, let M be a compact Riemannian manifold and let $\pi : P \to M$ be a principal bundle over M with structure group G and connection form ω. Let us choose on $\mathfrak{g}$ a G-invariant positive scalar product; we will denote by $\int_G f(g)\,dg$ the integral with respect to the corresponding Riemannian volume form; because the Riemannian metric is invariant, this is a Haar measure. For $p \in P$ above $x \in M$, the tangent space T_pP is the direct sum of the horizontal space H_pP, which is identified with T_xM by the projection $\pi_* : TP \to \pi^*TM$, and the vertical space identified with $\mathfrak{g}$. Taking H_pP and $\mathfrak{g}$ orthogonal to each other turns P into a Riemannian manifold, with metric depending on ω.

For $p \in P$, we denote by

$$J(p,a) : T_pP \to T_qP$$

the derivative of the map $\exp_p$ at the tangent vector $a \in T_pP$. The purpose of this section is to compute $J(p,a)$ when $a \in \mathfrak{g}$. This computation is very similar to the computation of the derivative of the exponential map in a Lie group.

If $a \in \mathfrak{g}$, let $j_{\mathfrak{g}}(a)$ equal

$$j_{\mathfrak{g}}(a) = \det_{\mathfrak{g}}\left(\frac{1-e^{-\operatorname{ad} a}}{\operatorname{ad} a}\right). \tag{5.1}$$

Note that $j_{\mathfrak{g}}(0) = 1$.

Proposition 5.1. *If dg is a Haar measure on G, then $d(\exp a) = j_{\mathfrak{g}}(a)\,da$ in exponential coordinates.*

Proof. If G is a compact Lie group with Lie algebra $\mathfrak{g}$, then

$$\frac{d}{d\varepsilon}\exp(-a)\exp(a+\varepsilon b)|_{\varepsilon=0} = \frac{1-e^{-\operatorname{ad} a}}{\operatorname{ad} a}b, \tag{5.2}$$

for a and $b \in \mathfrak{g}$. In a matrix group, this follows from Duhamel's formula (2.6); since G is compact, it has a faithful matrix representation. □

Let V be an Euclidean vector space with orthonormal basis e_i. The Euclidean scalar product on $\Lambda^2 V$ has orthonormal basis $\{e_i \wedge e_j \mid i < j\}$. Identify $\Lambda^2 V$ with $\mathfrak{so}(V)$ by the map $\tau : \Lambda^2 V \to \mathfrak{so}(V)$ of Proposition 3.7, defined by the formula $(\tau(\alpha)e_i, e_j) = 2(\alpha, e_i \wedge e_j)$.

We denote by Ω the curvature of the connection ω on P. It is a horizontal $\mathfrak{g}$-valued 2-form on P so, for $p \in P$, Ω_p is an element of $\Lambda^2 H_p^* P \otimes \mathfrak{g}$. If $a \in \mathfrak{g}$, the contraction of Ω_p with a is an element of $\Lambda^2 H_p^* P$ which we will denote by $\Omega_p \cdot a$.

If X is a vector field on M, we denote by X^P its horizontal lift on P. If $a \in \mathfrak{g}$, we denote by a the corresponding vertical vector field on P. Since X^P is invariant by G, we see that $[a, X^P] = 0$. From the definition of the curvature of a connection (1.6), we see that

$$[X^P, Y^P] = [X, Y]^P - \Omega(X^P, Y^P)_P,$$

where $\Omega(X^P, Y^P) \in \mathfrak{g}$ induces a vertical vector field $\Omega(X^P, Y^P)_P$ on P. We will usually write $\Omega(X^P, Y^P)$ instead of $\Omega(X^P, Y^P)_P$. The formula

$$\Omega_{pg}(X^P, Y^P) = g^{-1}\Omega_p(X^P, Y^P)$$

follows from the G-invariance of ω.

The skew-symmetric endomorphism $\tau(\Omega_p \cdot a) \in \operatorname{End}(T_x M)$ is given by

$$\tau(\Omega_p \cdot a)X = 2\sum_i (\Omega_p(X^P, e_i^P), a)e_i,$$

and it varies along the fibre containing p by the formula

$$\tau(\Omega_{pg} \cdot a) = \tau(\Omega_p \cdot (ga)). \tag{5.3}$$

Denote by ∇^M and ∇^P the Levi-Civita covariant derivatives on the tangent bundles TM and TP.

Lemma 5.2. *Let X and Y be vector fields on M, and let a and b be elements of $\mathfrak{g}$. Then*

1. $\nabla_a^P b = \frac{1}{2}[a, b]$.
2. $\nabla_a^P X^P = \frac{1}{4}\tau(\Omega_p \cdot a)X^P = \nabla_{X^P}^P a$
3. $\nabla_{X^P}^P Y^P = (\nabla_X^M Y)^P - \frac{1}{2}\Omega(X^P, Y^P)$

Proof. By the definition of ∇^P, if X, Y and Z are vector fields on P, $2(\nabla^P_X Y, Z)$ equals

$$X(Y,Z)+Y(Z,X)-Z(X,Y)+([X,Y],Z)-([Y,Z],X)+([Z,X],Y).$$

To prove (1), we observe that $(\nabla^P_a b, X^P)=0$, while

$$(\nabla^P_a b,c)=\tfrac{1}{2}([a,b],c)-\tfrac{1}{2}([b,c],a)+\tfrac{1}{2}([c,a],b)=\tfrac{1}{2}([a,b],c),$$

since the scalar product on $\mathfrak{g}$ is G–invariant.

To prove (2), we observe that

$$2(\nabla^P_a X^P, Y^P)=-([X^P,Y^P],a)=\tfrac{1}{2}(\tau(\Omega\cdot a)X^P,Y^P)$$

by the definition of τ, while $(\nabla^P_a X^P, b)=0$. A similar argument leads to the proof of (3). □

As a consequence of Lemma 5.2, we obtain a description of the horizontal and vertical geodesics in P.

Lemma 5.3. *1. For $a\in\mathfrak{g}$ and $p\in P$, the curve $s\mapsto p\exp(sa)$ is a geodesic, so that $\exp_p a = p\exp a$.*
2. If $x(s)$ is a geodesic in M, its horizontal lift is a geodesic in P.

Proof. By definition we have $\partial_s(p\exp sa)=a$, so (1) follows from the relation $\nabla^P_a a = 0$. Similarly (2) follows from Lemma 5.2 (3). □

Fix $x\in M$ and let $V=T_xM$. For $p\in P$ such that $\pi(p)=x$, identify H_pP with V and T_pP with $V\oplus\mathfrak{g}$. Then the map

$$J(p,a): T_pP\to T_{p\exp a}P$$

becomes a linear map from $V\oplus\mathfrak{g}$ to itself.

Theorem 5.4. *The map $J(p,a)$ preserves the subspaces V and $\mathfrak{g}$ of T_pP, and we have*

$$J(p,a)|_V=\frac{1-e^{-\tau(\Omega_p\cdot a)/2}}{\tau(\Omega_p\cdot a)/2},$$

$$J(p,a)|_{\mathfrak{g}}=\frac{1-e^{-\operatorname{ad}a}}{\operatorname{ad}a}.$$

Proof. The formula for $J(p,a)|_{\mathfrak{g}}$ follows from (5.2) and Lemma 5.3 (1). To calculate $J(p,a)|_V$, we need to compute $\partial_t \exp_p(a+tX)|_{t=0}$, for $X\in V$. Introduce $p(s,t)\doteq\exp_p s(a+tX)$ and $p(s)=\exp_p(sa)$. For every $t\in\mathbb{R}$, the curve $s\mapsto p(s,t)$ is a geodesic. Put $Y(s)=\partial_t p(s,t)|_{t=0}$, so that $Y(s)$ is a vector field along the curve $p(s)$, corresponding to the deformation of the geodesic $p(s)$ into the geodesic $p(s,\varepsilon)$; such a vector field is called a Jacobi vector field. We have $J(p,a)X=Y(s)|_{s=1}$. Let $R^P\in\mathscr{A}^2(P,\operatorname{End}(TP))$ be the curvature of the manifold P. The Jacobi vector field $Y(s)$ is determined by the differential equation

$$\nabla^P_{\partial_s}\nabla^P_{\partial_s}Y(s) = R^P(\partial_s p, Y(s))\cdot\partial_s p \tag{5.4}$$

with initial conditions $Y(0) = 0$ and $\nabla^P_{\partial_s}Y(s)|_{s=0} = X$. Let us recall how these equations are obtained: since ∇ is torsion free, taking the covariant derivative of $\nabla_{\partial_s}\partial_s p = 0$ gives

$$\begin{aligned} 0 = \nabla_{\partial_t}\nabla_{\partial_s}\partial_s p &= \nabla_{\partial_s}\nabla_{\partial_t}\partial_s p + R(\partial_t p, \partial_s p)\partial_s p \\ &= \nabla_{\partial_s}\nabla_{\partial_s}\partial_t p + R(\partial_t p, \partial_s p)\partial_s p, \end{aligned}$$

Let X be a vector field on M. Since $\nabla^P_a a = 0$ and $[a, X^P] = 0$, we see that

$$R^P(a, X^P)a = \nabla^P_a \nabla^P_{X^P} a.$$

Since $\exp(sa)a = a$, it follows from (5.3) that $\tau(\Omega_{p\exp(sa)}\cdot a) = \tau(\Omega_p \cdot a)$, hence is independent of s and

$$\left(\nabla^P_{X^P} a\right)_{p(s)} = \tfrac{1}{4}\left(\tau(\Omega_p\cdot a)X\right)^P.$$

It follows that

$$R^P(a, X^P)\cdot a = \left(\tfrac{1}{4}\tau(\Omega_p\cdot a)\right)^2 X^P.$$

If $V(s)$ is an horizontal vector field on P along the curve $p(s)$, the vectors $\nabla^P_a\nabla^P_a V(s)$ and $R^P(a, V(s))a$ are horizontal. Thus the differential equation (5.4) implies that $Y(s)$ remains horizontal, and identifies with an element $y(s)$ of V. In this context, the differential equation (5.4) says that

$$(\partial_s + \tau(\Omega_p\cdot a)/4)^2 y(s) = \left(\tau(\Omega_p\cdot a)/4\right)^2 y(s),$$

so that we obtain the theorem. □

For $X \in \mathrm{End}(V)$, recall the Definition 3.12 of j_V,

$$j_V(X) = \det\left(\frac{\sinh X/2}{X/2}\right).$$

The Hirzebruch $\hat{A}$-genus appears in the index formula because of the following corollary to the above theorem.

Corollary 5.5. $\det(J(p,a)) = j_{\mathfrak{g}}(a)\, j_V(\tau(\Omega_p\cdot a)/2))$

Proof. Let A be the matrix $\tau(\Omega_p\cdot a)/2$. Since A is antisymmetric, we see that $\det(e^{A/2}) = 1$, so that

$$\det\left(\frac{1-e^{-A}}{A}\right) = \det\left(\frac{\sinh A/2}{A/2}\right). \qquad \square$$

There is no conflict of notation between $j_{\mathfrak{g}}(a)$ and $j_V(X)$ since the compact group G is unimodular, so that $j_{\mathfrak{g}}(a)$ and $j_{\mathfrak{g}}(\mathrm{ad}\,a)$ are equal.

5.2 The Heat Kernel of a Principal Bundle

Let $P \to M$ be a principal bundle with compact structure group G and connection form ω. Let (ρ, E) be a representation of G in a vector space E and let $\mathcal{E} = P \times_G E$ be the associated vector bundle on M, with covariant derivative ∇ associated to the connection ω on P. Let $\Delta^{\mathcal{E}}$ be the Laplacian of $\mathcal{E}$; with respect to a local orthonormal frame e_i of TM,

$$\Delta^{\mathcal{E}} = -\sum_i (\nabla_{e_i}^2 - \nabla_{\nabla_{e_i} e_i}).$$

We can identify the space $\Gamma(M, \mathcal{E})$ with the subspace of $C^\infty(P) \otimes E$ invariant under the group G, denoted $C^\infty(P,E)^G$, as was explained in Proposition 1.7. In this section, we will relate the Laplacian $\Delta^{\mathcal{E}}$ and its heat kernel to the scalar Laplacian Δ^P on P, which acts on $C^\infty(P)$.

If $\phi \in C^\infty(P)$ is a function on P, we have the integration formula

$$\int_P \phi(p)dp = \int_M \Big(\int_G \phi(pg)dg \Big) dx.$$

In this formula, note that $\int_G \phi(pg)dg$ depends only on $x = \pi(p)$.

Let E_a be an orthonormal basis of the Lie algebra $\mathfrak{g}$ of G and let $\mathrm{Cas} = \sum_a \rho(E_a)^2 \in \mathrm{End}(E)$ be the **Casimir operator**; it is independent of the choice of the orthonormal basis E_a of $\mathfrak{g}$ and commutes with the action of G on E. In particular, if the representation ρ is irreducible, Cas is a scalar.

Proposition 5.6. *The Laplacian $\Delta^{\mathcal{E}}$ coincides with the restriction of the operator $\Delta^P \otimes 1 + 1 \otimes C$ to $(C^\infty(P) \otimes E)^G$.*

Proof. Let e_i^P be the horizontal lift of the vector field e_i on M to the principal bundle P. The vector fields e_i^P and E_a form a local orthonormal basis of TP. By (5.2) we see that

$$\Delta^P = -\sum_i ((e_i^P)^2 - (\nabla_{e_i} e_i)^P) - \sum_a E_a^2.$$

In the identification of $\Gamma(M, \mathcal{E})$ with $(C^\infty(P) \otimes E)^G$, the operator ∇_{e_i} corresponds to e_i^P, while, using the G-invariance of elements of $(C^\infty(P) \otimes E)^G$, the vertical vector field E_a may be replaced by the multiplication operator $-\rho(E_a)$; the proposition follows. □

Let us compare the semigroups of operators $e^{-t\Delta^{\mathcal{E}}}$ on $\Gamma(M, \mathcal{E})$ and $e^{-t\Delta^P}$ on $C^\infty(P)$. The Schwartz kernel of $e^{-t\Delta^{\mathcal{E}}}$ with respect to the Riemannian density $|dx|$ of M is denoted by $\langle x \mid e^{-t\Delta^{\mathcal{E}}} \mid y \rangle$. On pulling back to P, it becomes an element $(p_1, p_2) \mapsto \langle p_1 \mid e^{-t\Delta^{\mathcal{E}}} \mid p_2 \rangle$ of $C^\infty(P \times P) \otimes \mathrm{End}(E)$ which satisfies

$$\langle p_1 g_1 \mid e^{-t\Delta^{\mathcal{E}}} \mid p_2 g_2 \rangle = \rho(g_1)^{-1} \langle p_1 \mid e^{-t\Delta^{\mathcal{E}}} \mid p_2 \rangle \rho(g_2).$$

From Proposition 5.6, we see that

$$e^{-t\Delta^{\mathscr{E}}} = e^{-t\mathrm{Cas}} e^{-t\Delta^P}.$$

Thus we obtain an integral representation of the heat kernel of the Laplacian of an associated vector bundle $\mathscr{E}$ in terms of the scalar heat kernel on the manifold P.

Proposition 5.7. *If p_1 and p_2 are points in the principal bundle P, then*

$$\langle p_1 \mid e^{-t\Delta^{\mathscr{E}}} \mid p_2 \rangle = e^{-t\mathrm{Cas}} \int_G \langle p_1 \mid e^{-t\Delta^P} \mid p_2 g \rangle \rho(g)^{-1} dg.$$

Proof. We have

$$\begin{aligned}(e^{-t\Delta^{\mathscr{E}}} \phi)(p_1) &= e^{-t\mathrm{Cas}} \int_P \langle p_1 \mid e^{-t\Delta^P} \mid p_2 \rangle \phi(p_2) dp_2 \\ &= e^{-t\mathrm{Cas}} \int_M \Big(\int_G \langle p_1 \mid e^{-t\Delta^P} \mid p_2 g \rangle \phi(p_2 g) dg \Big) dx \\ &= e^{-t\mathrm{Cas}} \int_M \Big(\int_G \langle p_1 \mid e^{-t\Delta^P} \mid p_2 g \rangle \rho(g)^{-1} dg \Big) \phi(p_2) dx,\end{aligned}$$

from which the theorem follows. □

This theorem leads to an integral representation for the asymptotic expansion of $\langle x \mid e^{-t\Delta^{\mathscr{E}}} \mid x \rangle$ in terms of the asymptotic expansion of the heat kernel of P.

If $\phi \in C_c^\infty(\mathbb{R})$ is a smooth function with compact support, then by Proposition 2.13, the integral

$$\int_{\mathbb{R}} e^{-a^2/4t} \phi(a) da \sim (4\pi t)^{1/2} \sum_{k=0}^{\infty} \frac{(-t)^k}{k!} (\Delta^k \phi)(0)$$

has an asymptotic expansion in powers of $t^{1/2}$ near $t = 0$ which depends only on the Taylor series of ϕ at 0.

Let $\phi \in C_c^\infty(\mathfrak{g})$ be a smooth function with compact support on the vector space $\mathfrak{g}$. We will denote by

$$\int_{\mathfrak{g}}^{\mathrm{asymp}} e^{-\|a\|^2/4t} \phi(a) da \tag{5.5}$$

the asymptotic expansion in $t^{1/2}$ of the integral

$$\int_{\mathfrak{g}} e^{-\|a\|^2/4t} \phi(a) da.$$

This power series has the form $t^{\dim(\mathfrak{g})/2} \sum_{i=0}^{\infty} t^i \lambda_i$.

If ϕ is a smooth function defined only in a neighbourhood U of $0 \in \mathfrak{g}$, and ψ is a cut-off function on $\mathfrak{g}$ with support in U and equal to 1 near 0, then

$$\int_{\mathfrak{g}}^{\mathrm{asymp}} e^{-\|a\|^2/4t} \phi(a) \psi(a) da$$

is independent of the choice of ψ, so we will write it simply as $\int_{\mathfrak{g}}^{\text{asymp}} e^{-\|a\|^2/4t}\phi(a)da$. If $\Phi(t,a)=\sum_{i=0}^{\infty} t^i\Phi_i(a)$ is a formal power series in t with coefficients in $C^\infty(U)$, we will write

$$\int_{\mathfrak{g}}^{\text{asymp}} e^{-\|a\|^2/4t}\Phi(t,a)da$$

for the formal series

$$\sum_{i=0}^{\infty} t^i \int_{\mathfrak{g}}^{\text{asymp}} e^{-\|a\|^2/4t}\Phi_i(a)da.$$

To simplify the statement of the next theorem, we will suppose that the Casimir Cas of the representation ρ of G on E is a multiple of the identity.

Theorem 5.8. *There exist smooth functions Φ_i on $P\times\mathfrak{g}$ such that*

$$\langle p \mid e^{-t\Delta^{\mathcal{E}}} \mid p\rangle \sim (4\pi t)^{-(\dim(M)+\dim(\mathfrak{g}))/2}\int_{\mathfrak{g}}^{\text{asymp}} \sum_{i=0}^{\infty} t^i\Phi_i(p,a)e^{-\|a\|^2/4t}\rho(\exp a)^{-1}\, j_{\mathfrak{g}}(a)da.$$

Furthermore, denoting $T_{\pi(p)}M$ by V, we have

$$\begin{aligned}\Phi_0(p,a) &= \det{}^{-1/2}(J(p,a))\\ &= j_{\mathfrak{g}}^{-1/2}(a)\, j_V^{-1/2}(\tau(\Omega_p\cdot a)/2).\end{aligned}$$

Proof. Let $h_t(p_1,p_2)$ be the heat kernel of Δ^P on $C^\infty(P)$. By Theorem 5.8, we have

$$\langle p \mid e^{-t\Delta^{\mathcal{E}}} \mid p\rangle = e^{-t\mathrm{Cas}}\int_G h_t(p,pg)\rho(g)^{-1}dg.$$

Let $\psi(t)$ be a cut-off function which vanishes for t greater than half the square of the injectivity radius of M.

In Section 2.5, we showed how to obtain a formal series $\sum_{i=0}^{\infty} t^i\Psi_i$, whose coefficients Φ_i are smooth functions in the neighbourhood

$$\{(p_1,p_2)\mid d(p_1,p_2)<\varepsilon\}$$

of the diagonal of $P\times P$ such that if

$$h_t^N(p_1,p_2) = (4\pi t)^{-\dim(P)/2}\psi(d(p_1,p_2)^2)e^{-d(p_1,p_2)^2/4t}\sum_{i=0}^{N} t^i\Psi_i(p_1,p_2),$$

we have

$$\|h_t(p_1,p_2)-h_t^N(p_1,p_2)\| = O(t^{N-\dim(P)/2}).$$

By Theorem 2.30 and Corollary 5.5, we have, for a small,

$$h_t(p,p\exp a) = (4\pi t)^{-\dim(P)/2}e^{-\|a\|^2/4t}\sum_{i=0}^{N} t^i\Psi_i(p,a) + O(t^{N-\dim(P)/2}),$$

where $\Phi_0(p,a) = j_{\mathfrak{g}}^{-1/2}(a)j_V^{-1/2}(\tau(\Omega_p\cdot a)/2)$. The proof is completed by multiplying through by the formal power series $e^{-t\mathrm{Cas}}=\sum_{k=0}^{\infty}(-t\mathrm{Cas})^k/k!$, and inserting the formula $dg = j_{\mathfrak{g}}(a)da$. □

Let us extend this theorem to the case of a generalized Laplacian on a twisted bundle $\mathcal{W}\otimes\mathcal{E}$. Let $\mathcal{W}$ be a bundle over M, which we will call the **twisting bundle**. We can make the following identifications between space of sections:

$$\Gamma(M,\mathcal{W}\otimes\mathcal{E}) = (\Gamma(P,\pi^*\mathcal{W})\otimes E)^G,$$
$$\Gamma(M,\mathrm{End}(\mathcal{W}\otimes\mathcal{E})) = (\Gamma(P,\pi^*\mathrm{End}(\mathcal{W}))\otimes\mathrm{End}(E))^G.$$

If $F\in(\Gamma(P,\pi^*\mathrm{End}(\mathcal{W}))\otimes\mathfrak{g})^G$, then applying the map $\rho:\mathfrak{g}\to\mathrm{End}(E)$, we obtain a section

$$\rho(F)\in(\Gamma(P,\pi^*\mathrm{End}(\mathcal{W}))\otimes\mathrm{End}(E))^G\cong\Gamma(M,\mathrm{End}(\mathcal{W}\otimes\mathcal{E})),$$

which we will simply denote by F. We will also abbreviate our notation for the bundles $\pi^*\mathcal{W}$ and $\pi^*\mathrm{End}(\mathcal{W})$ on P to $\mathcal{W}$ and $\mathrm{End}(\mathcal{W})$. If $F\in(\Gamma(P,\mathrm{End}(\mathcal{W}))\otimes\mathfrak{g})^G$ and $a\in\mathfrak{g}$, then for all $p\in P$, the contraction $F_p\cdot a$ is a well-defined element of $\mathrm{End}(\mathcal{W}_x)$, where $x=\pi(p)$, and we have the invariance for $g\in G$,

$$F_{pg}\cdot a = F_p\cdot(ga). \tag{5.6}$$

We will consider generalized Laplacians on $\mathcal{W}\otimes\mathcal{E}$ of the form

$$H=\Delta^{\mathcal{W}\otimes\mathcal{E}}+F^0+F^1,$$

where

$$F^0\in\Gamma(M,\mathrm{End}(\mathcal{W}))\subset\Gamma(M,\mathrm{End}(\mathcal{W}\otimes\mathcal{E})),$$
$$F^1\in(\Gamma(P,\mathrm{End}(\mathcal{W}))\otimes\mathfrak{g})^G.$$

Here, $\mathcal{W}$ is a twisting bundle with connection $\nabla^{\mathcal{W}}$, and $\Delta^{\mathcal{W}\otimes\mathcal{E}}$ is the Laplacian on $\Gamma(M,\mathcal{W}\otimes\mathcal{E})$ with respect to the tensor product connection on $\mathcal{W}\otimes\mathcal{E}$.

The main result of this section will be an integral formula for the asymptotic expansion of the heat kernel of H. As before, we may view the kernel of e^{-tH} as a section

$$(p_1,p_2)\longmapsto\langle p_1\,|\,e^{-tH}\,|\,p_2\rangle$$

of the bundle $\mathcal{W}\boxtimes\mathcal{W}^*\otimes\mathrm{End}(E)$ over $P\times P$. Restricted to the diagonal, the kernel $\langle p\,|\,e^{-tH}\,|\,p\rangle$ gives a section of $(\Gamma(P,\mathrm{End}(\mathcal{W}))\otimes\mathrm{End}(E))^G$. We will give an asymptotic expansion of $\langle p\,|\,e^{-tH}\,|\,p\rangle$ compatible with this tensor product decomposition.

Theorem 5.9. *Let $H=\Delta^{\mathcal{W}\otimes\mathcal{E}}+F$, where $F=F^0+F^1$ is a potential such that $F^0\in\Gamma(M,\mathrm{End}(\mathcal{W}))$ and $F^1\in\Gamma(P,\mathrm{End}(\mathcal{W}))\otimes\mathfrak{g})^G$. Assume that the Casimir* Cas *of the representation ρ is scalar. Then there exist smooth sections $\Phi_j\in\Gamma(P\times\mathfrak{g},\mathrm{End}(\mathcal{W}))$ (where we denote by* $\mathrm{End}(\mathcal{W})$ *the pull-back of* $\mathrm{End}(\mathcal{W})$ *on M to $P\times\mathfrak{g}$) such that*

$$\langle p\,|\,e^{-tH}\,|\,p\rangle\sim$$
$$(4\pi t)^{-(\dim(M)+\dim(\mathfrak{g}))/2}\int_{\mathfrak{g}}^{\mathrm{asymp}} e^{-\|a\|^2/4t}\sum_{j=0}^{\infty}t^j\Phi_j(p,a)\otimes\rho(\exp a)^{-1}\,j_{\mathfrak{g}}(a)\,da.$$

Furthermore, letting $V = T_{\pi(p)}M$, we have

$$\Phi_0(p,a) = j_{\mathfrak{g}}^{-1/2}(a)\, j_V^{-1/2}(\tau(\Omega_p \cdot a)/2) \exp(F^1_p \cdot a/2).$$

Proof. The proof is very similar to the proof of Theorem 5.8. The first step is to write H as the restriction to $\Gamma(M, \mathscr{W} \otimes \mathscr{E}) = (\Gamma(P, \mathscr{W}) \otimes E)^G$ of a carefully chosen generalized Laplacian H^P on P. We will thus obtain an integral formula similar to Theorem 5.7 for the kernel $\langle p \mid e^{-tH} \mid p\rangle$.

The map $a \mapsto \frac{1}{2} F^1_p \cdot a \in \operatorname{End}(\mathscr{W}_x)$ defines a vertical one-form on P. We associate to the potential F a connection $\tilde{\nabla}$ on $\pi^*\mathscr{W}$ by adding to the pull-back $\pi^*\nabla^{\mathscr{W}}$ of the connection $\nabla^{\mathscr{W}}$ on $\mathscr{W}$ the vertical one-form $\frac{1}{2}F^1$:

$$\tilde{\nabla} = \nabla^{\mathscr{W}} + \tfrac{1}{2} F^1.$$

Let $\tilde{\Delta}$ be the Laplacian of the bundle $\pi^*\mathscr{W}$ associated to the Riemannian structure of P and the connection $\tilde{\nabla}$.

Lemma 5.10. *There exists a potential $\tilde{F} \in \Gamma(P, \operatorname{End}(\mathscr{W}))^G$ such that H is the restriction of the operator $(\tilde{\Delta} + \tilde{F}) \otimes 1 + 1 \otimes \mathrm{Cas}$ to $\Gamma(M, \mathscr{W} \otimes \mathscr{E}) = (\Gamma(P, \mathscr{W}) \otimes E)^G$.*

Proof. Let $\tilde{F} = \frac{1}{4}\sum_a (F^1 \cdot E_a)^2 + F^0$, where E_a is an orthonormal basis of $\mathfrak{g}$. We have

$$H = -\sum_i \tilde{\nabla}^2_{e_i^P} - \tilde{\nabla}_{(\nabla_{e_i} e_i)^P} + \sum_a F^1 \cdot E_a \otimes \rho(E_a) + F^0,$$

whereas

$$\tilde{\Delta} = -\sum_i \tilde{\nabla}^2_{e_i^P} - \tilde{\nabla}_{(\nabla_{e_i} e_i)^P} - \sum_a \tilde{\nabla}^2_{E_a}.$$

Observe that

$$\begin{aligned}\sum_a \tilde{\nabla}^2_{E_a} &= \sum_a \left(E_a + \tfrac{1}{2} F^1 \cdot E_a\right)^2 \\ &= \sum_a E_a^2 + \tfrac{1}{2}(F^1 \cdot E_a) E_a + \tfrac{1}{2} E_a (F^1 \cdot E_a) + \tfrac{1}{4}(F^1 \cdot E_a)^2.\end{aligned}$$

From the relation (5.6), we see that $F^1_{p\exp(sE_a)} \cdot E_a$ is independent of s, so that the action of the vector field E_a commutes with the endomorphism $F^1 \cdot E_a$, and we obtain

$$\sum_a \tilde{\nabla}^2_{E_a} = \sum_a E_a^2 + (F^1 \cdot E_a) E_a + \tfrac{1}{4}(F^1 \cdot E_a)^2.$$

As before, we can replace the second order operator $\sum_a E_a^2$ on $(\Gamma(P, \mathscr{W}) \otimes E)^G$, by Cas, while we can replace E_a by $-\rho(E_a)$; the lemma follows easily. □

For p_1 and $p_2 \in P$, let us denote by $h_t(p_1, p_2) \in \operatorname{Hom}(\mathscr{W}_{p_2}, \mathscr{W}_{p_1})$ the kernel of $e^{-t(\tilde{\Delta} + \tilde{F})}$. As in Theorem 5.7, we have

$$\langle p_1 \mid e^{-tH} \mid p_2 \rangle = e^{-t\mathrm{Cas}} \int_G h_t(p_1, p_2 g) \rho(g)^{-1} \, dg.$$

There exists smooth sections $(p_1,p_2)\mapsto \Phi_j(p_1,p_2)\in \mathrm{Hom}(\mathscr{W}_{p_2},\mathscr{W}_{p_1})$, defined in a neighbourhood of the diagonal of $P\times P$, such that if

$$h_t^N(p_1,p_2)=(4\pi t)^{-\dim(P)/2}\psi(d(p_1,p_2)^2)\,e^{-d(p_1,p_2)^2/4t}\sum_{i=0}^{N}t^i\Phi_i(p_1,p_2),$$

we have

$$\left\|h_t(p_1,p_2)-h_t^N(p_1,p_2)\right\|=O(t^{N-\dim(P)/2}).$$

Furthermore, the first term $\Phi_0(p_1,p_2)$ does not depend on $\tilde{F}$, and is given by the formula

$$\Phi_0(p_1,p_2)=\det{}^{-1/2}(J(p_1,\exp_{p_1}^{-1}p_2))\,\tilde{\tau}(p_1,p_2),$$

where $\tilde{\tau}(p_1,p_2)\in\mathrm{Hom}(\mathscr{W}_{p_2},\mathscr{W}_{p_1})$ is the geodesic parallel transport with respect to the connection $\tilde{\nabla}$. We will compute $\tilde{\tau}$ in the next lemma.

Lemma 5.11. *For $p\in P$ and $a\in\mathfrak{g}$, we have*

$$\tilde{\tau}(p,p\exp a)=\exp(\tfrac{1}{2}F_p^1\cdot a).$$

Proof. The function $t\mapsto\tilde{\tau}(t)=\tilde{\tau}(p,p\exp ta)$ is the solution of the differential equation

$$\frac{d}{dt}\tilde{\tau}(t)-\tfrac{1}{2}\left(F^1_{p\exp ta}\cdot a\right)\tilde{\tau}(t)=0,\quad \tilde{\tau}(0)=1,$$

and, by (5.6), we have $F^1_{p\exp ta}\cdot a=F^1_p\cdot a$. □

The rest of the proof of Theorem 5.9 is the same as that of Theorem 5.8, except that the Φ_0 term is multiplied by $\tilde{\tau}(p,p\exp a)$, calculated above. □

When we apply the above theorem to the heat kernel of D^2 in Section 4, the factor $j_V^{-1/2}(\tau(a\cdot\Omega_p)/2)$ in

$$\Phi_0(p,a)=j_{\mathfrak{g}}^{-1/2}(a)\,j_V^{-1/2}(\tau(\Omega_p\cdot a)/2)\exp(F_p^1\cdot a/2)$$

will gives rise to the $\hat{A}$-genus of M, while the factor $\exp(F_p^1\cdot a/2)$ will lead to the relative Chern character.

5.3 Calculus with Grassmann and Clifford Variables

Let ϕ be a continuous function on $\mathbb{R}$ slowly increasing at infinity. For $a\in\mathbb{R}$, we have:

$$\lim_{t\to0}(4\pi t)^{-1/2}\int_{\mathbb{R}}e^{-(x-a)^2/4t}\phi(x)dx=\phi(a),$$

as follows from the change of variable $x\mapsto t^{1/2}(a+x)$. We will prove a generalization of this formula, which allows a to lie in a more general algebra.

Let $\mathscr{A}$ be a finite dimensional supercommutative algebra with unit. (In practice, $\mathscr{A}$ will be an exterior algebra). Let $\xi = (\xi_1, \ldots, \xi_n)$ be a n-tuple of even nilpotent elements of $\mathscr{A}$. Let ϕ be an $\mathscr{A}$-valued smooth function on $\mathbb{R}^n$. We define:

$$\phi(x+\xi) = \sum_{J\in\mathbb{N}^n} \partial^J \phi(x)\xi^J/J!, \quad \text{for } x \in \mathbb{R}^n. \tag{5.7}$$

This sum is finite because the elements ξ are nilpotent. We will write $\|x-\xi\|^2$ for the function $\sum_{j=1}^n (x_j - \xi_j)^2$.

Lemma 5.12. *Let ϕ be an $\mathscr{A}$-valued smooth function on $\mathbb{R}^n$ slowly increasing at infinity (as well as all its derivatives). Let $\xi = (\xi_1, \ldots, \xi_n)$ be an n-tuple of even nilpotent elements of $\mathscr{A}$. Then*

$$\lim_{t\to 0}(4\pi t)^{-n/2} \int_{\mathbb{R}^n} e^{-\|x-\xi\|^2/4t} \phi(x)\, dx = \phi(\xi).$$

Proof. If Φ and its derivatives are rapidly decreasing functions on $\mathbb{R}^n$, we have

$$\int_{\mathbb{R}^n} (\Phi(x+\xi) - \Phi(x)) dx = \sum_{J\neq 0} \frac{\xi^J}{J!} \int_{\mathbb{R}^n} \partial^J \Phi(x)\, dx = 0.$$

Thus

$$\begin{aligned} \int_{\mathbb{R}^n} e^{-\|x-\xi\|^2/4t} \phi(x)\, dx &= \int_{\mathbb{R}^n} e^{-\|x\|^2/4t} \phi(x+\xi)\, dx \\ &= \sum_J \frac{\xi^J}{J!} \int_{\mathbb{R}^n} e^{-\|x\|^2/4t} \partial^J \phi(x)\, dx. \end{aligned}$$

In the limit $t \to 0$, we obtain the lemma. □

Let V be a Euclidean vector space. We denote by $\exp_\Lambda(\alpha)$ the exponential of $\alpha \in \Lambda V$, and by $\mathfrak{g} = C^2(V)$ the Lie algebra of $\mathrm{Spin}(V)$, which we can identify as a vector space with $\Lambda^2 V$ by the symbol map. By the universal property of the symmetric algebra $\mathbb{C}[\mathfrak{g}]$ of polynomial functions on $\mathfrak{g}$, the injection of $\mathfrak{g}$ into the even part of ΛV extends to an algebra homomorphism

$$A : \mathbb{C}[\mathfrak{g}] \to \Lambda V.$$

We can extend the map A to the algebra to $C^\infty(\mathfrak{g})$ by composing with the Taylor series expansion at 0. Let us describe this homomorphism in the case in which $V = \mathbb{R}^n$, with orthonormal basis e^i. An element f of $C^\infty(\mathfrak{g})$ is a smooth function $f(a_{ij})$ of the coordinates $\{a_{ij} \mid 1 \le i < j \le n\}$ on $\Lambda^2\mathbb{R}^n$, and $A(f)$ is the element $f(e^i \wedge e^j) \in \Lambda V$ defined in (5.7), which is obtained by replacing the variables a_{ij} by the elements $e^i \wedge e^j \in \Lambda^2\mathbb{R}^n$. Denote by $e \wedge e^*$ the antisymmetric matrix whose (i, j)-coefficient equals $e^i \wedge e^j \in \Lambda^2 V$. We will use the suggestive notation $f(e \wedge e^*)$ for $A(f)$, and will call it the **evaluation** of f at $e \wedge e^*$.

Since $e_i \wedge e_i = 0$, many polynomial expressions in the variables $e_i \wedge e_j$ will vanish, and consequently some functions on $\mathfrak{g}$ will have a vanishing evaluation at $e \wedge e^*$. In the following lemma, we collect a number of such functions. Recall that τ is the canonical representation $\mathfrak{g} \to \mathfrak{so}(V)$.

Lemma 5.13. *1. Let $v, w \in V$ and let $f^k_{v,w}(a) = (\tau(a)^k v, w)$ be the corresponding coefficient of the matrix $\tau(a)^k$. Then $f^k_{v,w}(e \wedge e^*) = 0$ for $k \geq 2$.*

2. Let $f_k(a) = \mathrm{Tr}((\mathrm{ad}\, a)^k)$, where ad *is the adjoint representation of $\mathfrak{g}$. Then $f_k(e \wedge e^*) = 0$ for $k \geq 1$.*

3. $j_{\mathfrak{g}}(e \wedge e^*) = 1$

Proof. The coefficients of the matrix $\tau(a)^2$ are linear combinations of monomials $a_{ik}a_{kj}$. Thus (1) follows from the relation $e_k \wedge e_k = 0$.

The adjoint representation can be written as $\mathrm{ad}(a) = \tau(a) \wedge 1 + 1 \wedge \tau(a)$ for $a \in \mathfrak{g}$. Hence it follows from (1) that the evaluation at $e \wedge e^*$ of any coefficient of $(\mathrm{ad}(a))^k$ vanishes, when $k > 2$. Since $\tau(a)$ is antisymmetric, we also have $\mathrm{Tr}(\tau(a) \wedge \tau(a)) = -\mathrm{Tr}(\tau(a)^2)$ and $\mathrm{Tr}(\mathrm{ad}\, a) = 0$. Thus, for any $k \geq 1$, the evaluation at $e \wedge e^*$ of $\mathrm{Tr}((\mathrm{ad}\, a)^k)$ vanishes.

Using the relation $\det(A) = \exp(\mathrm{Tr}(\log A))$, we can write $j_{\mathfrak{g}}(a)$ in the form

$$j_{\mathfrak{g}}(a) = \exp\Big(\mathrm{Tr}\Big(\sum_{k \geq 1} c_k (\mathrm{ad}\, a)^k\Big)\Big),$$

and (3) follows from (2). (Alternatively, (2) and (3) can be seen as a consequence of the fact that the constants are the only elements in ΛV which are invariants under the full orthogonal group $\mathrm{O}(V)$.) □

The following proposition is the crucial technical tool leading to the elimination of the singular part of the heat kernel index density.

Proposition 5.14. *If Φ is a smooth slowly increasing function on $\mathfrak{g}$, then*

$$\lim_{t \to 0} (4\pi t)^{-\dim(\mathfrak{g})/2} \int_{\mathfrak{g}} e^{-\|a\|^2/4t} \exp_\Lambda(a/2t)\, \Phi(a)\, da = \Phi(e \wedge e^*).$$

Proof. Since $(e^i \wedge e^j)^2 = 0$, we obtain by completion of squares the formula

$$e^{-\|a\|^2/4t} \exp_\Lambda(a/2t) = \exp_\Lambda\Big(-\frac{1}{4t} \sum_{i<j} (a_{ij} - e^i \wedge e^j)^2\Big).$$

Applying Lemma 5.12 gives the desired formula. □

For the proof of Theorem 4.1, we need to replace Grassmann variables by Clifford variables in Proposition 5.14. The next proposition shows that this is indeed possible when taking the limit $t \to 0$. Let δ_t be the automorphism of ΛV such that

$$\delta_t(\gamma) = t^{-j/2}\gamma \quad \text{for } \gamma \in \Lambda^j V.$$

As usual, we will identify the spaces $C(V)$ and ΛV by means of the symbol map.

Proposition 5.15. *If Φ is a smooth slowly increasing function on $\mathfrak{g}$, and $\gamma \in \Lambda^k V$, then*

$$\lim_{t \to 0} (4\pi t)^{-\dim(\mathfrak{g})/2} t^{k/2} \int_{\mathfrak{g}} e^{-\|a\|^2/4t} \delta_t(\gamma \cdot \exp_C(a/2))\Phi(a)\, da = \gamma \wedge \Phi(e \wedge e^*).$$

Proof. Recall the relation (3.13) between the exponentials in the exterior algebra and the Clifford algebra:

$$\exp_C a = j_V^{1/2}(a)\det(H_V^{1/2}(a))\left(H_V^{-1/2}(a)\cdot\exp_\Lambda a\right),$$

where $H_V(a) : \mathfrak{g} \to \mathrm{GL}(V)$ is given by the formula

$$H_V(a) = \frac{\tau(a)/2}{\tanh(\tau(a)/2)}.$$

In particular, the matrix entries of $H_V^{-1/2}(a)$ are analytic function of the coefficients of the matrix $\tau(a)^2$.

Let e_I be the orthonormal basis of ΛV obtained from an orthonormal basis e_i of V. We have

$$\delta_t(\exp_C(a/2))$$
$$= j_V^{1/2}(a)\det(H_V^{1/2}(a))\sum_{IJ}\langle\exp_\Lambda(a/2t), e_I\rangle\,\langle H_V^{-1/2}(a)\,e_I, e_J\rangle\, e_J.$$

By Lemma 5.13, the evaluation of $j_V^{1/2}(a)\det(H_V^{1/2}(a))\langle H_V^{-1/2}(a)\,e_I, e_J\rangle$ at $e\wedge e^*$ is equal to δ_{IJ}. Thus, applying Proposition 5.14, we obtain the result for $\gamma = 1$.

If $\gamma = e_1\wedge\ldots\wedge e_k$, we have

$$\gamma\cdot\exp_C(a/2) = (\varepsilon_1 - \iota_1)\cdots(\varepsilon_k - \iota_k)\exp_C(a/2).$$

Applying δ_t, we obtain

$$t^{k/2}\delta_t(\gamma\cdot\exp_C(a/2)) = (\varepsilon_1 - t\iota_1)\cdots(\varepsilon_k - t\iota_k)\delta_t(\exp_C(a/2)).$$

and taking the limit, we deduce the result from the case $\gamma = 1$. □

We restate the above result in a slightly different way.

Proposition 5.16. *Let Φ be a smooth function on $\mathfrak{g}$ defined in a neighbourhood of 0. Let $\gamma \in C_m(V)$, with symbol $\sigma_m(\gamma)\in\Lambda^m V$. Consider the formal power series*

$$(4\pi t)^{-\dim(\mathfrak{g})/2}\int_{\mathfrak{g}}^{\mathrm{asymp}} e^{-\|a\|^2/4t}\,\Phi(a)(\gamma\cdot\exp_C a)\,da \sim \sum_{i=0}^{\infty} t^i C_i,$$

where $C_i \in C(V)$. Then the element C_i is of Clifford degree less than $2i+m$, and we have

$$\sum_{i=0}^{n/2}\sigma_{2i+m}(C_i) = \gamma\wedge\Phi(2e\wedge e^*).$$

Proof. Proposition 5.15 asserts that the Laurent series $\sum_{i=0}^{\infty} t^{i+m/2}\delta_t(C_i)$ has no poles, and computes the constant term. □

5.4 The Index of Dirac Operators

We now give our second proof of Theorem 4.1. Thus M is an oriented compact Riemannian manifold of even dimension n. For simplicity, we assume that M admits a spin-structure; this is always true locally, and since the theorem is local, it is sufficient to prove the theorem in this case.

Let $V = \mathbb{R}^n$, with its standard orthonormal basis e_i. Let $G = \mathrm{Spin}(n)$, with Lie algebra $\mathfrak{g} = C^2(V)$, and let $\tau : \mathfrak{g} \to \mathfrak{so}(V)$ be the map defined in Proposition 3.7. We denote by $\pi : P \to M$ the principal bundle $\mathrm{Spin}(M)$ which defines the spin-structure, and provide it with its Levi-Civita connection ω. Let $\Omega \in \mathscr{A}^2(\mathrm{Spin}(M), \mathfrak{g})$ be the curvature of the principal bundle $\mathrm{Spin}(M)$. Then the Riemannian curvature R of M is equal to $\tau(\Omega) \in \mathscr{A}^2(M, \mathfrak{so}(TM))$. At a point $p \in P$, which corresponds under the two-fold covering $\mathrm{Spin}(M) \to \mathrm{SO}(M)$ to an orthonormal frame at the point $x = \pi(p) \in M$, we may identify both T_xM and T_x^*M with the vector space V.

Let $\rho : G \to \mathrm{End}(S)$ be the spinor representation, and let $\mathscr{S}$ be the associated superbundle of spinors, which carries the Levi-Civita connection $\nabla^{\mathscr{S}}$. The Casimir Cas of the representation ρ is given by

$$\mathrm{Cas} = \sum_{i<j} (e_i e_j)^2 = -\dim(\mathfrak{g}) = -n(n-1)/2,$$

so is a scalar.

We suppose given a Hermitian complex vector bundle $\mathscr{W} \to M$ and a Hermitian connection $\nabla^{\mathscr{W}}$ on $\mathscr{W}$ with curvature $F^{\mathscr{W}}$, and we let D be the corresponding twisted Dirac operator of the twisted spinor bundle $\mathscr{W} \otimes \mathscr{S}$. Let $k_t(x,y)$ be the heat kernel of the operator D^2; along the diagonal, it is a section of the bundle $\mathrm{End}(\mathscr{W} \otimes \mathscr{S}) \cong C(M) \otimes \mathrm{End}(\mathscr{W})$.

Since the operator D^2 satisfies the Lichnerowicz formula

$$\mathsf{D}^2 = \Delta^{\mathscr{W} \otimes \mathscr{S}} + \mathbf{c}(F^{\mathscr{W}}) + \tfrac{1}{4} r^M,$$

we may apply to it the analysis of Section 2. Indeed,

$$\mathsf{D}^2 = \Delta^{\mathscr{W} \otimes \mathscr{S}} + F^0 + F^1,$$

where $F^0_p = \frac{1}{4} r^M$ and

$$F^1_p = \sum_{i<j} F^{\mathscr{W}}_p(e^P_i, e^P_j) \otimes (e_i \wedge e_j) \in (\Gamma(P, \mathrm{End}(\mathscr{W})) \otimes \mathfrak{g})^G. \tag{5.8}$$

By Theorem 5.9, the asymptotic expansion of $\langle x \mid e^{-t\mathsf{D}^2} \mid x \rangle$ has the form

$$(4\pi t)^{-(\dim(\mathfrak{g})+n)/2} \int_{\mathfrak{g}}^{\mathrm{asymp}} e^{-\|a\|^2/4t} \sum_{j=0}^{\infty} t^j \, \Phi_j(a) \otimes \rho(\exp a) \, j_{\mathfrak{g}}(a) \, da,$$

where Φ_j form a series of smooth functions on $\mathfrak{g}$ with values in $\mathrm{End}(\mathscr{W}_x)$, and

$$\Phi_0(a) = j_{\mathfrak{g}}^{-1/2}(a)\, j_V^{-1/2}(\tau(\Omega_p \cdot a)/2) \exp(-F_p^1 \cdot a/2).$$

In the identification $\mathrm{End}(S) \cong C(V) \cong \Lambda V$ by the symbol map σ, the matrix $\rho(\exp a)$ corresponds to the element $\exp_C(a)$ of ΛV. The first assertion of Theorem 4.1 follows immediately from Proposition 5.15 with $\gamma = 1$.

The formula in Proposition 5.15 for the leading order of the above asymptotic expansion shows that

$$\sigma(k) = \Phi_0(2e \wedge e^*)\, j_{\mathfrak{g}}(2e \wedge e^*).$$

By Lemma 5.13, the function $j_{\mathfrak{g}}(2a)$ evaluated at $e \wedge e^*$ equals 1. Thus, it remains to calculate $\Phi_0(2e \wedge e^*)$.

Identifying V and V^*, we consider the curvature $F_p^{\mathcal{W}}$ as an element of $\Lambda^2 V \otimes \mathrm{End}(\mathcal{W}_x)$ and the Riemannian curvature R_p as an element of $\Lambda^2 V \otimes \mathrm{End}(T_x M)$. The function $a \mapsto F_p^1 \cdot a$ is a $\mathrm{End}(\mathcal{W}_x)$-valued linear function on $\mathfrak{g}$, thus its evaluation at $e \wedge e^*$ is an element of $\Lambda^2 V \otimes \mathrm{End}(\mathcal{W}_x)$. Similarly the function $a \mapsto \tau(\Omega_p \cdot a)$ is an $\mathrm{End}(T_x M)$-valued linear function on $\mathfrak{g}$, thus its evaluation at $e \wedge e^*$ is an element of $\Lambda^2 V \otimes \mathrm{End}(T_x M)$.

Lemma 5.17. *We have* $F_p^1 \cdot (e \wedge e^*) = F_p^{\mathcal{W}}$ *and* $\tau(\Omega_p \cdot (e \wedge e^*)) = R_p$.

Proof. The curvature $F_p^{\mathcal{W}}$ is the element

$$\sum_{i<j} e_i \wedge e_j \otimes F_p^{\mathcal{W}}(e_i^P, e_j^P)$$

of $\Lambda^2 V \otimes \mathrm{End}(\mathcal{W}_x)$. If $F_p^1 \in \mathrm{End}(\mathcal{W}_x) \otimes \mathfrak{g}$ is given by (5.8) and $a = \sum_{i<j} a_{ij} e_i \wedge e_j \in \mathfrak{g}$, we see that $F_p^1 \cdot a = \sum_{i<j} a_{ij} F_p^{\mathcal{W}}(e_i^P, e_j^P)$. Replacing the scalar variables a_{ij} by the Grassmann variables $e_i \wedge e_j$, we obtain the first formula.

The second formula is more subtle. Let us write $\Omega_p \in \Lambda^2 V \otimes \mathfrak{g}$ as

$$\Omega_p = \sum_{i<j;k<l} e_k \wedge e_l \otimes (\Omega_p(e_k, e_l), e_i \wedge e_j)\, e_i \wedge e_j.$$

Let $E_{ij} = \tau(e_i \wedge e_j)$, so that $\{E_{ij} \mid i < j\}$ is a basis of $\mathfrak{so}(V)$. Since the Riemannian curvature $R_p \in \Lambda^2 V \otimes \mathfrak{so}(V)$ is the image of the curvature Ω_p by the representation τ of $\mathfrak{g}$ in V, we see that

$$R_p = \sum_{i<j;k<l} e_k \wedge e_l \otimes (\Omega_p(e_k, e_l), e_i \wedge e_j)\, E_{ij}.$$

The fundamental symmetry of the Riemannian curvature shows that

$$R_p = \sum_{i<j;k<l} e_i \wedge e_j \otimes (\Omega_p(e_k, e_l), e_i \wedge e_j)\, E_{kl}.$$

For $a = \sum_{i<j} a_{ij} e_i \wedge e_j \in \mathfrak{g}$, the element $\Omega_p \cdot a$ of $\Lambda^2 V$ equals

$$\sum_{i<j;k<l} (\Omega_p(e_k,e_l),e_i\wedge e_j)\,a_{ij}e_k\wedge e_l,$$

so that

$$\tau(\Omega_p\cdot a)=\sum_{i<j;k<l} (\Omega_p(e_k,e_l),e_i\wedge e_j)\,a_{ij}E_{kl}.$$

Replacing the variable a_{ij} by the Grassmann variable $e_i\wedge e_j$, we obtain (2). □

The function $\Phi_0(2a)$ is the product of the real function $j_{\mathfrak{g}}^{-1/2}(2a)$, the real function $j_V^{-1/2}(\tau(\Omega_p\cdot a))$ and the $\mathrm{End}(\mathscr{W}_x)$-valued function $\exp(-F^1_p\cdot a)$. We know that $j_{\mathfrak{g}}^{-1/2}(2e\wedge e^*)=1$. By the preceding lemma, the evaluation of the function $j_V^{-1/2}(\tau(\Omega_p\cdot a))$ at $e\wedge e^*$ equals the $\hat{A}$-genus of M, while the evaluation of the $\mathrm{End}(\mathscr{W}_x)$-valued function $\exp(-F^1_p\cdot a)$ at $e\wedge e^*$ equals the element $\exp(-F^{\mathscr{W}})$ of $\mathscr{A}(M,\mathrm{End}(\mathscr{W}))$. This completes our second proof of Theorem 4.1.

Bibliographic Notes

The method used in this chapter first appeared in Berline-Vergne [24]. This work was influenced by the articles of Getzler [61] and Bismut [28].

6

The Equivariant Index Theorem

Let M be a compact oriented Riemannian manifold of even dimension n, and let H be a compact group of orientation preserving isometries acting on M. Let $\mathscr{E} \to M$ be a Clifford module with Clifford connection; if H acts on $\mathscr{E}$ compatibly with the Clifford action and Clifford connection, we call $\mathscr{E}$ an equivariant Clifford module. If D is the Dirac operator on $\mathscr{E}$ associated to the given data, then D commutes with the action of H; hence, the kernel of D is a finite-dimensional representation of H. The equivariant index is the virtual character of H given for $\gamma \in H$ by the formula

$$\operatorname{ind}_H(\gamma, \mathsf{D}) = \operatorname{Tr}(\gamma, \ker \mathsf{D}^+) - \operatorname{Tr}(\gamma, \ker \mathsf{D}^-).$$

In this chapter, we will prove a generalization of the local index theorem for D which enables us to calculate $\operatorname{ind}_H(\gamma, \mathsf{D})$ for arbitrary $\gamma \in H$. Let $\langle x \mid \gamma e^{-t\mathsf{D}^2} \mid y \rangle$ be the equivariant heat kernel of D, that is, the kernel of the operator $\gamma e^{-t\mathsf{D}^2}$. The restriction of $\langle x \mid \gamma e^{-t\mathsf{D}^2} \mid y \rangle$ to the diagonal of M is a section of $\operatorname{End}(\mathscr{E})$ which we denote by $k_t(\gamma, x)$.

Fix $\gamma \in H$, and denote its fixed point set by M^γ. The main result of this chapter, Theorem 6.11, is that $k_t(\gamma, x)$ has an asymptotic expansion, as $t \to 0$, of the form

$$(4\pi t)^{-\dim(M^\gamma)/2} \sum_{i=0}^{\infty} t^i \Phi_i(\gamma, x),$$

where the coefficients $\Phi_i(\gamma, x)$ are generalized sections of $\operatorname{End}(\mathscr{E})$ which are supported on M^γ, and such that $\Phi_i(\gamma, x)$ has Clifford filtration degree $2i + \dim(M) - \dim(M^\gamma)$. We will calculate $\sigma_{2i+\dim(M)-\dim(M^\gamma)} \Phi_i(\gamma, x)$ explicitly, obtaining a formula of a similar type to that of Chapter 4, which is the case where $H = \{1\}$; this formula involves the curvature of the normal bundle $\mathscr{N}$ of M^γ in M and the action of H on it.

We begin the chapter with some generalities on the equivariant index. In Section 2, we give a proof of the Atiyah-Bott fixed point formula for the equivariant index of the $\bar{\partial}$-operator when the fixed point set is discrete; this serves as a simple introduction to the equivariant index theorem. In Section 3 we state Theorem 6.11, while in Section 4, we show that it implies the equivariant index formula of Atiyah-Segal-Singer for $\operatorname{ind}_H(\gamma, \mathsf{D})$ as an integral over M^γ and the local formula of Gilkey. The rest of the chapter is devoted to the proof of Theorem 6.11, by a method generalizing that of Chapter 5.

6.1 The Equivariant Index of Dirac Operators

Let M be a compact oriented Riemannian manifold of even dimension n, on which a compact Lie group H acts by orientation-preserving isometries. There is induced an action of H on the Clifford bundle $C(M)$, which preserves the product.

Definition 6.1. *An* **equivariant Clifford module** $\mathcal{E}$ *over* M *is an* H*-equivariant bundle, with a Clifford module structure and Hermitian inner product preserved by the group action. We suppose that the action of* H *preserves the* $\mathbb{Z}_2$*-grading of* $\mathcal{E}$.

Let D be a Dirac operator associated to a Clifford connection $\nabla^{\mathcal{E}}$ on $\mathcal{E}$. The following lemma is evident.

Lemma 6.2. *The action of the group* H *commutes with the Dirac operator if and only if the Clifford connection* $\nabla^{\mathcal{E}}$ *is* H*-invariant.*

From now on, we only consider invariant Dirac operators D on $\mathcal{E}$. Since an element $\gamma \in H$ preserves the Dirac operator D, it must map $\ker(\mathsf{D})$ to itself, so that $\ker(\mathsf{D})$ becomes a $\mathbb{Z}_2$-graded representation of H. The character of this representation, that is, the supertrace

$$\operatorname{Str}(\gamma, \ker(\mathsf{D})) = \operatorname{Tr}(\gamma, \ker(\mathsf{D}^+)) - \operatorname{Tr}(\gamma, \ker(\mathsf{D}^-)),$$

is called the **equivariant index** of the Dirac operator D, and we will denote it by $\operatorname{ind}_H(\gamma, \mathsf{D})$; it is an element of the representation ring $R(H)$ of the compact group H. The purpose of this chapter is to calculate $\operatorname{ind}_H(\gamma, \mathsf{D})$ in terms of local data at the fixed point set of γ acting on M; for $\gamma = 1 \in H$, this is just the local index theorem of the Chapter 4. The other extreme occurs when the fixed point set of γ consists of isolated points; if D is associated to an elliptic complex, the formula for $\operatorname{ind}_H(\gamma, \mathsf{D})$ in this case, due to Atiyah and Bott, is particularly simple to derive, and we will present it in Section 2.

The foundation of the calculation of the equivariant index of a Dirac operator is the following generalization of the McKean-Singer formula.

Proposition 6.3. *Let* $k_t(\gamma, x)\,|dx| = \langle x \mid \gamma e^{-t\mathsf{D}^2} \mid x \rangle$ *be the restriction of the kernel of the operator* $\gamma e^{-t\mathsf{D}^2}$ *to the diagonal. Then for any* $t > 0$,

$$\operatorname{ind}_H(\gamma, \mathsf{D}) = \operatorname{Str}(\gamma e^{-t\mathsf{D}^2}) = \int_M \operatorname{Str}(k_t(\gamma, x))\,|dx|.$$

Proof. We will give two proof of this result, which are transcriptions of the two proofs of the McKean-Singer formula Theorem 3.50 in its simplest version.

If λ is a real number, the λ-eigenspace $\mathscr{H}_\lambda^\pm$ of the generalized Laplacian D^2, acting on $\Gamma(M,\mathscr{E}^\pm)$ is a finite dimensional representation of H, with character $\chi_\lambda^\pm$. We have the following formula for the supertrace of $\gamma e^{-t\mathsf{D}^2}$:

$$\operatorname{Str}(\gamma e^{-t\mathsf{D}^2}) = \sum_{\lambda\geq 0} (\chi_\lambda^+(\gamma) - \chi_\lambda^-(\gamma))e^{-t\lambda}.$$

The Dirac operator D induces an isomorphism between $\mathscr{H}_\lambda^+$ and $\mathscr{H}_\lambda^-$ when $\lambda \neq 0$. Since D commutes with H, this isomorphism respects the action of H, so that

$$\chi_\lambda^+(\gamma) - \chi_\lambda^-(\gamma) = 0 \quad \text{for } \lambda > 0,$$

and we are left with $\chi_0^+(\gamma) - \chi_0^-(\gamma)$, which is nothing but the equivariant index of D.

The second proof of Theorem 3.50 generalizes as follows. If $a(\gamma,t)$ is the function $\operatorname{Str}(\gamma e^{-t\mathsf{D}^2})$, we again see by Lemma 2.37 that for t large,

$$\left|\operatorname{Str}(\gamma e^{-t\mathsf{D}^2} - \gamma P_{\ker(\mathsf{D})})\right| \leq C e^{-t\lambda_1/2} \operatorname{vol}(M),$$

where λ_1 is the smallest non-zero eigenvalue of D^2. This shows that

$$a(\gamma,\infty) = \lim_{t\to\infty} \operatorname{Str}(\gamma e^{-t\mathsf{D}^2}) = \operatorname{ind}_H(\gamma,\mathsf{D}).$$

The proof is completed by showing that the function $a(t,\gamma)$ is independent of t, so that $a(\gamma,t) = a(\gamma,\infty) = \operatorname{ind}_H(\gamma,\mathsf{D})$. Differentiating with respect to t, we obtain

$$\begin{aligned}\frac{d}{dt}a(\gamma,t) &= -\operatorname{Str}(\gamma\mathsf{D}^2 e^{-t\mathsf{D}^2}) \\ &= -\tfrac{1}{2}\operatorname{Str}([\gamma\mathsf{D}, \mathsf{D}e^{-t\mathsf{D}^2}]),\end{aligned}$$

since D is odd and $[\mathsf{D},\gamma] = 0$. This last line vanishes by Lemma 3.49. □

Thus, the calculation of the equivariant index is reduced to the calculation of the limiting generalized section $\lim_{t\to 0} k_t(\gamma,x)$ of $\mathscr{E}$.

6.2 The Atiyah-Bott Fixed Point Formula

The Atiyah-Bott fixed point formula is a generalization of the Lefschetz fixed point formula, which gives the supertrace of the action of a diffeomorphism on the de Rham cohomology of a manifold. In some respects, it is more general than the equivariant index theorem which we will prove later, since the group H is allowed to be non-compact. However, it is more restrictive, in that it only gives a formula for elements $\gamma \in H$ whose fixed point set consists of non-degenerate isolated points, and D must be associated to an elliptic complex.

Let $D : \Gamma(M, \mathcal{E}^\bullet) \to \Gamma(M, \mathcal{E}^{\bullet+1})$ be a differential operator acting on the space of sections $\Gamma(M, \mathcal{E})$ of a $\mathbb{Z}$-graded finite dimensional vector bundle $\mathcal{E}$ over a compact manifold M. We say that $(\Gamma(M, \mathcal{E}), D)$ is a **complex** if $D^2 = 0$. Suppose that the cohomology space $H^\bullet(D) = \ker(D)/\operatorname{im}(D)$ is finite dimensional.

Consider a group H acting on $\mathcal{E} \to M$ and assume that the action of H on $\Gamma(M, \mathcal{E})$ commutes with D. The group H acts on the graded vector space $H^\bullet(D)$ and we define, for $\gamma \in H$,

$$\operatorname{ind}_H(\gamma, D) = \sum_i (-1)^i \operatorname{Tr}(\gamma, H^i(D)).$$

If γ acts on M with isolated nondegenerate fixed points, Atiyah and Bott gave a fixed point formula for the index of the transformation γ, even if H is a non-compact group.

In this section, we will give a proof of Atiyah-Bott formula for those complexes D for which there exist a Hermitian structure on $\mathcal{E}$ and a Riemannian metric on M such that the operator $D + D^*$ is a Dirac operator. Let $\mathcal{E} \to M$ be a H-equivariant vector bundle over a Riemannian manifold M. However, we do not need to assume that H acts by isometries on M; in particular, H need not be compact.

If x is an isolated fixed point of the action of an element $\gamma \in H$ on M, denote by γ_x the tangent action at the fixed point x, and by $\gamma_x^{\mathcal{E}}$ the endomorphism of the fibre $\mathcal{E}_x$ induced by γ. The point x is called a non-degenerate fixed point if the endomorphism $(1 - \gamma_x)$ of T_xM is invertible; for example, this condition is satisfied if x is an isolated fixed point and H is a compact group. Let L be a generalized Laplacian on $\mathcal{E}$ and let $k_t(\gamma, x)\,|dx|$ be the restriction of the kernel of γe^{-tL} to the diagonal; it is a section of $\mathcal{E} \otimes |\Lambda|$, where $|\Lambda|$ is the bundle of smooth densities on M.

Recall that a generalized section $s \in \Gamma^{-\infty}(M, \mathcal{E})$ of a bundle $\mathcal{E}$ is a continuous linear form on the space of smooth sections of $\mathcal{E}^* \otimes |\Lambda|$; thus, the space of smooth sections $\Gamma(M, \mathcal{E})$ of $\mathcal{E}$ may be embedded in $\Gamma^{-\infty}(M, \mathcal{E})$. We use the notation $\int_M \langle s(x), \phi(x)\rangle\,|dx|$ for $\langle s, \phi\,|dx|\rangle$. If $x \in M$, the delta distribution δ_x at x is the generalized section $\langle \delta_x, \phi\rangle = \phi(x)$ of the bundle of densities $|\Lambda|$.

Lemma 6.4. *Let $x_0 \in M$ be a non-degenerate isolated fixed point of $\gamma \in H$. If $\chi(x) \in C^\infty(M)$ is a function equal to 1 near x_0 and vanishing in a neighbourhood of all other fixed points of the action of γ on M, then as $t \to 0$, the section $\chi(x)k_t(\gamma, x)$ of* $\operatorname{End}(\mathcal{E}) \otimes |\Lambda|$ *has a limit in the space of generalized sections of* $\operatorname{End}(\mathcal{E}) \otimes |\Lambda|$, *given by*

$$\lim_{t\to 0} \chi(x)\,k_t(\gamma, x)\,|dx| = \frac{\gamma_{x_0}^{\mathcal{E}}\,\delta_{x_0}}{|\det(1 - \gamma_{x_0}^{-1})|} \in \Gamma^{-\infty}(M, \operatorname{End}(\mathcal{E}) \otimes |\Lambda|).$$

Proof. We must compute

$$\lim_{t\to 0} \int_M \chi(x)\,k_t(\gamma, x)\,\phi(x)\,|dx| \quad \text{for } \phi \in \Gamma(M, \mathcal{E}).$$

Since $k_t(\gamma, x) = \gamma^{\mathcal{E}}\langle \gamma^{-1}x \mid e^{-tL} \mid x\rangle$, we see that the function $\chi(x)k_t(\gamma, x)$ is rapidly decreasing as a function of t, except in a small neighbourhood U of x_0. Let $V = T_{x_0}M$, and let $F : U \to V$ be a diffeomorphism of U with a small neighbourhood of 0 in

V such that $d_{x_0}F = I$. We transport the integral over M to an integral on a small neighbourhood of 0 in V, and trivialize the bundle $\mathscr{E}$ on this neighbourhood.

The approximation of $\langle x \mid e^{-tL} \mid y \rangle$ shows that

$$\lim_{t\to 0} \int_M \chi(x) k_t(\gamma, x) \phi(x) |dx| = \lim_{t\to 0} (4\pi t)^{-n/2} \int_V e^{-d(\gamma_{x_0}^{-1}\mathbf{x},\mathbf{x})^2/4t} \psi(\mathbf{x}) \phi(\mathbf{x}) |d\mathbf{x}|,$$

where $\psi(\mathbf{x})$ is a smooth compactly supported section of $\mathrm{End}(\mathscr{E})$ equal to $\gamma_{x_0}^{\mathscr{E}}$ for $\mathbf{x} = 0$. Consider the change of variables $\mathbf{x} \to t^{1/2}\mathbf{x}$ on V. We see by Proposition 1.28 that

$$\lim_{t\to 0} t^{-1} d(\gamma_{x_0}^{-1} t^{1/2}\mathbf{x}, t^{1/2}\mathbf{x})^2 = \|\gamma_{x_0}^{-1}\mathbf{x} - \mathbf{x}\|^2,$$

where $\|\cdot\|$ is the Euclidean norm on V, and the lemma follows. □

If D is a complex on a $\mathbb{Z}$-graded Hermitian vector bundle $\mathscr{E} \to M$ over a Riemannian manifold M such that $D + D^*$ is a Dirac operator, then $L = DD^* + D^*D = (D + D^*)^2$ is a generalized Laplacian which commutes with D:

$$(DD^* + D^*D)D = DD^*D = D(DD^* + D^*D).$$

It follows from Hodge's Theorem 3.54 that $D + D^*$ is essentially self-adjoint, that the cohomology $H(D)$ is a finite-dimensional vector space and that $\ker(D + D^*) = \ker(D) \cap \ker(D^*)$ is a subspace of $\ker(D)$ isomorphic to $H(D)$.

Lemma 6.5. *Let L be a generalized Laplacian of the form $DD^* + D^*D$ with $D^2 = 0$. Assume that the action of H on $\Gamma(M, \mathscr{E})$ commutes with D. Then, for every $t > 0$,*

$$\mathrm{ind}_H(\gamma, D) = \mathrm{Str}(\gamma e^{-tL}).$$

Proof. The proof is very similar to that of the equivariant McKean-Singer formula of Section 1. First, note that the right-hand side of the equation is independent of t; indeed, its derivative with respect to t is

$$\begin{aligned}\frac{d}{dt} \mathrm{Str}(\gamma e^{-tL}) &= -\mathrm{Str}(\gamma (DD^* + D^*D) e^{-tL}) \\ &= -\mathrm{Str}([\gamma D, D^* e^{-tL}]) = 0.\end{aligned}$$

Here, we have used the fact that the supertrace vanishes on supercommutators and that L and γ commute with D.

When $t \to \infty$, the semigroup e^{-tL} tends to the projection P_0 onto $\ker(D + D^*)$. This subspace need not be fixed by the action of γ, since γ does not commute with D^*. However, we have $\ker(D) = P_0(\ker(D)) \oplus \mathrm{Im}(D)$, thus the supertrace of the operator $P_0 \gamma P_0$ is the index of γ, and we obtain

$$\mathrm{ind}_H(\gamma, D) = \mathrm{Str}(P_0 \gamma P_0) = \mathrm{Str}(\gamma e^{-tL}).$$ □

From the above two lemmas, we obtain the following special case of the Atiyah-Bott fixed point formula.

Theorem 6.6 (Atiyah-Bott). *Let $\mathscr{E} \to M$ be a $\mathbb{Z}$-graded Hermitian vector bundle on a Riemannian manifold M, and let D be a first-order differential operator on $\Gamma(M,\mathscr{E})$ such that $D^2 = 0$, and such that the operator $DD^* + D^*D$ is a generalized Laplacian.*

Let γ be a bundle map of the bundle $\mathscr{E}$ covering the action of γ on M, and commuting with D. Assume that the action of γ on M has only isolated non-degenerate fixed points. Then

$$\operatorname{ind}_H(\gamma, D) = \sum_{x_0 \in M^\gamma} \frac{\operatorname{Str}(\gamma_{x_0}^{\mathscr{E}})}{|\det(1-\gamma_{x_0}^{-1})|},$$

where M^γ is the fixed point set of the action of γ on M.

Let us state the formulas that this theorem gives for the d and $\bar{\partial}$-complexes.

Corollary 6.7. *If M is a compact manifold, and γ is a diffeomorphism of M with isolated non-degenerate fixed points, then*

$$\sum_i (-1)^i \operatorname{Tr}(\gamma, H^i(M)) = \sum_{x_0 \in M^\gamma} \varepsilon(x_0, \gamma),$$

where $\varepsilon(x_0,\gamma)$ is the sign of the determinant $\det(1-\gamma_{x_0}^{-1})$.

Proof. The operator d acts on sections of the $\mathbb{Z}$-graded bundle $\mathscr{E} = \Lambda T^*M$ and satisfies the condition of the above theorem for any choice of Riemannian metric on M. If γ is an endomorphism of a real vector space V, then $\sum_i (-1)^i \operatorname{Tr}(\gamma, \Lambda^i V^*) = \det(1-\gamma^{-1}, V)$. □

Note that when γ lies in a compact group, the sign of $\det(1-\gamma_{x_0})$ is always positive, since the real eigenvalues of γ are ± 1.

Next, we turn to the $\bar{\partial}$-complex.

Corollary 6.8. *If M is a compact complex manifold with holomorphic vector bundle $\mathscr{W} \to M$, and γ is a holomorphic transformation of $\mathscr{W} \to M$, then γ acts on the $\bar{\partial}$-cohomology spaces $H^{0,i}(M,\mathscr{W})$. If the action of γ on M has only isolated non-degenerate fixed points, then*

$$\sum_i (-1)^i \operatorname{Tr}(\gamma, H^{0,i}(M,\mathscr{W})) = \sum_{x_0 \in M^\gamma} \frac{\operatorname{Tr}(\gamma_{x_0}^{\mathscr{W}})}{\det_{T_{x_0}^{1,0}M}(1-\gamma_{x_0}^{-1})}$$

Proof. The operator $\bar{\partial}$ acts on sections of the $\mathbb{Z}$-graded bundle $\Lambda(T^{0,1}M)^* \otimes \mathscr{W}$. We have

$$\sum_i (-1)^i \operatorname{Tr}_{\Lambda^i(T_{x_0}^{0,1}M)^* \otimes \mathscr{W}_{x_0}}(\gamma) = \operatorname{Tr}(\gamma_{x_0}^{\mathscr{W}}) \cdot \det_{T_{x_0}^{0,1}M}(1-\gamma_{x_0}^{-1}).$$

Since $|\det(1-\gamma_{x_0}^{-1})| = \det_{T_{x_0}^{1,0}M}(1-\gamma_{x_0}^{-1}) \det_{T_{x_0}^{0,1}M}(1-\gamma_{x_0}^{-1})$, the corollary follows.

□

6.3 Asymptotic Expansion of the Equivariant Heat Kernel

Let M be a compact oriented Riemannian manifold of even dimension n, and let H be a compact group of orientation preserving isometries of M. Since an element γ of H is an isometry of M, its fixed point set M^γ is a submanifold of M, which may consist of several components of different dimension. Using the Riemannian metric on M, we may decompose the tangent bundle along M^γ as

$$TM|_{M^\gamma} \cong TM^\gamma \oplus \mathcal{N},$$

where $\mathcal{N}$ is the normal bundle along M^γ.

Since γ is an isometry of M, the Levi-Civita connection preserves the decomposition $TM|_{M^\gamma} = TM^\gamma \oplus \mathcal{N}$, so induces connections on TM^γ and $\mathcal{N}$ which preserve the metrics on these bundles; the connection on TM^γ is in fact the Levi-Civita connection of M^γ. Let $R \in \mathscr{A}^2(M, \mathfrak{so}(M))$ be the Riemannian curvature of TM, and let $R^0 \in \mathscr{A}^2(M^\gamma, \mathfrak{so}(M^\gamma))$ and $R^1 \in \mathscr{A}^2(M^\gamma, \mathfrak{so}(\mathcal{N}))$ be the curvatures of the induced connections on TM^γ and $\mathcal{N}$; thus, R^0 is the Riemannian curvature of the manifold M^γ. We have the following relation:

$$R|_{M^\gamma} = R^0 \oplus R^1. \tag{6.1}$$

Since R^0 is the Riemannian curvature of the manifold M^γ, we see that

$$\hat{A}(M^\gamma) = \det{}^{1/2}\left(\frac{R^0/2}{\sinh(R^0/2)}\right)$$

is the $\hat{A}$-genus of M^γ.

The tangent map $d\gamma$ restricted to M^γ defines a section of the bundle $\mathrm{SO}(M)|_{M^\gamma}$, which preserves the orthogonal decomposition $TM|_{M^\gamma} = TM^\gamma \oplus \mathcal{N}$. Furthermore, TM^γ is exactly the eigen-bundle with eigenvalue one of $d\gamma$: if $v \in T_{x_0}M$ is a tangent vector fixed by $d_{x_0}\gamma$ for some $x_0 \in M^\gamma$, the curve $\exp_{x_0} tv$ is fixed by γ so that $v \in T_{x_0}M^\gamma$.

Fix a component M_0 of M^γ and let γ_1 be the transformation induced on $\mathcal{N}|_{M_0}$. Since the only possible real eigenvalue of γ_1 is -1, we see that $\det(1-\gamma_1) > 0$. Furthermore, since $\det(\gamma_1) = 1$, we see that $\dim(\mathcal{N}_{x_0}) = \dim(M) - \dim(M_0)$ must be even. We denote by ℓ_0 and ℓ_1 the locally constant functions on M^γ such that $\dim(M_0) = 2\ell_0$ and $\dim(\mathcal{N}_{x_0}) = 2\ell_1$ for all $x_0 \in M_0$; it is clear that $n = 2\ell_0 + 2\ell_1$. Denote by $\det(\mathcal{N})$ the line bundle over M^γ which over the component M_0 of M^γ is equal to $\Lambda^{2\ell_1}\mathcal{N}^*$; this line bundle is contained in $\Lambda T^*M|_{M^\gamma}$.

Lemma 6.9. *The function* $\det(1-\gamma_1)$ *is constant on each component of* M^γ.

Proof. If M_0 is a component of M^γ, each fibre $\mathcal{N}_x$, $x \in M_0$, of the bundle $\mathcal{N}$ carries a finite-dimensional representation of the closure $G(\gamma)$ of the group generated by γ in the group of isometries of M. Since M is compact, this is a compact group. Since the dual of the compact group $G(\gamma)$ is discrete, the representation must be independent of $x \in M_0$, from which the lemma follows. □

Let $\gamma \in \mathrm{SO}(V)$ and $A \in \mathfrak{so}(V)$. In Definition 3.25, we have defined the analytic square root $\det^{1/2}(1-\gamma\exp A)$. Applying this with $\gamma_1 \in \mathrm{SO}(\mathcal{N}_{x_0})$ and $A = -R_1 \in \mathfrak{so}(\mathcal{N}_{x_0}) \otimes \Lambda^2 T^*_{x_0}M^\gamma$, we obtain an element of $\Lambda T^*_{x_0}M^\gamma$; as the point x_0 varies, we obtain a differential form $\det^{1/2}(1-\gamma_1\exp(-R_1)) \in \mathcal{A}(M^\gamma)$. It follows from the theory of characteristic classes that this form is closed. Its zero-degree component is equal to $|\det(1-\gamma_1)|^{1/2}$ so that $\det^{1/2}(1-\gamma_1\exp(-R^1))$ is a well-defined invertible element of $\mathcal{A}(M^\gamma)$.

We may trivialize the density bundles $|\Lambda_M|$ and $|\Lambda_{M^\gamma}|$ by means of the canonical sections $|dx|$ and $|dx_0|$ derived from the Riemannian metrics on M and M^γ respectively. Thus, the delta-function δ_{M^γ} of M^γ is the generalized function on M defined by

$$\int_M \delta_{M^\gamma}(x)\,\phi(x)\,|dx| = \int_{M^\gamma} \phi(x_0)\,|dx_0|,$$

where $|dx|$ and $|dx_0|$ are the Riemannian densities of M and M^γ. If $\mathcal{F}$ is a bundle over M and $v \in \Gamma(M^\gamma,\mathcal{F})$, then $v\delta_{M^\gamma}$ is a generalized section of $\mathcal{F}$ over M.

From now on, $\mathcal{E}$ will be a H-equivariant Clifford module on M, with H-invariant Clifford action and Clifford connection. As usual, we write $\mathrm{End}(\mathcal{E}) \cong C(M) \otimes \mathrm{End}_{C(M)}(\mathcal{E})$. If M has a spin-structure, then $\mathcal{E}$ is isomorphic to $\mathcal{W} \otimes \mathcal{S}$ for some bundle $\mathcal{W}$, and $\mathrm{End}_{C(M)}(\mathcal{E}) \cong \mathrm{End}(\mathcal{W})$.

Lemma 6.10. *Along M^γ, the action of γ on $\mathcal{E}$ may be identified with a section $\gamma^{\mathcal{E}}$ of $C(\mathcal{N}^*) \otimes \mathrm{End}_{C(M)}(\mathcal{E})$.*

Proof. We have $\gamma c(\alpha)\gamma^{-1} = c(\gamma\alpha)$ for $\alpha \in \Gamma(M,T^*M)$. Thus the action on $\mathcal{E}_{x_0}$ of γ at a point $x_0 \in M^\gamma$ commutes with the Clifford action of $C(T^*_{x_0}M^\gamma)$. It is clear that the commutant of $C(T^*_{x_0}M^\gamma)$ in $\mathrm{End}(\mathcal{E}_{x_0})$ is $C(\mathcal{N}^*_{x_0}) \otimes \mathrm{End}_{C(M)}(\mathcal{E})_{x_0}$. □

By this lemma, we may form the highest degree symbol $\sigma_{\dim(\mathcal{N})}(\gamma^{\mathcal{E}})$, which is a section of $\mathrm{End}_{C(M)}(\mathcal{E}) \otimes \det(\mathcal{N}) \subset \mathrm{End}_{C(M)}(\mathcal{E}) \otimes \Lambda T^*M|_{M^\gamma}$, since $\det(\mathcal{N}) = \Lambda^{\dim(\mathcal{N})}\mathcal{N}^*$.

Let $F^{\mathcal{E}/S}$ be the twisting curvature of the Clifford module $\mathcal{E}$. Recall that if M has a spin-structure with spinor bundle $\mathcal{S}$, we may write $\mathcal{E} = \mathcal{W} \otimes \mathcal{S}$, and then $F^{\mathcal{E}/S} = F^{\mathcal{W}}$ is the curvature of the twisting bundle $\mathcal{W}$. We denote the restriction of $F^{\mathcal{E}/S}$ to M^γ by $F_0^{\mathcal{E}/S} \in \Gamma(M^\gamma, \Lambda T^*M^\gamma \otimes \mathrm{End}_{C(M)}(\mathcal{E}))$.

Let D be the Dirac operator of the Clifford module $\mathcal{E}$; it is an H-invariant operator on $\Gamma(M,\mathcal{E})$. Let

$$k_t(\gamma,x)\,|dx| \in \Gamma(M, C(M) \otimes \mathrm{End}_{C(M)}(\mathcal{E}) \otimes |\Lambda|)$$

be the restriction of the kernel of the operator $\gamma e^{-t\mathsf{D}^2}$ to the diagonal. The main result of this chapter is to show that $k_t(\gamma,x)$ has an asymptotic expansion when $t \to 0$ in the space of generalized sections $\Gamma^{-\infty}(M, C(M) \otimes \mathrm{End}_{C(M)}(\mathcal{E}) \otimes |\Lambda|)$, and to calculate the leading order of the expansion. This theorem is a generalization of a result of Gilkey.

Theorem 6.11. *The section $k_t(\gamma)$ of the bundle $C(M)\otimes \mathrm{End}_{C(M)}(\mathscr{E})\otimes|\Lambda|$ has an asymptotic expansion as $t\to 0$ in the space of generalized sections of $C(M)\otimes \mathrm{End}_{C(M)}(\mathscr{E})\otimes|\Lambda|$ of the form*

$$k_t(\gamma)\sim(4\pi t)^{-\dim(M^\gamma)/2}\sum_{i=0}^{\infty} t^i\Phi_i(\gamma),$$

where $\Phi_i(\gamma)$ is a generalized section of $C_{2i+\dim(\mathscr{N})}(M)\otimes \mathrm{End}_{C(M)}(\mathscr{E})\otimes|\Lambda|$ supported on M^γ. The symbol of $\Phi_i(\gamma)$ is given by the formula

$$\sum_{i=0}^{\dim(M^\gamma)/2}\sigma_{2i+\dim(\mathscr{N})}(\Phi_i(\gamma))=I(\gamma)\cdot\delta_{M^\gamma},$$

where $I(\gamma)\in\Gamma(M^\gamma,\Lambda T^*M\otimes \mathrm{End}_{C(M)}(\mathscr{E}))$ equals

$$I(\gamma)=\frac{\hat{A}(M^\gamma)\exp(-F_0^{\mathscr{E}/S})\,\sigma_{\dim(\mathscr{N})}(\gamma^{\mathscr{E}})}{\det^{1/2}(1-\gamma_1)\det^{1/2}(1-\gamma_1\exp(-R^1))}.$$

Observe that when γ is the identity, this theorem reduces to the local index theorem of the last chapter, Theorem 4.1.

From the formula

$$k_t(\gamma,x)=\gamma^{\mathscr{E}}\langle\gamma^{-1}x\mid e^{-t\mathrm{D}^2}\mid x\rangle,$$

we see that if $\gamma x\neq x$, the function $t\to k_t(\gamma,x)$ is rapidly decreasing as $t\to 0$; thus only generalized sections with support in M^γ will appear in the asymptotics of the generalized section $k_t(\gamma)$, and we need only study $k_t(\gamma)$ in a neighbourhood of M^γ.

The map $(x_0,v)\in\mathscr{N}\mapsto \exp_{x_0}v$ defines a diffeomorphism between a neighbourhood of the zero section of $\mathscr{N}$ and a neighbourhood U of M^γ. We may identify $\mathscr{E}_{\exp_{x_0}v}$ with $\mathscr{E}_{x_0}$ by parallel transport along the geodesic $s\in[0,1]\mapsto \exp_{x_0}sv$; this respects the filtration on $\mathrm{End}(\mathscr{E})\cong C(M)\otimes \mathrm{End}_{C(M)}(\mathscr{E})$. Thus $k_t(\gamma,\exp_{x_0}v)$ may be thought of as lying in the fixed filtered algebra $\mathrm{End}(\mathscr{E}_{x_0})$. If ϕ is a smooth function on M with support in U, the integral

$$I(t,\gamma,\phi,x_0)=\int_{\mathscr{N}_{x_0}}k_t(\gamma,\exp_{x_0}v)\,\phi(\exp_{x_0}v)\,|dv|$$

defines an element of $\mathrm{End}(\mathscr{E}_{x_0})$, where $|dv|$ is the Euclidean density of $\mathscr{N}_{x_0}$. Using the rapid decrease of $k_t(\gamma,x)$ as $t\to 0$ for $x\notin M^\gamma$ and a partition of unity argument, we see that Theorem 6.11 is implied by the following result.

Proposition 6.12. *Let M_0 be the component of the point x_0 in M^γ, and let ϕ be a smooth function on M supported in a small neighbourhood of M_0. Then*

1. *$I(t,\gamma,\phi,x_0)$ has an asymptotic expansion as $t\to 0$ of the form*

$$I(t,\gamma,\phi,x_0)\sim(4\pi t)^{-\ell_0}\sum_{i=0}^{\infty}t^i\Phi_i(\phi,x_0),$$

where $x_0\mapsto\Phi_i(\phi,x_0)\in\Gamma(M_0,C_{2(i+\ell_1)}(M)\otimes \mathrm{End}_{C(M)}(\mathscr{E}))$.

2. *If* $\sigma(k(\gamma,\phi,x_0))$ *is given by the formula* $\sum_{i=0}^{\ell_0}\sigma_{2(i+\ell_1)}(\Phi_i(\phi,x_0))$, *then*
$$\sigma(k(\gamma,\phi,x_0)) = I(\gamma,x_0)\,\phi(x_0).$$

We will prove this proposition later in this chapter. However, if x_0 is an isolated point of M^γ, the proof is much the same as that of Lemma 6.4, which we proved in the course of proving the Atiyah-Bott fixed point theorem in the last section. Indeed we see, using the change of variables $v \to t^{1/2}v$, that

$$\lim_{t\to 0} I(t,\gamma,\phi,x_0) = \frac{\gamma_{x_0}^{\mathcal{E}}\,\phi(x_0)}{\det\left(1-\gamma_{x_0}^{T_{x_0}M}\right)}.$$

6.4 The Local Equivariant Index Theorem

The Atiyah-Segal-Singer fixed point formula for $\operatorname{ind}_H(\gamma,\mathrm{D})$ is a formula for the equivariant index $\operatorname{ind}_H(\gamma,\mathrm{D})$ of an equivariant Dirac operator D as an integral over the fixed point manifold M^γ. In this section, we will show how combining Theorem 6.11 with the equivariant McKean-Singer formula leads to a local version of this theorem.

Let $\mathcal{W}$ be an H-equivariant superbundle on M. Fix $\gamma\in H$. If $\mathbb{A}$ is an H-invariant superconnection on $\mathcal{W}$, denote by F_0 the restriction to M^γ of the curvature $F=\mathbb{A}^2$ of $\mathbb{A}$. Define the differential form $\mathrm{ch}_H(\gamma,\mathbb{A})\in\mathcal{A}(M^\gamma)$, by the formula

$$\mathrm{ch}_H(\gamma,\mathbb{A}) = \mathrm{Str}_{\mathcal{W}}(\gamma\cdot\exp(-F_0)).$$

This is a closed differential form on M^γ whose cohomology class is independent of the H-invariant superconnection $\mathbb{A}$, by a proof which is identical to that for the ordinary Chern character. In this section, we denote $\mathrm{ch}_H(\gamma,\mathbb{A})$ by $\mathrm{ch}_H(\gamma,\mathcal{W})$.

If $\mathcal{E}$ is an H-equivariant Clifford module over M with H-invariant superconnection $\mathbb{A}$, we will define a differential form

$$\mathrm{ch}_H(\gamma,\mathcal{E}/S)\in\mathcal{A}(M^\gamma,\det(\mathcal{N})) = \Gamma(M^\gamma,\Lambda T^*M^\gamma\otimes\det(\mathcal{N})),$$

which agrees with $\mathrm{ch}_H(\gamma,\mathcal{W})$ up to a sign when $\mathcal{E}=\mathcal{W}\otimes\mathcal{S}$ is an equivariant twisted spinor bundle. Recall that we have defined a map

$$\mathrm{Str}_{\mathcal{E}/S}:\Gamma\left(M^\gamma,\mathrm{End}_{C(M)}(\mathcal{E})\right)\to C^\infty(M^\gamma).$$

If $\mathcal{E}=\mathcal{W}\otimes\mathcal{S}$ and $a\in\Gamma(M^\gamma,\mathrm{End}(\mathcal{W}))$, $\mathrm{Str}_{\mathcal{E}/S}(a)=\mathrm{Str}_{\mathcal{W}}(a)$. We extend $\mathrm{Str}_{\mathcal{E}/S}$ to a map

$$\mathrm{Str}_{\mathcal{E}/S}:\Gamma\left(M^\gamma,\mathrm{End}_{C(M)}(\mathcal{E})\otimes\det(\mathcal{N})\right)\to\Gamma(M^\gamma,\det(\mathcal{N})).$$

Let $F_0^{\mathcal{E}/S}$ be the restriction to M^γ of the twisting curvature of $\mathbb{A}$. Then

$$\mathrm{Str}_{\mathcal{E}/S}\left(\sigma_{\dim(\mathcal{N})}(\gamma^{\mathcal{E}})\exp(-F_0^{\mathcal{E}/S})\right)$$

is a differential form on M^γ with values in the line bundle $\det(\mathcal{N})$.

Definition 6.13. *Define the* **localized relative Chern character** *form*

$$\mathrm{ch}_H(\gamma,\mathcal{E}/S)=\frac{2^{\ell_1}}{\det^{1/2}(1-\gamma_1)}\,\mathrm{Str}_{\mathcal{E}/S}\big(\sigma_{\dim(\mathcal{N})}(\gamma^{\mathcal{S}})\exp(-F_0^{\mathcal{E}/S})\big)$$
$$\in\mathcal{A}(M^\gamma,\det(\mathcal{N})).$$

Obviously, this rather peculiar definition requires some justification; we will rewrite it in terms of more familiar objects when M has an H-equivariant spin-structure; this is by definition a spin-structure determined by a principal bundle $\mathrm{Spin}(M)$ which carries an action of the group H, such that the projection map $\mathrm{Spin}(M)\to\mathrm{SO}(M)$ is equivariant.

Proposition 6.14. *If M has an H-equivariant spin-structure, then for any $\gamma\in H$, the manifold M^γ is naturally oriented.*

Proof. The action of γ on the fibre of the spinor bundle $\mathcal{S}_{x_0}$ at the point $x_0\in M^\gamma$ gives an element $\tilde\gamma\in\mathrm{End}(\mathcal{S}_{x_0})\cong C(T^*_{x_0}M)$ such that $\tilde\gamma c(\alpha)=c(\gamma_1\alpha)\tilde\gamma$ for all $\alpha\in\mathcal{N}^*_{x_0}$. With respect to the decomposition $T_{x_0}M=V_0\oplus V_1$, where $V_0=T_{x_0}M^\gamma$ and $V_1=\mathcal{N}_{x_0}$, we see that $\tilde\gamma$ is an element of $C(V_1^*)$.

Let e^i, $1\le i\le 2\ell_0$, be an orthonormal basis of V_0^*. If

$$\Gamma_0=\exp_C\Big(\frac{\pi}{2}\sum_{i=1}^{\ell_0}e^{2i-1}\wedge e^{2i}\Big)=\mathbf{c}(e^1\wedge\ldots\wedge e^{2\ell_0})\in\mathrm{Spin}(V_0^*),$$

then $\Gamma_0\cdot\tilde\gamma\in\mathrm{Spin}(V^*)$ maps to $(-1)\times\gamma_1\in\mathrm{SO}(V_0^*)\times\mathrm{SO}(V_1^*)$.

Let $T:\Lambda T^*_{x_0}M\to\mathbb{R}$ be the Berezin integral. By Lemma 3.28,

$$|T(\Gamma_0\cdot\tilde\gamma)|=2^{-\ell_1}\det{}^{1/2}(1-\gamma_1).$$

Thus $\sigma_{\dim(\mathcal{N})}(\tilde\gamma)$ is a non-vanishing section of $\det(\mathcal{N})$ such that

$$|\sigma_{\dim(\mathcal{N})}(\tilde\gamma)|=2^{-\ell_1}\det{}^{1/2}(1-\gamma_1).$$

Thus $\sigma_{\dim(\mathcal{N})}(\tilde\gamma)$ defines a trivialization of $\det(\mathcal{N})$, which is the same thing as an orientation of $\mathcal{N}$. But given an orientation of M and $\mathcal{N}^*$, there is a unique orientation on M^γ compatible with these data. □

Fix an orientation of M^γ. Sometimes, there is a natural orientation on M^γ (for example, if M^γ is discrete), which may not agree with the orientation which we obtained above. We will write $\varepsilon(\gamma)(x_0)=1$ if the two orientations agree, and $\varepsilon(\gamma)(x_0)=-1$ if they do not; thus, for fixed γ, $\varepsilon(\gamma)$ is a function from M^γ to $\{\pm1\}$ which is constant on each component.

This discussion implies the following result.

Proposition 6.15. *Let M be an H-equivariant even-dimensional spin manifold, and let $\mathcal{W}$ be an H-equivariant vector bundle with H-equivariant superconnection $\mathbb{A}$. Consider the H-equivariant Clifford connection $\nabla^{\mathcal{W}}\otimes1+1\otimes\nabla^{\mathcal{S}}$ on the Clifford module $\mathcal{E}=\mathcal{W}\otimes\mathcal{S}$. Then*

$$\mathrm{ch}_H(\gamma,\mathcal{E}/S)=\varepsilon(\gamma)\,\mathrm{ch}_H(\gamma,\mathcal{W}).$$

Proof. In this case, $\gamma^{\mathscr{E}} = \gamma^{\mathscr{W}} \otimes \tilde{\gamma}$, and

$$\sigma_{\dim(\mathscr{N})}(\gamma^{\mathscr{E}}) = \varepsilon(\gamma) 2^{-\ell_1} \det{}^{1/2}(1-\gamma_1)\gamma^{\mathscr{W}}. \qquad \square$$

We now return to the general case, in which M does not necessarily have a spin-structure, and state the equivariant local index theorem. Consider the section of ΛT^*M over M^γ given by

$$\frac{\hat{A}(M^\gamma)\,\mathrm{ch}_H(\gamma, \mathscr{E}/S)}{\det^{1/2}(1-\gamma_1\exp(-R_1))}.$$

Since M is orientable, we may take the Berezin integral of this section, to obtain a function on M^γ, which we denote by

$$T_M\left(\frac{\hat{A}(M^\gamma)\,\mathrm{ch}_H(\gamma, \mathscr{E}/S)}{\det^{1/2}(1-\gamma_1\exp(-R^1))}\right) \in C^\infty(M^\gamma).$$

We can now state the equivariant index theorem.

Theorem 6.16 (Atiyah-Segal-Singer). *The equivariant index* $\mathrm{ind}_H(\gamma, \mathsf{D})$ *of an equivariant Dirac operator* D *associated to an ordinary connection is given by the formula*

$$\mathrm{ind}_H(\gamma,\mathsf{D}) = i^{-\dim(M)/2} \int_{M^\gamma} (2\pi)^{-\dim(M_\gamma)/2}\, T_M\left(\frac{\hat{A}(M^\gamma)\,\mathrm{ch}_H(\gamma, \mathscr{E}/S)}{\det^{1/2}(1-\gamma_1\exp(-R^1))}\right) |dx_0|.$$

Proof. We will show that this theorem is a consequence of Theorem 6.11. Let U be a tubular neighbourhood of M^γ and let ψ be a smooth function with support in U identically equal to 1 in a neighbourhood of M^γ. Using a partition of unity, we see that

$$\begin{aligned}\int_M \mathrm{Str}(k_t(\gamma,x))\,dx &\sim \int_U \mathrm{Str}(k_t(\gamma,x))\,\psi(x)\,|dx| \\ &\sim \int_{\mathscr{N}} \mathrm{Str}(k_t(\gamma, \exp_{x_0} v))\,\psi(\exp_{x_0} v) b(x_0,v)\,dx_0\,|dv|.\end{aligned}$$

Here, we have rewritten the Riemannian volume $|dx|$ in terms of the Riemannian volume on $\mathscr{N}$ and a smooth function $b(x_0,v)$ satisfying $b(x_0,0)=1$, by the formula

$$\int_M \phi(x)\,|dx| = \int_{M^\gamma}\int_{\mathscr{N}_{x_0}} \phi(\exp_{x_0} v) b(x_0,v)\,|dv|\,|dx_0|.$$

Applying Proposition 6.12 with $\phi(v) = \psi(\exp_{x_0} v)\, b(x_0,v)$, we see that

$$I(t,\gamma,\phi,x_0) = \int_{\mathscr{N}_{x_0}} k_t(\gamma, \exp_{x_0} v)\,\phi(v)\,|dv|$$

has an asymptotic expansion of the form

$$I(t,\gamma,\phi,x_0) \sim (4\pi t)^{-\dim(M^\gamma)/2} \sum_{i=0}^{\infty} t^i \Phi_i(x_0),$$

where the Clifford degree of $\Phi_i(x_0)$ is less than or equal to $2(i+\ell_1)$. We now use the fact that the supertrace vanishes on elements of Clifford degree strictly less than n. It follows from Theorem 6.11 that $\operatorname{Str}(I(t,\gamma,x_0))$ has an asymptotic expansion without singular part, so that

$$\begin{aligned}
\operatorname{ind}_H(\gamma,\mathsf{D}) &= \lim_{t\to 0}\int_M \operatorname{Str}(k_t(\gamma,x))\,|dx| \\
&= \int_{M^\gamma} \lim_{t\to 0} \operatorname{Str} I(t,\gamma,\phi,x_0)\,|dx_0| \\
&= \int_{M^\gamma} (4\pi)^{-\dim(M^\gamma)/2} \operatorname{Str}\left(\Phi_{\dim(M^\gamma)/2}(x_0)\right)|dx_0|,
\end{aligned}$$

from which the theorem follows on substituting the formula for $\sigma_n\left(\Phi_{\dim(M^\gamma)/2}\right)$. □

There are two special cases of this formula in which the geometry is easier to understand. The first case is that in which M has an H-equivariant spin-structure, with spinor bundle $\mathcal{S}$. Thus, the Clifford module $\mathcal{E}$ is a twisted spinor bundle $\mathcal{W}\otimes\mathcal{S}$, where $\mathcal{W} = \operatorname{Hom}_{C(M)}(\mathcal{S},\mathcal{E})$, and D is the twisted Dirac operator associated to an invariant connection $\nabla^{\mathcal{W}}$ on $\mathcal{W}$ with curvature $F^{\mathcal{W}}$. Then Proposition 6.15 shows that

$$\operatorname{ind}_H(\gamma,\mathsf{D}) = \int_{M^\gamma} (2\pi i)^{-\ell_0} i^{-\ell_1} \varepsilon(\gamma) \frac{\hat{A}(M^\gamma)\operatorname{ch}_H(\gamma,\mathcal{W})}{\det^{1/2}(1-\gamma_1\exp(-R^1))}.$$

A further simplification of the formula occurs if in addition the fixed point set M^γ has a spin-structure. In this case, the normal bundle $\mathcal{N}$ has associated to it a spinor bundle $\mathcal{S}(\mathcal{N})$, defined by the formula

$$\mathcal{S}(\mathcal{N}) = \operatorname{Hom}_{C(M^\gamma)}(\mathcal{S}_{M^\gamma}, \mathcal{S}_M|_{M^\gamma}).$$

The actions of γ on the spinor bundles $\mathcal{S}_{M^\gamma}$ and $\mathcal{S}_M$ induces an automorphism $\tilde{\gamma}_1$ of the vector bundle $\mathcal{S}(\mathcal{N})$. Choosing compatible orientations on M^γ and $\mathcal{N}$, we have

$$\begin{aligned}
\operatorname{ch}_H(\gamma,\mathcal{S}(\mathcal{N})) &= \operatorname{ch}_H(\gamma,\mathcal{S}^+(\mathcal{N})) - \operatorname{ch}_H(\gamma,\mathcal{S}^-(\mathcal{N})) \\
&= i^{-\ell_1}\varepsilon(\gamma,M^\gamma)\det^{1/2}(1-\gamma_1\exp(-R^1)),
\end{aligned}$$

and the equivariant index formula may be rewritten as follows:

$$\operatorname{ind}_H(\gamma,\mathsf{D}) = \int_{M^\gamma} (2\pi i)^{-\ell_0}(-1)^{\ell_1} \frac{\hat{A}(M^\gamma)\operatorname{ch}_H(\gamma,\mathcal{W})}{\operatorname{ch}_H(\gamma,\mathcal{S}(\mathcal{N}))}.$$

6.5 Geodesic Distance on a Principal Bundle

In this section, we prove a technical result which is basic to the proof of Proposition 6.12. Let $\pi : P \to M$ be a principal bundle over a Riemannian manifold, with compact structure group G, connection form $\omega \in \mathscr{A}^1(P)\otimes \mathfrak{g}$, and curvature Ω.

We denote by $d^2(p_1,p_2)$ the square of the geodesic distance between two points p_1 and p_2 of P. If p_1 and p_2 belong to the same fibre $\pi^{-1}(x)$, the horizontal tangent spaces at p_1 and p_2 may be identified with T_xM. Given v_1, v_2 in T_xM, we are going to compute the Taylor expansion up to second order in $t \in \mathbb{R}$ of $d^2(\exp_{p_1} tv_1, \exp_{p_2} tv_2)$. We have $p_2 = p_1 \exp a$ for some $a \in \mathfrak{g}$. Recall from Theorem 5.4 that

$$\exp_{p_1}(a+tv) = \exp_{p_2}(tJ(p_1,a)v + o(t)) \quad \text{for } v \in T_xM,$$

where

$$J(p_1,a)|_{T_xM} = \frac{1-e^{-\tau(\Omega_{p_1}\cdot a)/2}}{\tau(\Omega_{p_1}\cdot a)/2}. \tag{6.2}$$

Proposition 6.17. *For small a, $d^2(\exp_{p_1} tv_1, \exp_{p_2} tv_2)$ equals*

$$\|a\|^2 + t^2\left((v_1, J(p_1,a)^{-1}v_1) - 2(v_1, J(p_1,a)^{-1}v_2) + (v_2, J(p_1,a)^{-1}v_2)\right) + o(t^2).$$

Proof. Let $p_i(t_i) = \exp_{p_i}(t_i v_i)$, $i = 1, 2$, and define $X(t_1,t_2) \in T_{p_1(t_1)}P$ by $p_2(t_2) = \exp_{p_1(t_1)} X(t_1,t_2)$. Thus, we must compute $\|X(t_1,t_2)\|^2$ to second order with respect to (t_1,t_2).

We identify a neighbourhood of p_1 in P with a neighbourhood of the origin in $V \times \mathfrak{g}$ by means of the exponential map; here, $V = T_xM$. Thus, for t small, $T_{p_1(t)}P$ may be identified with $V \oplus \mathfrak{g}$. Under this identification, the metric on $T_{p_1(t)}P$ coincides with the orthogonal sum metric on $V \oplus \mathfrak{g} \cong T_{p_1}P$ up to second order.

Let us start by computing $X(t_1,t_2)$ up to first order. We have

$$X(t) = X(t,t) = a + tX_1 + O(t^2).$$

Differentiating both sides of the equation $\exp_{p_1(t)} X(t) = p_2(t)$ with respect to t, we obtain

$$v_1 + J(p_1,a)X_1 = v_2;$$

here, we have used the fact that $\exp_{p_1(t)} a = \exp_{p_2}(tv_1)$, which follows because both sides of this equation are geodesic curves starting at p_2 with tangent vector v_1. Thus, we obtain

$$X(t_1,t_2) = a + J(p_1,a)^{-1}(t_2v_2 - t_1v_1) + o(|t_1| + |t_2|).$$

In particular, since the horizontal and vertical tangent spaces are orthogonal for $t_1 = t_2 = 0$, this implies that $\|X(t_1,t_2)\|^2 - \|a\|^2$ vanishes to at least second order.

Denote by ∇^P the Riemannian covariant differentiation on P, and by R^P its curvature. Consider the map $p : \mathbb{R}^3 \to P$ given by

$$p(s,t_1,t_2) = \exp_{p_1(t_1)}(sX(t_1,t_2)).$$

This map satisfies the equations

$$\nabla^P_{\partial_s}\partial_s p = 0,\ \|\partial_s p\|^2 = \|X(t_1,t_2)\|^2, \text{ and } \partial_s p|_{s=0} = X(t_1,t_2).$$

Thus, for $i = 1, 2$,

$$\begin{aligned}
\partial_s^2(\partial_{t_i} p, \partial_s p) &= (\nabla^P_{\partial_s}\nabla^P_{\partial_s}\partial_{t_i} p, \partial_s p),\\
&= (\nabla^P_{\partial_s}\nabla^P_{\partial_{t_i}}\partial_s p, \partial_s p), \text{ by the vanishing of the torsion,}\\
&= (R^P(\partial_s p, \partial_{t_i} p)\partial_s p, \partial_s p),\\
&= 0, \quad \text{since } R^P \text{ is antisymmetric.}
\end{aligned}$$

From this, we see that $(\partial_{t_i} p, \partial_s p)$ is an affine function of s. Furthermore, the vector $\partial_{t_1} p|_{s=1}$ is tangent to the curve

$$t_1 \longmapsto \exp_{p_1(t_1)} X(t_1,t_2) = p_2(t_2).$$

Thus $\partial_{t_1} p|_{s=1} = 0$, from which it follows that

$$(\partial_{t_1} p, \partial_s p) = (1-s)(\partial_{t_1} p, \partial_s p)\big|_{s=0}.$$

Writing

$$\begin{aligned}
\partial_{t_1}\|\partial_s p\|^2 &= 2(\nabla^P_{\partial_{t_1}}\partial_s p, \partial_s p)\\
&= 2\partial_s(\partial_{t_1} p, \partial_s p)\\
&= -2(\partial_{t_1} p, \partial_s p)\big|_{s=0},
\end{aligned}$$

we obtain the differential equation

$$\begin{aligned}
\partial_{t_1}\|X(t_1,t_2)\|^2 &= -2(\partial_{t_1} p, X(t_1,t_2))_{p_1(t_1)}\\
&= 2(v_1, J(p_1,a)^{-1}(t_1 v_1 - t_2 v_2)) + o(|t_1| + |t_2|).
\end{aligned}$$

Let $\|X(t_1,t_2)\|^2 = \|a\|^2 + t_1^2 x_{11} + 2t_1 t_2 x_{12} + t_2^2 x_{22} + o(t_1^2 + t_2^2)$ be the Taylor expansion of $\|X(t_1,t_2)\|^2$. We see that

$$\begin{aligned}
x_{11} &= (v_1, J(p_1,a)^{-1} v_1), \quad \text{and}\\
x_{12} &= -(v_1, J(p_1,a)^{-1} v_2).
\end{aligned}$$

Now $d^2(p_1(t_1), p_2(t_2)) = d^2(p_2(t_2), \exp_{p_2 \exp(-a)} t_1 v_1)$. Thus, exchanging the roles of (p_1,t_1) and (p_2,t_2) we obtain

$$x_{22} = (v_2, J(p_2,-a)^{-1} v_2) = (J(p_2,a)^{-1} v_2, v_2).$$

Since $J(p_1,a) = J(p_2,a)$, this completes the proof. □

6.6 The heat kernel of an equivariant vector bundle

As in Chapter 5, let $\pi : P \to M$ be a principal bundle with compact structure group G, and let $\mathscr{E} = P \times_G E$ be an associated vector bundle corresponding to the representation ρ of G on the vector space E. Let ω be a connection one-form on P, and let ∇ be the corresponding covariant derivative on $\mathscr{E}$; we will always consider P with the Riemannian structure associated to the connection ω. In this section, we will generalize the results of Section 5.2 to the situation in which a compact group H acts on the left on P by orientation-preserving isometries, and leaves the connection one-form ω invariant.

Also, suppose we are given an auxiliary bundle $\mathscr{W}$ with an action of H, endowed with an H-invariant Hermitian metric and an H-invariant Hermitian connection. Let F be a potential of the form $F = F^0 + F^1$, where $F^0 \in \Gamma(M, \operatorname{End}(\mathscr{W}))$ and $F^1 \in \Gamma(P, \operatorname{End}(\mathscr{W}) \otimes \mathfrak{g})^G$, as in Section 5.2. Assume that F commutes with the action of γ.

The group H acts on $\mathscr{E}$, hence on the space of sections $\Gamma(M,\mathscr{E})$ by the formula $(\gamma \cdot s)(x) = \gamma s(\gamma^{-1}x)$, where $\gamma \in H$ and $x \in M$. If $\Delta^{\mathscr{W}\otimes\mathscr{E}}$ is the Laplacian on the bundle $\mathscr{W}\otimes\mathscr{E}$ associated to the connection $\nabla^{\mathscr{W}} \otimes 1 + 1 \otimes \nabla^{\mathscr{E}}$ on $\mathscr{W}\otimes\mathscr{E}$, we will interested in the smooth kernel

$$\langle x_0 \mid \gamma e^{-t(\Delta^{\mathscr{W}\otimes\mathscr{E}}+F)} \mid x_1 \rangle.$$

We can lift this kernel to the principal bundle P, obtaining there the kernel

$$\langle p_0 \mid \gamma e^{-t(\Delta^{\mathscr{W}\otimes\mathscr{E}}+F)} \mid p_1 \rangle \in \Gamma(P \times P, \mathscr{W} \boxtimes \mathscr{W}^*) \otimes \operatorname{End}(E).$$

Fix $\gamma \in H$, and let $\mathscr{N}$ be the normal bundle to M^γ in M. The exponential map $(x_0, v) \mapsto \exp_{x_0} v$, for $x_0 \in M^\gamma$ and $v \in \mathscr{N}_{x_0}$, gives an isomorphism between a neighbourhood of the zero section in $\mathscr{N}$ and a neighborhood of M^γ in M, and if $p \in \pi^{-1}(x_0)$, the point $\exp_p v$ projects to $\exp_{x_0} v$ under π. We identify $\mathscr{W}_{\exp_{x_0} v}$ and $\mathscr{W}_{x_0}$ by parallel transport along the geodesic. Let ϕ be a smooth function on $\mathscr{N}_{x_0}$ with small support. Our aim is to study the asymptotic behaviour of the time-dependent element of $\operatorname{End}(\mathscr{W}_{x_0}) \otimes \operatorname{End}(E)$

$$I(t,\gamma,\phi,p) = \int_{\mathscr{N}_{x_0}} \langle \exp_p v \mid \gamma \exp(-t(\Delta^{\mathscr{W}\otimes\mathscr{E}} + F)) \mid \exp_p v \rangle \, \phi(v) \, dv.$$

Generalizing Theorem 4.1, we will prove that $I(t,\gamma,\phi,p)$ has an asymptotic expansion as a Laurent power series in $t^{1/2}$ when $t \to 0$, and give an integral formula for this expansion.

Consider the quadratic form Q_a on $T_{x_0}M \oplus T_{x_0}M$ defined for v_1, $v_2 \in T_{x_0}M$ by the formula

$$Q_a(v_1,v_2) = (v_1, J(p,a)^{-1}v_1) - 2(v_1, J(p,a)^{-1}v_2) + (v_2, J(p,a)^{-1}v_2).$$

By Proposition 6.17, Q_a is the Hessian of the function

$$(v_1,v_2) \longmapsto d^2(\exp_p v_1, \exp_{p \exp a} v_2)$$

at the critical point $v_1 = v_2 = 0$, and for $a = 0$, we have that $Q_a(v_1, v_2)|_{a=0} = \|v_1 - v_2\|^2$. Let $Q_1(a, \gamma)$ be the quadratic form on $\mathcal{N}_{x_0}$ given by

$$Q_1(a,\gamma)(v) = Q_a(v, \gamma v) \quad \text{for } v \in \mathcal{N}_{x_0};$$

for $a = 0$, we have $\det(Q_1(0,\gamma)) = \det(1-\gamma_1)^2 \neq 0$. Finally, let $\tilde{\gamma}$ be the element of G such that $\gamma p = p\tilde{\gamma}$.

Theorem 6.18. *Assume that the Casimir* Cas *of the representation ρ is a scalar. There exist smooth sections $\Phi_i \in \Gamma(P \times \mathfrak{g}, \mathrm{End}(\mathcal{W}))$ such that as $t \to 0$, $I(t, \gamma, \phi, p)$ is asymptotic to the Laurent series*

$$(4\pi t)^{-\dim(P)/2+\ell_1} \int_{\mathfrak{g}}^{\text{asymp}} e^{-\|a\|^2/4t} \sum_{i=0}^{\infty} t^i \Phi_i(p,a)\, \rho((\exp a)^{-1}\tilde{\gamma})\, j_{\mathfrak{g}}(a)\, da.$$

Furthermore, $\Phi_0(p,a)$ is given by the formula

$$\Phi_0(p,a) = \frac{\gamma_{x_0}^{\mathcal{W}} \exp(F_p^1 \cdot a/2)\, \phi(0)}{|\det(Q_1(a,\gamma))|^{1/2}\, j_{\mathfrak{g}}^{1/2}(a)\, j_{T_{x_0}M}^{1/2}(\tau(\Omega_p \cdot a)/2)}.$$

Proof. The proof is a generalization of that of Theorem 5.9. However, we will only give the details in the scalar case, in which $\mathcal{W}$ is the trivial bundle, since the extension to non-zero potential F is similar to the corresponding extension in the non-equivariant case, apart from the occurrence of $\gamma_{x_0}^{\mathcal{W}}$ in the formula for Φ_0. Thus, from now on, we study the heat kernel of the operator $\Delta^{\mathcal{E}}$.

Let $h_t(p_0, p_1)$ be the scalar heat kernel of P and let Cas be the Casimir of the representation (ρ, E) of G. The following lemma is the equivariant version of Proposition 5.7.

Lemma 6.19.

$$\langle p_0 \mid \gamma e^{-t\Delta^{\mathcal{E}}} \mid p_1 \rangle = e^{-t\mathrm{Cas}} \int_G h_t(p_0, \gamma p_1 g) \rho(g)^{-1}\, dg$$

Proof. In the identification of $\Gamma(M, \mathcal{E})$ with $C^\infty(P, E)^G$, the action of H on $\Gamma(M, \mathcal{E})$ corresponds to the action on $C^\infty(P)$ given by $(\gamma s)(x) = s(\gamma^{-1}x)$. Now the operator $\gamma e^{-t\Delta^P}$ on $C^\infty(P)$ has kernel $h_t(\gamma^{-1}p_0, p_1) = h_t(p_0, \gamma p_1)$; the rest of the proof is as in that of Proposition 5.7. □

Identifying $v \in T_{x_0}M$ with a horizontal tangent vector at $p \in P$, the geodesic $\exp_p tv$ is a horizontal curve which projects to the curve $\exp_{x_0} tv$ in M. It follows from the above lemma that

$$\langle \exp_p v \mid \gamma e^{-t\Delta^{\mathcal{E}}} \mid \exp_p v \rangle = e^{-t\mathrm{Cas}} \int_G h_t(\exp_p v, \gamma \exp_p(v) \cdot g) \rho(g)^{-1}\, dg.$$

The curve $\gamma \exp_p(tv)$ is a geodesic in P starting at $\gamma p = p\tilde{\gamma}$ with a horizontal tangent vector which projects onto $\gamma_1 v$. Thus $\gamma \exp_p v = \exp_p(\gamma_1 v) \cdot \tilde{\gamma}$ and, replacing g by $\tilde{\gamma}^{-1} g$, we obtain

$$\langle \exp_p v \mid \gamma e^{-t\Delta^{\mathscr{E}}} \mid \exp_p v \rangle = \int_G h_t(\exp_p v, \exp_p(\gamma_1 v)\cdot g) e^{-t\mathrm{Cas}} \rho(g^{-1}\tilde{\gamma})\, dg.$$

When g is far from the identity and v is small enough, the function

$$t \longmapsto h_t(\exp_p v, \exp_p(\gamma_1 v)\cdot g)$$

decreases rapidly as $t \to 0$. Thus, since only a small neighbourhood of the identity contributes asymptotically, we can work in exponential coordinates on G. Let $\psi \in C_c^\infty(\mathfrak{g})$ be a smooth cut-off function equal to 1 on a small neighbourhood of 0. Since $|d(\exp a)| = j_{\mathfrak{g}}(a)\,|da|$, we see that $I(t,\gamma,\phi,x_0)$ is asymptotic to

$$e^{-t\mathrm{Cas}} \int_{\mathfrak{g}\times\mathscr{N}_{x_0}}^{\mathrm{asymp}} h_t(\exp_p v, \exp_p(\gamma_1 v)\cdot e^a) \rho(e^{-a}\tilde{\gamma})\, \phi(v)\, \psi(a)\, j_{\mathfrak{g}}(a)\, |da|\, |dv|.$$

Applying Theorem 2.30, we may replace $h_t(\exp_p v, \exp_p(\gamma_1 v)\cdot \exp a)$ by its approximation

$$(4\pi t)^{-\dim(P)/2} e^{-f(a,v)/4t} \sum_{i=0}^{\infty} t^i \Phi_i(a,v),$$

where $f(a,v) = d^2(\exp_p v, \exp_p(\gamma_1 v)\cdot \exp a)$ and

$$\Phi_i(a,v) = U_i(\exp_p v, \exp_p(\gamma_1 v)\cdot \exp a).$$

Furthermore, Theorem 5.8 shows that

$$\Phi_0(a,0) = j_{\mathfrak{g}}(a)^{-1/2}\, j_{T_{x_0}M}^{-1/2}(\tau(\Omega_p\cdot a)/2)).$$

By Proposition 6.17, the function $v \mapsto f(a,v)$ on $\mathscr{N}_{x_0}$ has its only critical point at $v=0$, where its Hessian is the quadratic form $v \mapsto Q_1(a,\gamma)(v)$. When $a=0$, we have $Q_1(0,\gamma)(v) = \|(1-\gamma_1)v\|^2$. Thus, for small a, the quadratic form $Q_1(a,\gamma)$ is not degenerate on $\mathscr{N}_{x_0}$. By Morse's Lemma, there exists a local diffeomorphism $v \mapsto w = F_a(v)$ of $\mathscr{N}_{x_0}$ such that $F_a(0)=0$ and $f(a,v) = \|a\|^2 + \|w\|^2$. Note that

$$\left.\frac{dv}{dw}\right|_{w=0} = \left|\det(Q_1(a,\gamma))\right|^{-1/2}.$$

Thus, we see that $I(t,\gamma,\phi,x_0)$ is asymptotic to

$$(4\pi t)^{-\dim(P)/2} e^{-t\mathrm{Cas}} \int_{\mathfrak{g}\times\mathscr{N}_{x_0}}^{\mathrm{asymp}} e^{-(\|a\|^2+\|w\|^2)/4t}$$
$$\times \sum_{i=0}^{\infty} t^i \Phi_i(a, F_a^{-1}(w))\, \phi(F_a^{-1}(w))\, \psi(a)\, \rho(\exp(a)^{-1}\tilde{\gamma})\, j_{\mathfrak{g}}(a) \left|\frac{dv}{dw}\right| |da|\, |dw|.$$

The change of variable $w \to t^{-1/2}w$ gives us the asymptotic expansion of the right-hand side that we wished to prove, with

$$\Phi_0(p,a) = \frac{\phi(0)}{|\det(Q_1(a,\gamma))|^{1/2}\, j_{\mathfrak{g}}^{1/2}(a)\, j_{T_{x_0}M}^{1/2}(\tau(\Omega_p\cdot a)/2)}.$$

In the case of a twisting bundle $\mathcal{W}$, we consider the heat kernel h_t of the operator $\tilde{\Delta}+\tilde{F}$ of Theorem 5.9, and we use the formula

$$\langle p_0 \mid \gamma e^{-t(\Delta^{\mathcal{W}\otimes\mathcal{E}}+F)} \mid p_1 \rangle = \int_G h_t(p_0, \gamma p_1 \cdot g)\, \gamma_x^{\mathcal{W}} \otimes e^{-t\mathrm{Cas}} \rho(g^{-1})\, dg. \quad \square$$

6.7 Proof of Proposition 6.13

In this section, we will prove Proposition 6.12, and hence Theorem 6.11. The proof is a generalization of the method of Chapter 5. As in that chapter, we will take advantage of the fact that the theorem is local to require that the manifolds M has an equivariant spin structure, with spinor bundle $\mathcal{S}$. Thus, the Clifford module $\mathcal{E}$ is isomorphic to $\mathcal{W}\otimes\mathcal{S}$, where $\mathcal{W}$ is a Hermitian vector bundle with connection $\nabla^{\mathcal{W}}$.

Let $P = \mathrm{Spin}(M)$ be the double-cover of the orthonormal frame bundle $\mathrm{SO}(M)$ corresponding to the spin-structure on M. A point $p \in \mathrm{Spin}(M)$ projects to an orthonormal frame at the point $x_0 = \pi(p) \in M$, which we will denote by $p : V \to T_{x_0}M$, where V is the vector space $\mathbb{R}^n$. The theorem will follow from an application of Theorem 6.18, with group $G = \mathrm{Spin}(n)$.

Let M_0 be a component of the fixed point set γ. Decompose $V = \mathbb{R}^n$ into the direct sum of $V_0 = \mathbb{R}^{2\ell_0}$ and $V_1 = \mathbb{R}^{2\ell_1}$. If x_0 is a point in M_0, we will denote by p a point in the fibre of the bundle $\mathrm{Spin}(M)$ at x_0 such p restricted to V_0 is a frame of $T_{x_0}M_0 \subset T_{x_0}M$, while p restricted to V_1 is a frame of $\mathcal{N}_{x_0}$.

Let $\tilde{\gamma}$ be the element of $\mathrm{Spin}(V)$ such that $\gamma u = u\tilde{\gamma}$. The element $\tilde{\gamma}$ induces an endomorphism of $\mathcal{S}_{x_0}$. Since $\tilde{\gamma}$ lies above a transformation $\gamma_1 \in \mathrm{SO}(\mathcal{N}_{x_0})$ such that $\det(1-\gamma_1) \neq 0$, we see that $\tilde{\gamma}$ belongs to $C(V_1)$, and that its symbol $\sigma_{2\ell_1}(\tilde{\gamma})$ is non-zero.

We may identify the curvature $F_{x_0}^{\mathcal{W}}$ of the connection $\nabla^{\mathcal{W}}$ with an element of $\Lambda^2 V \otimes \mathrm{End}(\mathcal{W}_{x_0})$. Lichnerowicz's formula (Theorem 3.52) shows that $\mathrm{D}^2 = \Delta^{\mathcal{W}\otimes\mathcal{S}} + F^0 + F^1$, where $F_p^0 = r_{x_0}^M/4$ and $F_p^1 = F_{x_0}^{\mathcal{W}}$. Thus, by Theorem 6.18, we obtain the following result.

Lemma 6.20. *The section $I(t,\gamma,\phi,x_0)$ has the asymptotic expansion*

$$(4\pi t)^{-\dim(\mathfrak{g})/2-\ell_0} \int_{\mathfrak{g}}^{\text{asymp}} e^{-\|a\|^2/4t} \sum_{i=0}^{\infty} t^i \Phi_i(a)\, \rho(\exp(a)\tilde{\gamma})\, j_{\mathfrak{g}}(a)\, da,$$

where $\Phi_0(a)$ equals

$$\frac{\exp(-F_{x_0}^{\mathcal{W}}\cdot a/2)\, \gamma_{x_0}^{\mathcal{W}}\, \phi(x_0)}{|\det(Q_1(-a,\gamma_1))|^{1/2}\, j_{\mathfrak{g}}(a)^{1/2}\, j_V^{-1/2}(\tau(\Omega_p\cdot a/2))}.$$

Proposition 5.16 implies the first assertion of Theorem 6.12, as well as giving us the formula

$$\sigma(k(\gamma,\phi,x_0)) = j_{\mathfrak{g}}(2e\wedge e^*)\, \Phi_0(2e\wedge e^*) \wedge \sigma_{2\ell_1}(\tilde{\gamma}).$$

Let us consider the decomposition $V = V_0 \oplus V_1 = \mathbb{R}^{2\ell_0} \oplus \mathbb{R}^{2\ell_1}$. Denote by $e_0 \wedge e_0^*$ the antisymmetric $n \times n$-matrix whose (i,j)-coefficient equals $e^i \wedge e^j$ if $i \leq 2\ell_0$ and $j \leq 2\ell_0$, and such that all other coefficients vanish. If Φ is a function on $\mathfrak{g}$, we denote by $\Phi(e_0 \wedge e_0^*)$ its evaluation at this element of $\mathfrak{g} \otimes \Lambda^2 V_0^*$. It is clear that $\Phi(e_0 \wedge e_0^*)$ depends only on the restriction of Φ to the subspace $\mathfrak{g}_0 = \Lambda^2 V_0$ of $\mathfrak{g} = \Lambda^2 V$.

Since $\sigma_{2\ell_1}(\tilde{\gamma}) \in \Lambda^{2\ell_1} V_1$, it is clear that

$$\Phi(2e \wedge e^*) \wedge \sigma_{2\ell_1}(\tilde{\gamma}) = \Phi(2e_0 \wedge e_0^*) \wedge \sigma_{2\ell_1}(\tilde{\gamma}).$$

We now have the following results:

1. $j_{\mathfrak{g}}(2e_0 \wedge e_0^*) = 1$
2. $F^{\mathcal{W}} \cdot (e_0 \wedge e_0^*)$ equals the restriction to M^γ of the curvature $F^{\mathcal{W}}$;
3. if $R = R^0 \oplus R^1$ is the decomposition of the curvature of the bundle $TM|_{M_0}$ of (6.1), corresponding to the decomposition $TM|_{M_0} = TM_0 \oplus \mathcal{N}$, it follows that for $a \in \mathfrak{g}_0 = \Lambda^2 V_0$,
$$j_V(\tau(\Omega_p \cdot a)/2)) = j_{V_0}(\tau(\Omega^0_{p_0} \cdot a)/2)\, j_{V_1}(\tau(\Omega^1_p \cdot a)/2).$$

It only remains to calculate $Q_1(2e_0 \wedge e_0^*, \gamma_1)$.

Lemma 6.21. *For $v \in V_1$ and $a \in \mathfrak{g}_0$, we have*

$$(Q_1(a,\gamma)v,v) = 2(v, J_{V_1}(\tau(\Omega^1_p \cdot a)/2)^{-1}(1-\gamma_1)v).$$

Proof. By Proposition 6.17, we have

$$(Q_1(a,\gamma_1)v,v) = (v, J_V(p,a)^{-1}v) - 2(v, J_V(p,a)^{-1}\gamma_1 v) + (\gamma_1 v, J_V(p,a)^{-1}\gamma_1 v).$$

For $a \in \mathfrak{g}_0$, the endomorphism $J_V(p,a)$ preserves the decomposition $V = V_0 \oplus V_1$, and

$$J_V(p,a)|_{V_1} = J_{V_1}(\tau(\Omega^1_p \cdot a)/2).$$

The lemma follows from the fact that $J_{V_1}(\tau(\Omega^1_p \cdot a)/2)$ commutes with γ_1. □

Let $A = J_{V_1}(\tau(\Omega^1_p \cdot a)/2)^{-1}$. The matrix $A \cdot (1-\gamma_1)$ is not symmetric, so we must rewrite $(Q_1(a,\gamma)v,v)$ in the form (Bv,v), where B is a symmetric matrix. Thus, we write

$$(Q_1(a,\gamma)v,v) = (v, A \cdot (1-\gamma_1)v) + (v, (1-\gamma_1)^* \cdot A^* v).$$

We have $A^* = e^{\tau(\Omega^1_p \cdot a)/2} A$, and hence

$$A \cdot (1-\gamma_1) + (1-\gamma_1)^* \cdot A^* = A \cdot (1-\gamma_1) \cdot (1-\gamma_1^* e^{\tau(\Omega^1_p \cdot a)/2}).$$

Thus for $a \in \mathfrak{g}_0$, we find that

$$\begin{aligned} &j_V^{-1/2}(\tau(\Omega_p, a)/2)\,|\det(Q_1(a,\gamma_1))|^{-1/2} \\ &= j_{V_0}^{-1/2}(\tau(\Omega^0_p \cdot a/2)) \det(1-\gamma_1)^{-1/2} \det(1-\gamma_1 \exp(-\tau(\Omega^1_p \cdot a)/2)^{-1/2}. \end{aligned}$$

We now replace a by $2e_0 \wedge e_0^*$. Using the fundamental symmetry of the Riemannian curvature as in Lemma 5.17 we see that

$$j_{V_0}^{-1/2}(\tau(\Omega^0_{p_0}, e_0 \wedge e_0^*)) = \hat{A}(M^\gamma),$$

which completes the proof of Proposition 6.12.

Bibliographic Notes

The fixed point formula Theorem 6.16 for elliptic operators invariant under a compact Lie group G was proved by Atiyah-Segal [12]; the proof uses the Atiyah-Singer index theorem [14] and the localization result for the ring $K_G(M)$. The Atiyah-Bott fixed point formula was proved in [5].

The generalization of the local index theorem to a fixed point formula in the presence of a compact group action, was announced by Patodi [91] in the case of the $\bar{\partial}$-operator, and was proved by Donnelly-Patodi [54] for the G-signature. It was first proved for the twisted G-signature operator by Gilkey [66]. Our proof of Theorem 6.11 is taken from Berline-Vergne [24], which was inspired by Bismut [29]. For a proof more in the style of Chapter 4, see the article of Lafferty-Yu-Zhang [79].

7 Equivariant Differential Forms

Although it is rarely possible to calculate the integral of a differential form exactly, it was a beautiful discovery of Bott that it is sometimes possible to localize the calculation of such integrals to the zero set of a vector field on the manifold. In this chapter, we will describe a generalization of this, a localization formula for equivariant differential forms. Only the results of Chapter 1 are a prerequisite to reading this chapter.

Let M be an n-dimensional manifold acted on by a Lie group G with Lie algebra $\mathfrak{g}$. A G-equivariant differential form on M is defined to be a polynomial map $\alpha : \mathfrak{g} \to \mathscr{A}(M)$ such that $\alpha(gX) = g \cdot \alpha(X)$ for $g \in G$. The equivariant exterior differential $d_{\mathfrak{g}}$ is the operator defined by the formula

$$(d_{\mathfrak{g}}\alpha)(X) = (d - \iota(X_M))\alpha(X),$$

where X_M is the vector field on M which generates the action of the one-parameter group $\exp tX$. The space of equivariant differential forms with this differential is a complex, with cohomology the equivariant cohomology of M, denoted $H_G^\bullet(M)$.

The condition that $X \mapsto \alpha(X)$ is equivariantly closed means that for each $X \in \mathfrak{g}$, the homogeneous components of the differential form

$$\alpha(X) = \alpha(X)_{[0]} + \alpha(X)_{[1]} + \dots$$

satisfy the series of relations

$$\iota(X_M)\alpha(X)_{[i]} = d\alpha(X)_{[i-2]}.$$

These relations imply that $\alpha(X)_{[n]}$ is exact outside the set M_0 of zeroes of the vector field X_M, if the one-parameter group of diffeomorphism generated by X is a circle. The localization formula then reduces the integral $\int_M \alpha(X)$ to a certain integral over M_0.

Equivariantly closed differential forms arise naturally in a variety of situations. Bott's formulas for characteristic numbers, the exact stationary phase formula of Duistermaat-Heckmann for a Hamiltonian action, Harish-Chandra's formula for the

Fourier transform of orbits of the coadjoint representation of a compact Lie group, and Chevalley's theorem on the structure of the algebra of invariant functions on a semi-simple Lie algebra are all consequences of the localization formula. These results will not be needed in the rest of the book, but are included for their intrinsic interest.

We will also use equivariant differential forms to give a generalization of the Chern-Weil map, thereby relating the equivariant cohomology of a G-manifold M to the cohomology of fibre bundles with M as fibre. As an illustration, we consider in Section 7 the simple case of a Euclidean vector bundle with an action of the orthogonal group, which leads to the Mathai-Quillen universal Thom form of a vector bundle.

7.1 Equivariant Characteristic Classes

Let M be a C^∞-manifold with an action of a Lie group G, and let $\mathfrak{g}$ be the Lie algebra of G. The group G acts on $C^\infty(M)$ by the formula $(g\cdot\phi)(x) = \phi(g^{-1}x)$. For $X \in \mathfrak{g}$, we denote by X_M (or sometimes simply X) the vector field on M given by

$$(X_M\cdot\phi)(x) = \frac{d}{d\varepsilon}\phi(\exp(-\varepsilon X)x)\Big|_{\varepsilon=0}$$

The minus sign is there so that $X \mapsto X_M$ is a Lie algebra homomorphism. On the other hand, if G acts on the right on a manifold P then we must write

$$(X_P\cdot\phi)(p) = \frac{d}{d\varepsilon}\phi(p\exp\varepsilon X)\Big|_{\varepsilon=0}$$

Let $\mathbb{C}[\mathfrak{g}]$ denote the algebra of complex valued polynomial functions on $\mathfrak{g}$. We may view the tensor product $\mathbb{C}[\mathfrak{g}]\otimes\mathscr{A}(M)$ as the algebra of polynomial maps from $\mathfrak{g}$ to $\mathscr{A}(M)$. The group G acts on an element $\alpha \in \mathbb{C}[\mathfrak{g}]\otimes\mathscr{A}(M)$ by the formula

$$(g\cdot\alpha)(X) = g\cdot(\alpha(g^{-1}\cdot X)) \quad \text{for all } g \in G \text{ and } X \in \mathfrak{g}.$$

Let $\mathscr{A}_G(M) = (\mathbb{C}[\mathfrak{g}]\otimes\mathscr{A}(M))^G$ be the subalgebra of G-invariant elements; an element α of $\mathscr{A}_G(M)$ satisfies the relation $\alpha(g\cdot X) = g\cdot\alpha(X)$, and will be called an **equivariant differential form**.

The algebra $\mathbb{C}[\mathfrak{g}]\otimes\mathscr{A}(M)$ has a $\mathbb{Z}$-grading, defined by the formula

$$\deg(P\otimes\alpha) = 2\deg(P) + \deg(\alpha)$$

for $P \in \mathbb{C}[\mathfrak{g}]$ and $\alpha \in \mathscr{A}(M)$. We define the equivariant exterior differential $d_{\mathfrak{g}}$ on $\mathbb{C}[\mathfrak{g}]\otimes\mathscr{A}(M)$ by the formula

$$(d_{\mathfrak{g}}\alpha)(X) = d(\alpha(X)) - \iota(X)(\alpha(X)),$$

where $\iota(X)$ denotes contraction by the vector field X_M; from now on, we will frequently write X instead of X_M. Thus, $d_{\mathfrak{g}}$ increases by one the total degree on

$\mathbb{C}[\mathfrak{g}] \otimes \mathscr{A}(M)$, and preserves $\mathscr{A}_G(M)$. The homotopy formula $\iota(X)d + d\iota(X) = \mathscr{L}(X)$ (see (1.4)) implies that

$$(d_{\mathfrak{g}}^2 \alpha)(X) = -\mathscr{L}(X)\alpha(X),$$

for any $\alpha \in \mathbb{C}[\mathfrak{g}] \otimes \mathscr{A}(M)$, hence $(\mathscr{A}_G(M), d_{\mathfrak{g}})$ is a complex. The elements of $\mathscr{A}_G(M)$ such that $d_{\mathfrak{g}}\alpha = 0$ are called equivariantly closed forms; those of the form $\alpha = d_{\mathfrak{g}}\beta$ are called equivariantly exact forms. This complex was introduced by H. Cartan.

Definition 7.1. *The* **equivariant cohomology** $H_G^\bullet(M)$ *of* M *is the cohomology of the complex* $(\mathscr{A}_G(M), d_{\mathfrak{g}})$.

We will also consider equivariant differential forms $X \mapsto \alpha(X)$ which depend smoothly on X in a neighbourhood of $0 \in \mathfrak{g}$, and not polynomially; the algebra of all of these is written $\mathscr{A}_G^\infty(M)$. Although $\mathscr{A}_G^\infty(M)$ is not a $\mathbb{Z}$-graded algebra, it has a differential $d_{\mathfrak{g}}$, defined by the same formula as before, which is odd with respect to the $\mathbb{Z}_2$-grading.

If $H \to G$ is a homomorphism of Lie groups, there is a pull-back map $\mathscr{A}_G(M) \to \mathscr{A}_H(M)$ on equivariant forms, defined using the restriction map $\mathbb{C}[\mathfrak{g}] \to \mathbb{C}[\mathfrak{h}]$. This is a homomorphism of differential graded algebras; that is, it sends exterior product to exterior product and equivariant exterior differential $d_{\mathfrak{g}}$ to exterior differential $d_{\mathfrak{h}}$, and hence induces a map $H_G^\bullet(M) \to H_H^\bullet(M)$.

It is clear that when H is the trivial group $\{1\}$, the equivariant cohomology of M is the ordinary de Rham cohomology. Thus, letting $H = \{1\}$ and pulling back by the inclusion of the identity $\{1\} \to G$, we obtain a map $\mathscr{A}_G(M) \to \mathscr{A}(M)$, given explicitly by evaluation $\alpha \mapsto \alpha(0)$ at $X = 0$.

If $\phi : N \to M$ is a map of G-manifolds which intertwines the actions of G, then pull-back by ϕ induces a homomorphism of differential graded algebras $\phi^* : \mathscr{A}_G(M) \to \mathscr{A}_G(N)$. In particular, if N is a G-invariant submanifold of M, the restriction map maps $\mathscr{A}_G(M)$ to $\mathscr{A}_G(N)$.

If M is a compact oriented manifold, we can integrate equivariant differential forms over M, obtaining a map

$$\int_M : \mathscr{A}_G(M) \to \mathbb{C}[\mathfrak{g}]^G,$$

by the formula $(\int_M \alpha)(X) = \int_M \alpha(X)$; here, if α is a non-homogeneous differential form, the integral $\int_M \alpha$ is understood to mean the integral of the homogeneous component of top degree. If α is equivariantly exact, that is, $\alpha = d_{\mathfrak{g}}\nu$ for some $\nu \in \mathscr{A}_G(M)$, and $\dim(M) = n$, then

$$\alpha(X)_{[n]} = d(\nu(X)_{[n-1]}),$$

so that $\int_M \alpha(X) = 0$. Thus, if α is an equivariantly closed form, $\int_M \alpha$ is an element of $\mathbb{C}[\mathfrak{g}]^G$ which only depends on the equivariant cohomology class of α.

We will now associate to an equivariant superbundle certain equivariantly closed forms whose classes in equivariantly cohomology will be called equivariant characteristic classes. This construction follows closely the construction of non-equivariant characteristic forms in Chapter 1.

Recall the definition of a G-equivariant vector bundle Definition 1.5.

Definition 7.2. *If $\mathscr{E}$ is a G-equivariant vector bundle, the space of equivariant differential forms with values in $\mathscr{E}$ is the space*

$$\mathscr{A}_G(M,\mathscr{E}) = \big(\mathbb{C}[\mathfrak{g}] \otimes \mathscr{A}(M,\mathscr{E})\big)^G,$$

with $\mathbb{Z}$-grading defined by a formula analogous to that on $\mathscr{A}_G(M)$.

A G-equivariant superbundle $\mathscr{E} = \mathscr{E}^+ \oplus \mathscr{E}^-$ is a superbundle such that $\mathscr{E}^+$ and $\mathscr{E}^-$ are G-equivariant vector bundles. If $\mathbb{A}$ is a superconnection on a G-equivariant superbundle $\mathscr{E}$ which commutes with the action of G on $\mathscr{A}(M,\mathscr{E})$, we see that

$$[\mathbb{A}, \mathscr{L}^{\mathscr{E}}(X)] = 0$$

for all $X \in \mathfrak{g}$. We say that $\mathbb{A}$ is a **G-invariant superconnection.** If G is compact, then every G-equivariant bundle admits an invariant connection, constructed as in (1.10).

Definition 7.3. *The* **equivariant superconnection** $\mathbb{A}_{\mathfrak{g}}$ *corresponding to a G-invariant superconnection $\mathbb{A}$ is the operator on $\mathbb{C}[\mathfrak{g}] \otimes \mathscr{A}(M,\mathscr{E})$ defined by the formula*

$$(\mathbb{A}_{\mathfrak{g}}\alpha)(X) = (\mathbb{A} - \iota(X))(\alpha(X)), \quad X \in \mathfrak{g},$$

where $\iota(X)$ denotes the contraction operator $\iota(X_M)$ on $\mathscr{A}(M,\mathscr{E})$.

The justification for this definition is that

$$\mathbb{A}_{\mathfrak{g}}(\alpha \wedge \theta) = d_{\mathfrak{g}}\alpha \wedge \theta + (-1)^{|\alpha|}\alpha \wedge \mathbb{A}_{\mathfrak{g}}\theta$$

for all $\alpha \in \mathbb{C}[\mathfrak{g}] \otimes \mathscr{A}(M)$ and $\theta \in \mathbb{C}[\mathfrak{g}] \otimes \mathscr{A}(M,\mathscr{E})$. The operator $\mathbb{A}_{\mathfrak{g}}$ preserves the subspace $\mathscr{A}_G(M,\mathscr{E}) \subset \mathbb{C}[\mathfrak{g}] \otimes \mathscr{A}(M,\mathscr{E})$, and we will also denote its restriction to this space by $\mathbb{A}_{\mathfrak{g}}$.

Bearing in mind the formula

$$d_{\mathfrak{g}}^2\alpha(X) + \mathscr{L}(X)\alpha(X) = 0,$$

we are motivated to define the equivariant curvature $F_{\mathfrak{g}}$ of the equivariant superconnection $\mathbb{A}_{\mathfrak{g}}$ by the formula

$$\varepsilon(F_{\mathfrak{g}}(X)) = \mathbb{A}_{\mathfrak{g}}(X)^2 + \mathscr{L}^{\mathscr{E}}(X).$$

Given $\theta \in \mathscr{A}_G(M,\mathrm{End}(\mathscr{E}))$, let $\mathbb{A}_{\mathfrak{g}}\theta$ be the element of $\mathscr{A}_G(M,\mathrm{End}(\mathscr{E}))$ such that $\varepsilon(\mathbb{A}_{\mathfrak{g}}\theta) = [\mathbb{A}_{\mathfrak{g}}, \varepsilon(\theta)]$.

Proposition 7.4. *The* **equivariant curvature** $F_{\mathfrak{g}}$ *is in $\mathscr{A}_G(M,\mathrm{End}(\mathscr{E}))$, and satisfies the equivariant Bianchi formula*

$$\mathbb{A}_{\mathfrak{g}}F_{\mathfrak{g}} = 0.$$

Proof. To prove that $F_{\mathfrak{g}} \in \mathscr{A}_G(M, \mathrm{End}(\mathscr{E}))$, we must show that $\mathbb{A}_{\mathfrak{g}}(X)^2 + \mathscr{L}^{\mathscr{E}}(X)$ commutes with multiplication by any $\alpha \in \mathbb{C}[\mathfrak{g}] \otimes \mathscr{A}(M)$:

$$\begin{aligned}[\mathbb{A}_{\mathfrak{g}}(X)^2 + \mathscr{L}^{\mathscr{E}}(X), \varepsilon(\alpha(X))] &= [\mathbb{A}_{\mathfrak{g}}(X), [\mathbb{A}_{\mathfrak{g}}(X), \varepsilon(\alpha(X))]] + [\mathscr{L}^{\mathscr{E}}(X), \varepsilon(\alpha(X))] \\ &= [\mathbb{A}_{\mathfrak{g}}(X), \varepsilon(d_{\mathfrak{g}}\alpha(X))] + \varepsilon(\mathscr{L}(X)\alpha(X)) \\ &= \varepsilon(d_{\mathfrak{g}}^2\alpha(X) + \mathscr{L}(X)\alpha(X)) = 0.\end{aligned}$$

The equivariant Bianchi identity is just another way of writing the obvious identity $[\mathbb{A}_{\mathfrak{g}}(X), \mathbb{A}_{\mathfrak{g}}(X)^2 + \mathscr{L}^{\mathscr{E}}(X)] = 0$. □

If we expand the definition of $F_{\mathfrak{g}}$, we see (identifying $F_{\mathfrak{g}}$ with the operator $\varepsilon(F_{\mathfrak{g}})$ on $\mathscr{A}_G(M, \mathrm{End}(\mathscr{E}))$) that

$$\begin{aligned}F_{\mathfrak{g}}(X) &= (\mathbb{A} - \iota(X))^2 + \mathscr{L}^{\mathscr{E}}(X) \\ &= F - [\mathbb{A}, \iota(X)] + \mathscr{L}^{\mathscr{E}}(X).\end{aligned}$$

where $[\iota(X), \mathbb{A}] = \iota(X)\mathbb{A} + \mathbb{A}\iota(X)$ is the supercommutator, and $F = \mathbb{A}^2$ is the curvature of $\mathbb{A}$. In particular, $F_{\mathfrak{g}}(0) = F$. Motivated by this formula for $F_{\mathfrak{g}}(X)$, we make the following definition.

Definition 7.5. *The* **moment** *of $X \in \mathfrak{g}$ (relative to a superconnection $\mathbb{A}$) is the element of $\mathscr{A}^+(M, \mathrm{End}(\mathscr{E}))$ given by the formula*

$$\mu(X) = \mathscr{L}^{\mathscr{E}}(X) - [\iota(X), \mathbb{A}].$$

Observe that $\mu = F_{\mathfrak{g}} - F$ is an element of $\mathscr{A}_G(M, \mathrm{End}(\mathscr{E}))$, so that

$$F_{\mathfrak{g}}(X) = F + \mu(X).$$

The following formula is a restatement of the equivariant Bianchi identity,

$$\mathbb{A}\mu(X) = \iota(X)F. \tag{7.1}$$

The use of the word moment is justified by the similarity between this equation and the definition of the moment of a Hamiltonian vector field in symplectic geometry, as we will see later.

In the special case in which the invariant superconnection $\mathbb{A}$ on the equivariant bundle $\mathscr{E}$ is a connection ∇, the above formulas take a simplified form. Since $[\nabla, \iota(X)] = \nabla_X$, we see that the moment $\mu(X)$, given by the formula

$$\mu(X) = \mathscr{L}^{\mathscr{E}}(X) - \nabla_X,$$

lies in $\Gamma(M, \mathrm{End}(\mathscr{E}))$, and hence

$$\mu \in (\mathfrak{g}^* \otimes \Gamma(M, \mathrm{End}(\mathscr{E})))^G \subset \mathscr{A}_G^2(M, \mathrm{End}(\mathscr{E})).$$

If the vector field X_M vanishes at a point $x_0 \in M$, the endomorphism $\mu(X)(x_0)$ coincides with the infinitesimal action $\mathscr{L}(X)(x_0)$ of $\exp tX$ in $\mathscr{E}_{x_0}$.

We can give a geometric interpretation of the moment $\mu(X)$ in this case. Denote by $\pi : \mathscr{E} \to M$ the projection onto the base. The action of G on $\mathscr{E}$ determines vector fields $X_{\mathscr{E}}$ on the total space $\mathscr{E}$ corresponding with $X \in \mathfrak{g}$. The connection ∇ on $\mathscr{E}$ determines a splitting of the tangent space $T\mathscr{E}$,

$$T\mathscr{E} = \pi^*\mathscr{E} \oplus \pi^*TM,$$

into a vertical bundle isomorphic to $\pi^*\mathscr{E}$ and a horizontal bundle isomorphic to π^*TM.

Proposition 7.6. *Let* $\mathbf{x}$ *be the tautological section of the bundle* $\pi^*\mathscr{E}$ *over* $\mathscr{E}$. *The vertical component of* $X_{\mathscr{E}}$ *may be identified with* $-\mu(X)\mathbf{x}$.

Proof. This follows from Proposition 1.20, once we observe that $\mathscr{L}^{\pi^*\mathscr{E}}(X)\mathbf{x} = 0$. □

We now construct the equivariant generalization of characteristic forms. If $f(z)$ is a polynomial in the indeterminate z, then $f(F_{\mathfrak{g}})$ is an element of $\mathscr{A}_G^+(M, \mathrm{End}(\mathscr{E}))$. When we apply the supertrace map

$$\mathrm{Str} : \mathscr{A}_G^+(M, \mathrm{End}(\mathscr{E})) \to \mathscr{A}_G^+(M),$$

we obtain an element of $\mathscr{A}_G^+(M)$, which we call an **equivariant characteristic form**.

Theorem 7.7. *The equivariant differential form* $\mathrm{Str}(f(F_{\mathfrak{g}}))$ *is equivariantly closed, and its equivariant cohomology class is independent of the choice of the G-invariant superconnection* $\mathbb{A}$.

Proof. If $\alpha \in \mathscr{A}_G(M, \mathrm{End}(\mathscr{E}))$, it follows from Lemma 1.42 that

$$d_{\mathfrak{g}} \mathrm{Str}(\alpha) = \mathrm{Str}(\mathbb{A}_{\mathfrak{g}}\alpha).$$

The equation $d_{\mathfrak{g}}(\mathrm{Str}(f(F_{\mathfrak{g}}))) = 0$ now follows from the equivariant Bianchi identity $\mathbb{A}_{\mathfrak{g}}F_{\mathfrak{g}} = 0$. Similarly, if $\mathbb{A}^t$ is a one-parameter family of G-invariant superconnections, with equivariant curvature $F_{\mathfrak{g}}^t$, we have

$$\begin{aligned}
\frac{d}{dt}\mathrm{Str}(f(F_{\mathfrak{g}}^t)) &= \mathrm{Str}\Big(\Big[\mathbb{A}_{\mathfrak{g}}^t, \frac{d\mathbb{A}_{\mathfrak{g}}^t}{dt}\Big] f'(F_{\mathfrak{g}}^t)\Big) \\
&= d_{\mathfrak{g}} \mathrm{Str}\Big(\frac{d\mathbb{A}_{\mathfrak{g}}^t}{dt} f'(F_{\mathfrak{g}}^t)\Big)
\end{aligned}$$

so that the difference of the two equivariant characteristic forms

$$\mathrm{Str}(f(F_{\mathfrak{g}}^1)) - \mathrm{Str}(f(F_{\mathfrak{g}}^0))$$

is the equivariant coboundary of

$$\int_0^1 \mathrm{Str}\Big(\frac{d\mathbb{A}_{\mathfrak{g}}^t}{dt} f'(F_{\mathfrak{g}}^t)\Big)\, dt. \qquad \square$$

As in the non-equivariant case discussed in Section 1.4, we can also allow $f(z)$ to be a power series with infinite radius of convergence. In particular, the **equivariant Chern character** form $\mathrm{ch}_{\mathfrak{g}}(\mathbb{A})$ of an equivariant superbundle $\mathscr{E}$ with superconnection $\mathbb{A}$ is the element of the analytic completion $\mathscr{A}_G^{\omega}(M)$ of $\mathscr{A}_G(M)$ defined by

$$\mathrm{ch}_{\mathfrak{g}}(\mathbb{A}) = \mathrm{Str}(\exp(-F_{\mathfrak{g}})), \tag{7.2}$$

in other words, $\mathrm{ch}_{\mathfrak{g}}(\mathbb{A})(X) = \mathrm{Str}(\exp(-F_{\mathfrak{g}}(X)))$.

In the case of a G-invariant connection ∇, the component of exterior degree zero of the equivariant curvature $F_{\mathfrak{g}}(X) = \mu(X) + F$ is $\mu(X)$, which satisfies $\mu(X) = O(|X|)$. Thus we can define $f(F_{\mathfrak{g}}(X)) = f(\mu(X) + F)$ even if f is a germ of an analytic function at $0 \in \mathfrak{g}$. As an example, we have the **equivariant $\hat{A}$-genus** $\hat{A}_{\mathfrak{g}}(\nabla)$ of a vector bundle $\mathscr{E}$ over a compact manifold M,

$$\hat{A}_{\mathfrak{g}}(\nabla)(X) = \det{}^{1/2}\left(\frac{F_{\mathfrak{g}}(X)/2}{\sinh(F_{\mathfrak{g}}(X)/2)}\right).$$

Finally, we define the **equivariant Euler form.** Let $\mathscr{E}$ be a G-equivariant oriented vector bundle with G-invariant metric and G-invariant connection ∇ compatible with the metric. The curvature F and the moment μ are both elements of $\mathscr{A}_G(M, \mathfrak{so}(\mathscr{E}))$. We define the equivariant Euler form of $\mathscr{E}$ by

$$\chi_{\mathfrak{g}}(\nabla)(X) = \mathrm{Pf}(-F_{\mathfrak{g}}(X)) = \det{}^{1/2}(-F_{\mathfrak{g}}(X)),$$

where the notations are those of Definition 1.35. An argument similar to that of Theorem 7.7 shows that $\chi_{\mathfrak{g}}(\nabla)$ is an equivariantly closed form, and that its class in cohomology depends neither on the connection nor on the Euclidean structure of $\mathscr{E}$, but only on the orientation of $\mathscr{E}$.

We close this section with two important examples of the equivariant moment map.

Example 7.8. Let us compute the moment $\mu^M(X) \in \Gamma(M, \mathfrak{so}(TM))$ of an infinitesimal isometry X acting on the tangent bundle of a Riemannian manifold; this is called the **Riemannian moment** of M. The group G of isometries of M preserves the Levi-Civita connection ∇ on TM. By the definition of the moment, we see that

$$\mu^M(X)Y = [X,Y] - \nabla_X Y,$$

and from the vanishing of the torsion we obtain

$$\mu^M(X)Y = -\nabla_Y X, \tag{7.3}$$

in other words, $\mu^M(X) = -\nabla X$. Since $\mathscr{L}(X)$ and ∇_X preserve the metric on M, $\mu^M(X)$ is skew-symmetric. In this example, (7.1) states that the covariant derivative of the moment μ^M is given by the formula

$$[\nabla_Y, \mu^M(X)] = R(X,Y), \tag{7.4}$$

where R is the Riemannian curvature of M.

The equivariant curvature of a Riemannian manifold M acted on isometrically by a compact group G is defined by the formula $R_{\mathfrak{g}}(X) = R + \mu^M(X) \in \mathcal{A}^2_G(M, \mathfrak{so}(TM))$.

Example 7.9. The next example of a moment map is due to Kostant and Souriau; it is fundamental to the theory of geometric quantization. Let M be a **symplectic manifold** with symplectic two-form Ω, that is, $\Omega \in \mathcal{A}^2(M)$ is a closed two-form, $d\Omega = 0$, such that the bilinear form $\Omega_x(X,Y)$ on T_xM is non-degenerate for each $x \in M$. If $f \in C^\infty(M)$, then the **Hamiltonian vector field** generated by f is defined as the unique vector field H_f such that

$$df = \iota(H_f)\Omega.$$

Assume that a Lie group G acts on M and that the action is Hamiltonian: this means that, for every $X \in \mathfrak{g}$, there is given a function $\mu(X)$ on M such that

1. $\mu(X)$ depends linearly on X;
2. the vector field X_M is the Hamiltonian vector field generated by $\mu(X)$, that is

$$d\mu(X) = \iota(X_M)\Omega;$$

3. $\mu(X)$ is equivariant, that is, $g \cdot \mu(X) = \mu(g \cdot X)$ for $g \in G$.

The **symplectic moment map** of the action is the C^∞ map $\mu : M \to \mathfrak{g}^*$ defined by $(X, \mu(m)) = \mu(X)(m)$. One sees easily that

$$X \longmapsto \Omega_{\mathfrak{g}}(X) = \mu(X) + \Omega$$

is an equivariantly closed form on M such that $\Omega_{\mathfrak{g}}(0) = \Omega$.

This equivariant differential form may in some cases be identified with i times the equivariant curvature of an equivariant line bundle. Let $\mathcal{L}$ be a complex line bundle on M. Suppose that $\mathcal{L}$ carries a connection $\nabla^{\mathcal{L}}$ whose curvature

$$(\nabla^{\mathcal{L}})^2 = i\Omega \in \mathcal{A}^2(M)$$

equals i times the symplectic form Ω on M. Furthermore, suppose that the Lie group G acts on the manifold M and on the line bundle $\mathcal{L}$, in such a way as to preserve the connection $\nabla^{\mathcal{L}}$. It follows immediately that the symplectic form Ω is preserved by the action of G.

We may define the moment of this action, $\mu^{\mathcal{L}}(X) = \mathcal{L}^{\mathcal{L}}(X) - \nabla^{\mathcal{L}}_X$, and the equivariant curvature

$$F_{\mathfrak{g}} = \mu^{\mathcal{L}} + i\Omega \in \mathcal{A}^2_{\mathfrak{g}}(M).$$

In fact, $\mu^{\mathcal{L}}$ equals $\sqrt{-1}$ times the moment μ defined above, since by (7.1),

$$i\iota(X_M)\Omega = d\mu^{\mathcal{L}}(X).$$

It follows that $F_{\mathfrak{g}} = i\Omega_{\mathfrak{g}}$.

7.2 The Localization Formula

If M is a manifold acted on by a compact group G, there is a Riemannian metric $(\cdot,\cdot)$ on M which is invariant under the action of G. For example, such a metric may be formed by averaging any metric on M with respect to the Haar measure of G. In this section, we will show how such a metric may be used to study equivariant differential forms.

Proposition 7.10. *Let G be a compact Lie group and $X \to \alpha(X)$ be an equivariantly closed differential form on M. If $X \in \mathfrak{g}$, let $M_0(X)$ be the set of zeroes of the vector field X_M. Then for each $X \in \mathfrak{g}$, the differential form $\alpha(X)_{[n]}$ is exact outside $M_0(X)$.*

Proof. Fix $X \in \mathfrak{g}$ and write $d_X = d - \iota(X)$. Let θ be a differential form on M such that $\mathscr{L}(X)\theta = 0$ and such that $d_X\theta$ is invertible outside $M_0(X)$. Such a differential form may be constructed using a G-invariant Riemannian metric $(\cdot,\cdot)$ on M. We define $\theta(\xi) = (X_M, \xi)$ for $\xi \in \Gamma(M, TM)$, then θ is a one-form on M invariant under the action of X such that $d_X\theta = |X|^2 + d\theta$, which is clearly invertible outside the set $M_0(X) = \{|X|^2 = 0\}$.

On the set $M - M_0(X)$, we have

$$\alpha(X)_{[n]} = d\left(\frac{\theta \wedge \alpha(X)}{d_X\theta}\right)_{[n-1]}$$

for every equivariantly closed differential form α on M. Indeed, since $d_X^2\theta = 0$, we have

$$\alpha(X) = d_X\left(\frac{\theta \wedge \alpha(X)}{d_X\theta}\right),$$

and the result follows by taking the highest degree piece of each side. □

Note that it is essential to assume that G is compact. For example, consider $M = S^1 \times S^1$ with coordinates $x, y \in \mathbb{R}/2\pi\mathbb{Z}$. Let X be the nowhere-vanishing vector field $(1 + \frac{1}{2}\sin x)\,\partial_y$, and let

$$\alpha(X) = -\tfrac{1}{2}(7\cos x + \sin 2x) + (1 - 4\sin x)dx \wedge dy.$$

It is easy to verify that $d_X\alpha(X) = 0$. However, $\int_M \alpha(X) = (2\pi)^2$, so that $\alpha(X)_{[2]}$ is not exact.

Proposition 7.10 strongly suggests that when G is a compact Lie group and M is compact, the integral of an equivariantly closed form $\alpha(X)$ depends only on the restriction of $\alpha(X)$ to $M_0(X)$. In the rest of this section, we will prove the localization formula, which expresses the integral $\int_M \alpha(X)$ of an equivariantly closed differential form α as an integral over the set of zeroes of the vector field X_M.

We will first state and prove the localization formula in the important special case where X_M has isolated zeroes. Here, at each point $p \in M_0(X)$, the Lie action $\mathscr{L}(X)\xi = [X_M, \xi]$ on $\Gamma(M, TM)$ gives rise to an invertible transformation L_p of T_pM. This can be proved using the exponential map with respect to a G-invariant Riemannian metric: if $\xi \in T_pM$ was annihilated by L_p, all the points of the curve $\exp_p(s\xi)$

would be fixed by $\exp(tX)$. The transformation L_p being the Lie derivative of an action of a compact Lie group, it has only imaginary eigenvalues. Thus the dimension of M is even and there exists an oriented basis e_i, $1 \leq i \leq n$, of T_pM such that for $1 \leq i \leq \ell = n/2$,

$$L_p e_{2i-1} = \lambda_i e_{2i},$$
$$L_p e_{2i} = -\lambda_i e_{2i-1}.$$

We have $\det(L_p) = \lambda_1^2 \lambda_2^2 \dots \lambda_\ell^2$, and it is natural to take the following square root (dependent only on the orientation of the manifold):

$$\det{}^{1/2}(L_p) = \lambda_1 \dots \lambda_\ell.$$

Theorem 7.11. *Let G be a compact Lie group with Lie algebra $\mathfrak{g}$ acting on a compact oriented manifold M, and let α be an equivariantly closed differential form on M. Let $X \in \mathfrak{g}$ be such that X_M has only isolated zeroes. Then*

$$\int_M \alpha(X) = (-2\pi)^\ell \sum_{p \in M_0(X)} \frac{\alpha(X)(p)}{\det^{1/2}(L_p)},$$

where $\ell = \dim(M)/2$, and by $\alpha(X)(p)$, we mean the value of the function $\alpha(X)_{[0]}$ at the point $p \in M$.

Proof. Replacing G by the closed group generated by X, we may assume that it is abelian. Let $p \in M_0(X)$. Using a G-invariant metric and the exponential map, the vector field X can be linearized around p. Thus there exist local coordinates $x_1, \dots, x_n$ around p such that X_M is the vector field

$$X_M = \lambda_1(x_2\partial_1 - x_1\partial_2) + \dots + \lambda_\ell(x_n\partial_{n-1} - x_{n-1}\partial_n),$$

and $\det^{1/2}(L_p) = \lambda_1 \dots \lambda_\ell$.

Let θ^p be the one-form in a neighbourhood U_p of p given by

$$\theta^p = \lambda_1^{-1}(x_2 dx_1 - x_1 dx_2) + \dots + \lambda_\ell^{-1}(x_n dx_{n-1} - x_{n-1} dx_n).$$

Then θ^p is such that $\mathcal{L}(X)\theta^p = 0$ and $\theta^p(X_M) = \sum_i x_i^2 = \|x\|^2$. Using a G-invariant partition of unity subordinate to the covering of M by the G-invariant open sets U_p and $M - M_0(X)$ (which may be constructed by averaging any partition of unity with respect to the action of G), we can construct a one-form θ such that $\mathcal{L}(X)\theta = 0$, $d_X\theta$ is invertible outside $M_0(X)$, and such that θ coincides with θ^p in a neighbourhood of p.

Consider the neighbourhood B_ε^p of p in M given by $B_\varepsilon^p = \{x \mid \|x\|^2 \leq \varepsilon\}$. If $S_\varepsilon^p = \{x \mid \|x\|^2 = \varepsilon\}$, then

$$\begin{aligned}\int_M \alpha(X) &= \lim_{\varepsilon \to 0} \int_{M - \cup_p B_\varepsilon^p} \alpha(X) \\ &= \lim_{\varepsilon \to 0} \int_{M - \cup_p B_\varepsilon^p} d\left(\frac{\theta \wedge \alpha(X)}{d_X\theta}\right) \\ &= -\sum_p \lim_{\varepsilon \to 0} \int_{S_\varepsilon^p} \frac{\theta \wedge \alpha(X)}{d_X\theta}.\end{aligned}$$

(The sign change comes from exchanging the interior for the exterior orientation of S^p_ε). Let us fix a point p. Near p, $\theta = \theta^p$. Rescaling the variable x by a factor of $\varepsilon^{1/2}$, the sphere S_ε becomes the unit sphere S_1, while $\theta(d_X\theta)^{-1}$ being homogeneous of degree zero is invariant. Then

$$\int_{S_\varepsilon} \frac{\theta \wedge \alpha(X)}{d_X\theta} = \int_{S_1} \frac{\theta \wedge \alpha_\varepsilon(X)}{d_X\theta},$$

where $\alpha_\varepsilon(X)(x,dx) = \alpha(X)(\varepsilon^{1/2}x, \varepsilon^{1/2}dx)$. When $\varepsilon \to 0$, $\alpha_\varepsilon(X)$ tends to the constant $\alpha(X)(p)$. To prove the proposition, it remains to compute

$$-\int_{S_1} \theta(d_X\theta)^{-1} = \int_{S_1} \theta(1-d\theta)^{-1} = \int_{S_1} \theta(d\theta)^{\ell-1} = \int_{B_1} (d\theta)^\ell$$

But

$$(d\theta)^\ell = (-2)^\ell \ell!(\lambda_1 \dots \lambda_\ell)^{-1} dx_1 \wedge \dots \wedge dx_n.$$

Since the volume of the 2ℓ-dimensional unit ball equals $\pi^\ell/\ell!$, we obtain the theorem. □

We now turn to the localization formula in the general case. Fix an element $X \in \mathfrak{g}$, and let M_0 denote the set of zeroes of X.

Proposition 7.12. *The set M_0 of zeroes of the vector field X_M is a submanifold of M, which may have several components of different dimension. The normal bundle $\mathcal{N}$ of M_0 in M is an even-dimensional orientable vector bundle.*

Proof. Fix a G-invariant Riemannian structure on M. If x_0 is a zero of X_M, the one parameter group $\exp tX$ acts on $T_{x_0}M$. If a tangent vector at x_0 is fixed by $\exp tX$, then the geodesic tangent to it is contained in M_0, which proves that M_0 is a submanifold of M.

The Lie derivative $Y \mapsto \mathcal{L}(X)Y = [X_M, Y]$ is an endomorphism of the vector space $\mathcal{N}_{x_0} = T_{x_0}M/T_{x_0}M_0$ which is invertible and antisymmetric. This implies that the real vector bundle $\mathcal{N}$ can be given a complex structure and is therefore even-dimensional and orientable. Indeed, observe that the eigenvalues of $\mathcal{L}(X)$ in $\mathcal{N}^{\mathbb{C}}_{x_0} = \mathcal{N}_{x_0} \otimes_{\mathbb{R}} \mathbb{C}$ come in pairs $\pm i\lambda_j$, with $\lambda_j > 0$. Therefore, we have a decomposition

$$\mathcal{N}^{\mathbb{C}}_{x_0} = \mathcal{N}^+_{x_0} \oplus \overline{\mathcal{N}^+_{x_0}}$$

where $\mathcal{N}^+_{x_0}$ is the sum of the eigenspaces such that $\lambda_j > 0$. □

Let $\mathrm{rk}(\mathcal{N}) : M_0 \to \mathbb{Z}$ be the locally constant function on M_0 which gives the codimension of each component. Let $\mathfrak{g}_0 = \{Y \in \mathfrak{g} \mid [Y,X] = 0\}$ be the centralizer of $X \in \mathfrak{g}$, and let G_0 be the connected Lie subgroup of G with Lie algebra $\mathfrak{g}_0$. Then G_0 preserves the submanifold M_0 and acts on the normal bundle $\mathcal{N}$. We choose a G_0-invariant Euclidean structure and a G_0-invariant metric connection $\nabla^{\mathcal{N}}$ on $\mathcal{N}$. The normal moment of $Y \in \mathfrak{g}_0$ is the endomorphism $\mu^{\mathcal{N}}(Y)$ of $\mathcal{N}$ such that

$$\mathcal{L}^{\mathcal{N}}(Y) = \nabla^{\mathcal{N}}_Y + \mu^{\mathcal{N}}(Y).$$

In particular, since the vector field X_M vanishes on M_0, we have

$$\mu^{\mathscr{N}}(X)\xi = [X_M, \xi] \quad \text{for } \xi \in \mathscr{N},$$

and $\mu^{\mathscr{N}}(X)$ is an invertible transformation of $\mathscr{N}$. Choose an orientation on $\mathscr{N}$ and consider the equivariant Euler form $\chi_{\mathfrak{g}_0}(\mathscr{N}) \in \mathscr{A}^+(M_0)$. Its degree zero piece equals $\det(-\mu^{\mathscr{N}})^{1/2}$, and therefore $\chi_{\mathfrak{g}_0}(\mathscr{N})(Y)$ is invertible in $\mathscr{A}^+(M_0)$ when Y is sufficiently close to X. We can now state the general localization formula.

Theorem 7.13. *Let G be a compact Lie group acting on a compact manifold M. Let α be an equivariantly closed form on M. Let $X \in \mathfrak{g}$, let M_0 be the zero set of the vector field X_M and let $\mathscr{N}$ be the normal bundle of M_0 in M. Choose an orientation on $\mathscr{N}$ and impose the corresponding orientation on M_0. Then for Y in the centralizer $\mathfrak{g}_0$ of X in $\mathfrak{g}$ and sufficiently close to X, we have*

$$\int_M \alpha(Y) = \int_{M_0} (2\pi)^{\mathrm{rk}(\mathscr{N})/2} \frac{\alpha(Y)}{\chi_{\mathfrak{g}_0}(\mathscr{N})(Y)}$$

where $\chi_{\mathfrak{g}_0}(\mathscr{N})$ is the equivariant Euler form of the normal bundle. In particular,

$$\int_M \alpha(X) = \int_{M_0} (-2\pi)^{\mathrm{rk}(\mathscr{N})/2} \frac{\alpha(X)}{\det^{1/2}(\mathscr{L}^{\mathscr{N}}(X) + R^{\mathscr{N}})}.$$

The proof follows from several lemmas. As before, let θ be the G_0-invariant one-form on M dual to the vector field X_M with respect to the G-invariant Riemannian metric on M,

$$\theta(Z) = (X_M, Z) \quad \text{for } Z \in \Gamma(M, TM).$$

Then we have

$$(d_{\mathfrak{g}}\theta)(Y) = d\theta - (X_M, Y_M), \quad \text{for } Y \in \mathfrak{g}_0.$$

Lemma 7.14. *For all $t \in \mathbb{R}$ and $Y \in \mathfrak{g}_0$,*

$$\int_M \alpha(Y) = \int_M e^{t(d_{\mathfrak{g}}\theta)(Y)} \alpha(Y).$$

Proof. We give two arguments, the second of which extends better to non-compact manifolds. The first proof follows from the fact that $e^{td_{\mathfrak{g}}\theta} - 1$ is equivariantly exact:

$$e^{td_{\mathfrak{g}}\theta} - 1 = d_{\mathfrak{g}}\left(\frac{\theta(e^{td_{\mathfrak{g}}\theta} - 1)}{d_{\mathfrak{g}}\theta}\right).$$

The other argument is analogous to the proof of the McKean-Singer formula Theorem 3.50: we observe that

$$\begin{aligned}\frac{d}{dt}\int_M e^{t(d_{\mathfrak{g}}\theta)(Y)}\alpha(Y) &= \int_M (d_{\mathfrak{g}}\theta)(Y) e^{t(d_{\mathfrak{g}}\theta)(Y)} \alpha(Y) \\ &= \int_M d_{\mathfrak{g}}(\theta e^{td_{\mathfrak{g}}\theta}\alpha)(Y) = 0,\end{aligned}$$

since α is equivariantly closed. Thus, we may set $t = 0$ without changing the integral, whence the result. □

Thus, we must compute the limit when $t \to 0$ of

$$\int_M e^{(d_{\mathfrak{g}}\theta)(Y)/t}\alpha(Y) = \int_M e^{-(X,Y)/t}\alpha(Y)\sum_{k=0}^{n/2} t^{-k}\frac{(d\theta)^k}{k!}.$$

To do this, we study the form θ in the neighbourhood of M_0.

By orthogonal projection, the Levi-Civita connection ∇ gives a connection $\nabla^{\mathcal{N}}$ on the normal bundle $\mathcal{N}$ which is compatible with the induced metric. Identifying M_0 with the zero section of $\mathcal{N}$, we obtain a canonical isomorphism

$$T\mathcal{N}|_{M_0} \cong TM|_{M_0}.$$

Consider the moment map $\mu^M(Y)Z = -\nabla_Z Y$ of TM. Since $\mu^M(Y)$ commutes with the operator $\mathcal{L}(X)$ on M_0, it preserves the decomposition

$$TM|_{M_0} = TM_0 \oplus \mathcal{N};$$

thus, the restriction of $\mu^M(Y)$ to $\mathcal{N}$ coincides with the moment endomorphism $\mu^{\mathcal{N}}(Y)$. The endomorphism $\mu^{\mathcal{N}}(X)$ of $\mathcal{N}$ coincides with the infinitesimal action $\mathcal{L}(X)$. The action of G_0 on the manifold $\mathcal{N}$ determines vector fields on $\mathcal{N}$. The vector field $X_{\mathcal{N}}$ is vertical and is given at the point $(x,\mathbf{y}) \in M_0 \times \mathcal{N}_x$ by the vector $-\mu^M(X)\mathbf{y} \in \mathcal{N}_x$.

The connection $\nabla^{\mathcal{N}}$ determines a splitting of the tangent space to the total space $\mathcal{N}$ into horizontal and vertical tangent spaces. We denote by $(v_1,v_2)_0$ the Euclidean structure of the vector bundle $\mathcal{N}$. Consider the one-form θ_0 on the total space $\mathcal{N}$ given by

$$\theta_0(Z) = (X_{\mathcal{N}}, Z_V)_0 \tag{7.5}$$

where Z_V is the vertical part of a vector field Z on $\mathcal{N}$.

To prove Theorem 7.13, we could proceed as in the proof of Theorem 7.11, multiplying the one-form θ_0 by a cut-off function to obtain a one-form θ which coincides with θ_0 in a neighbourhood of M_0. However, we will see in the next lemma that this is not necessary, since θ_0 arises naturally as a scaling limit of the one-form θ.

Let ψ be a diffeomorphism of a neighbourhood U of the zero section in $\mathcal{N}$ with a neighbourhood of M_0 in M, such that

$$\begin{aligned} \psi(x,0) &= x \quad \text{for } x \in M_0, \\ d\psi|_{M_0} &= I \quad \text{on } T\mathcal{N}|_{M_0} \cong TM|_{M_0}. \end{aligned}$$

An example of such a map is the exponential map $\psi(x,\mathbf{y}) = \exp_x \mathbf{y}$, but the exponential map is difficult to compute explicitly in most cases, so that it is interesting to permit a different choice of ψ. Transport the metric $(\cdot,\cdot)$ and the one-form θ to U, by means of the diffeomorphism ψ. Consider the dilation δ_t, $t > 0$, of $\mathcal{N}$ given by the formula

$$\delta_t(x,\mathbf{y}) = (x, t^{1/2}\mathbf{y}), \quad \text{where } x \in M_0 \text{ and } \mathbf{y} \in \mathcal{N}_x.$$

We assume that U is preserved by δ_t for $t \leq 1$.

Lemma 7.15. *1.* $d\theta(Z_1,Z_2) = -2(\mu^M(X)Z_1,Z_2)$ *for* $Z_1, Z_2 \in \Gamma(M,TM)$
2. $\lim_{t\to 0} t^{-1}\delta_t^*\theta = \theta_0$
3. For $Y \in \mathfrak{g}_0$, $\lim_{t\to 0} t^{-1}(X,Y)_{(x,t^{1/2}\mathbf{y})} = (\mu^{\mathcal{N}}(X)\mathbf{y},\mu^{\mathcal{N}}(Y)\mathbf{y})_0$.

Proof. It is easy to calculate $d\theta$, using the explicit formula for the exterior differential of a one-form: if Z_1 and Z_2 are vector fields on M, then

$$\begin{aligned} d\theta(Z_1,Z_2) &= Z_1(X,Z_2) - Z_2(X,Z_1) - (X,[Z_1,Z_2]) \\ &= (\nabla_{Z_1}X,Z_2) - (\nabla_{Z_2}X,Z_1) \\ &= -2(\mu^M(X)Z_1,Z_2). \end{aligned}$$

Since $\mu^M(X)Z_1 = 0$ if Z_1 is tangent to M_0, we see that $d\theta(Z_1,Z_2)$ vanishes if Z_1 or Z_2 is tangent to M_0. Consider local coordinates $(x,\mathbf{y})$ on U such that M_0 is the set $\mathbf{y}=0$. Since θ vanishes on M_0, as does $\iota(Z_1)d\theta$ if Z_1 is tangential to M_0, we see that

$$\theta = \sum_{ijk} a_{ijk}(x,\mathbf{y})\mathbf{y}_i\mathbf{y}_j dx_k + \sum_{ij} b_{ij}(x,\mathbf{y})\mathbf{y}_i d\mathbf{y}_j,$$

for some smooth coefficients $a_{ijk}(x,\mathbf{y})$ and $b_{ij}(x,\mathbf{y})$. This shows that

$$t^{-1}\delta_t^*\theta = \sum_{ijk} a_{ijk}(x,t^{1/2}\mathbf{y})\mathbf{y}_i\mathbf{y}_j dx_k + \sum_{ij} b_{ij}(x,t^{1/2}\mathbf{y})\mathbf{y}_i d\mathbf{y}_j$$

has a limit when $t \to 0$:

$$\lim_{t\to 0} t^{-1}\delta_t^*\theta = \sum_{ijk} a_{ijk}(x,0)\mathbf{y}_i\mathbf{y}_j dx_k + \sum_{ij} b_{ij}(x,0)\mathbf{y}_i d\mathbf{y}_j.$$

In order to obtain (2), it remains to show that

$$\partial_{\mathbf{y}_i}\theta(\partial_{\mathbf{y}_j})|_{\mathbf{y}=0} = \partial_{\mathbf{y}_i}\theta_0(\partial_{\mathbf{y}_j})|_{\mathbf{y}=0} \tag{7.6}$$

and

$$\partial_{\mathbf{y}_i}\partial_{\mathbf{y}_j}\theta(\partial_{x_k})|_{\mathbf{y}=0} = \partial_{\mathbf{y}_i}\partial_{\mathbf{y}_j}\theta_0(\partial_{x_k})|_{\mathbf{y}=0}. \tag{7.7}$$

Denote by Θ the one-form on M_0 which represents the connection $\nabla^{\mathcal{N}}$ in the local basis $\partial_{\mathbf{y}_j}$ of the normal bundle $\mathcal{N}$. Thus

$$\nabla^{\mathcal{N}}_{\partial_{x_k}}\partial_{\mathbf{y}_i} = \Theta(\partial_{x_k})\partial_{\mathbf{y}_i}.$$

By Proposition 1.20, we have

$$\theta_0 = -(\mu^{\mathcal{N}}(X)\mathbf{y}, d\mathbf{y} + \Theta\mathbf{y})_0.$$

Since X vanishes on M_0, we have on M_0 the formula

$$\begin{aligned} \partial_{\mathbf{y}_i}(X,\partial_{\mathbf{y}_j})|_{\mathbf{y}=0} &= (\nabla_{\partial_{\mathbf{y}_i}}X,\partial_{\mathbf{y}_j})|_{M_0} \\ &= -(\mu^{\mathcal{N}}(X)\partial_{\mathbf{y}_i},\partial_{\mathbf{y}_j})_0, \end{aligned}$$

which proves (7.6).

Let us prove 7.7. We have

$$\partial_{y_i}\partial_{y_j}\theta_0(\partial_{x_k})|_{y=0} = -(\mu^M(X)\partial_{y_i},\Theta(\partial_{x_k})\partial_{y_j})_0 - (\mu^M(X)\partial_{y_j},\Theta(\partial_{x_k})\partial_{y_i})_0.$$

By (7.4), we have

$$\nabla_{Y_1}\nabla_{Y_2}X = -\nabla_{Y_1}(\mu^M(X)Y_2) = -\mu^M(X)\nabla_{Y_1}Y_2 - R(X,Y_1)Y_2.$$

If Z is tangential to M_0, so that $\mu^M(X)Z$ and $R(X,Y_1)$ vanish on M_0, we see that

$$(\nabla_{Y_1}\nabla_{Y_2}X,Z)|_{M_0} = (\nabla_{Y_1}Y_2,\mu^M(X)Z)|_{M_0} - (R(X,Y_1)Y_2,Z)|_{M_0} = 0.$$

It follows that

$$Y_1Y_2(X,Z)|_{M_0} = (\nabla_{Y_1}X,\nabla_{Y_2}Z)|_{M_0} + (\nabla_{Y_2}X,\nabla_{Y_1}Z)|_{M_0} \tag{7.8}$$

We apply this with $Z = \partial_{x_k}$, $Y_1 = \partial_{y_i}$, and $Y_2 = \partial_{y_j}$. Using the relation $\nabla_{\partial_{y_i}}\partial_{x_k} = \nabla_{\partial_{x_k}}\partial_{y_i}$ and the fact that $\mu^M(X)$ preserves the decomposition $TM|M_0 = TM_0 \oplus \mathcal{N}$, we obtain

$$\begin{aligned}(\nabla_{\partial_{y_i}}X,\nabla_{\partial_{y_j}}\partial_{x_k})|_{M_0} &= -(\mu^{\mathcal{N}}(X)\partial_{y_i},\nabla^{\mathcal{N}}_{\partial_{x_k}}\partial_{y_j})_0\\ &= -(\mu^{\mathcal{N}}(X)\partial_{y_i},\Theta(\partial_{x_k})\partial_{y_j})_0,\end{aligned}$$

and hence

$$\begin{aligned}\partial_{y_i}\partial_{y_j}\theta(\partial_{x_k})|_{y=0} &= -(\mu^{\mathcal{N}}(X)\partial_{y_i},\Theta(\partial_{x_k})\partial_{y_j})_0 - (\mu^{\mathcal{N}}(X)\partial_{y_j},\Theta(\partial_{x_k})\partial_{y_i})_0\\ &= \partial_{y_i}\partial_{y_j}\theta_0(\partial_{x_k})|_{y=0},\end{aligned}$$

proving the second part of the lemma.

The vector field $X_{\mathcal{N}}$ is vertical and is given at the point $(x,\mathbf{y})$, $x\in M$, $\mathbf{y}\in\mathcal{N}_x$, by the vector $-\mu^{\mathcal{N}}(X)\mathbf{y}\in\mathcal{N}_x$. If $Y\in\mathfrak{g}_0$, the vector field Y_M commutes with X_M, and hence is tangent to M_0, so that the function (X,Y) vanishes to second order on M_0. Equation (7.8) shows that

$$Y_1Y_2(X,Y)|_{M_0} = (\mu^{\mathcal{N}}(X)Y_1,\mu^{\mathcal{N}}(Y)Y_2)_0 + (\mu^{\mathcal{N}}(X)Y_2,\mu^{\mathcal{N}}(Y)Y_1)_0,$$

which proves the third part of the lemma. □

Using the diffeomorphism ψ, we can identify the neighbourhood U of the zero section in $\mathcal{N}$ with a neighbourhood U of M_0 in M. Outside the neighbourhood U, the function (X,X) is strictly positive, so we can find $\varepsilon>0$ such that if $Y\in\mathfrak{g}_0$ is sufficiently close to X, $(X,Y)_{\mathbf{y}}\geq\varepsilon$ for all $\mathbf{y}\notin U$. Since $d\theta$ is nilpotent, a partition of unity argument now shows that

$$\lim_{t\to0}\int_M e^{-(X,Y)/t}e^{d\theta/t}\alpha(Y) = \lim_{t\to0}\int_U e^{-(X,Y)/t}e^{d\theta/t}\alpha(Y)\phi,$$

if ϕ is a compactly supported function on U equal to one on a neighbourhood of M_0.

Let us consider $\alpha(Y)\phi$ as a differential form on $\mathcal{N}$. If we denote the one-form $\alpha(Y)$ on M_0 and its pull-back to $\mathcal{N}$ by the same notation, we see that $\lim_{t\to 0}\delta_t^*(\alpha(Y)\phi) = \alpha(Y)|_{M_0}$. Thus, it follows from Lemma 7.15 that

$$\lim_{t\to 0}\delta_t^*(e^{-(X,Y)/t}e^{d\theta/t}\alpha(Y)\phi) = e^{-(\mu^{\mathcal{N}}(X)\mathbf{y},\mu^{\mathcal{N}}(Y)\mathbf{y})_0}e^{d\theta_0}\alpha(Y)|_{M_0}.$$

Furthermore, for Y near X and $\mathbf{y}$ in a neighbourhood of M_0, we have a bound $(X,Y)_{(x,\mathbf{y})} \geq c\|\mathbf{y}\|^2$ with $c > 0$, so that by dominated convergence and change of variables $(x,\mathbf{y}) \mapsto (x, t^{1/2}\mathbf{y})$, we see that

$$\lim_{t\to 0}\int_{\mathcal{N}} e^{-(X,Y)/t}e^{d\theta/t}\alpha(Y)\phi = \int_{\mathcal{N}} e^{-(\mu^{\mathcal{N}}(X)\mathbf{y},\mu^{\mathcal{N}}(Y)\mathbf{y})_0}e^{d\theta_0}\alpha(Y)|_{M_0}.$$

Consider the differential form defined by integration over the fibre,

$$\int_{\mathcal{N}/M_0} e^{-(\mu^{\mathcal{N}}(X)\mathbf{y},\mu^{\mathcal{N}}(Y)\mathbf{y})_0}e^{d\theta_0}.$$

We have

$$\theta_0 = -(\mu^{\mathcal{N}}(X)\mathbf{y}, \nabla^{\mathcal{N}}\mathbf{y})_0.$$

Recall that $\nabla^{\mathcal{N}}$ is invariant under $\mathcal{L}(X)$, so that $[\nabla^{\mathcal{N}},\mu^{\mathcal{N}}(X)] = 0$. We see that $d\theta_0$ equals

$$-(\mu^{\mathcal{N}}(X)\nabla^{\mathcal{N}}\mathbf{y},\nabla^{\mathcal{N}}\mathbf{y})_0 - (\mu^{\mathcal{N}}(X)\mathbf{y},R^{\mathcal{N}}\mathbf{y})_0.$$

It follows that

$$e^{-(\mu^{\mathcal{N}}(X)\mathbf{y},\mu^{\mathcal{N}}(Y)\mathbf{y})_0}e^{d\theta_0} = e^{-(\mu^{\mathcal{N}}(X)\mathbf{y},(\mu^{\mathcal{N}}(Y)+R^{\mathcal{N}})\mathbf{y})_0-(\mu^{\mathcal{N}}(X)\nabla^{\mathcal{N}}\mathbf{y},\nabla^{\mathcal{N}}\mathbf{y})_0}.$$

The proof of the theorem is completed by the following lemma. Here, the element $\chi_{\mathfrak{g}_0}(\mathcal{N})(Y) = \det^{1/2}(-(\mu^{\mathcal{N}}(Y)+R^{\mathcal{N}}))$ is an invertible element of $\mathcal{A}(M_0)$, for Y sufficiently near X.

Lemma 7.16.

$$\int_{\mathcal{N}/M_0} e^{-(\mu^{\mathcal{N}}(X)\mathbf{y},(\mu^{\mathcal{N}}(Y)+R^{\mathcal{N}})\mathbf{y})_0-(\mu^{\mathcal{N}}(X)\nabla^{\mathcal{N}}\mathbf{y},\nabla^{\mathcal{N}}\mathbf{y})_0} = \frac{(2\pi)^{\mathrm{rk}(\mathcal{N})/2}}{\chi_{\mathfrak{g}_0}(\mathcal{N})(Y)}.$$

Proof. Choose a point $x_0 \in M_0$ and let $V = \mathcal{N}_{x_0}$. Given a local oriented orthonormal frame of $\mathcal{N}$ around x_0, we may write $\nabla^{\mathcal{N}} = d + \Theta$. We must show that

$$\int_V e^{-(\mu(X)\mathbf{y},(\mu(Y)+\mathrm{R})\mathbf{y})-(\mu(X)d\mathbf{y},d\mathbf{y})} = (-2\pi)^{\mathrm{rk}(\mathcal{N})/2}\frac{\det^{1/2}(\mu(X))}{\det(\mu(X)(\mu(Y)+\mathrm{R}))^{1/2}},$$

where $\mu = \mu_{x_0}^{\mathcal{N}} \in \mathfrak{g}^*\otimes\mathrm{End}(V)$ and $\mathrm{R} = R_{x_0}^{\mathcal{N}} \in \Lambda^2 T^*_{x_0}M_0\otimes\mathrm{End}(V)$.

It is a simple matter to evaluate this integral. First we apply the definition of the Pfaffian, which shows that

$$\int_V e^{-(\mu(X)y,(\mu(Y)+R)y)-(\mu(X)dy,dy)}$$

$$= (-2)^{n/2} \int_V e^{-(\mu(X)y,(\mu(Y)+R)y)} \det{}^{1/2}(\mu(X)) dy_1 \wedge \ldots \wedge dy_n.$$

Recall the formula for a Gaussian integral: if A is a positive-definite endomorphism, then

$$\int_V e^{-(y,Ay)}\, dy = \pi^{n/2} \det(A)^{-1/2}.$$

Since $\mu(X)$ and $\mu(Y)+R$ are antisymmetric and commute, the matrix

$$-\mu(X)(\mu(Y)+R)$$

is symmetric. Its zero-degree component $-\mu(X)\mu(Y)$ is positive definite when $X = Y$, and hence for all Y sufficiently close to X. Hence, we obtain

$$\int_V e^{-(\mu(X)y,(\mu(Y)+R)y)}\, dy_1 \wedge \ldots \wedge dy_n = \pi^{n/2} \det(-\mu(X)(\mu(Y)+R))^{-1/2}$$

$$= \pi^{n/2} \det{}^{-1/2}(\mu(X)) \det{}^{-1/2}(\mu(Y)+R). \qquad \square$$

7.3 Bott's Formulas for Characteristic Numbers

In the next three sections, we will give some applications of the localization formula. The first of these predates this formula, and is due to Bott.

If $\Phi \in \mathbb{C}[\mathfrak{so}(n)]^{O(n)}$ is an invariant polynomial function on the Lie algebra $\mathfrak{so}(n)$, where $n = 2\ell$ is even, then Φ is uniquely determined by its restriction to the Cartan subalgebra $\mathfrak{t} \subset \mathfrak{so}(n)$ of matrices X of the form

$$X e_{2i-1} = x_i e_{2i},$$
$$X e_{2i} = -x_i e_{2i-1}.$$

Let p_k be the k-th elementary symmetric polynomials in x_i^2, given by the formula

$$p_k = \sum_{1 \le i_1 < \cdots < i_k \le \ell} x_{i_1}^2 \ldots x_{i_k}^2.$$

(In particular, $p_\ell = \det(X)$.) If Φ is $O(n)$-invariant, the restriction of Φ to $\mathfrak{t}$ is a symmetric function of the variables x_i^2, so belongs to the polynomial ring generated by p_i, $1 \le i \le \ell$, that is, $\mathbb{C}[\mathfrak{so}(n)]^{O(n)} = \mathbb{C}[p_1, \ldots, p_\ell]$.

The function $\det^{1/2}(X) = x_1 \ldots x_\ell = p_\ell^{1/2}$ is only invariant under the group $SO(n)$, and is the additional generator of the ring of $SO(n)$-invariant polynomials over the ring of $O(n)$-invariant polynomials; it is nothing but the classical Pfaffian function $\mathrm{Pf}(X)$.

Let M be a compact oriented Riemannian manifold of dimension $n = 2\ell$, and let R be the curvature of a connection on the tangent bundle TM compatible with the metric. The Chern-Weil map defines a homomorphism

$$\phi : \mathbb{C}[\mathfrak{so}(n)]^{SO(n)} = \mathbb{C}[p_1,\ldots,p_\ell,p_\ell^{1/2}] \to \mathscr{A}(M),$$

by the formula $\phi(\Phi) = \Phi(R)$, and $\Phi(R)$ is a closed form on M whose cohomology class is independent of the choice of connection. Define

$$\Phi(M) = (-2\pi)^{-\ell} \int_M \Phi(R).$$

The number $\Phi(M)$ is called a **characteristic number** of M; it is a classical theorem of algebraic topology that if $\Phi \in \mathbb{Z}[p_i,p_\ell^{1/2}]$, $\Phi(M)$ is an integer.

Suppose that M admits a circular symmetry with isolated fixed points. At each fixed point p, the action of the infinitesimal generator X of the circle group $G = \{e^{2\pi i\theta}\}$ on the tangent space T_pM determines an endomorphism L_p, given in an appropriate oriented basis by

$$L_p e_{2i-1} = -\lambda_i e_{2i},$$
$$L_p e_{2i} = \lambda_i e_{2i-1},$$

where λ_i are integers called the exponents of the action at p. Recall that by definition $\det^{1/2}(L_p) = \lambda_1 \ldots \lambda_\ell$, and depends on the orientation of T_pM but not on its metric.

Theorem 7.17 (Bott). *Let M be a manifold on which the circle acts with isolated fixed points. If Φ is a homogeneous polynomial of degree $k \leq \ell$,*

$$\sum_p \frac{\Phi(L_p)}{\det^{1/2}(L_p)} = \begin{cases} \Phi(M) & k = \ell, \\ 0, & k < \ell. \end{cases}$$

Proof. Identify the Lie algebra of G with $\{uX \mid u \in \mathbb{R}\}$. Choose a G-invariant metric, and let $u\mu^M(X) + R$ be the equivariant curvature of the associated Levi-Civita connection. If Φ is a polynomial, $u \to \Phi(u\mu^M(X) + R)$ is an equivariantly closed form on M, and we see from Theorem 7.11 that

$$(-2\pi)^{-\ell} \int_M \Phi(u\mu^M(X) + R) = u^{-\ell} \sum_p \frac{\Phi(uL_p)}{\det^{1/2}(L_p)}.$$

The left hand side of this equation is a polynomial in u with value at zero the characteristic number $\Phi(M)$, while the right hand side is a Laurent polynomial in u. The equality of the two sides implies remarkable cancellations properties over fixed points of the values of $\Phi(L_p)$ if $\deg(\Phi) < \ell$, while the equality of the constant term of this Laurent polynomial gives the formula when $\deg(\Phi) = \ell$. □

When applied to the Pfaffian function $\Phi(X) = \mathrm{Pf}(X)$ on $\mathfrak{so}(n)$, the characteristic number $(-2\pi)^{-\ell} \int_M \mathrm{Pf}(R)$ is $(-1)^\ell$ times the Euler characteristic of M, by the Gauss-Bonnet-Chern theorem, and we see that the number of fixed points of the circle action is the Euler characteristic of M. Since the vector field X generating the circle action is an infinitesimal isometry, it always has index $\nu(p,X) = 1$ at an isolated zero

$X(p) = 0$. Thus, this result is a very particular case of the Poincaré-Hopf theorem, Theorem 1.58.

The general localization theorem allows us to generalize the above formula to situations in which $M_0(X)$ is not zero-dimensional; this extension was made by Baum and Cheeger. Let us merely mention the case of the Pffafian, in which we obtain the following result.

Proposition 7.18. *Let M be a compact oriented even-dimensional Riemannian manifold, and let X be an isometry of M with fixed point set M_0. Then the Euler characteristic of M equals the Euler characteristic of M_0.*

Proof. Observe that

$$\det{}^{1/2}(-R_{\mathfrak{g}}(X))|_{M_0} = \det{}^{1/2}(-R_{\mathfrak{g}}^{\mathcal{N}}(X))\det{}^{1/2}(-R^0),$$

where R^0 is the Riemannian curvature of M_0. By the localization theorem, we see that

$$(2\pi)^{-n/2}\int_M \det{}^{1/2}(-R_{\mathfrak{g}}(X)) = \int_{M_0}(2\pi)^{-\dim(M_0)/2}\det{}^{1/2}(-R^0).$$

Setting $X = 0$ and applying the Gauss-Bonnet-Chern theorem to both sides, the result follows. □

7.4 Exact Stationary Phase Approximation

One important application of the localization formula is the "exact stationary phase approximation" of Duistermaat-Heckmann.

Let M be a compact manifold of dimension $n = 2\ell$, f a smooth function on M and let dx be a smooth density on M. Let $t \in \mathbb{R}$ and consider the function

$$F(t) = \int_M e^{itf}dx.$$

The major contribution to the value of this integral when t tends to infinity arises from the neighbourhood of stationary points, that is, points where the differential df of the phase function f vanishes. Assume that the phase function f is non degenerate, which means the set M_0 where the differential of f vanishes consists in a finite number of points and at these points the Hessian $H_p = \nabla_p df$ of f is a non-degenerate quadratic form on T_pM. Let $\sigma(H_p)$ be the signature of the quadratic form H_p. If $T_pM = T_p^+ \oplus T_p^-$ is an orthogonal splitting for H_p such that H_p is positive definite on T^+ and negative definite on T^- then $\sigma(H_p) = \dim(T^+) - \dim(T^-)$.

If e_i is a basis of T_pM such that $(|dx|_p, e_1 \wedge \ldots \wedge e_n) = 1$, let

$$a_p(f,dx) = |\det(H_p(e_i,e_j))|^{-1/2}.$$

It is not hard to show that when t tends to infinity,

$$F(t) = \sum_{p \in M_0} \left(\frac{2\pi}{t}\right)^{\ell} e^{\pi i \sigma(H_p)/4} a_p(f, dx) e^{itf(p)} + O(t^{-\ell-1}).$$

Duistermaat and Heckman discovered a class of examples of "exact stationary phase approximation", where the error term in the above formula vanishes.

Let (M, Ω) be a compact symplectic manifold of dimension n, and let G be a compact group of Hamiltonian transformations of M. For $X \in \mathfrak{g}$, let $\mu(X)$ be the symplectic moment of X, defined in (7.9). Since $d\mu(X) = \iota(X_M)\Omega$, the set of points where the one-form $d\mu(X)$ vanishes coincides with the zero set $M_0(X)$ of the vector field X_M. For $p \in M_0(X)$, we denote by $\mathcal{L}_p(X)$ the infinitesimal action of X on T_pM. The **Liouville form** of the symplectic manifold M is the form

$$d\beta = \left(e^{\Omega/2\pi}\right)_{[n]} = \frac{\Omega^{\ell}}{(2\pi)^{\ell}\ell!};$$

this is a volume form which defines a canonical orientation on M.

Theorem 7.19 (Duistermaat-Heckman). *Let (M, Ω) be a compact symplectic manifold, with compact Hamiltonian symmetry group G. If X is an element of the Lie algebra of G such that $M_0(X)$ consists of a finite number of points, then*

$$\int_M e^{i\mu(X)} d\beta = i^{\ell} \sum_{p \in M_0(X)} \frac{e^{i\mu(X)(p)}}{\det^{1/2}(\mathcal{L}_p(X))}.$$

(The square root of $\det(\mathcal{L}_p(X))$ is computed with respect to the canonical orientation of T_pM.)

Proof. The equivariant symplectic form $\Omega_{\mathfrak{g}}(X) = \mu(X) + \Omega$ is an equivariantly closed form on M, as is

$$e^{i\Omega_{\mathfrak{g}}(X)} = e^{i\mu(X)} e^{i\Omega}.$$

The theorem follows from the application of the localization theorem, Theorem 7.11 to the integral

$$\int_M e^{i\mu(X)} d\beta = (2\pi i)^{-\ell} \int_M e^{i\Omega_{\mathfrak{g}}(X)}.$$

□

Corollary 7.20. *Let X be a Hamiltonian vector field with Hamiltonian f on a compact symplectic manifold M, such that the flow generated by X is periodic, and X has discrete zeroes. Then the error term in the stationary-phase approximation to the integral $\int_M e^{itf} d\beta$ vanishes:*

$$\int_M e^{itf} d\beta = \sum_{p \in M_0} \left(\frac{2\pi}{t}\right)^{\ell} e^{\pi i \sigma(H_p)/4} a_p(f, dx) e^{itf(p)}.$$

Proof. We need the following lemma.

Lemma 7.21. *The Hessian $H_p(Y, Z)$ of the function f at the point $p \in M_0(X)$ is given by the formula*

$$H_p(Y, Z) = -\Omega(\mathcal{L}_p(X)Y, Z).$$

Proof. Since $\mathscr{L}(Z)f = \iota(Z)\iota(X)\Omega$, we have

$$\begin{aligned}\mathscr{L}(Y)\mathscr{L}(Z)f &= \mathscr{L}(Y)(\iota(Z)\iota(X)\Omega)\\ &= \iota(\mathscr{L}(Y)Z)\iota(X)\Omega + \iota(Z)\iota(\mathscr{L}(Y)X)\Omega + \iota(Z)\iota(X)(\mathscr{L}(Y)\Omega).\end{aligned}$$

Since X vanishes at p, we obtain $H_p(Y,Z) = -\Omega([X,Y],Z)$, proving the lemma. □

By the lemma, we see that

$$a_p(f,d\beta) = (2\pi)^{-\ell}|\det(\mathscr{L}_p(X))|^{-1/2}.$$

Thus, we must prove that for each critical point p of f,

$$\operatorname{sgn}\big(\det{}^{1/2}(\mathscr{L}_p(X))\big) = i^{-n/2}e^{\pi i\sigma(H_p)/4} \tag{7.9}$$

Let (V,Ω) be a symplectic vector space, and let A be an invertible semisimple endomorphism of V with purely imaginary eigenvalues. The quadratic form $H(v,w) = -\Omega(Av,w)$ is non-degenerate, and A is a skew-adjoint endomorphism of V with respect to H. Choose a basis for V of vectors p_i and q_i, $1 \le i \le \ell$, such that $\Omega(p_i,q_j) = \delta_{ij}$, $\Omega(p_i,p_j) = \Omega(q_i,q_j) = 0$, $Ap_i = \lambda_i q_i$, and $Aq_i = -\lambda_i p_i$. We see that

$$\operatorname{sgn}(H) = 2\sum_{i=1}^{\ell}\operatorname{sgn}(\lambda_i),$$

while

$$\det{}^{1/2}(A) = \prod_{i=1}^{\ell}\lambda_i.$$

It is now easy to check (7.9). □

Of course, using the general localization formula, it is possible to remove the condition in the above theorem that $M_0(X)$ is zero-dimensional, at the cost of introducing the curvature of the normal bundle to $M_0(X)$.

7.5 The Fourier Transform of Coadjoint Orbits

Let G be a Lie group with Lie algebra $\mathfrak{g}$, and let $\mathfrak{g}^*$ be the dual vector space to the vector space $\mathfrak{g}$. The group G acts on $\mathfrak{g}$ by the adjoint action, and the dual action of G on $\mathfrak{g}^*$ is called the coadjoint action. One of the most important examples of a Hamiltonian action arises in this situation.

Let M be an orbit of the coadjoint representation; thus, for some $f \in \mathfrak{g}^*$, $M = G\cdot f$. The vector fields X_M, $X \in \mathfrak{g}$, are sections of TM and span the tangent space at each point, and we have $(X_M)_f = -X\cdot f$, where $-(X\cdot f)(Y) = \langle f,[X,Y]\rangle$. Let $\mu(X)$ be the restriction to M of the linear form $f \mapsto f(X)$ on $\mathfrak{g}^*$:

Lemma 7.22. *The form $\Omega(X_M,Y_M)_f = -\langle f,[X,Y]\rangle$ defines a G-invariant symplectic form on M. Furthermore the action of G on M is Hamiltonian and the symplectic moment of X is the function $\mu(X)$.*

Proof. The form Ω is clearly non-degenerate and G-invariant. The function $\mu(X)$ being the restriction of a linear function obviously satisfies

$$d\mu(X)(Y_M)_f = (Y_M)_f(X) = -\langle f, [X,Y]\rangle,$$

and hence $d\mu(X) = \iota(X_M)\Omega$, which shows that $\mu(X)$ is the Hamiltonian for the vector field X_M. Furthermore, it follows that

$$0 = d^2\mu(X) = d\iota(X_M)\Omega = -\iota(X_M)d\Omega,$$

where in the last equality, we used the invariance of Ω, $\mathscr{L}(X_M)\Omega = 0$. Since the vector fields X_M span the tangent space, we see that $d\Omega = 0$. □

Let $d\beta$ be the Liouville measure on M. By the Fourier transform of an orbit, we mean the integral

$$F_M(X) = \int_M e^{i\langle f,X\rangle} d\beta;$$

this is a generalized function on $\mathfrak{g}$ if the orbit M is sufficiently well-behaved. Such integrals are very important in representation theory. The above lemma shows that such a Fourier transform is the integral of an equivariantly closed differential form, since

$$F_M(X) = (2\pi i)^{-n/2} \int_M e^{i\Omega_\mathfrak{g}(X)}.$$

When G is a compact Lie group, the localization theorem will lead to a formula for the Fourier transform of the Liouville measure of M, due to Harish-Chandra. This formula in turn implies Chevalley's theorem on the structure of $\mathbb{C}[\mathfrak{g}]^G$. To state Harish-Chandra's formula precisely, we first recall a few results on the structure of compact Lie groups.

Let $\mathfrak{g}_\mathbb{C}$ be a reductive Lie algebra over $\mathbb{C}$; in other words,

$$\mathfrak{g}_\mathbb{C} = \mathfrak{z} \oplus [\mathfrak{g}_\mathbb{C}, \mathfrak{g}_\mathbb{C}],$$

where $\mathfrak{z}$ is the centre of $\mathfrak{g}_\mathbb{C}$ and $[\mathfrak{g}_\mathbb{C}, \mathfrak{g}_\mathbb{C}]$ is semisimple. Consider the collection of all abelian subalgebras $\mathfrak{a}$ of $\mathfrak{g}_\mathbb{C}$ such that the transformations

$$\{\operatorname{ad} X \mid X \in \mathfrak{a}\} \subset \operatorname{End}(\mathfrak{g}_\mathbb{C})$$

are simultaneously diagonalizable; a maximal subalgebra among this set is called a Cartan subalgebra, and as is well known, any two Cartan subalgebras are conjugate under the action of the adjoint group $\operatorname{Ad}(\mathfrak{g}_\mathbb{C})$.

Choose a Cartan subalgebra $\mathfrak{h}$ of $\mathfrak{g}_\mathbb{C}$. If $\alpha \in \mathfrak{h}^*$, define

$$(\mathfrak{g}_\mathbb{C})_\alpha = \{X \in \mathfrak{g}_\mathbb{C} \mid [H,X] = \langle \alpha, H\rangle X\}, \quad \text{for all } H \in \mathfrak{h}.$$

If $\alpha \neq 0$ and $(\mathfrak{g}_\mathbb{C})_\alpha \neq 0$, α is called a root of $\mathfrak{h}$ in $\mathfrak{g}_\mathbb{C}$. The set of roots of $\mathfrak{h}$ in $\mathfrak{g}_\mathbb{C}$ is denoted by $\Delta(\mathfrak{g}_\mathbb{C}, \mathfrak{h})$, or Δ when $\mathfrak{h}$ is fixed. If $\alpha \in \Delta$, then $\dim(\mathfrak{g}_\mathbb{C})_\alpha = 1$, $-\alpha \in \Delta$ and $\dim[(\mathfrak{g}_\mathbb{C})_\alpha, (\mathfrak{g}_\mathbb{C})_{-\alpha}] = 1$. Furthermore, $[(\mathfrak{g}_\mathbb{C})_\alpha, (\mathfrak{g}_\mathbb{C})_{-\alpha}] \subset \mathfrak{h}$, and there exists a unique element $H_\alpha \in [(\mathfrak{g}_\mathbb{C})_\alpha, (\mathfrak{g}_\mathbb{C})_{-\alpha}]$ such that $\langle \alpha, H_\alpha\rangle = 2$.

For $\alpha \in \Delta$, we may consider the reflection of $\mathfrak{h}$ about the plane orthogonal to H_α, $s_\alpha(H) = H - \langle \alpha, H \rangle H_\alpha$, which satisfies $s_\alpha^2 = 1$. The subgroup $W = W(\mathfrak{g}_\mathbb{C}, \mathfrak{h})$ of transformations of $\mathfrak{h}$ generated by the reflections s_α is a finite group, called the Weyl group. We denote by $\varepsilon(w) = \pm 1$ the determinant of the transformation $w \in W$.

The Killing form $B(X,Y) = \mathrm{Tr}(\mathrm{ad}X \cdot \mathrm{ad}Y)$ is positive definite on the real span $\sum_\alpha \mathbb{R} H_\alpha$. The vector $\hat{H}_\alpha = 2B(H_\alpha, H_\alpha)^{-1} H_\alpha \in \mathfrak{h}$ is the unique element of $\mathfrak{h}$ such that for all H in $\mathfrak{h}$, $B(\hat{H}_\alpha, H) = \langle \alpha, H \rangle$.

A positive system P of roots is a subset of Δ such that for $\alpha \in \Delta$, either α or $-\alpha$, but not both, belongs to P, and such that if $\alpha \in P$, $\beta \in P$ and $\alpha + \beta \in \Delta$ then $\alpha + \beta \in P$. We define

$$\rho_P = \tfrac{1}{2} \sum_{\alpha \in P} \alpha.$$

A simple root with respect to a fixed positive system P is a root which cannot be written as the sum of two elements of P. The set $\{s_\alpha \mid \alpha \text{ simple}\}$ is a set of generators for W. If $w \in W$, we denote by $|w|$ the length of w with respect to this set of generators.

A compact connected abelian Lie group T is called a torus; it is the quotient of its Lie algebra $\mathfrak{t}$ by a lattice. If T acts on a finite dimensional real vector space V, then $V = V_0 \oplus \sum_{k \in \mathfrak{t}^* \setminus \{0\}} V_{[k]}$; where V_0 is the subspace of V fixed by T and $V_{[k]}$ is an even-dimensional real vector space on which the spectrum of the action of $\mathfrak{t}$ is $\pm ik$.

If G is a compact connected Lie group, every element of G belongs to a torus and any two maximal tori of G are conjugate under the adjoint action of G. The complexification $\mathfrak{g}_\mathbb{C}$ of the Lie algebra $\mathfrak{g}$ of G is a complex reductive Lie algebra. If T is a maximal torus in G with Lie algebra $\mathfrak{t}$, we have $\mathfrak{g} = \mathfrak{t} \oplus \mathfrak{r}$, where $\mathfrak{r} = [\mathfrak{t}, \mathfrak{g}]$, and thus $\mathfrak{g}^* = \mathfrak{t}^* \oplus \mathfrak{r}^*$, and $\mathfrak{t}_\mathbb{C}$ is a Cartan subalgebra of $\mathfrak{g}_\mathbb{C}$. Every element of $\mathfrak{g}$ is conjugate to an element of $\mathfrak{t}$ by the action of an element of G, and every element of $\mathfrak{g}^*$ is conjugate to an element $\lambda \in \mathfrak{t}^*$. Let $\Delta = \Delta(\mathfrak{g}_\mathbb{C}, \mathfrak{t}_\mathbb{C})$. Roots $\alpha \in \Delta$ take imaginary values on $\mathfrak{t}$, and $iH_\alpha \in \mathfrak{t}$ for all $\alpha \in \Delta$. The group T acts on $\mathfrak{r}$ and we write $\mathfrak{r}_{[i\alpha]}$ for the two-dimensional subspace of $\mathfrak{r}$ on which $X \in \mathfrak{t}$ acts by the infinitesimal rotation of angle $i\alpha(X)$.

If T is a maximal torus in a compact Lie group G, let

$$N(T) = \{g \in G \mid gTg^{-1} = T\}$$

be the normalizer of T in G. The group $N(T)$ acts on $\mathfrak{t}$ by the restriction of the adjoint action. Since T acts trivially on $\mathfrak{t}$, this defines an action of the quotient group $W(G,T) = N(T)/T$, which can be identified with the Weyl group $W(\mathfrak{g}_\mathbb{C}, \mathfrak{t}_\mathbb{C})$. We will denote it by W.

For a compact Lie group G, every coadjoint orbit is of the form $M_\lambda = G \cdot \lambda$, with $\lambda \in \mathfrak{t}^*$. We will give an explicit formula for the Liouville measure of an orbit M_λ and for its Fourier transform $F_\lambda(X)$. For $\lambda \in \mathfrak{t}^*$, let P_λ be the set of roots

$$P_\lambda = \{\alpha \in \Delta \mid \langle \lambda, iH_\alpha \rangle > 0\},$$

and let $\mathfrak{r}_\lambda = \sum_{\alpha \in P_\lambda} \mathfrak{r}_{[i\alpha]}$. The orbit M_λ may be identified with the homogeneous space $G/G(\lambda)$, where $G(\lambda)$ is the stabiliser of λ. The space $\mathfrak{r}_\lambda$ is isomorphic to the tangent space $T_e(G/G(\lambda))$ at the base point $e = G(\lambda)$. In particular, if $\langle \lambda, iH_\alpha \rangle \neq 0$

for every $\alpha \in \Delta$, then $G(\lambda) = T$ and $\mathfrak{t}_\lambda = \mathfrak{t}$. Such a point λ is called regular and its orbit has maximal dimension. The restriction of minus the Killing form to $\mathfrak{t}_\lambda$ is positive definite and determines a $G(\lambda)$-invariant inner product on $T_e(G/G(\lambda))$. Thus the homogeneous space $G/G(\lambda)$ is a Riemannian manifold, with a G-invariant metric which coincides with $-B|_{\mathfrak{t}_\lambda}$ at the point e. Denote by $d\bar{g}$ the corresponding G-invariant measure on $G/G(\lambda)$. This measure depends only on the subgroup $G(\lambda)$, which varies over a finite number of subgroups of G.

Lemma 7.23. *The Liouville measure $d\beta_\lambda$ of M_λ is given by the formula*

$$\int_{M_\lambda} \phi \, d\beta_\lambda = \prod_{\alpha \in P_\lambda} \frac{\langle \lambda, i\hat{H}_\alpha \rangle}{2\pi} \int_{G/G(\lambda)} \phi(g\lambda) \, d\bar{g}.$$

Proof. We can choose $X_{\pm\alpha}$ such that $\{e_\alpha, f_\alpha\}$ is an orthonormal basis of $\mathfrak{t}_{[i\alpha]}$, where

$$e_\alpha = \frac{X_\alpha - X_{-\alpha}}{B(H_\alpha, H_\alpha)^{1/2}}, \text{ and } f_\alpha = \frac{iX_\alpha + iX_{-\alpha}}{B(H_\alpha, H_\alpha)^{1/2}}$$

Since $\Omega_\lambda(f_\alpha \cdot \lambda, e_\alpha \cdot \lambda) = i\lambda(\hat{H}_\alpha)$, we easily obtain the lemma. □

The Fourier transform of the measure $d\beta_\lambda$ is a G-invariant analytic function $F_\lambda(X)$ on $\mathfrak{g}$, which is clearly determined by its restriction to $\mathfrak{t}$.

Theorem 7.24 (Harish-Chandra). *Given $\lambda \in \mathfrak{t}^*$, let $W_\lambda = \{w \in W \mid w\lambda = \lambda\}$ be the stabilizer of λ in the Weyl group W. For $X \in \mathfrak{t}$, X regular, the Fourier transform $F_\lambda(X) = \int_{M_\lambda} e^{if(X)} d\beta_\lambda(f)$ is given by the formula*

$$F_\lambda(X) = \sum_{w \in W/W_\lambda} \frac{e^{i\langle w\lambda, X\rangle}}{\prod_{\alpha \in P_\lambda} \langle w\alpha, X\rangle}.$$

Proof. Let X be a regular element of $\mathfrak{t}$. In this case the zero set of the vector field generated by the action of $\exp tX$ on $\mathfrak{g}^*$ is the subspace $\mathfrak{t}^*$ of $\mathfrak{g}^*$ which is fixed by the whole of T. Thus the zero set of the vector field X_{M_λ} is the finite set

$$M_\lambda \cap \mathfrak{t}^* = \{w\lambda \mid w \in W/W_\lambda\}.$$

Applying Theorem 7.11, we see that it suffices to compute $\det^{1/2}(\mathscr{L}_{w\lambda}(X))$. The element $\prod_{\alpha \in P_\lambda} f_\alpha \wedge e_\alpha$ determines the orientation of $\mathfrak{t}_\lambda \cong T_e M_\lambda$. Since $\mathscr{L}_\lambda(X) f_\alpha = i\alpha(X) e_\alpha$, we find that $\det^{1/2}(\mathscr{L}_\lambda(X)) = \prod_{\alpha \in P_\lambda} (i\alpha(X))$. □

If λ is regular, P_λ is "half" of Δ, and up to a sign, the denominator

$$\det{}^{1/2}_{w\lambda}(\mathscr{L}(X)) = \prod_{\alpha \in P_\lambda} (w\alpha, iX)$$

is the same at each fixed point. Thus we obtain the following corollary.

Corollary 7.25 (Harish-Chandra). *If M_λ is a regular orbit of the coadjoint representation, then for $X \in \mathfrak{t}$, X regular,*

$$F_{M_\lambda}(X) = \prod_{\alpha\in P_\lambda} \langle \alpha, X\rangle^{-1} \sum_{w\in W} \varepsilon(w) e^{i\langle w\lambda, X\rangle}.$$

When λ is regular, we can compute the volume $\mathrm{vol}(M_\lambda)$ of the symplectic manifold (M, Ω_λ) as the limit when X tends to 0 of $F_{M_\lambda}(X)$.

Proposition 7.26. *Let λ be regular and $\Lambda = i\lambda$. Then*

$$\mathrm{vol}(M_\lambda) = \frac{\prod_{\alpha>0}\langle \alpha, \Lambda\rangle}{\prod_{\alpha>0}\langle \alpha, \rho\rangle}.$$

Proof. We will also write ρ for the element of $\mathfrak{t}_{\mathbb{C}}$ dual to $\rho \in \mathfrak{t}^*$, which is half the sum of the roots in the positive system $P_\lambda \subset \Delta$. The volume of M_λ is the limit when t tends to 0 of $F_{M_\lambda}(t\rho)$. Since

$$\begin{aligned} t^{-n/2}\sum_{w\in W}\varepsilon(w)e^{i\langle w\lambda, t\rho\rangle} &= t^{-n/2}\sum_{w\in W}\varepsilon(w)e^{t\langle \Lambda, w^{-1}\rho\rangle} \\ &= t^{-n/2}\prod_{\alpha>0}(e^{t\langle\Lambda,\alpha\rangle/2} - e^{-t\langle\Lambda,\alpha\rangle/2}) \end{aligned}$$

tends to $\prod_{\alpha>0}\langle \alpha, \Lambda\rangle$, when $t \to 0$, we obtain the proposition. □

In particular, for $i\lambda = \rho$ the symplectic volume is 1. Using Lemma 7.23, which relates the symplectic volume and the Riemannian volume, we obtain the following result.

Corollary 7.27. *The Riemannian volume of the flag manifold G/T equals*

$$\prod_{\alpha>0} 2\pi\langle \alpha, \rho\rangle^{-1}.$$

Using the above results, we will now analyse the structure of the algebra of G-invariant functions on $\mathfrak{g}$. If $j : \mathfrak{t} \to \mathfrak{g}$ is the inclusion map, then we obtain a restriction map

$$j^* : C^\infty(\mathfrak{g})^G \to C^\infty(\mathfrak{t})^W.$$

There are similar maps where the space of functions $C^\infty(\mathfrak{g})$ is replaced by analytic functions $C^\omega(\mathfrak{g})$ or polynomials $\mathbb{C}[\mathfrak{g}]$. We will prove that the restriction map is an isomorphism in all of these cases, by constructing an explicit inverse for it.

Choose a system $P \subset \Delta$ of positive roots, and define

$$\omega(\lambda) = \prod_{\alpha\in P} \frac{\langle \lambda, i\alpha\rangle}{\langle \rho, \alpha\rangle} \in \mathbb{C}[\mathfrak{t}^*].$$

Let us denote by ∂_ω the constant coefficient differential operator on $\mathfrak{t}$, given by

$$\partial_\omega = \prod_{\alpha\in P}\langle \rho, \alpha\rangle^{-1} \prod_{\alpha\in P} \partial_{iH_\alpha},$$

where $\partial_{i\hat{H}_\alpha}$ is differentiation in the direction of the vector $i\hat{H}_\alpha$.

Let pr be the projection from $\mathfrak{g} = \mathfrak{t} \oplus \mathfrak{r}$ to $\mathfrak{t}$. Let $\pi_{\mathfrak{g}/\mathfrak{t}}$ be the real polynomial on $\mathfrak{t}$

$$\pi_{\mathfrak{g}/\mathfrak{t}}(X) = \prod_{\alpha \in P} i\alpha(X) \in \mathbb{C}[\mathfrak{t}].$$

All integrals over G are with respect to the Haar measure of volume one.

Theorem 7.28 (Chevalley). *If $\phi \in C^\infty(\mathfrak{t})$, let $c(\phi)$ be the function on $\mathfrak{g}$ defined by the formula*

$$(c(\phi))(X) = \frac{1}{|W|} \int_G (\partial_\omega(\pi_{\mathfrak{g}/\mathfrak{t}}\phi))(\mathrm{pr}(g \cdot X))\, dg.$$

Then c is an isomorphism of $C^\infty(\mathfrak{t})^W$, $C^\omega(\mathfrak{t})^W$ and $\mathbb{C}[\mathfrak{t}^*]^W$ to $C^\infty(\mathfrak{g})^G$, $C^\omega(\mathfrak{g})^G$ and $\mathbb{C}[\mathfrak{g}]^G$, and is the inverse of the restriction map j^*.

Proof. The theorem follows from the following formula: if X is a regular element of $\mathfrak{t}$, then

$$\int_G (\partial_\omega \phi)(\mathrm{pr}(gX))\, dg = \sum_{w \in W} \frac{\phi(wX)}{\pi_{\mathfrak{g}/\mathfrak{t}}(wX)}. \tag{7.10}$$

To see that $j^* c(\phi) = \phi$ if ϕ is invariant under the Weyl group, we apply this formula to the function $\pi_{\mathfrak{g}/\mathfrak{t}}\phi$

It suffices to prove (7.10) for $\phi(X) = e^{i\langle\lambda, X\rangle}$, $\lambda \in \mathfrak{t}^*$ regular. Indeed, this proves it by continuity for all λ, hence for all $\phi \in C_c^\infty(\mathfrak{t})$, and finally, since G is compact, for all $\phi \in C^\infty(\mathfrak{t})$.

For the function $\phi(X) = e^{i\langle\lambda, X\rangle}$, $(\partial_\omega \phi)(X) = \omega(i\lambda) e^{i\langle\lambda, X\rangle}$ and the left-hand side of (7.10) becomes

$$i^{|P|}\omega(\lambda) \int_G e^{i\langle\lambda, gX\rangle}\, dg.$$

If λ is regular, this is equal to $F_{M_\lambda}(X)$ up to a constant independent of X. To evaluate this constant, we compare both sides with $X = 0$; Proposition 7.26 shows that

$$\int_G (\partial_\omega \phi)(\mathrm{pr}(gX))\, dg = i^{|P|} F_{M_\lambda}(X).$$

Applying Corollary 7.25, we obtain (7.10). □

When G is a non-compact Lie group and M is a coadjoint orbit of G, the Fourier transform $F_M(X) = \int_M e^{if(X)} d\beta$ may often still be defined as a generalized function on $\mathfrak{g}$. If K is a compact subgroup of G, the action of K on M is the Hamiltonian action of a compact Lie group, and so, by Proposition 7.10, for $X \in \mathfrak{k}$ the form $e^{if(X)} d\beta$ is exact outside the zeroes of X_M. Thus we can hope that when F_M admits a restriction as a generalized function on the Lie algebra $\mathfrak{k}$ of K, the localization formula will allow us to compute this restriction $F_M|_{\mathfrak{k}}$. We state here a result of this type, where G is a real semisimple Lie group with maximal compact subgroup K. This result can be proved using Stokes's theorem and estimates at infinity on the noncompact manifold M.

Let G be a connected real semisimple Lie group with Lie algebra $\mathfrak{g}$, and let K be a maximal compact subgroup of G with Lie algebra $\mathfrak{k}$. Let $\mathfrak{g} = \mathfrak{k} \oplus \mathfrak{p}$ be the Cartan decomposition of $\mathfrak{g}$, and let T be a maximal torus of K with Lie algebra $\mathfrak{t}$. If the Cartan subalgebra $\mathfrak{t}_{\mathbb{C}}$ of $\mathfrak{k}_{\mathbb{C}}$ is also a Cartan subalgebra of $\mathfrak{g}_{\mathbb{C}}$, we say that G and K have equal rank. We have then $\mathfrak{g} = \mathfrak{t} \oplus \mathfrak{r}$, where $\mathfrak{r} = [\mathfrak{t}, \mathfrak{g}]$ and we identify $\mathfrak{t}^*$ with a subset of $\mathfrak{g}^*$.

Let $\Delta = \Delta(\mathfrak{g}_{\mathbb{C}}, \mathfrak{t}_{\mathbb{C}})$, and $\Delta_{\mathfrak{p}} = \{\alpha \mid (\mathfrak{g}_{\mathbb{C}})_\alpha \subset \mathfrak{p}_{\mathbb{C}}\}$; a root belonging to $\Delta_{\mathfrak{p}}$ is called noncompact. For $\lambda \in \mathfrak{t}^*$, let $P_\lambda = \{\alpha \in \Delta \mid (\lambda, iH_\alpha) > 0\}$, and let n_λ be the number of non-compact roots contained in P_λ. If

$$\mathfrak{t}_r = \{H \in \mathfrak{t} \mid \langle \alpha, H\rangle \neq 0 \text{ for all } \alpha \in \Delta\}$$

is the set of regular elements in $\mathfrak{t}$, then $G \cdot \mathfrak{t}_r$ is an open set of $\mathfrak{g}$. The function $F_M(X)$ is a generalized function on $\mathfrak{g}$ and is analytic on $G \cdot \mathfrak{t}_r$. The restriction of F_M to $G \cdot \mathfrak{t}_r$ is thus determined by its restriction to $\mathfrak{t}_r$. When M is of maximal dimension, the following formula is due to Rossmann.

Theorem 7.29 (Rossmann). *Let M be a closed orbit of the coadjoint representation of a real semisimple Lie group G such that G and K have equal rank. Let $W = W(\mathfrak{k}_{\mathbb{C}}, \mathfrak{t}_{\mathbb{C}})$ be the compact Weyl group. Then for $X \in \mathfrak{t}_r$, we have the following results:*

1. *If $M \cap \mathfrak{t}^* = \emptyset$, then $F_M(X) = 0$.*
2. *If $M = G \cdot \lambda$ with $\lambda \in \mathfrak{t}^*$, and W_λ is the subgroup of W stabilizing λ,*

$$F_M(X) = (-1)^{n(\lambda)} \sum_{w \in W/W_\lambda} \frac{e^{i\langle w\lambda, X\rangle}}{\prod_{\alpha \in P_\lambda} \langle w\alpha, X\rangle}.$$

7.6 Equivariant Cohomology and Families

In this section, we will define the equivariant Chern-Weil homomorphism, which maps the equivariant de Rham complex of a G-manifold M to the ordinary de Rham complex of a fibre bundle with fibre M and structure group G, which we do not need to assume is compact. This is an extension of the theory of characteristic classes in Section 1.5, although here we only consider the case of connections, and not of superconnections.

Let $P \to B$ be a principal bundle with structure group G, with connection one-form $\omega \in \mathscr{A}^1(P, \mathfrak{g})^G$ and curvature $\Omega = d\omega + \frac{1}{2}[\omega, \omega]$. The decomposition $TP = HP \oplus VP$ of the tangent bundle TP into a horizontal subbundle $HP = \ker(\omega)$ and a vertical subbundle VP determines a projection operator h from $\mathscr{A}(P)$ onto the subalgebra of horizontal forms

$$\mathscr{A}(P)_{\text{hor}} = \{\alpha \mid \iota(X)\alpha = 0 \text{ for all } X \in \mathfrak{g}\}.$$

With respect to a basis X_i, $1 \le i \le m$, of $\mathfrak{g}$, we may write

$$\omega = \sum_{i=1}^m \omega^i X_i \text{ and } \Omega = \sum_{i=1}^m \Omega^i X_i,$$

where ω^i are one-forms on P, and Ω^i are horizontal two-forms.

Lemma 7.30. *The projection h onto the algebra $\mathscr{A}(P)_{\mathrm{hor}}$ of horizontal forms is given by the explicit formula*

$$h = \prod_{i=1}^{m}(I - \omega^i \iota(X_i))$$
$$= \sum_{1 \le i_1 < \cdots < i_r \le m} (-1)^{r(r+1)/2} \omega^{i_1} \ldots \omega^{i_r} \iota(X_{i_1}) \ldots \iota(X_{i_r}).$$

Proof. If $p_i = I - \omega^i \iota(X_i)$, the relations

$$[\omega^i, \omega^j] = [\iota(X_i), \iota(X_j)] = 0 \text{ and } [\iota(X_i), \omega^j] = \delta_{ij},$$

imply that $\iota(X_i)p_i = [p_i, p_j] = 0$ and $p_i^2 = p_i$. It is also clear that p_i is an algebra homomorphism and that $(\prod_{i=1}^m p_i)\omega_k = 0$. Thus $\prod_{i=1}^m p_i$ is the projection h on the space of horizontal forms. □

Let D denote the operator $h \cdot d \cdot h$ on $\mathscr{A}(P)$; under the identification of $\mathscr{A}(P)_{\mathrm{bas}}$ with $\mathscr{A}(B)$, the restriction of D to $\mathscr{A}(P)_{\mathrm{bas}}$ corresponds to d, since h acts as the identity on $\mathscr{A}(P)_{\mathrm{hor}}$ and d preserves $\mathscr{A}(P)_{\mathrm{bas}}$.

Lemma 7.31. *1.* $D = h \cdot (d - \sum_{i=1}^m \Omega^i \iota(X_i))$, *and*
2. $D^2 + h \cdot (\sum_{i=1}^m \Omega^i \mathscr{L}(X_i)) = 0$

Proof. Since d is a derivation and $h\omega^i = 0$, $hd\omega^i = \Omega^i$, we have for $\alpha \in \mathscr{A}(P)$,

$$h \cdot d \cdot h(\alpha) = h \cdot d\Big(\alpha - \sum_{i=1}^m \omega^i \iota(X_i)\alpha\Big)$$
$$= h\Big(d\alpha - \sum_{i=1}^m \Omega^i \iota(X_i)\alpha\Big).$$

To calculate D^2, we observe that

$$D^2 = h \cdot d \cdot h \cdot d \cdot h$$
$$= h \cdot d \cdot h \cdot \Big(d - \sum_{i=1}^m \Omega^i \iota(X_i)\Big)$$
$$= h\Big(d - \sum_{i=1}^m \Omega^i \iota(X_i)\Big)^2$$
$$= h\Big(-\sum_{i=1}^m d\Omega^i \iota(X_i) - \sum_{i=1}^m \Omega^i \mathscr{L}(X_i)\Big)$$

since $h(d\Omega^i) = 0$. □

The following result shows how the covariant derivative $\nabla^{\mathscr{V}}$ on an associated vector bundle $\mathscr{V} = P \times_G V$ is related to the horizontal projection.

Proposition 7.32. *The covariant derivative $\nabla^{\mathcal{V}}$ on $\mathcal{V}$ coincides with the restriction of the operator $D \otimes I$ on $\mathscr{A}(P,V)$ to $\mathscr{A}(B,\mathcal{V}) = \mathscr{A}(P,V)_{\mathrm{bas}}$.*

Proof. If $\alpha \in \mathscr{A}(P,V)_{\mathrm{bas}}$, we have $\iota(X)d\alpha = (\mathscr{L}(X) \otimes I)\alpha = -\rho(X)\alpha$. Hence $\iota(X_i)\iota(X_j)d\alpha = 0$ and

$$h(dh\alpha) = h(d\alpha) = d\alpha + \sum \omega^i \rho(X_i)\alpha = \nabla^{\mathcal{V}}\alpha. \qquad \square$$

If M is a G-manifold, let $\mathcal{M} = P \times_G M$ be the associated fibre bundle, with base B and typical fibre M. Since $\mathcal{M}$ is the quotient of $P \times M$ by a free action of G, we may identify $\mathscr{A}(\mathcal{M})$ with the space $\mathscr{A}(P \times M)_{\mathrm{bas}}$ of forms α on $P \times M$ which are basic with respect to the action of G. The form ω is a connection form for the action of G on $P \times M$. Thus, if we write simply $\iota(X)$ instead of $\iota(X_{P\times M}) = \iota(X_P) + \iota(X_M)$, the projection h onto the algebra of G-horizontal forms $\mathscr{A}(P \times M)_{\mathrm{hor}}$ is given by the formula

$$h = \prod_{i=1}^{m} (I - \omega^i \iota(X_i))$$

and the operator $D = h \cdot d \cdot h$ restricts to $d_{\mathcal{M}}$ on the space $\mathscr{A}(\mathcal{M}) = \mathscr{A}(P \times M)_{\mathrm{bas}}$.

Consider the differential graded algebra $(\mathbb{C}[\mathfrak{g}] \otimes \mathscr{A}(M), d_{\mathfrak{g}})$ of $\mathscr{A}(M)$-valued polynomial functions on $\mathfrak{g}$. For $\alpha = f \otimes \beta \in \mathbb{C}[\mathfrak{g}] \otimes \mathscr{A}(M)$, we define $\alpha(\Omega) \in \mathscr{A}(P) \otimes \mathscr{A}(M)$ by $\alpha(\Omega) = f(\Omega) \otimes \beta$.

Definition 7.33. *The map $\phi_\omega : \mathbb{C}[\mathfrak{g}] \otimes \mathscr{A}(M) \to \mathscr{A}(P \times M)_{\mathrm{hor}}$ defined by*

$$\phi_\omega(\alpha) = h(\alpha(\Omega))$$

is called the **Chern-Weil homomorphism.**

Observe that the restriction of ϕ_ω to $\mathscr{A}_G(M) = \left(\mathbb{C}[\mathfrak{g}] \otimes \mathscr{A}(M)\right)^G$ sends $\mathscr{A}_G(M)$ into $\mathscr{A}(\mathcal{M}) \cong \mathscr{A}(P \times M)_{\mathrm{bas}}$.

Theorem 7.34. *Let G be a Lie group and let $P \to B$ be a principal bundle with structure group G and connection form ω. Let M be a manifold with smooth action of G. Then the Chern-Weil homomorphism induces a homomorphism of differential graded algebras*

$$\phi_\omega : (\mathscr{A}_G(M), d_{\mathfrak{g}}) = ((\mathbb{C}[\mathfrak{g}] \otimes \mathscr{A}(M))^G, d_{\mathfrak{g}}) \to (\mathscr{A}(\mathcal{M}), d)$$

Proof. Since ϕ_ω clearly preserves products, we have only to show that it intertwines the differential $d_{\mathfrak{g}}$ on $\mathscr{A}_G(M)$ and the differential d on $\mathscr{A}(\mathcal{M})$. In fact, we will prove that the map $\phi_\omega : \mathbb{C}[\mathfrak{g}] \otimes \mathscr{A}(M) \to (\mathscr{A}(P \times M))_{\mathrm{hor}}$ satisfies the formula

$$D \cdot \phi_\omega = \phi_\omega \cdot d_{\mathfrak{g}}.$$

If $f \otimes \alpha \in \mathbb{C}[\mathfrak{g}] \otimes \mathscr{A}(M)$, we see by Lemma 7.31 that

$$D\phi_\omega(f\otimes\alpha) = h\Big(d(f(\Omega)\otimes\alpha) - \sum_{i=1}^{m}\Omega^i\iota(X_i)(f(\Omega)\otimes\alpha)\Big)$$
$$= h\Big(f(\Omega)\otimes(d\alpha - \sum_{i=1}^{m}\Omega^i\iota(X_i)\alpha)\Big)$$

since $h(d\Omega^i) = 0$ and $\iota(X_i)\Omega^j = 0$,

$$= \phi_\omega \cdot d_{\mathfrak{g}}(f\otimes\alpha).$$ □

Note that when M is a point, the above Chern-Weil homomorphism becomes a map
$$\phi_\omega : \mathbb{C}[\mathfrak{g}]^G \to \mathscr{A}(B),$$
which is just the ordinary Chern-Weil homomorphism. In the general case, the map ϕ_ω still has a geometric interpretation. The connection on P determines a horizontal subbundle $H\mathscr{M}$ of the tangent bundle $T\mathscr{M}$, where $H\mathscr{M}$ is defined as the image of HP under the projection $P\times M \to \mathscr{M}$. We have $T\mathscr{M} = H\mathscr{M}\oplus V\mathscr{M}$, where $V\mathscr{M}$ is the vertical tangent bundle. Thus, if $(p,m)\in P\times M$ projects to $y=[p,m]\in\mathscr{M}$, we have the isomorphism
$$j_{(p,m)} : H_p\mathscr{M}\oplus T_mM \cong T_y\mathscr{M},$$
which induces an isomorphism between $\Lambda H_p^*P\otimes\Lambda T_m^*M$ and $\Lambda T_y^*\mathscr{M}$.

If $f\otimes\beta\in\mathbb{C}[\mathfrak{g}]\otimes\mathscr{A}(M)$, then substituting Ω for $X\in\mathfrak{g}$ in f, we obtain a horizontal form $f(\Omega)_p\in\Lambda H_p^*P$, while $\beta_m\in\Lambda T_m^*M$ may be considered as a vertical form on $P\times M$. If we define $(f\otimes\beta)(\Omega)$ to be $f(\Omega)\otimes\beta$, then by linearity we can extend this to define $\alpha(\Omega)$ for any $\alpha\in\mathbb{C}[\mathfrak{g}]\otimes\mathscr{A}(M)$. It is easy to see that if α is G-invariant, then $j_{(p,m)}(\alpha(\Omega)_{(p,m)})$ depends only on $y\in\mathscr{M}$. Thus, we obtain a differential form on $\mathscr{M}$ which corresponds to $h(\alpha(\Omega))$, since $h(\alpha(\Omega))$ is a basic form on $P\times M$ whose restriction to the subspace $\Lambda HP\otimes\Lambda TM$ coincides with that of $\alpha(\Omega)$.

The Chern-Weil homomorphism has the following functoriality. A morphism $f : N\to M$ of G-manifolds gives rise to a G-invariant map of fibre bundles $f_P : \mathscr{N} = P\times_G N \to \mathscr{M} = P\times_G M$, and to the commutative diagram

$$\begin{array}{ccc} \mathscr{A}_{\mathfrak{g}}(M) & \xrightarrow{\phi_\omega} & \mathscr{A}(\mathscr{M}) \\ {\scriptstyle f^*}\downarrow & & \downarrow{\scriptstyle f_P^*} \\ \mathscr{A}_{\mathfrak{g}}(N) & \xrightarrow{\phi_\omega} & \mathscr{A}(\mathscr{N}) \end{array}$$

Proposition 7.35. *Let G be a Lie group, and let M be a compact oriented manifold with smooth action of G. We obtain the commutative diagram*

$$\begin{array}{ccc} \mathscr{A}_{\mathfrak{g}}(M) & \xrightarrow{\phi_\omega} & \mathscr{A}(\mathscr{M}) \\ \downarrow & & \downarrow \\ \mathbb{C}[\mathfrak{g}]^G & \xrightarrow{\phi_\omega} & \mathscr{A}(B) \end{array}$$

where the left vertical arrow is the integration over M and the right vertical arrow is the integration over the fibres of $\mathcal{M} \to B$.

We will now relate the equivariant curvature with the ordinary curvature. Let $E = E^+ \oplus E^-$ be a G-equivariant superbundle on M, with invariant superconnection $\mathbb{A}$. Let $F^E = \mathbb{A}^2 \in \mathcal{A}(M, \mathrm{End}(E))$ be the curvature of $\mathbb{A}$ and let μ^E be its moment. Consider the associated family of vector bundles

$$\mathcal{E} = P \times_G E \to \mathcal{M} = P \times_G M.$$

Then $\mathcal{A}(\mathcal{M}, \mathcal{E})$ may be identified with the space of basic forms $\mathcal{A}(P \times M, E)_{\mathrm{bas}}$, where we consider E as a vector bundle on $P \times M$. The operator

$$h^E = \prod_{i=1}^m (I - \omega^i \iota(X_i))$$

is a projection from $\mathcal{A}(P \times M, E)$ onto $\mathcal{A}(P \times M, E)_{\mathrm{hor}}$. Consider the operator

$$D^{\mathcal{E}} = h^E \cdot (d_P \otimes 1 + 1 \otimes \mathbb{A}) \cdot h^E$$

on $\mathcal{A}(P \times M, E)$, where $d_P \otimes 1 + 1 \otimes \mathbb{A}$ is the pull-back of the superconnection $\mathbb{A}$ on $P \times M$. Clearly $D^{\mathcal{E}}$ preserves $\mathcal{A}(\mathcal{M}, \mathcal{E})$ and induces on it a superconnection.

Definition 7.36. *The superconnection $\mathbb{A}^{\mathcal{E}}$ on $\mathcal{A}(M, \mathcal{E})$ is the restriction of the operator $D^{\mathcal{E}}$ to $\mathcal{A}(P \times M, \mathcal{E})_{\mathrm{bas}}$.*

Let us compute the curvature $F^{\mathcal{E}} = (\mathbb{A}^{\mathcal{E}})^2$ of the superconnection $\mathbb{A}^{\mathcal{E}}$. The operator $D^{\mathcal{E}}$ on $\mathcal{A}(P \times M, \mathcal{E})$ satisfies

$$D^{\mathcal{E}}(\alpha s) = (D\alpha)s + (-1)^{|\alpha|} h(\alpha) D^{\mathcal{E}} s$$

for all $\alpha \in \mathcal{A}(P \times M)$ and $s \in \mathcal{A}(P \times M, \mathcal{E})$, where $D = h \cdot d_{P \times M} \cdot h$. By the formula $D^2 + h(\sum_{i=1}^m \Omega^i \mathcal{L}(X_i)) = 0$ of Lemma 7.31, we see that the operator

$$(D^{\mathcal{E}})^2 + h^E \cdot \sum_{i=1}^m \Omega^i \mathcal{L}(X_i)$$

commutes with exterior multiplication by any horizontal differential form, and hence equals $h^E \cdot F^{\mathcal{E}}$. We will now compute $F^{\mathcal{E}}$ as a function of the curvature F^E of $\mathbb{A}$ and the moment μ^E.

Lemma 7.37. *1. $D^{\mathcal{E}} = h^E \cdot (d_P + \mathbb{A} - \sum_{i=1}^m \Omega^i \iota(X_i))$*
2. The curvature $F^{\mathcal{E}}$ equals $F^E + \sum_{i=1}^m \Omega^i \mu^E(X_i)$, and

$$(D^{\mathcal{E}})^2 + h^E \cdot \Big(\sum_{i=1}^m \Omega^i \mathcal{L}(X_i) \Big) = h^E \cdot F^{\mathcal{E}}.$$

Proof. The proof of this lemma is almost the same as that of Lemma 7.31, except that the operator $d = d_P + d_M$ on $\mathscr{A}(P\times M)$ is replaced by $d_P + \mathbb{A}$ on $\mathscr{A}(P\times M, E)$. Since $\mathbb{A}$ is a superconnection and $h(\omega_i) = 0$,

$$\begin{aligned} D^{\mathscr{E}} &= h^E \cdot (d_P + \mathbb{A}) \cdot h^E \\ &= h^E \left(d_P + \mathbb{A} - \textstyle\sum_{i=1}^m \Omega^i \iota(X_i)\right). \end{aligned}$$

From this, it follows that

$$(D^{\mathscr{E}})^2 = h^E \left(d_P + \mathbb{A} - \textstyle\sum_{i=1}^m \Omega^i \iota(X_i)\right)^2.$$

Using the relation

$$[d_P + \mathbb{A}, \iota(X_i)] = \mathscr{L}^E(X_i) - \mu^E(X_i) + \mathscr{L}((X_i)_P),$$

we see that

$$[d_P + \mathbb{A}, \Omega^i \iota(X_i)] = d\Omega^i \cdot \iota(X_i) + \Omega^i \mathscr{L}(X_i) - \Omega^i \mu^E(X_i).$$

It follows that

$$\begin{aligned} &\left(d_P + \mathbb{A} - \textstyle\sum_{i=1}^m \Omega^i \iota(X_i)\right)^2 \\ &\qquad = F^E + \sum_{i=1}^m \Omega^i \mu^E(X_i) - \sum_{i=1}^m d\Omega^i \cdot \iota(X_i) - \sum_{i=1}^m \Omega^i \mathscr{L}(X_i), \end{aligned}$$

and hence that

$$(D^{\mathscr{E}})^2 + h^E \cdot \Big(\sum_{i=1}^m \Omega^i \mathscr{L}(X_i)\Big) = h^E \cdot \Big(F^E + \sum_{i=1}^m \Omega^i \mu^E(X_i)\Big),$$

since $h(d\Omega^i) = 0$. □

Recall that the equivariant superconnection $\mathbb{A}_{\mathfrak{g}}$ is the operator on $\mathbb{C}[\mathfrak{g}] \otimes \mathscr{A}(M,E)$ defined by

$$(\mathbb{A}_{\mathfrak{g}} \alpha)(X) = \mathbb{A}(\alpha(X)) - \iota(X)(\alpha(X))$$

and that the operator $\mathbb{A}_{\mathfrak{g}}(X)^2 + \mathscr{L}(X)$ is given by exterior multiplication by the equivariant curvature $F_{\mathfrak{g}}(X) = F^E + \mu^E(X)$.

Theorem 7.38. *Let G be a Lie group and let $P \to B$ be a principal bundle with structure group G and connection form ω. Let M be a manifold with smooth action of G, and let $E \to M$ be a G-equivariant bundle on M with G-invariant superconnection $\mathbb{A}$.*

1. The map

$$\phi_\omega : \mathbb{C}[\mathfrak{g}] \otimes \mathscr{A}(M,E) \to \mathscr{A}(P \times M, E)_{\mathrm{hor}}$$

given by $\phi_\omega(\alpha) = h^E(\alpha(\Omega))$ satisfies the formula $\phi_\omega \cdot \mathbb{A}_{\mathfrak{g}} = D^{\mathscr{E}} \cdot \phi_\omega$.

2. *If we also denote by* ϕ_ω *the map*

$$\phi_\omega : \mathbb{C}[\mathfrak{g}] \otimes \mathscr{A}(M,\mathrm{End}(E)) \to \mathscr{A}(P \times M, \mathrm{End}(E))_{\mathrm{hor}},$$

then $\phi_\omega(F_\mathfrak{g}) = F^{\mathscr{E}}$.

3. *Let* $\mathrm{ch}_\mathfrak{g}(\mathbb{A})$ *be the equivariant Chern character of the G-equivariant superbundle* $E \to M$*, with G-invariant superconnection* $\mathbb{A}$*, and let* $\mathrm{ch}(\mathbb{A}^{\mathscr{E}}) = \mathrm{Str}(e^{-(\mathbb{A}^{\mathscr{E}})^2})$ *be the Chern character form of the associated family of superbundles* $\mathscr{E} \to \mathscr{M}$ *with superconnection* $\mathbb{A}^{\mathscr{E}}$*. Then the image of* $\mathrm{ch}_\mathfrak{g}(\mathbb{A})$ *by the Chern-Weil homomorphism is the Chern character form of* $\mathbb{A}^{\mathscr{E}}$

$$\phi_\omega(\mathrm{ch}_\mathfrak{g}(\mathbb{A})) = \mathrm{ch}(\mathbb{A}^{\mathscr{E}}).$$

Proof. The proof of (1) follows the proof in Theorem 7.34 that $\nabla \cdot \phi_\omega = \phi_\omega \cdot d$, except that the operator $d = d_P + d_M$ on $\mathscr{A}(P \times M)$ is replaced by $d_P + \mathbb{A}$.

To prove (2), we must show that

$$\phi_\omega(\mathscr{L}(X)) = \sum_{i=1}^m \Omega^i \mathscr{L}(X_i) \cdot \phi_\omega.$$

Let $f^i \in \mathfrak{g}^*$ be the dual basis to the basis X_i of $\mathfrak{g}$. If $f \in \mathbb{C}[\mathfrak{g}]$ and $\alpha \in \mathscr{A}(M)$, we have

$$\begin{aligned}\phi_\omega(\mathscr{L}(X)(f \otimes \alpha)) &= \phi_\omega\Big(\sum_{i=1}^m f^i f \otimes \mathscr{L}(X_i)\alpha\Big) \\ &= h\Big(\sum_{i=1}^m \Omega^i f(\Omega) \otimes \mathscr{L}(X_i)\alpha\Big) \\ &= h\Big(\sum_{i=1}^m \Omega^i \mathscr{L}(X_i)(f(\Omega) \otimes \alpha)\Big),\end{aligned}$$

since $\sum_{i=1}^m \Omega^i \mathscr{L}(X_i) f(\Omega) = 0$.

Part (3) now follows immediately from (2). □

Let us see what these formulas become in the case of a trivial principal bundle $P = B \times G$. A connection one-form on P is just a $\mathfrak{g}$-valued one-form on B,

$$\omega = \sum_{i=1}^m \omega^i X_i.$$

Let M be a manifold with G action, and E be a G-equivariant vector bundle over M with invariant connection ∇ and moment $\mu^E \in \mathfrak{g}^* \otimes \Gamma(M, \mathrm{End}(E))$.

The manifold $\mathscr{M} = P \times_G M$ is the direct product $B \times M$, and the bundle $\mathscr{E} = P \times_G E$ is the pull-back of E by the projection of $B \times M$ onto M. Using the connection form ω, we obtain a connection $\nabla^{\mathscr{E}}$ on $\mathscr{E}$. The following formula is easily shown.

Proposition 7.39. *Let* $\omega \cdot \mu^E = \sum_{i=1}^m \omega^i \mu^E(X_i)$ *be the contraction of* ω *with* μ^E*. Then* $\nabla^{\mathscr{E}} = d + \nabla + \omega \cdot \mu^E$.

7.7 The Bott Class

Let $\mathscr{V} \to M$ be an oriented Euclidean vector bundle of even rank with spin-structure, let $\mathscr{S} \to M$ be the corresponding spin superbundle, and $\mathscr{S}_{\mathscr{V}} \to \mathscr{V}$ be the pull-back of $\mathscr{S}$ to $\mathscr{V}$. In this section, we will describe the "Riemann-Roch" formula of Mathai and Quillen relating the Chern character of a superconnection on $\mathscr{S}_{\mathscr{V}}$ and of the Thom class $U(\mathscr{V})$ of the bundle $\mathscr{V}$. This result is an application of the functorial properties of the equivariant Chern-Weil differential forms proved in the last section.

Consider the case where M is a point. Let V be an oriented Euclidean vector space of dimension $n = 2\ell$. Let $G = \mathrm{Spin}(V)$, with Lie algebra $\mathfrak{g} = \Lambda^2 V \subset C(V)$, and let $\tau : \mathfrak{g} \to \mathfrak{so}(V)$ be the action of $\mathfrak{g}$ on V defined in (3.4). Let S be the spinor space of V and let ρ be the representation of G in S. The trivial bundle $S_V = V \times S$ is a G-equivariant vector bundle with action $g(\mathbf{x}, s) = (g \cdot \mathbf{x}, \rho(g)s)$.

Let $c : C(V) \to \mathrm{End}(S)$ be the spin representation of the Clifford algebra of V, and consider the odd endomorphism $c(\mathbf{x}) \in \Gamma(V, \mathrm{End}^-(S_V))$, where $\mathbf{x} \in \Gamma(V, V \times V)$ is the tautological section of the trivial bundle with fibre V. From the section $c(\mathbf{x}) = \sum_k \mathbf{x}_k c_k$, we can construct a G-invariant superconnection $\mathbb{A} = d + ic(\mathbf{x})$ on $S_V \to V$ with curvature

$$F_V = \|\mathbf{x}\|^2 + i\,dc(\mathbf{x}) = \|\mathbf{x}\|^2 + i\sum_k d\mathbf{x}_k c_k \in \mathscr{A}(V, \mathrm{End}(S)) \cong \mathscr{A}(V, C(V)).$$

The vector bundle $S_V \to V$, with superconnection $d + ic(\mathbf{x})$, is called the **Bott class**; by Bott periodicity, it represents an element of the K-theory of V which generates $K(V)$ as a free module, but we will not discuss this point of view.

If $X = \sum_{i<j} X_{ij} e_i \wedge e_j$ is an element of $\mathfrak{g}$, the vector field on V defined by the action of G on V is

$$X_V = -2\sum_{i<j} X_{ij}(x_i \partial_j - x_j \partial_i).$$

If $X \in \mathfrak{g}$, the moment $\mu(X) = \mathscr{L}(X) - \nabla_X$ of X is equal to $c(X)$, hence the equivariant curvature of the superconnection is the $\mathrm{End}(S)$-valued form on V

$$F_V(X) = \|\mathbf{x}\|^2 + i\,dc(\mathbf{x}) + c(X) = \sum_k \mathbf{x}_k^2 + i\sum_k d\mathbf{x}_k c_k + \sum_{k<l} X_{kl} c_k c_l.$$

In this context, the Bianchi identity $[\mathbb{A} - \iota(X), F_{\mathfrak{g}}(X)] = 0$ becomes

$$\big(d - \iota(X_V)\big)F_V(X) + i[c(\mathbf{x}), F_V(X)] = 0, \tag{7.11}$$

which is also easy to check directly.

The equivariant Chern character A_V of the trivial bundle $S_V \to V$ with superconnection $\mathbb{A} = d + ic(\mathbf{x})$ is the equivariantly closed form on V given by the formula

$$\begin{aligned} A_V(X) &= \mathrm{Str}\big(\exp(-F_V(X))\big) \\ &= \mathrm{Str}\big(e^{-(\|\mathbf{x}\|^2 + i\,dc(\mathbf{x}) + c(X))}\big) \\ &= \mathrm{Str}\big(e^{-(\|\mathbf{x}\|^2 - ic(d\mathbf{x}) + c(X))}\big). \end{aligned}$$

This differential form decays rapidly at infinity. Such a behaviour would not have been possible if we had taken instead the Chern character of a connection. This illustrates one of the applications of superconnections: the construction of Chern characters which decrease rapidly at infinity on non-compacts manifolds.

Consider now an oriented Euclidean vector bundle $\mathscr{V} \to M$ of even rank $n = 2\ell$ over a compact manifold M, and assume that $\mathscr{V}$ has a spin structure. Thus $\mathscr{V}$ is associated to a principal bundle $P \to M$ with structure group $G = \mathrm{Spin}(V)$, in other words, $\mathscr{V} = P \times_G V$. Let ω be a connection form on P with curvature form Ω. The connection ω defines a Chern-Weil homomorphism $\phi_\omega : \mathscr{A}_G(V) \to \mathscr{A}(\mathscr{V})$.

Let $\mathscr{S} = P \times_G S$ be the corresponding spinor bundle over M, and let $\mathscr{S}_\mathscr{V} = P \times_G S_V$ be a bundle over $\mathscr{V}$ defined using the trivial bundle $S_V = V \times S$ over V; $\mathscr{S}_\mathscr{V}$ is the pull-back of the bundle $\mathscr{S} \to M$ to $\mathscr{V}$ by the projection $\mathscr{V} \to M$. The bundle map $c(\mathbf{x}) : S_V \to S_V$ defines a bundle map $c(\mathbf{x}) : \mathscr{S}_\mathscr{V} \to \mathscr{S}_\mathscr{V}$. Consider the G-invariant superconnection $\mathbb{A} = d + ic(\mathbf{x})$ on $S_V \to V$. As in the preceding section, the connection ω allows us to lift the superconnection $\mathbb{A}$ to a superconnection $\mathbb{A}^{\mathscr{S}_\mathscr{V}}$ on $\mathscr{S}_\mathscr{V} \to \mathscr{V}$. (Beware that the base of our principal bundle P is M, and the fibre of the associated bundle is V, whereas in the last section, the base was B and the fibre was M.) If we denote by ∇_ω the connection induced on the pull-back $\mathscr{S}_\mathscr{V} \to \mathscr{V}$ by the connection on $\mathscr{S} \to M$ with connection form ω, then

$$\mathbb{A}^{\mathscr{S}_\mathscr{V}} = \nabla_\omega + ic(\mathbf{x}).$$

If $d\mathbf{x} \in \mathscr{A}^1(V,V)$ is the canonical one-form on V, Lemma 7.30 shows that

$$\phi_\omega(d\mathbf{x}) = d\mathbf{x} + \omega \cdot \mathbf{x} \in \mathscr{A}^1(P \times V, V).$$

Thus the Chern-Weil homomorphism $\phi_\omega : \mathbb{C}[\mathfrak{g}] \otimes \mathscr{A}(V) \to \mathscr{A}(P \times V)_{\mathrm{hor}}$ is obtained by substituting Ω for $X \in \mathfrak{g}$, and the one-form $d\mathbf{x} + \omega \cdot \mathbf{x}$ on $P \times V$ for the one-form $d\mathbf{x}$ on V. It follows that

$$\phi_\omega(A_V) = \mathrm{Str}\big(e^{-(\|\mathbf{x}\|^2 - ic(d\mathbf{x} + \omega \cdot \mathbf{x}) + c(\Omega))}\big).$$

The following result follows from the functorial properties of the Chern-Weil homomorphism.

Proposition 7.40. *The closed form $\phi_\omega(A_V) \in \mathscr{A}(\mathscr{V})$ is the Chern character form $\mathrm{ch}(\mathbb{A}^{\mathscr{S}_\mathscr{V}})$ of the bundle $\mathscr{S}_\mathscr{V} \to \mathscr{V}$ with superconnection $\mathbb{A}^{\mathscr{S}_\mathscr{V}}$.*

The differential form $\phi_\omega(A_V)$ has the important property of decaying rapidly in the fibre directions.

In Section 1.6, we defined the Thom form of an oriented Euclidean vector bundle $\mathscr{V} \to M$. This may be obtained by a construction analogous to that used above; we will restrict attention to the case of even-dimensional $\mathscr{V}$, although a similar construction works in the odd-dimensional case. The equivariant Thom form U_V of the vector space V may be defined by a formula analogous to that of A_V, except that we work within the Grassman algebra of V instead of its Clifford algebra. The analogy between the Clifford algebra and the Grassmann algebra leads us replace the $C(V)$-valued equivariant curvature F_V of S_V by the ΛV-valued form f_V on V, given by the formula

$$f_V(X) = \|\mathbf{x}\|^2 + i d\mathbf{x} + X = \sum_k \mathbf{x}_k^2 + i \sum_k d\mathbf{x}_k e_k + \sum_{k<l} X_{kl} e_k \wedge e_l.$$

This is the analogue of the differential form which we denoted Ω in Section 1.6.

If $e \in V$, we denote by $\iota_\Lambda(e)$ the derivation of $\mathscr{A}(V, \Lambda V)$ defined by

$$\iota_V(\alpha \otimes \xi) = (-1)^{|\alpha|} \alpha \otimes \iota_\Lambda(\xi) \quad \text{for } \alpha \in \mathscr{A}^k(V) \text{ and } \xi \in \Lambda V,$$

and by $\iota_\Lambda(\mathbf{x})$ the operator

$$\iota_\Lambda(\mathbf{x}) = \sum_{k=1}^m \mathbf{x}_k \iota_\Lambda(e_k).$$

It is easy to verify that for every $X \in \mathfrak{g}$,

$$(d - \iota(X_V)) f_V(X) - 2i\iota_\Lambda(\mathbf{x}) f_V(X) = 0.$$

This formula is the analogue of Part (1) of Proposition 1.51.

Let $\exp(-f_V(X))$ be the exponential of $f_V(X)$ in the algebra $\mathscr{A}(V, \Lambda V)$. Since the operators d, $\iota(X_V)$ and $\iota_\Lambda(\mathbf{x})$ are derivations, we see that

$$(d - \iota_\Lambda(X_V) - 2i\iota_\Lambda(\mathbf{x})) \exp(-f_V(X)) = 0. \tag{7.12}$$

The equivariant Thom form U_V on V is defined using the Berezin integral $T : \mathscr{A}(V, \Lambda V) \to \mathscr{A}(V)$ in place of the supertrace $\mathrm{Str} : \mathscr{A}(V, C(V)) \to \mathscr{A}(V)$:

$$U_V(X) = T(\exp(-f_V(X))) = T(e^{-(\|\mathbf{x}\|^2 + i d\mathbf{x} + X)}).$$

More explicitly, $U_V(X)$ is given by a sum over multi-indices $I \subset \{1, \ldots, n\}$ of the form

$$U_V(X) = e^{-\|\mathbf{x}\|^2} \sum_{|I| \text{ even}} P_{I'}(X) d\mathbf{x}_I$$

where $P_I(X)$ are homogeneous polynomials of degree $(n - |I|)/2$ in X, I' is the complement of I, and $P_{I'}(X)$ coincides up to a sign with the Pfaffian of

$$X_{I'} = \sum_{\{(j,k) \in I' \times I' | j<k\}} X_{jk} e_j \wedge e_k \in \Lambda^2 \mathbb{R}^{I'}.$$

In particular,

$$U_V(X) = \mathrm{Pf}(-X) e^{-\|\mathbf{x}\|^2} + \sum_{i=1}^{\ell-1} \alpha_i + e^{-\|\mathbf{x}\|^2} d\mathbf{x}_1 \wedge \ldots \wedge d\mathbf{x}_n,$$

where $\alpha_i \in \mathscr{A}^{2i}(V) \otimes \mathbb{C}[\mathfrak{g}]_{\ell-i}$.

Theorem 7.41. *1. The form U_V is an equivariantly closed form on V, and for all $X \in \mathfrak{g}$, we have*

$$\pi^{-\ell} \int_V U_V(X) = 1.$$

2. If $i : \{0\} \to V$ is the injection of the origin into V,

$$i^*(U_V(X)) = \mathrm{Pf}(-X).$$

Proof. It is clear that $T(\iota_\Lambda(\mathbf{x})\alpha) = 0$ for every $\alpha \in \mathscr{A}(V, \Lambda V)$. Applying the Berezin integral to $\exp(-f_V(X))$, we see from (7.12) that

$$(d - \iota(X_V))T(\exp(-f_V(X))) = 0.$$

The other properties of $U_V(X)$ are obvious. □

Definition 7.42. *Define $\mathscr{U}(\mathscr{V}) \in \mathscr{A}(\mathscr{V})$ by the formulas*

$$U(\mathscr{V}) = 2^\ell \phi_\omega(U_V) = T\left(e^{-(\|\mathbf{x}\|^2 + i(d\mathbf{x} + \omega\cdot\mathbf{x}) + \Omega)}\right).$$

The differential form $\mathscr{U}(\mathscr{V})$ is the Thom form of Section 1.6, for the metric $\|\mathbf{x}\|^2/2$. We obtain another proof that $U(\mathscr{V})$ is closed, from the properties of the Chern-Weil map ϕ_ω. We also see that

$$(2\pi)^{-\ell} \int_{\mathscr{V}/M} U(\mathscr{V}) = 1.$$

Recall the function $j_V(X)$, which we have already met a number of times in this book:

$$j_V(X) = \det\left(\frac{\sinh(\tau(X)/2)}{\tau(X)/2}\right).$$

By definition,

$$\phi_\omega(j_V^{1/2}) = \det{}^{1/2}\left(\frac{\sinh \Omega/2}{\Omega/2}\right) = \hat{A}(\mathscr{V})^{-1}.$$

The following proposition is a generalization of Proposition 3.16.

Proposition 7.43. *The equivariant differential forms A_V and U_V are related by the formula*

$$(-2i)^{-\ell} A_V(X) = j_V^{1/2}(X)\, U_V(X).$$

Proof. Choose an oriented orthonormal basis e_i of V such that

$$X = \sum_{i=1}^{\ell} \lambda_i e_{2i-1} \wedge e_{2i}.$$

From this, we see that it suffices to consider the case in which $\dim(V) = 2$ and $X = \lambda e_1 \wedge e_2$. Then

$$\begin{aligned}-F_V(X) &= -\|\mathbf{x}\|^2 - id\mathbf{x}_1c_1 - id\mathbf{x}_2c_2 - \lambda c_1c_2\\ &= -\|\mathbf{x}\|^2 - \lambda(c_1 + i\lambda^{-1}d\mathbf{x}_2)(c_2 - i\lambda^{-1}d\mathbf{x}_1) - \lambda^{-1}d\mathbf{x}_1d\mathbf{x}_2.\end{aligned}$$

If we let $\xi_1 = c_1 + i\lambda^{-1}d\mathbf{x}_2$ and $\xi_2 = c_2 - i\lambda^{-1}d\mathbf{x}_1$, then $\xi_1^2 = \xi_2^2 = -1$ and $\xi_1\xi_2 + \xi_2\xi_1 = 0$, so that $e^{-\lambda\xi_1\xi_2} = \cos\lambda - \xi_1\xi_2\sin\lambda$. Thus, we see that

$$\begin{aligned}e^{-F_V(X)} = e^{-\|\mathbf{x}\|^2}\Big(\cos\lambda - \sin\lambda\, c_1c_2 - i\frac{\sin\lambda}{\lambda}(d\mathbf{x}_1c_1 + d\mathbf{x}_2c_2)\\ + \frac{\sin\lambda - \lambda\cos\lambda}{\lambda^2}d\mathbf{x}_1d\mathbf{x}_2 + \frac{\sin\lambda}{\lambda}d\mathbf{x}_1d\mathbf{x}_2c_1c_2\Big).\end{aligned}$$

Since $\mathrm{Str}(c_1c_2) = -2i$, we obtain

$$\mathrm{Str}\big(e^{-F_V(X)}\big) = 2i\sin\lambda\, e^{-\|\mathbf{x}\|^2}(1 - \lambda^{-1}d\mathbf{x}_1d\mathbf{x}_2).$$

On the other hand,

$$-f_V(X) = -\|\mathbf{x}\|^2 - id\mathbf{x}_1e_1 - id\mathbf{x}_2e_2 - \lambda e_1\wedge e_2$$

and hence

$$e^{-f_V(X)} = e^{-\|\mathbf{x}\|^2}\Big(1 - id\mathbf{x}_1e_1 - id\mathbf{x}_2e_2 - \lambda e_1\wedge e_2 + d\mathbf{x}_1d\mathbf{x}_2e_1\wedge e_2\Big).$$

It follows that $T\big(e^{-f_V(X)}\big) = -\lambda\, e^{-\|\mathbf{x}\|^2}(1 - \lambda^{-1}d\mathbf{x}_1d\mathbf{x}_2)$.

The matrix $\tau(X)$ has eigenvalues $\pm 2i\lambda$, so that $j_V^{1/2}(X) = \lambda^{-1}\sin\lambda$. Thus, it follows that

$$\mathrm{Str}\big(e^{-F_V(X)}\big) = -2i\frac{\sin\lambda}{\lambda}T\big(e^{-f_V(X)}\big). \qquad \square$$

Thus, we obtain the following result, which is a refinement at the level of differential forms of the well known "Riemann-Roch" relation between the Thom classes in K-theory and in cohomology. In cohomology, this formula is due to Atiyah-Hirzebruch [10].

Theorem 7.44. *Let $\mathcal{V} \to M$ be an even-dimensional oriented Euclidean vector bundle with spin-structure over a manifold M. The Chern character $\mathrm{ch}(\mathbb{A}^{\mathcal{S}_\mathcal{V}})$ of the Thom bundle $\mathcal{S}_\mathcal{V} \to \mathcal{V}$ with superconnection $\mathbb{A}^{S_V}$ and the Thom form $U(\mathcal{V})$ are related by the formula*

$$i^\ell\,\mathrm{ch}(\mathbb{A}^{\mathcal{S}_\mathcal{V}}) = \hat{A}(\mathcal{V})^{-1}U(\mathcal{V}).$$

Bibliographic Notes

Section 1

The complex of equivariant differential forms was introduced in H. Cartan [47]. He shows that if G is compact, the cohomology $H_G^\bullet(M)$ of this complex agrees with the topological definition of equivariant cohomology $H^\bullet(M \times_G BG, \mathbb{R})$. The map between the two theories may be described using the Chern-Weil map of Section 6, with base the universal classifying space BG (or a finite-dimensional approximation).

The definitions of the moment of an equivariant connection, the equivariant curvature and equivariant characteristic classes are taken from Berline and Vergne [22], except that here we allow superconnections instead of just connections.

Sections 2, 3 and 4

Theorem 7.13 is due to Berline-Vergne [20] when $X = Y$, and to Bismut [31] as stated. The original proof used an argument due to Bott, of approximating M_0 by a small tubular neighbourhood, similar to that which we give here for the proof of Theorem 7.11. Here, we use an argument by Gaussian approximation; it is interesting to observe its close analogy to the proof of the McKean-Singer formula. Special cases which were proved earlier include Bott's formulas for characteristic numbers, when the zero-set of the vector field is discrete [43], corresponding formulas when the zero-set may have arbitrary codimension proved in the Riemannian case by Baum and Cheeger [18] and in the holomorphic case by Bott [44], Duistermaat and Heckman's theorem on the exactness of stationary phase for functions whose Hamiltonian vector field generates a circle action Corollary [58], and the localization theorem for equivariant differential forms, proved by Berline and Vergne [20], and by Atiyah and Bott [7]. A generalization of Theorem 7.11 to algebraic manifolds with singularities has been given by Rossmann [99].

A localization theorem for equivariant K-theory was proved by Segal [102], and by Quillen for equivariant generalized cohomology theories [92]. The similarity between the localization properties of the complex of equivariant differential forms and the localization theorem in K-theory were observed by Berline and Vergne [20] and Witten [106].

Witten has proposed an interesting analogy between the local index theorem for the Dirac operator on a manifold M and the localization theorem in equivariant cohomology for the action of the circle group on on the loop space LM (see Atiyah [4]). Bismut generalizes their results to twisted Dirac operators in [31].

Section 5

The homogeneous Hamiltonian actions of Section 5 were introduced by Kostant [76]. A general relationship between the Fourier transform of a coadjoint orbit and the character of the corresponding representation was conjectured by Kirillov [73], [74], based on his work on his formulas for the characters of compact and nilpotent groups. Further conjectures on this formula are stated in Vergne [103].

Theorem 7.24 is proved in Harish-Chandra [69].

Further results on the relationship between equivariant cohomology and Fourier transforms of orbits may be found in Berline and Vergne [21], Duflo-Heckman-Vergne [55], Duflo-Vergne [56]. The proof we give of Chevalley's Theorem 7.28 is in the last of these references. Theorem 7.29 is due to Rossmann [98].

Section 7

The results of this section are due to Mathai and Quillen [82], although we have made greater use of the Berezin integral in the exposition than they did. As explained in Atiyah-Singer [15], it allows one to pass from the Atiyah-Singer Theorem for the index of an elliptic operator, expressed in terms of K-theory, to formulas in terms of characteristic classes.

Additional remarks

Let M be a compact manifold on which a compact group G acts differentiably. If $g \in G$, denote by $G(g) = \{h \in G \mid gh = hg\}$ the centralizer of g, by $\mathfrak{g}(g)$ its Lie algebra, and by $M(g)$ the fixed point set of g acting on M, considered as a manifold with differentiable $G(g)$-action.

In Block-Getzler [38] and Duflo-Vergne [57], a de Rham model $\mathscr{K}_G(M)$ for the G-equivariant K-theory of a G-manifold M is introduced. (In the first of these articles, the relationship with equivariant cyclic homology is also described.) An element of $\mathscr{K}_G(M)$ is represented by a collection of equivariant differential forms $\{\omega_g(X)\}_{g\in G}$, called a **bouquet**, where $\omega_g(X) \in \mathscr{A}^\infty_{G(g)}(M(g))$ is a $G(g)$-equivariant differential form on $M(g)$. Axioms of invariance and compatibility for bouquets are given in [38] and [57].

By analogy with the case of equivariant differential forms, one can define an integral for bouquets. If M has a G-invariant spin-structure, Proposition 6.14 induces a family of orientations on the fixed point sets $M(g)$. Let $R^\infty(G) = C^\infty(G)^G$ be the ring of smooth functions on G invariant under the adjoint action of G. The integral $\int_M^b : \mathscr{K}_G(M) \to R^\infty(G)$ is defined by the formula

$$\int_M^b \omega(g\exp X) = \int_{M(g)} (2\pi)^{\dim(\mathscr{N}(g))/2} \frac{\omega_g(X)\hat{A}_{\mathfrak{g}(g)}(X,M(g))}{\det^{1/2}(1-g^{\mathscr{N}(g)}\exp R^{\mathscr{N}(g)}_{\mathfrak{g}(g)}(X))}.$$

Here, $\hat{A}_{\mathfrak{g}(g)}(X,M(g))$ is the equivariant $\hat{A}$-genus of $M(g)$, and $R^{\mathscr{N}}_{\mathfrak{g}(g)}$ is the equivariant curvature of the normal bundle $\mathscr{N}(g)$ of the embedding $M(g) \hookrightarrow M$. The fact that these germs of smooth functions on G glue together to form an element of $R^\infty(G)$ is a corollary of the Localization Theorem 7.13 (see [57]). There is also a generalization of this formula to Spin$_c$-manifolds.

8

The Kirillov Formula for the Equivariant Index

The character of a finite-dimensional irreducible representation of a compact Lie group G can be described by the Weyl character formula, which is a special case of the fixed point formula for the equivariant index of Chapter 6. However, there is another formula, the universal character formula of Kirillov[73], which presents the character not as a sum over fixed points but as an integral over a certain orbit of G in its coadjoint representation on $\mathfrak{g}^*$; this second formula is in principle much more general than the first, since it applies to many cases other than that of compact groups.

Let M be a compact oriented even-dimensional Riemannian G-manifold, where G is a compact Lie group, and let $\mathcal{E}$ be a G-equivariant Clifford module over M. In this chapter we will give an analogue of Kirillov's formula which computes the equivariant index of a Dirac operator on the bundle $\mathcal{E}$ over M as the integral over M of an equivariantly closed differential form. Let $\mathfrak{g}$ be the Lie algebra of G, let $\hat{A}_{\mathfrak{g}}(X,M)$, $X \in \mathfrak{g}$, be the equivariant $\hat{A}$-genus of the tangent bundle of M, and let $\operatorname{ch}_{\mathfrak{g}}(X,\mathcal{E}/S)$ be the equivariant relative Chern character of $\mathcal{E}$ (see (8.2)). If D is a Dirac operator on $\mathcal{E}$ associated to a G-invariant connection on $\mathcal{E}$, we call the Kirillov formula the formula for the equivariant index of D, which holds for $X \in \mathfrak{g}$ sufficiently small,

$$\operatorname{ind}_G(e^{-X},\mathsf{D}) = (2\pi i)^{-\dim(M)/2} \int_M \hat{A}_{\mathfrak{g}}(X,M) \operatorname{ch}_{\mathfrak{g}}(X,\mathcal{E}/S).$$

In Section 1, we will prove this formula by combining the fixed point formula for the index with the localization formula of the last chapter. We discuss the special case in which D is an invariant Dirac operator on a homogeneous space of a compact Lie group in Section 2.

In Section 3, we introduce Bismut's "quantized equivariant differential", namely, we replace the Dirac operator D by the operator $\mathsf{D} + \frac{1}{4} c(X_M)$, where X_M is the vector field on M corresponding to the Lie algebra element $X \in \mathfrak{g}$. Following Bismut, we rewrite the general Kirillov formula as a local index theorem for this operator; the proof of this theorem follows closely the proof of the local index theorem of Chapter 4.

8.1 The Kirillov Formula

In Chapter 6, we proved for the equivariant index of the Dirac operator $\operatorname{ind}_G(\gamma, \mathsf{D})$ the fixed point formula (where $n = \dim(M)$)

$$\operatorname{ind}_G(\gamma, \mathsf{D}) = \int_{M^\gamma} \frac{(2\pi)^{\dim(\mathcal{N})/2}}{(2\pi i)^{n/2}} T_M \left(\frac{\hat{A}(M^\gamma)\operatorname{ch}_G(\gamma, \mathcal{E}/S)}{\det^{1/2}(1 - \gamma^{\mathcal{N}} \exp(-R^{\mathcal{N}}))} \right) |dx_0|, \quad (8.1)$$

which expresses the equivariant index as an integral over the fixed point set M^γ of γ in M. In this section, we will rewrite this for γ near the identity in G, in terms of the equivariant cohomology of M; we call the resulting formula the **Kirillov formula**, by analogy with Kirillov's formulas for characters of Lie groups.

Let M be a compact oriented Riemannian manifold of even dimension n, and let G be a Lie group, with Lie algebra $\mathfrak{g}$, acting on M by positively oriented isometries. Let R be the Riemannian curvature of M, and let

$$\mu^M \in (\mathfrak{g}^* \otimes \Gamma(M, \mathfrak{so}(M)))^G \subset \mathcal{A}_G^2(M, \mathfrak{so}(M))$$

be the Riemannian moment of X defined in (7.3), that is, the skew-endomorphism of TM given by

$$\mu^M(X)Y = -\nabla_Y X.$$

Let $R_\mathfrak{g} = \mu^M + R \in \mathcal{A}_G^2(M, \mathfrak{so}(M))$ be the equivariant Riemannian curvature of M. Let $X \mapsto \hat{A}_\mathfrak{g}(X, M)$ be the equivariant $\hat{A}$-genus of the tangent bundle of M,

$$\hat{A}_\mathfrak{g}(X, M) = \det{}^{1/2}\left(\frac{R_\mathfrak{g}(X)/2}{\sinh(R_\mathfrak{g}(X)/2)} \right).$$

Let $\mathcal{E}$ be a G-equivariant Clifford module over M with G-invariant Hermitian metric, and invariant Clifford connection $\nabla^\mathcal{E}$. Let $\mathcal{L}^\mathcal{E}(X)$ be the Lie derivative of the action of $X \in \mathfrak{g}$ on $\Gamma(M, \mathcal{E})$, and let

$$\mu^\mathcal{E}(X) = \mathcal{L}^\mathcal{E}(X) - \nabla_X^\mathcal{E} \in \Gamma(M, \operatorname{End}(\mathcal{E}))$$

be the moment of X with respect to the connection $\nabla^\mathcal{E}$; thus, $\mu^\mathcal{E}$ is an element of $(\mathfrak{g}^* \otimes \Gamma(M, \operatorname{End}(\mathcal{E})))^G \subset \mathcal{A}_G^2(M, \operatorname{End}(\mathcal{E}))$.

Recall from (3.7) the canonical map

$$\tau : \Lambda^2 T^*M \to \mathfrak{so}(M);$$

if $a \in \mathcal{A}^2(M)$ and $\xi \in \mathcal{A}^1(M)$, then as operators on $\mathcal{E}$,

$$[c(a), c(\xi)] = c(\tau(a)\xi).$$

If μ^M is the Riemannian moment of M, then $c(\tau^{-1}(\mu^M)) \in \mathcal{A}_G^2(M, \operatorname{End}(\mathcal{E}))$ is given by Mehler's formula

$$c(\tau^{-1}(\mu^M(X))) = \tfrac{1}{2} \sum_{i<j} (\mu^M(X)e_i, e_j) c(e^i) c(e^j)$$

where e^i, $(1 \leq i \leq n)$, is an orthonormal frame of T^*M. Thus

$$[c(\tau^{-1}(\mu^M(X))), c(\alpha)] = c(\mu^{T^*M}(X)\alpha)$$

for all $\alpha \in \mathscr{A}^1(M)$, where μ^{T^*M} is the moment of the bundle T^*M, equal to minus the adjoint of μ^M.

Define the **twisting moment** $\mu^{\mathscr{E}/S}(X)$ for the action of X on $\mathscr{E}$ by the formula

$$\mu^{\mathscr{E}/S}(X) = \mu^{\mathscr{E}}(X) - c(\tau^{-1}(\mu^M(X))).$$

Lemma 8.1. *The twisting moment $\mu^{\mathscr{E}/S}$ commutes with $c(\alpha)$ for all $\alpha \in \mathscr{A}^1(M)$, and hence lies in $(\mathfrak{g}^* \otimes \Gamma(M, \mathrm{End}_{C(M)}(\mathscr{E})))^G \subset \mathscr{A}^2_G(M, \mathrm{End}_{C(M)}(\mathscr{E}))$.*

Proof. Since $\mu^{\mathscr{E}}(X) = \mathscr{L}^{\mathscr{E}}(X) - \nabla^{\mathscr{E}}_X$ and $\mu^{T^*M}(X) = \mathscr{L}^{T^*M}(X) - \nabla^{T^*M}_X$, we see, since $\nabla^{\mathscr{E}}$ is a G-invariant Clifford connection, that

$$\begin{aligned}[\mu^{\mathscr{E}}(X), c(\alpha)] &= c(\mathscr{L}^{T^*M}(X)\alpha) - c(\nabla^{T^*M}_X \alpha) \\ &= c(\mu^{T^*M}(X)\alpha) \\ &= [c(\tau^{-1}(\mu^M(X)), c(\alpha)].\end{aligned}$$

□

We now define the **equivariant twisting curvature** of $\mathscr{E}$ to be

$$F^{\mathscr{E}/S}_{\mathfrak{g}}(X) = \mu^{\mathscr{E}/S}(X) + F^{\mathscr{E}/S} \in \mathscr{A}^2_G(M, \mathrm{End}_{C(M)}(\mathscr{E})).$$

Note that if M has an equivariant spin-structure with spinor bundle $\mathscr{S}$, so that $\mathscr{E}$ is isomorphic to the equivariant twisted spinor bundle $\mathscr{W} \otimes \mathscr{S}$, then we may identify $F^{\mathscr{E}/S}_{\mathfrak{g}} \in \mathscr{A}_G(M, \mathrm{End}_{C(M)}(\mathscr{E}))$ with $F^{\mathscr{W}}_{\mathfrak{g}} \in \mathscr{A}_G(M, \mathrm{End}(\mathscr{W}))$ under the isomorphism of bundles $\mathrm{End}_{C(M)}(\mathscr{E}) \cong \mathrm{End}(\mathscr{W})$. Define the **equivariant relative Chern character** form of $\mathscr{E}$ to be

$$\mathrm{ch}_{\mathfrak{g}}(X, \mathscr{E}/S) = \mathrm{Str}_{\mathscr{E}/S}(\exp(-F^{\mathscr{E}/S}_{\mathfrak{g}}(X))). \tag{8.2}$$

In this chapter, we will study the following reformulation of the equivariant index theorem near the identity in G. We will explain in the next section why we call this the Kirillov formula.

Theorem 8.2 (Kirillov formula). *For $X \in \mathfrak{g}$ sufficiently close to zero,*

$$\mathrm{ind}_G(e^{-X}, \mathsf{D}) = (2\pi i)^{-n/2} \int_M \hat{A}_{\mathfrak{g}}(X, M)\, \mathrm{ch}_{\mathfrak{g}}(X, \mathscr{E}/S).$$

Let us first comment on some differences between the fixed point formula (8.1) and the Kirillov formula for $\mathrm{ind}_G(\gamma, \mathsf{D})$. In the first formula, it is not *a priori* clear, and in fact quite remarkable that the right-hand side of the fixed point formula depends analytically on γ near $\gamma = 1$, and in particular that it has a limit when $\gamma \to 1$ equal to

$$\mathrm{ind}(\mathsf{D}) = (2\pi i)^{-n/2} \int_M \hat{A}(M)\, \mathrm{ch}(\mathscr{E}/S).$$

In contrast the analytic behaviour of $\operatorname{ind}_G(e^{-X}, \mathsf{D})$ near $X = 0$ is exhibited in the Kirillov formula. However, the fact that this analytic function of X is the restriction to a neighbourhood of zero of an analytic function of e^{-X} is not apparent. This fact is an analogue of the integrality property of the $\hat{A}$-genus of a spin manifold.

The Kirillov formula can be deduced from the localization formula, since the integrals over M of an equivariantly closed form $\alpha(X)$ localize on the set of zeroes of X_M.

We now start the proof of Theorem 8.2. Choose $X \in \mathfrak{g}$ sufficiently close to zero that the zero set of X_M coincides with the fixed point set M_0 of $\gamma = \exp(-X)$. By Proposition 7.12, the bundle $\mathcal{N}$ is orientable, and hence so is M_0. Fix compatible orientations on M_0 and its normal bundle $\mathcal{N}$. Let $T_{\mathcal{N}}$ be the Berezin integral

$$T_{\mathcal{N}} : \Gamma(M_0, \Lambda\mathcal{N}^*) \to C^\infty(M_0).$$

Since $\gamma^{\mathcal{E}} \in \Gamma(M_0, C(\mathcal{N}^*) \otimes \operatorname{End}_{C(M)}(\mathcal{E})) \cong \Gamma(M_0, \Lambda\mathcal{N}^* \otimes \operatorname{End}_{C(M)}(\mathcal{E}))$, we see that

$$T_{\mathcal{N}}(\gamma^{\mathcal{E}}) \in \Gamma(M_0, \operatorname{End}_{C(M)}(\mathcal{E})).$$

The fixed point formula for the equivariant index states that

$$\operatorname{ind}_G(\gamma, \mathsf{D}) = (2\pi i)^{-n/2} \int_{M_0} (2\pi)^{\dim(\mathcal{N})/2} \frac{\hat{A}(M_0)\operatorname{ch}_G(\gamma, \mathcal{E}/S)}{\det^{1/2}(1 - \gamma^{\mathcal{N}} \exp(-R^{\mathcal{N}}))},$$

where $R^{\mathcal{N}}$ is the curvature of the normal bundle $\mathcal{N} \to M_0$, $\gamma^{\mathcal{N}}$ is the induced action of γ on $\mathcal{N}$, and $\operatorname{ch}_G(\gamma, \mathcal{E}/S)$ is the differential form on M_0 given by the formula

$$\operatorname{ch}_G(\gamma, \mathcal{E}/S) = \frac{2^{\dim(\mathcal{N})/2}}{\det^{1/2}(1 - \gamma^{\mathcal{N}})} \operatorname{Str}_{\mathcal{E}/S}(T_{\mathcal{N}}(\gamma^{\mathcal{E}}) \cdot \exp(-F_0^{\mathcal{E}/S})).$$

By the localization theorem, we have

$$\int_M \hat{A}_{\mathfrak{g}}(X, M)\operatorname{ch}_{\mathfrak{g}}(X, \mathcal{E}/S) = \int_{M_0} (2\pi)^{\dim(\mathcal{N})/2} \frac{(\hat{A}_{\mathfrak{g}}(X, M)\operatorname{ch}_{\mathfrak{g}}(X, \mathcal{E}/S))|_{M_0}}{\det_{\mathcal{N}}^{1/2}(-R_{\mathfrak{g}}^{\mathcal{N}}(X))}.$$

Thus, we need only prove that, for $\gamma = e^{-X}$ with X sufficiently small,

$$\frac{\hat{A}(M_0)\operatorname{ch}_G(\gamma, \mathcal{E}/S)}{\det^{1/2}(1 - \gamma^{\mathcal{N}} \exp(-R^{\mathcal{N}}))} = \frac{(\hat{A}_{\mathfrak{g}}(X, M)\operatorname{ch}_{\mathfrak{g}}(X, \mathcal{E}/S))|_{M_0}}{\det_{\mathcal{N}}^{1/2}(-R_{\mathfrak{g}}^{\mathcal{N}}(X))}.$$

Observe that the differential forms $\det_{\mathcal{N}}^{1/2}(-R_{\mathfrak{g}}^{\mathcal{N}}(X))$ and $\operatorname{ch}_G(\gamma, \mathcal{E}/S)$ both depend on the orientation of $\mathcal{N}$; choose the orientation such that $\det_{\mathcal{N}}^{1/2}(\mathcal{L}^{\mathcal{N}}(X)) > 0$. Under the splitting

$$TM|_{M_0} = TM_0 \oplus \mathcal{N},$$

the Riemannian curvature restricted to M_0 splits as $R = R_0 \oplus R^{\mathcal{N}}$, where R_0 is the Riemannian curvature of M_0. The Riemannian moment $\mu^M(X)$ of X restricted to M_0 is the infinitesimal action of X on TM, and equals the moment $\mu^{\mathcal{N}}(X)$. Thus,

$$R_{\mathfrak{g}}(X) = \begin{pmatrix} R_0 & 0 \\ 0 & R_{\mathfrak{g}}^{\mathcal{N}}(X) \end{pmatrix}.$$

It follows that

$$\begin{aligned} \hat{A}_{\mathfrak{g}}(X,M) &= \det{}^{1/2}\left(\frac{R_{\mathfrak{g}}(X)/2}{\sinh(R_{\mathfrak{g}}(X)/2)}\right) \\ &= \det{}^{1/2}\left(\frac{R_0/2}{\sinh(R_0/2)}\right)\det{}^{1/2}\left(\frac{R_{\mathfrak{g}}^{\mathcal{N}}(X)/2}{\sinh(R_{\mathfrak{g}}^{\mathcal{N}}(X)/2)}\right). \end{aligned}$$

The exponential of $R_{\mathfrak{g}}^{\mathcal{N}}(X)$ equals $(\gamma^{\mathcal{N}})^{-1}\exp(R^{\mathcal{N}})$ where $\gamma = e^{-X}$, since $\mu^{\mathcal{N}}(X)$ may be identified with the action of the vector field X on the bundle $\mathcal{N}$, and $\mu^{\mathcal{N}}(X)$ and $R^{\mathcal{N}}$ commute. Since $R_{\mathfrak{g}}^{\mathcal{N}} \in \mathcal{A}_G(M_0, \mathfrak{so}(\mathcal{N}))$ is antisymmetric and $\mathcal{N}$ is even-dimensional, we see that

$$\det{}^{-1/2}(\sinh(R_{\mathfrak{g}}^{\mathcal{N}}(X)/2)) = 2^{\dim(\mathcal{N})/2}\det{}^{-1/2}(1-\gamma^{\mathcal{N}}\exp(-R^{\mathcal{N}})),$$

and hence

$$\frac{\hat{A}_{\mathfrak{g}}(X,M)}{\det_{\mathcal{N}}^{1/2}(R_{\mathfrak{g}}^{\mathcal{N}}(X))} = \frac{\hat{A}(M_0)}{\det^{1/2}(1-\gamma^{\mathcal{N}}\exp(-R^{\mathcal{N}}))}.$$

It remains to show that

$$\mathrm{ch}_{\mathfrak{g}}(X, \mathcal{E}/S)|_{M_0} = (-1)^{\dim(\mathcal{N})/2}\,\mathrm{ch}_G(e^{-X}, \mathcal{E}/S),$$

Consider the decomposition

$$\mu^{\mathcal{E}}(X) = c(\tau^{-1}(\mu^M(X))) + \mu^{\mathcal{E}/S}(X),$$

where $\tau^{-1}(\mu^M(X)) \in \Lambda^2\mathcal{N}^*$. Since the two terms on the right-hand side commute, we see on taking the exponential of both sides that

$$\gamma^{\mathcal{E}} = \exp(-c(\tau^{-1}(\mu^M(X)))) \cdot \exp(-\mu^{\mathcal{E}/S}(X))$$

Let us write a for $\exp(c(\tau^{-1}(\mu^M(X)))) \in C(\mathcal{N}^*)$. With our choice of orientation of $\mathcal{N}$, $T_{\mathcal{N}}(a) > 0$, and

$$T_{\mathcal{N}}(a) = 2^{-\dim(\mathcal{N})/2}\det{}^{1/2}(1-\gamma^{\mathcal{N}}),$$

so that

$$\begin{aligned} \mathrm{ch}_G(\gamma, \mathcal{E}/S) &= \frac{2^{\dim(\mathcal{N})/2}}{\det^{1/2}(1-\gamma^{\mathcal{N}})}\mathrm{Str}_{\mathcal{E}/S}(T_{\mathcal{N}}(\gamma^{\mathcal{E}})\exp(-F_0^{\mathcal{E}/S})) \\ &= (-1)^{\dim(\mathcal{N})/2}\,\mathrm{Str}_{\mathcal{E}/S}(\exp(-\mu^{\mathcal{E}/S}(X))\exp(-F_0^{\mathcal{E}/S})) \\ &= \mathrm{ch}_{\mathfrak{g}}(X, \mathcal{E}/S). \end{aligned}$$

This completes the proof of Theorem 8.2.

8.2 The Weyl and Kirillov Character Formulas

Let G be a connected compact Lie group with maximal torus T. The irreducible finite dimensional representations of G were classified by E. Cartan, and their characters were calculated by H. Weyl. In the notation of Section 7.5, let L_T be the lattice $\{X \in \mathfrak{t} \mid e^X = 1\}$, so that $T = \mathfrak{t}/L_T$, and denote by L_T^* the dual lattice

$$L_T^* = \{l \in \mathfrak{t}^* \mid l(X) \in 2\pi\mathbb{Z} \text{ for every } X \in L_T\} \subset \mathfrak{t}^*.$$

Choose a system $P \subset \Delta$ of positive roots. The subset $X_G = i\rho_P + L_T^*$ of $\mathfrak{t}^*$ is independent of the choice of P and W-invariant. In the notation of Section 7.5, we write Weyl's character formula as follows.

Theorem 8.3 (Weyl). *If λ is a regular element of X_G, there exists a unique finite-dimensional irreducible representation T_λ of G, such that for $X \in \mathfrak{t}$,*

$$\operatorname{Tr}(T_\lambda(e^X)) = \frac{\sum_{w\in W} \varepsilon(w) e^{i\langle w\lambda, X\rangle}}{\prod_{\alpha\in P_\lambda}(e^{\langle\alpha,X\rangle/2} - e^{-\langle\alpha,X\rangle/2})}$$

Let us now describe Kirillov's formula for the character of the representation T_λ. As usual, for $X \in \mathfrak{g}$, let

$$j_{\mathfrak{g}}(X) = \det\left(\frac{\sinh(\operatorname{ad}X/2)}{\operatorname{ad}X/2}\right).$$

We can define an analytic square root of $j_{\mathfrak{g}}(X)$ on the whole of $\mathfrak{g}$. Indeed, if $X \in \mathfrak{t}$,

$$j_{\mathfrak{g}}(X) = \prod_{\alpha\in\Delta} \frac{e^{\langle\alpha,X\rangle/2} - e^{-\langle\alpha,X\rangle/2}}{\langle\alpha,X\rangle} = \left(\prod_{\alpha\in P} \frac{e^{\langle\alpha,X\rangle/2} - e^{-\langle\alpha,X\rangle/2}}{\langle\alpha,X\rangle}\right)^2.$$

Thus, by Chevalley's Theorem 7.28, there exists a unique analytic G-invariant function $j_{\mathfrak{g}}^{1/2}$ on $\mathfrak{g}$, which on $\mathfrak{t}$ is equal to

$$j_{\mathfrak{g}}^{1/2}(X) = \prod_{\alpha\in P} \frac{e^{\langle\alpha,X\rangle/2} - e^{-\langle\alpha,X\rangle/2}}{\langle\alpha,X\rangle}.$$

Theorem 8.4 (Kirillov). *Let λ be a regular element of X_G and let M_λ be the symplectic manifold $G\cdot\lambda$ with its canonical symplectic structure Ω_λ. For $X \in \mathfrak{g}$, we have*

$$j_{\mathfrak{g}}^{1/2}(X)\operatorname{Tr}(T_\lambda(e^X)) = \int_{M_\lambda} e^{if(X)}\, dm(f)$$

where dm is the Liouville measure of M_λ.

Proof. Both sides of the equation are G-invariant functions on $\mathfrak{g}$. Thus it is sufficient to check this equality when $X \in \mathfrak{t}$. The equality then follows immediately from Corollary 7.25. □

In this section, we will explain how these formulas are special cases of two forms of the equivariant index theorem, respectively the fixed point formula Theorem 6.16 and the Kirillov formula Theorem 8.2. To do this, we must find an equivariant Dirac operator whose equivariant index is the character of T_λ. As underlying space, we take the flag manifold G/T. Consider G/T as a Riemannian manifold, with the metric induced by the restriction of the opposite of the Killing form to $\mathfrak{r} = T_e(G/T)$. Choosing a system $P \subset \Delta$ of positive roots, let

$$e_\alpha = \frac{X_\alpha - X_{-\alpha}}{\|H_\alpha\|} \quad , \quad f_\alpha = \frac{iX_\alpha + iX_{-\alpha}}{\|H_\alpha\|}.$$

Choose the orientation on G/T such that $\prod_{\alpha \in P} e_\alpha \wedge f_\alpha$ is the orientation of $\mathfrak{r} = T_e(G/T)$. The manifold G/T is an even-dimensional oriented Riemannian manifold with spin-structure.

Lemma 8.5. *Let $\hat{A}_\mathfrak{g}(X, G/T)$ be the equivariant $\hat{A}$-genus of the Riemannian manifold G/T. Then $\hat{A}_\mathfrak{g}(X, G/T)$ and $j_\mathfrak{g}^{-1/2}(X)$ represent the same class in the equivariant de Rham cohomology $H^\bullet_\mathfrak{g}(G/T)$.*

Proof. Consider the direct sum decomposition $\mathfrak{g} = \mathfrak{t} \oplus \mathfrak{r}$ of $\mathfrak{g}$. From this, we obtain a direct sum decomposition of the bundle $G \times_T \mathfrak{g}$,

$$G \times_T \mathfrak{g} = (G \times_T \mathfrak{t}) \oplus T(G/T).$$

Since T acts trivially on $\mathfrak{t}$, the bundle $G \times_T \mathfrak{t}$ is isomorphic to the trivial bundle $G/T \times \mathfrak{t}$. The trivial connection on this bundle has vanishing equivariant curvature, so the equivariant $\hat{A}$-genus equals 1. Thus, the bundle $G \times_T \mathfrak{g}$, with connection the direct sum of the trivial connection on $G/T \times \mathfrak{t}$ and the Levi-Civita connection on $T(G/T)$, has equivariant $\hat{A}$-genus equal to $\hat{A}_\mathfrak{g}(X, G/T)$.

However, since $\mathfrak{g}$ is a G-module, the bundle $G \times_T \mathfrak{g}$ may be trivialized to $G/T \times \mathfrak{g}$ by the map

$$[(g, X)] \in G \times_T \mathfrak{g} \longmapsto ([g], \operatorname{ad} g \cdot X) \in G/T \times \mathfrak{g}.$$

The trivial connection on $G/T \times \mathfrak{g}$ has equivariant curvature $F_\mathfrak{g}(X) = \operatorname{ad} X$, so that for the trivial connection,

$$\hat{A}_\mathfrak{g}(X, G \times_T \mathfrak{g}) = j_\mathfrak{g}^{-1/2}(X).$$

Since the equivariant cohomology class of the equivariant $\hat{A}$-genus is independent of the connection used in its definition, we obtain the lemma. □

Of course, $\hat{A}_\mathfrak{g}(X, G/T)$ and $j_\mathfrak{g}^{-1/2}(X)$ are not in general equal as differential forms, since $j_\mathfrak{g}^{-1/2}(X)$ is a function whereas the differential form $\hat{A}(G/T) = \hat{A}_\mathfrak{g}(0, G/T)$ has non-vanishing higher degree terms in general.

If $\lambda \in L_T^*$, there exists a unique character of T with differential $i\lambda$ on $\mathfrak{t}$. Denote this character by $e^{i\lambda}$ and by $\mathcal{L}_\lambda$ the homogeneous line bundle $G \times_T \mathbb{C}_\lambda$, where $\mathbb{C}_\lambda$ is the one-dimensional vector space $\mathbb{C}$ with the representation $e^{i\lambda}$ of T. Let us denote by $\mu^\lambda(X)$ the restriction to M_λ of the linear function $f \mapsto (f,X)$. If λ is regular, the map $g \mapsto g\cdot\lambda$ is an isomorphism of G/T with $M_\lambda = G\cdot\lambda$. Thus, $\mathcal{L}_\lambda$ is a line bundle over M_λ.

Proposition 8.6 (Kostant). *The homogeneous line bundle $\mathcal{L}_\lambda$ has a G-invariant canonical Hermitian connection. The equivariant curvature of $\mathcal{L}_\lambda$ is the equivariantly closed form $i(\mu^\lambda(X)+\Omega_\lambda)$.*

Proof. For $X \in \mathfrak{g}$, define an operator ∇_X on sections of $\mathcal{L}_\lambda$ by the equation

$$\mathcal{L}(X) = \nabla_X + i\mu^\lambda(X).$$

Since the vector fields X_{M_λ} span the tangent bundle of M_λ, it is only necessary to verify the following condition in order for ∇_X to be a connection: if X_{M_λ} vanishes at a point $f \in M_\lambda$, in other words, if $X \in \mathfrak{g}(f)$, and if ϕ is a section of $\mathcal{L}_\lambda$, then $\nabla_X\phi(f) = 0$. But if $X \in \mathfrak{g}(f)$,

$$(\mathcal{L}^\lambda(X)\phi)(f) = i(f,X)\phi(f) = i\mu^\lambda(X)(f)\,\phi(f)$$

by definition of the homogeneous vector bundle $\mathcal{L}_\lambda$. Thus ∇ is a connection with moment map $i\mu^\lambda(X)$, from which it follows that its equivariant curvature equals $i(\mu^\lambda(X)+\Omega_\lambda)$. □

If G is simply connected, the element $i\rho_P \in L_T^*$, so that if $X_G = L_T^*$.

Consider the G-invariant twisted Dirac operator D_λ associated to the twisting bundle $\mathcal{L}_\lambda$:

$$\mathsf{D}_\lambda : \Gamma(G/T, \mathcal{S}\otimes\mathcal{L}_\lambda) \to \Gamma(G/T, \mathcal{S}\otimes\mathcal{L}_\lambda).$$

Its equivariant index

$$\operatorname{ind}_G(g,\mathsf{D}_\lambda) = \operatorname{Tr}_{\ker(\mathsf{D}_\lambda^+)}(g) - \operatorname{Tr}_{\ker(\mathsf{D}_\lambda^-)}(g)$$

is the character of a virtual representation of G. Denote by $\varepsilon(P,P_\lambda) = \pm 1$ the quotient of the two orientations on $\mathfrak{r}$ determined by P and P_λ. The following theorem identifies the Weyl and Kirillov character formulas with the equivariant index theorem for the Dirac operator D_λ.

Theorem 8.7. *The fixed point formula for the equivariant index of D_λ is given for $X \in \mathfrak{t}$ by the formula*

$$\operatorname{ind}_G(e^X,\mathsf{D}_\lambda) = \frac{\sum_{w\in W}\varepsilon(w)e^{i\langle w\lambda,X\rangle}}{\prod_{\alpha\in P}(e^{\langle\alpha,X\rangle/2} - e^{-\langle\alpha,X\rangle/2})}.$$

Thus, the index $\operatorname{ind}_G(g,\mathsf{D}_\lambda)$ equals the character of the representation T_λ up to a sign.

The Kirillov formula Theorem 8.2 for the equivariant index of D_λ at $X \in \mathfrak{g}$ small is

$$\operatorname{ind}_G(e^X,\mathsf{D}_\lambda) = \varepsilon(P,P_\lambda)\int_{M_\lambda} j_{\mathfrak{g}}^{-1/2}(X)e^{if(X)}\,dm(f),$$

where dm is the Liouville measure of M_λ.

Proof. Let us start with the fixed point formula. The action of T on the flag manifold leaves stable the points

$$\{wT \mid w \in W(G,T) = N(G,T)/T\} \subset G/T,$$

and for generic $t \in T$, the action of t on G/T has a finite number of fixed points indexed by the Weyl group W. Thus it is easy to write the fixed point formula for the character of the virtual representation $\operatorname{ind} \mathsf{D}_\lambda$: it equals

$$\sum_w (-i)^{|P|} \varepsilon(w,t,P) \frac{e^{iw\lambda}(t)}{|\det_{\mathfrak{r}}^{1/2}(1 - wtw^{-1})|},$$

where the sign $\varepsilon(w,t,P)$ is the quotient of the orientations of $\mathfrak{r}$ determined respectively by P and by the element $\rho(wtw^{-1})$ in the Clifford algebra $C(\mathfrak{r})$ of the Euclidean space $\mathfrak{r}$. We have, for $t = e^X$, $X \in \mathfrak{t}$,

$$\begin{aligned} \rho(e^X) &= \exp\Big(\tfrac{1}{2} \sum_{\alpha \in P} (Xe_\alpha, f_\alpha) e_\alpha f_\alpha\Big) \\ &= \exp\Big(-i \sum_{\alpha \in P} \langle \alpha, X/2 \rangle e_\alpha f_\alpha\Big) \\ &= \prod_{\alpha \in P} (\cosh\langle \alpha, X/2 \rangle - i \sinh\langle \alpha, X/2 \rangle e_\alpha f_\alpha). \end{aligned}$$

It follows that

$$\varepsilon(w,P,e^X) = (-1)^{|P|} \varepsilon(w) \prod_{\alpha \in P} \operatorname{sgn}(i \sinh\langle \alpha, X/2 \rangle).$$

Since

$$\det_{\mathfrak{r}}(1 - e^X) = \prod_{\alpha \in P} (2i \sinh\langle \alpha, X/2 \rangle)^2,$$

the fixed point formula follows.

Since the Clifford module $\mathcal{S} \otimes \mathcal{L}_\lambda$ that we are considering is a twisted spinor bundle, the Kirillov formula may be rewritten as

$$\operatorname{ind}(\mathsf{D}_\lambda)(e^{-X}) = (2\pi i)^{-\dim(G/T)/2} \int_{G/T} \hat{A}_{\mathfrak{g}}(X, G/T) \operatorname{ch}_{\mathfrak{g}}(X, \mathcal{L}_\lambda).$$

By Lemma 8.6, the equivariant Chern character of $\mathcal{L}_\lambda$ equals

$$\operatorname{ch}_{\mathfrak{g}}(X, \mathcal{L}_\lambda) = e^{-i(f,X)} e^{-i\Omega_\lambda}.$$

The integral on G/T of an equivariantly closed differential form depends only on its equivariant cohomology class. By Lemma 8.5, we can choose as a representative for $\hat{A}_{\mathfrak{g}}(X, G/T)$ the function $j_{\mathfrak{g}}^{-1/2}(X)$. It only remains to compute the highest degree term of $\operatorname{ch}_{\mathfrak{g}}(X, \mathcal{L}_\lambda)$, which is $(-2\pi i)^{\dim(G/T)/2} e^{-i(f,X)} dm$. The replacement of X by $-X$ and a careful comparison of orientations proves the second formula. □

Bott has given another realization of the representation T_λ, as the $\bar{\partial}$-cohomology space of a line bundle on the flag manifold G/T. Define $\mathfrak{r}^+ = \sum_{\alpha \in P} (\mathfrak{g}_\mathbb{C})_\alpha$. Then $\mathfrak{r}^+$ is a subalgebra of $\mathfrak{g}_\mathbb{C}$ and $\mathfrak{r}$ has a complex structure J such that $\mathfrak{r}^+ = \{X \in \mathfrak{r}_\mathbb{C} \mid JX = iX\}$. There exists a unique G-invariant complex structure on G/T such that $T_e^{1,0}(G/T) = \mathfrak{r}^+$. Furthermore, for any $\mu \in L_T^*$, the line bundle $\mathcal{L}_\mu = G \times_T \mathbb{C}_\mu$ has a structure of an holomorphic line bundle over G/T. It is easy to write down the fixed point formula for the character of the virtual representation $\sum (-1)^i H^{0,i}(G/T, \mathcal{L}_\mu)$ on the $\bar{\partial}$-cohomology spaces of the line bundle $\mathcal{L}_\mu$.

Theorem 8.8. *If $\mu \in L_T^*$, then for $X \in \mathfrak{t}$,*

$$\sum (-1)^p \operatorname{Tr}_{H^{0,p}(G/T,\mathcal{L}_\mu)}(e^X) = \frac{\sum_{w \in W} \varepsilon(w) e^{\langle w(i\mu + \rho_P), X \rangle}}{\prod_{\alpha \in P} (e^{\langle \alpha, X \rangle/2} - e^{-\langle \alpha, X \rangle/2})}.$$

This theorem shows that the virtual representation $\sum (-1)^p H^{0,p}(G/T, \mathcal{L}_\mu)$ is zero if $\lambda = \mu - i\rho_P$ is not a regular element of $\mathfrak{t}^*$. If λ is regular, let w_λ be the element of W such that $w(P_\lambda) = P$. Then the virtual representation $\sum (-1)^p H^{0,p}(G/T, \mathcal{L}_\mu)$ is the representation $\varepsilon(w_\lambda) T_\lambda$. In fact a more refined result, the **Borel-Weil-Bott theorem**, is true:

$$H^{0,p}(G/T, \mathcal{L}_\mu) = \begin{cases} T_\lambda, & |w_\lambda| = p, \\ 0, & \text{if } |w_\lambda| \neq p, \end{cases}$$

where $|w|$ is the length of the Weyl group element $w \in W(G,T)$.

8.3 The Heat Kernel Proof of the Kirillov Formula

Let M be a compact oriented Riemannian manifold of even dimension n, and let G be a compact Lie group with Lie algebra $\mathfrak{g}$ acting on M by orientation-preserving isometries. Let $\mathcal{E}$ be a G-equivariant Clifford module over M with G-equivariant Hermitian structure and G-invariant Clifford connection $\nabla^{\mathcal{E}}$.

Definition 8.9. *The* **Bismut Laplacian** *is the second-order differential operator on $\Gamma(M, \mathcal{E})$ given by the formula*

$$H(X) = \left(\mathsf{D} + \tfrac{1}{4} c(X)\right)^2 + \mathcal{L}(X),$$

where we write $c(X)$ for the Clifford action of the dual of the vector field X_M on $\mathcal{E}$.

The operator $H(X)$ is a generalized Laplacian depending polynomially on $X \in \mathfrak{g}$. It may be thought of as a quantum analogue of the equivariant Riemannian curvature

$$R_\mathfrak{g}(X) = (\nabla - \iota(X))^2 + \mathcal{L}(X).$$

The restriction $k_t(x,x,X)\,|dx|$ of the heat kernel of the semigroup $e^{-tH(X)}$ to the diagonal is a section of the bundle $\operatorname{End}(\mathcal{E}) \cong C(M) \otimes \operatorname{End}_{C(M)}(\mathcal{E}) \otimes |\Lambda|$. The asymptotic expansion of $k_t(x,x,X)$ as a function of t for t small has the form

$$k_t(x,x,X) \sim (4\pi t)^{-n/2} \sum_{i=0}^{\infty} t^i k_i(x,X),$$

where $k_i(x,X)$ depends polynomially on X (as we will show later). Denoting by $\mathbb{C}[[\mathfrak{g}]]$ the algebra of formal power series on $\mathfrak{g}$, we may identify k_i with a section of the equivariant Clifford algebra $\mathbb{C}[[\mathfrak{g}]] \otimes C(M) \otimes \mathrm{End}_{C(M)}(\mathscr{E})$. The associated graded algebra of this filtered algebra is $\mathbb{C}[[\mathfrak{g}]] \otimes \Lambda T^*M \otimes \mathrm{End}_{C(M)}(\mathscr{E})$.

If V is a Euclidean vector space with Clifford algebra $C(V^*)$, consider the algebra $\mathbb{C}[[\mathfrak{g}]] \otimes C(V^*)$ with filtration

$$F_i(\mathbb{C}[[\mathfrak{g}]] \otimes C(V^*)) = \sum_{k \leq i/2} \mathbb{C}[\mathfrak{g}]_k \otimes C_{i-2k}(V^*),$$

where $\mathbb{C}[\mathfrak{g}]_k$ is the space of homogeneous polynomials on $\mathfrak{g}$ of degree k. This filtration is analogous to the grading of the equivariant cohomology complex, and will be called the equivariant Clifford filtration. The associated graded algebra is isomorphic to $\mathbb{C}[\mathfrak{g}] \otimes \Lambda V^*$, with equivariant grading

$$\deg(P \otimes a) = 2\deg(P) + \deg(a) \quad \text{for } P \in \mathbb{C}[\mathfrak{g}] \text{ and } a \in \Lambda V^*.$$

The next theorem is the generalization of Theorem 4.1 to the equivariant context. It generalizes Bismut's infinitesimal Lefschetz formula [31].

Theorem 8.10. *Let $k_t(x,x,X) = \langle x \mid e^{-tH(X)} \mid x \rangle$ be the restriction of the heat kernel of the semigroup $e^{-tH(X)}$ to the diagonal. Consider the asymptotic expansion*

$$k_t(x,x,X) \sim (4\pi t)^{-n/2} \sum_{i=0}^{\infty} t^i k_i(x,X).$$

1. The equivariant Clifford degree of k_i is less or equal than $2i$.
2. Let $\sigma(k) = \sum_{i=0}^{\infty} \sigma_{[2i]}(k_i) \in \mathbb{C}[[\mathfrak{g}]] \otimes \mathscr{A}(M, \mathrm{End}_{C(M)}(\mathscr{E}))$. Then

$$\sigma(k)(X) = \det{}^{1/2}\left(\frac{R_{\mathfrak{g}}(X)/2}{\sinh(R_{\mathfrak{g}}(X)/2)}\right) \exp(-F_{\mathfrak{g}}^{\mathscr{E}/S}(X)).$$

Before proving this theorem, let us show how it implies the Kirillov formula for the equivariant index. Recall that the heat equation computation of the index is based on the McKean-Singer formula Proposition 6.3: for every $t > 0$,

$$\mathrm{ind}_G(\gamma, \mathsf{D}) = \mathrm{Str}(\gamma e^{-t\mathsf{D}^2}).$$

For $u \in \mathbb{C}$, we consider the operator

$$\mathsf{D}_u = \mathsf{D} + uc(X).$$

Since the operators D and $c(X)$ commute with the action of e^{tX} on $\Gamma(M, \mathscr{E})$, so does D_u.

Proposition 8.11. *For all $t > 0$ and all $u \in \mathbb{C}$, we have*

$$\operatorname{ind}_G(\exp X, \mathrm{D}) = \operatorname{Str}\left(e^X e^{-t\mathrm{D}_u^2}\right).$$

Proof. The result is true for $u = 0$; we will show that the supertrace on the right-hand side does not depend on u. By Proposition 2.50, and the fact e^X and the heat operator $e^{-t\mathrm{D}_u^2}$ commute, we have

$$\frac{d}{du}\operatorname{Str}\left(e^X e^{-t\mathrm{D}_u^2}\right) = -t\operatorname{Str}\left(\frac{d\mathrm{D}_u^2}{du} e^{-t\mathrm{D}_u^2} e^X\right).$$

Since e^X commutes with $c(X)$ and D, we obtain

$$\begin{aligned}\frac{d}{du}\operatorname{Str}\left(e^X e^{-t\mathrm{D}_u^2}\right) &= -t\operatorname{Str}\left(\frac{d\mathrm{D}_u^2}{du} e^X e^{-t\mathrm{D}_u^2}\right)\\ &= -t\operatorname{Str}\left((\mathrm{D}_u c(X) + c(X)\mathrm{D}_u) e^X e^{-t\mathrm{D}_u^2}\right)\\ &= -t\operatorname{Str}\left(\left[\mathrm{D}_u, c(X) e^X e^{-t\mathrm{D}_u^2}\right]\right).\end{aligned}$$

This vanishes by Lemma 3.49 □

Let us now set $u = \frac{1}{4}$. It follows from this proposition that for every $t > 0$,

$$\operatorname{ind}_G(e^{-tX}, \mathrm{D}) = \operatorname{Str}\left(e^{-t\mathcal{L}(X)} e^{-t(\mathrm{D}+c(X)/4)^2}\right) = \operatorname{Str}\left(e^{-tH(X)}\right).$$

Thus we obtain the formula

$$\operatorname{ind}_G(e^{-tX}, \mathrm{D}) = \int_M \operatorname{Str}(k_t(x,x,X))\,|dx|.$$

Consider the asymptotic expansion in t of both sides of this formula. We have

$$\operatorname{Str}(k_t(x,x,X)) \sim (4\pi t)^{-n/2} \sum_{j=0}^{\infty} t^j \operatorname{Str}(k_j(x,X)).$$

It follows from the first assertion of Theorem 8.10 that if $j < n/2$, the total degree of k_j is strictly less than n. In particular, its exterior degree is strictly less than the top degree and the supertrace of k_j vanishes. Thus there is no singular part in the asymptotic development of $\operatorname{Str}(k_t(x,x,X))$. Using the formula $\operatorname{Str}(a) = (-2i)^{n/2} \operatorname{Str}_{\mathcal{E}/S}(a)_{[n]}$, we obtain

$$\operatorname{ind}_G(e^{-tX}, \mathrm{D}) \sim (2\pi i)^{-n/2} \sum_{j=0}^{\infty} t^j \int_M \operatorname{Str}_{\mathcal{E}/S}\left(k_{n/2+j}(x,X)_{[n]}\right) |dx|$$

The first assertion of Theorem 8.10 also implies that $k_{n/2+j}(X,x))_{[n]}$ is a polynomial in X of degree less or equal to j. The asymptotic expansion of the left-hand side $\operatorname{ind}_G(e^{-tX}, \mathrm{D})$ of the above equation is a convergent series:

$$\operatorname{ind}_G(e^{-tX}, \mathsf{D}) = \sum_{j=0}^{\infty} \frac{(-t)^j}{j!} \left(\operatorname{Tr}_{\ker(\mathsf{D}^+)}(\mathscr{L}(X)^j) - \operatorname{Tr}_{\ker(\mathsf{D}^-)}(\mathscr{L}(X)^j) \right).$$

Comparing the coefficient of t^j in these two power series in t, we see that only the component of $\sigma(k_{n/2+j}(X,x))$ lying in $\mathbb{C}[\mathfrak{g}]_j \otimes \Lambda^n T_x^* M$ contributes to the equivariant index. Using the explicit formula of the theorem for $\sigma(k)$, we obtain

$$\operatorname{ind}_G(e^{-X}, \mathsf{D}) = (2\pi i)^{-n/2} \int_M \hat{A}_{\mathfrak{g}}(X,M) \operatorname{ch}_{\mathfrak{g}}(X, \mathscr{E}/S).$$

which is the Kirillov formula.

Let us now prove Theorem 8.10. Following a plan which should be familiar by now, we start by proving a generalization of Lichnerowicz's formula. Let $\theta_X \in \mathscr{A}^1(M)$ be the one-form associated to X,

$$\theta_X(\xi) = (X, \xi) \in \mathscr{A}^1(M).$$

Let $\nabla^{\mathscr{E},X}$ be the Clifford connection $\nabla^{\mathscr{E}} - \frac{1}{4}\theta_X$ on the bundle $\mathscr{E}$; since θ_X is scalar-valued, it is clear that $\nabla^{\mathscr{E},X}$ is again a Clifford connection. Let us show how Bismut's Laplacian $H(X)$ can be written as a generalized Laplacian associated to the connection $\nabla^{\mathscr{E},X}$ on $\mathscr{E}$. Let Δ_X be the Laplacian on $\mathscr{E}$ associated to $\nabla^{\mathscr{E},X}$.

Proposition 8.12. *If $F^{\mathscr{E}/S}$ and $\mu^{\mathscr{E}/S}(X)$ are, respectively, the twisting curvature and the twisting moment of X for the connection $\nabla^{\mathscr{E}}$ and r_M is the scalar curvature of M, then*

$$H(X) = \Delta_X + \tfrac{1}{4} r_M + \mathbf{c}(F^{\mathscr{E}/S}) + \mu^{\mathscr{E}/S}(X).$$

Proof. Recall from Lemma 7.15 that

$$(\mu^M(X)Y, Z) = -\tfrac{1}{2} d\theta_X(Y,Z), \quad Y, Z \in \Gamma(M, TM).$$

Since θ_X is a scalar-valued one-form, the twisting curvature $F^{\mathscr{E}/S,X}$ of the connection $\nabla^{\mathscr{E},X}$ is equal to $F^{\mathscr{E}/S} - \frac{1}{4} d\theta_X$. Let $\mathsf{D}_X = \mathsf{D} + \frac{1}{4} c(X)$; since D_X is the Dirac operator associated to the connection $\nabla^{\mathscr{E},-X}$, Lichnerowicz's formula shows that

$$\begin{aligned} \mathsf{D}_X^2 &= \Delta_{-X} + \tfrac{1}{4} r_M + \mathbf{c}(F^{\mathscr{E}/S,-X}) \\ &= \Delta_{-X} + \tfrac{1}{4} r_M + \mathbf{c}(F^{\mathscr{E}/S}) + \tfrac{1}{4}\mathbf{c}(d\theta_X). \end{aligned}$$

By this formula and Proposition 3.45, we see that

$$\begin{aligned} \Delta_{-X} - \Delta_X &= \mathsf{D}_X^2 - \mathsf{D}_{-X}^2 - \tfrac{1}{2}\mathbf{c}(d\theta_X) \\ &= \tfrac{1}{2}[\mathsf{D}, \mathbf{c}(\theta_X)] - \tfrac{1}{2}\mathbf{c}(d\theta_X) \\ &= -\nabla_X^{\mathscr{E}} + \tfrac{1}{2} d^*\theta_X. \end{aligned}$$

But $d^*\theta_X = \operatorname{Tr}(\mu^M(X)) \in C^\infty(M)$ vanishes, since μ^M is antisymmetric; this implies that

$$\Delta_X - \Delta_{-X} = \nabla_X^{\mathscr{E}}.$$

Since $\mathcal{L}^{\mathcal{E}}(X) = \nabla^{\mathcal{E}}_X + \mu^{\mathcal{E}}(X)$, we see that

$$\begin{aligned} H(X) &= \mathsf{D}^2_X + \mathcal{L}^{\mathcal{E}}(X) \\ &= \Delta_{-X} + \tfrac{1}{4} r_M + \mathbf{c}(F^{\mathcal{E}/S}) + \nabla^{\mathcal{E}}_X + \mu^{\mathcal{E}}(X) + \tfrac{1}{4}\mathbf{c}(d\theta_X). \end{aligned}$$

Since $\Delta_X = \Delta_{-X} + \nabla^{\mathcal{E}}_X$ and $\mu^{\mathcal{E}/S}(X) = \mu^{\mathcal{E}}(X) + \frac{1}{4}\mathbf{c}(d\theta_X)$ (by definition of the twisting moment), the result is proved. □

We could model the proof of Theorem 8.10 on the proofs of the simple index theorem of either Chapter 4 or Chapter 5. We will follow the former. Thus, as in Section 4.3, let $V = T_{x_0}M$, $E = \mathcal{E}_{x_0}$ and $U = \{\mathbf{x} \in V \mid \|\mathbf{x}\| < \varepsilon\}$, where ε is smaller than the injectivity radius of the manifold M at x_0. We identify U by means of the exponential map $\mathbf{x} \mapsto \exp_{x_0}\mathbf{x}$ with a neighbourhood of x_0 in M.

For $x = \exp_{x_0}\mathbf{x}$, the fibre $\mathcal{E}_x$ and E are identified by the parallel transport map $\tau(x_0,x) : \mathcal{E}_x \to E$ along the geodesic $x_s = \exp_{x_0} s\mathbf{x}$. Thus the space $\Gamma(U,\mathcal{E})$ of sections of $\mathcal{E}$ over U may be identified with the space $C^\infty(U,E)$.

Choose an orthonormal basis ∂_i of $V = T_{x_0}M$, with dual basis $d\mathbf{x}^i$ of $T^*_{x_0}M$, and let $c^i = c(d\mathbf{x}^i) \in \mathrm{End}(E)$. Let e_i be the local orthonormal frame obtained by parallel transport along geodesics from the orthonormal basis ∂_i of $T_{x_0}M$, and let e^i be the dual frame of T^*M.

Let S be the spinor space of V^* and let $W = \mathrm{Hom}_{C(V^*)}(S,E)$ be the auxiliary vector space such that $E = S \otimes W$, so that $\mathrm{End}(E) \cong \mathrm{End}(S) \otimes \mathrm{End}(W) \cong C(V^*) \otimes \mathrm{End}(W)$, and $\mathrm{End}_{C(V^*)}(E) = \mathrm{End}(W)$.

We must study the operator $H(X)$ in the above coordinate system, but with respect to the frame of the bundle $\mathcal{E}$ obtained by parallel transport along geodesics emerging from x_0 with respect to the connection $\nabla^{\mathcal{E},X}$. Define $\alpha : U \times \mathfrak{g} \to \mathbb{C}^\times$ by the formula

$$\alpha_X(\mathbf{x}) = -\frac{1}{4}\int_0^1 (\iota(\mathcal{R})\theta_X)(t\mathbf{x})t^{-1}\,dt,$$

so that $\mathcal{R}\alpha_X = -\frac{1}{4}\iota(\mathcal{R})\theta_X$. Let $\rho(X,\mathbf{x}) = e^{\alpha_X(\mathbf{x})}$. Then $\rho(X,\mathbf{x})$ is the parallel transport map on the trivial line bundle over M with respect to the connection $d - \frac{1}{4}\theta_X$, along the geodesic leading from x and x_0.

Lemma 8.13. *The conjugate*

$$\rho(X,\mathbf{x})(\nabla^{\mathcal{E},X}_{\partial_i})\rho(X,\mathbf{x})^{-1}$$

of the covariant derivative $\nabla^{\mathcal{E},X}_{\partial_i} = \nabla^{\mathcal{E}}_{\partial_i} - \frac{1}{4}\theta_X(\partial_i)$ *is given on elements of* $C^\infty(U,E)$ *by the formula*

$$\partial_i + \tfrac{1}{4}\sum_{j;k<l} R_{ijkl}\mathbf{x}^j c^k c^l - \sum_j \tfrac{1}{4}\mu^M_{ij}(X)\mathbf{x}^j + \sum_{j<k} f_{ijk}(\mathbf{x})c^j c^k + \langle g_i(\mathbf{x}),X\rangle + h_i(\mathbf{x}).$$

Here, $R_{ijkl} = (R(\partial_k,\partial_l)_{x_0}\partial_j,\partial_i)_{x_0}$ *and* $\mu^M_{ij}(X) = (\mu^M(X)_{x_0}\partial_i,\partial_j)_{x_0}$ *are the Riemannian curvature and moment at* x_0, *and* $f_{ijk}(\mathbf{x}) \in C^\infty(U)$, $g_i(\mathbf{x}) \in C^\infty(U)\otimes\mathfrak{g}^*$ *and* $h_i(\mathbf{x}) \in C^\infty(U,\mathrm{End}_{C(V^*)}(E))$ *are error terms which satisfy*

$$f_{ijk}(\mathbf{x}) = O(|\mathbf{x}|^2),\ g_i(\mathbf{x}) = O(|\mathbf{x}|)^2,\ \text{and } h_i(\mathbf{x}) = O(|\mathbf{x}|).$$

Proof. Let us write

$$\rho(X,\mathbf{x})(d - \tfrac{1}{4}\theta_X)\rho(X,\mathbf{x})^{-1} = d + \omega_X,$$

where

$$\omega_X = -(d\alpha_X + \tfrac{1}{4}\theta_X) = \sum_i \omega_i(X,\mathbf{x})\,d\mathbf{x}^i$$

satisfies $\omega_X(\mathscr{R}) = 0$ and depends linearly on $X \in \mathfrak{g}$. As a special case of (1.12), we have

$$\mathscr{L}(\mathscr{R})\omega = d\iota(\mathscr{R})\omega + \iota(\mathscr{R})d\omega = -\tfrac{1}{4}\iota(\mathscr{R})d\theta_X.$$

The one-form $-\frac{1}{4}\iota(\mathscr{R})d\theta_X$ is dual to $\frac{1}{2}\mu^M(X)\mathscr{R}$. Taking the Taylor expansion of both sides of this formula, we see that

$$\omega_i(X,\mathbf{x}) = \tfrac{1}{4}\sum_j (\mu^M(X)\partial_j, \partial_i)_{x_0}\mathbf{x}^j + \langle X, h_i(\mathbf{x})\rangle,$$

where $h_i(\mathbf{x}) = O(|\mathbf{x}|^2)$. Since $\rho(X,\mathbf{x})\big(\nabla^{\mathscr{E},X}_{\partial_i}\big)\rho(X,\mathbf{x})^{-1} = \nabla^{\mathscr{E}}_{\partial_i} + \omega_i$, the lemma follows from Lemma 4.15. □

Let $\langle x \mid e^{-tH(X)} \mid x_0\rangle$ be the heat kernel of the operator $H(X)$; by Theorem 2.48, $\langle x \mid e^{-tH(X)} \mid x_0\rangle$ depends smoothly on $X \in \mathfrak{g}$. We transfer this kernel to the neighbourhood U of $0 \in V$, thinking of it as taking values in $\mathrm{End}(E)$, by writing

$$k(t,\mathbf{x},X) = \rho(X,\mathbf{x})\tau(x_0,x)\langle x \mid e^{-tH(X)} \mid x_0\rangle,$$

where $x = \exp_{x_0}\mathbf{x}$. Since we may identify $\mathrm{End}(E) = C(V^*)\otimes\mathrm{End}(W)$ with $\Lambda V^*\otimes\mathrm{End}(W)$ by means of the full symbol map σ, $k(t,\mathbf{x},X)$ is identified with a $\Lambda V^*\otimes\mathrm{End}(W)$-valued function on $U\times\mathfrak{g}$. Consider the space $\Lambda V^*\otimes\mathrm{End}(W)$ as a $C(V^*)\otimes\mathrm{End}(W)$ module, where the action of $C(V^*)$ on ΛV^* is the usual one $c(\alpha) = \varepsilon(\alpha) - \iota(\alpha)$. The following lemma shows that $k(t,\mathbf{x},X)$ may be characterized as a solution to a heat equation on the open manifold U; the proof is analogous to that of Lemma 4.16.

Lemma 8.14. *Let $L(X)$ be the differential operator on U, with coefficients in $C(V^*)\otimes\mathrm{End}(W)$, defined by the formula*

$$\rho(X,\mathbf{x})^{-1}L(X)\rho(X,\mathbf{x}) = -\sum_i\big((\nabla^{\mathscr{E}}_{e_i} - \tfrac{1}{4}\theta_X(e_i))^2 - \nabla^{\mathscr{E}}_{\nabla_{e_i}e_i} + \tfrac{1}{4}\theta_X(\nabla_{e_i}e_i)\big)$$
$$+ \tfrac{1}{4}r_M + \sum_{i<j}F^{\mathscr{E}/S}(e_i,e_j)c^ic^j + \mu^{\mathscr{E}/S}(X).$$

The function $k(t,\mathbf{x},X) \in C^\infty(\mathbb{R}_+\times U, \Lambda V^*\otimes\mathrm{End}(W))$ satisfies the differential equation

$$(\partial_t + L(X))k(t,\mathbf{x},X) = 0.$$

Using the explicit recurrence formula of Theorem 2.26 and the fact that the coefficients of $L(X)$ are polynomial in X, we see that the coefficients of the asymptotic expansion of $k(t,\mathbf{x},X)$ are polynomial in X.

To prove the Kirillov formula, we will rescale the kernel in a way similar to that used in Section 4.3, except that we will perform an additional rescaling on the Lie algebra variable $X \in \mathfrak{g}$. If $\alpha \in C^\infty(\mathbb{R}_+ \times U \times \mathfrak{g}, \Lambda V^* \otimes \operatorname{End}(W))$, let $\delta_u\alpha$, $(0 < u \leq 1)$, equal

$$(\delta_u\alpha)(t,\mathbf{x},X) = \sum_{i=0}^{n} u^{-i/2}\alpha(ut, u^{1/2}\mathbf{x}, u^{-1}X)_{[i]}.$$

Define the rescaled heat kernel $r(u,t,\mathbf{x},X)$ by

$$r(u,t,\mathbf{x},X) = u^{n/2}(\delta_u k)(t,\mathbf{x},X).$$

We define an operator $L(u,X)$ by the formula $u\delta_u L(X)\delta_u^{-1}$; it is clear that $r(u,t,\mathbf{x},X)$ satisfies the heat equation

$$(\partial_t + L(u,X))r(u,t,\mathbf{x},X) = 0.$$

Using Lemma 8.13, we can give an asymptotic expansion for $L(u,X)$. We will show that

$$L(u,X) = K(X) + O(u^{1/2}),$$

where $K(X)$ is a harmonic oscillator similar to that which entered in Section 4.3.

In order to state the formula for $K(X)$, recall the matrix R with nilpotent entries equal to $\mathsf{R}_{ij} = (R_{x_0}\partial_i, \partial_j) \in \Lambda^2 V^*$ where R_{x_0} is the Riemannian curvature of M at x_0, and the element F of $\Lambda^2 V^* \otimes \operatorname{End}(W)$ obtained by evaluating the twisting curvature $F^{\mathcal{E}/S}$ at the point x_0. Let $\mu^M(X) \in \operatorname{End}(V)$ be the evaluation of the moment of the manifold M at the point x_0 and Lie algebra element $X \in \mathfrak{g}$, and let $\mu^{\mathcal{E}/S}(X) \in \operatorname{End}(W)$ be the evaluation of the relative moment of the Clifford module $\mathcal{E}$ at x_0 and Lie algebra element $X \in \mathfrak{g}$.

Proposition 8.15. *The family of differential operators*

$$L(u,X) = u\delta_u L(X)\delta_u^{-1}$$

acting on $C^\infty(U \times \mathfrak{g}, \Lambda V^* \otimes \operatorname{End}(W))$ *has a limit* $K(X)$ *when* u *tends to* 0, *given by the formula*

$$K(X) = -\sum_i \Big(\partial_i - \tfrac{1}{4}\sum_j (\mathsf{R}_{ij} + \mu^M_{ij}(X))\mathbf{x}_j\Big)^2 + \mathsf{F} + \mu^{\mathcal{E}/S}(X).$$

Proof. Using the fact that $u^{1/2}\delta_u c^i \delta_u^{-1} = \varepsilon(e^i) - u\iota(e^i)$, we see that the differential operator

$$\nabla^{\mathcal{E},X,u}_{\partial_i} = u^{1/2}\delta_u\Big(\rho(X,\mathbf{x})\nabla^{\mathcal{E},X}_{\partial_i}\rho(X,\mathbf{x})^{-1}\Big)\delta_u^{-1} = A_1(u) + A_2(u),$$

where

$$A_1(u) = \partial_i + \tfrac{1}{4} \sum_{j;k<l} R_{ijkl} \mathbf{x}^j (\varepsilon^k - u\iota^k)(\varepsilon^l - u\iota^l) - \sum_j \tfrac{1}{4} \mu^M_{ij}(X) \mathbf{x}^j,$$
$$A_2(u) = u^{-1/2} \sum_{j<k} f_{ijk}(u^{1/2}\mathbf{x})(\varepsilon^j - u\iota^j)(\varepsilon^k - u\iota^k)$$
$$+ u^{1/2} g_i(u^{1/2}\mathbf{x}) + u^{-1/2} \langle h_i(u^{1/2}\mathbf{x}), X \rangle.$$

The functions $f_{ijk}(\mathbf{x})$, $g_i(\mathbf{x})$ and $h_i(\mathbf{x})$ vanish to sufficiently high order in $\mathbf{x}$ that $\lim_{u\to 0} A_2(u) = 0$. Thus, we see that $\nabla^{\mathscr{E},X,u}_{\partial_i}$ has a limit as $u \to 0$, equal to

$$\partial_i + \tfrac{1}{4} \sum_{j;k<l} R_{ijkl} \mathbf{x}^j \varepsilon^k \varepsilon^l - \tfrac{1}{4} \sum_j \mu^M_{ij}(X) \mathbf{x}^j.$$

Since $\mathsf{R}_{ij} = \sum_{k<l} R_{jikl} \varepsilon^k \varepsilon^l$, the above limit is seen to equal

$$\partial_i - \tfrac{1}{4} \sum_j (\mathsf{R}_{ij} + \mu^M_{ij}(X)) \mathbf{x}^j.$$

Let $c^i(u) = u^{1/2} \delta_u c^i \delta_u^{-1} = \varepsilon^i - u\iota^i$. The operator $L(u,X)$ equals the sum of

$$-\sum_i (\nabla^{\mathscr{E},X,u}_{e_i})^2 + \sum_{i<j} F^{\mathscr{E}/S}(e_i,e_j)(u^{1/2}\mathbf{x}) c^i(u) c^j(u) + (\mu^{\mathscr{E}/S}(X))(u^{1/2}\mathbf{x})$$

and

$$u^{1/2} \sum_i \nabla^{\mathscr{E},X,u}_{\nabla_{e_i} e_i} + \tfrac{1}{4} u r_M(u^{1/2}\mathbf{x}).$$

Clearly, these last two terms vanish in the limit $u \to 0$. The first term has the limit

$$-\sum_i \Big(\partial_i - \tfrac{1}{4} \sum_j (\mathsf{R}_{ij} + \mu^M_{ij}(X)) \mathbf{x}^j \Big)^2,$$

while the second and third terms converge to

$$\sum_{i<j} \mathsf{F}(\partial_i, \partial_j) \varepsilon^i \varepsilon^j + \mu^{\mathscr{E}/S}(X)|_{\mathbf{x}=0} = \mathsf{F} + \mu^{\mathscr{E}/S}(X),$$

since the vector fields e_i coincide with ∂_i when $\mathbf{x} = 0$. □

The operator $L(0,X) = K(X)$ is a generalized harmonic oscillator, in the sense of Definition 4.11, associated to the $n \times n$ antisymmetric curvature matrix $\mathsf{R}_{ij} + \mu^M_{ij}(X)$ and the $N \times N$-matrix $\mathsf{F} + \mu^{\mathscr{E}/S}(X)$ (where $N = \dim W$), with coefficients in the commutative algebra $\Lambda^+ V^*$. It follows that there is a unique solution of the heat equation

$$(\partial_t + K(X)) r(t,\mathbf{x},X) = 0$$

such that $\lim_{t\to 0} r(t,\mathbf{x},X) = \delta(\mathbf{x})$, given by the formula

$$(4\pi t)^{-n/2} \det\left(\frac{t(\mathsf{R}+\mu^M(X))/2}{\sinh t(\mathsf{R}+\mu^M(X))/2}\right)^{1/2} \exp(-t(\mathsf{F}+\mu^{\mathscr{E}/S}(X)))$$
$$\times \exp\left(-\frac{1}{4t}\left\langle \mathbf{x} \left| \frac{t(\mathsf{R}+\mu^M(X))}{2} \coth\left(\frac{t(\mathsf{R}+\mu^M(X))}{2}\right) \right| \mathbf{x} \right\rangle\right).$$

If J is an integer, let $\mathbb{C}[\mathfrak{g}]_{(J)}$ be the quotient of the algebra of polynomials on the vector space $\mathfrak{g}$ by the ideal of polynomials of order $J+1$. If $a \in \mathbb{C}[\mathfrak{g}]_{(J)}$, it has a unique polynomial representative of the form $a(X) = \sum_{|\alpha|\leq J} a_\alpha X^\alpha$; we define the norm of a to be $\sup_\alpha |a_\alpha|$.

We can think of $r(u,t,\mathbf{x},X)$ as defining an element

$$r_J(u,t,\mathbf{x},X) \in C^\infty(\mathbb{R}_+ \times U, \Lambda V^* \otimes \operatorname{End}(W)) \otimes \mathbb{C}[\mathfrak{g}]_{(J)}$$

by taking the Taylor series expansion of $r(u,t,\mathbf{x},X)$ at $X=0$ up to order J, in other words,

$$r_J(u,t,\mathbf{x},X) = \sum_{|\alpha|\leq J} \frac{1}{\alpha!} (\partial_X^\alpha r(u,t,\mathbf{x},X))_{X=0} X^\alpha.$$

Lemma 8.16. *There exist polynomials $\gamma_{i,J}(t,\mathbf{x})$ on $\mathbb{R}\times V$, with values in $\Lambda V^* \otimes \operatorname{End}(W) \otimes \mathbb{C}[\mathfrak{g}]_{(J)}$, such that for every integer N, the function*

$$r_J^N(u,t,\mathbf{x},X) = q_t(\mathbf{x}) \sum_{i=-n-2J}^{2N} u^{i/2} \gamma_{i,J}(t,\mathbf{x},X)$$

approximates $r_J(u,t,\mathbf{x},X)$ in the following sense:

$$\left\|\partial_t^j \partial_{\mathbf{x}}^\alpha (r_J(u,t,\mathbf{x},X) - r_J^N(u,t,\mathbf{x},X))\right\|_{\Lambda V^*\otimes \operatorname{End}(W)\otimes\mathbb{C}[\mathfrak{g}]_{(J)}} \leq C(J,N,j,\alpha) u^N,$$

for $0 < u \leq 1$ and $(t,\mathbf{x})$ lying in a compact subset of $(0,1)\times U$. Furthermore, $\gamma_{i,J}(0,0) = 0$ if $i \neq 0$, while $\gamma_{0,J}(0,0) = 1$.

Proof. By Theorem 2.48, there exist $\Lambda V^* \otimes \operatorname{End}(W)$-valued smooth functions ψ_i on $U \times \mathfrak{g}$ such that

$$\left\| k(t,\mathbf{x},X) - q_t(\mathbf{x}) \sum_{i=0}^N t^i \psi_i(\mathbf{x},X) \right\|_{\Lambda V^*\otimes\operatorname{End}(W)} \leq C(N) t^{N-n/2}$$

for $t \in (0,1)$, $\mathbf{x} \in U$, and X in a bounded neighbourhood of $0 \in \mathfrak{g}$. Furthermore, we can differentiate this asymptotic expansion with respect to the parameter X. It follows that there exist $\Lambda V^* \otimes \operatorname{End}(W) \otimes \mathbb{C}[\mathfrak{g}]_{(J)}$-valued smooth functions $\psi_{i,J}$ on U such that

$$\left\| k_J(t,\mathbf{x},X) - q_t(\mathbf{x}) \sum_{i=0}^N t^i \psi_{i,J}(\mathbf{x},X) \right\|_{\Lambda V^*\otimes\operatorname{End}(W)\otimes\mathbb{C}[\mathfrak{g}]_{(J)}} \leq C(N,J) t^{N-n/2}$$

for $t \in (0,1)$ and $\mathbf{x} \in U$, where $k_J(t,\mathbf{x},X)$ is the Taylor expansion of $k(t,\mathbf{x},X)$ around $X=0$ to order J. As in the proof of Proposition 4.17, we can suppose that $\psi_{i,J}$ is polynomial in $\mathbf{x}$.

Rescaling by $u \in (0,1)$, we obtain

$$\left\| r_J(u,t,\mathbf{x},X)_{[p]} - u^{-p/2} q_t(\mathbf{x}) \sum_{i=0}^{N} (ut)^i \Psi_{i,J}(u^{1/2}\mathbf{x}, u^{-1}X)_{[p]} \right\| \leq C(N,J) u^{N-p/2-J} t^{N-n/2}.$$

Thus, we see that the function $u \mapsto r_J(u,t,\mathbf{x},X)$ has an asymptotic expansion of the desired form, where $\gamma_{i,J}(t,\mathbf{x},X)_{[p]}$ is the coefficient of $u^{i/2}$ in

$$u^{-p/2} \sum_{j=0}^{\infty} (ut)^j \Psi_{j,J}(u^{1/2}\mathbf{x}, u^{-1}X)_{[p]}.$$

Since $\psi_{j,J}$ is a polynomial in X of order at most J, we see that $(\gamma_{i,J})_{[p]} = 0$ if $i \leq -p/2 - J$. The argument for the derivatives is similar. Since

$$\sum_i u^{i/2} \gamma_{i,J}(0,0,X)_{[p]} = u^{-p/2} \Psi_{0,J}(0, u^{-1}X)_{[p]},$$

we see that $\gamma_{i,J}(0,0,X)_{[p]} = 0$ unless $i = p = 0$, and that $\gamma_{0,J}(0,0,X) = 1$. □

It is clear that $r_J(u,t,\mathbf{x},X)$ satisfies the differential equation

$$(\partial_t + L_J(u,X)) r_J(u,t,\mathbf{x},X) = 0, \tag{8.3}$$

where $L_J(u,X)$ is the action induced by the differential operator $L(u,X)$ on $C^\infty(U) \otimes \Lambda V^* \otimes \mathrm{End}(W) \otimes \mathbb{C}[\mathfrak{g}]_{(J)}$. Let us show how this implies that $r_J(u,t,\mathbf{x},X)$ has a limit as $u \to 0$, using the fact that $L(u,X) = K(X) + O(u^{1/2})$. Let ℓ be the largest integer for which $\gamma_{-\ell,J}(t,\mathbf{x},X)$ is non-zero. Expanding (8.3) in a Laurent series in $u^{1/2}$, we see that the leading term $u^{-\ell/2} q_t(\mathbf{x}) \gamma_{-\ell,J}(t,\mathbf{x},X)$ of the asymptotic expansion of $r_J(u,t,\mathbf{x},X)$ satisfies the heat equation $(\partial_t + K(X))(q_t(\mathbf{x})\gamma_{-\ell,J}(t,\mathbf{x},X)) = 0$. Since formal solutions of the heat equation $(\partial_t + K(X))(q_t(\mathbf{x})\gamma(t,\mathbf{x},X)) = 0$ for the harmonic oscillator are uniquely determined by $\gamma(0,0,X)$, and since $\gamma_{i,J}(0,0,X) = 0$ for $i < 0$, we see that $\gamma_{-\ell,J}$ must equal zero unless $\ell \leq 0$. In particular, we see that there are no poles in the Laurent expansion of $r_J(u,t,\mathbf{x},X)$ in powers of $u^{1/2}$.

We also learn from this argument that the leading term of the expansion of $r_J(u,t,\mathbf{x})$, namely $r_J(0,t,\mathbf{x},X) = q_t(\mathbf{x})\gamma_{0,J}(t,\mathbf{x},X)$, satisfies the heat equation for the operator $L(0,X) = K(X)$, with initial condition

$$\gamma_{0,J}(t = 0, \mathbf{x} = 0, X) = 1.$$

Applying Theorem 4.13 to the operator $K(X)$, we obtain the following result.

Proposition 8.17. *The limit $\lim_{u \to 0} r_J(u,t,\mathbf{x},X)$ exists, and is given by reduction modulo X^{J+1} of the formula*

$$(4\pi t)^{-n/2} \det\left(\frac{t(\mathsf{R}+\mu^M(X))/2}{\sinh t(\mathsf{R}+\mu^M(X))/2}\right)^{1/2} \times$$
$$\times \exp\left(-\frac{1}{4t}\left\langle \mathbf{x} \left| \frac{t(\mathsf{R}+\mu^M(X))/2}{\coth(t(\mathsf{R}+\mu^M(X))/2)} \right| \mathbf{x} \right\rangle - t(\mathsf{F}+\mu^{\mathscr{E}/S}(X))\right).$$

Since this is true for arbitrary J, we have only to set $t = 1$ and $\mathbf{x} = 0$ to obtain Theorem 8.10.

Bibliographic Notes

The Kirillov formula Theorem 8.2 was proved in Berline-Vergne [23], using the localization theorem for equivariant differential forms and the equivariant index theorem of Atiyah-Segal-Singer. The local version of the theorem proved in Section 3 is a generalization of the results of Bismut [32]; in particular, the definition of the "quantized equivariant curvature" $H(X)$, and the proofs of Propositions 8.11 and 8.12 are due to him.

In Sections 9.4 and 10.7, we will explain the relationship between the Kirillov formula and the index theorem for families with compact structure group; the relationship comes about by considering the family $P \times_G M$, where P is an approximation of the universal bundle EG for G. We will see that the Kirillov formula is equivalent to Bismut's local family index theorem, in the case of families with compact structure group.

Using the language of bouquets (see the Additional Remarks at the end of Chapter 7) we may formulate a theorem which combines the equivariant index formula of Chapter 6 with the Kirillov formula Theorem 8.2 (Berline-Vergne [23]). Introducing a bouquet $\mathrm{bch}(\mathscr{E}/S)$ which collects the equivariant relative Chern characters of the bundle $\mathscr{E}$ over the fixed point sets $M(g)$, we have

$$\mathrm{ind}_G(\mathsf{D}) = (2\pi i)^{-\dim(M)/2} \int_M^b \mathrm{bch}(\mathscr{E}/S).$$

This type of formula may also be used to express the index of a transversally elliptic operator (Berline-Vergne [26]; see also Duflo [46]), and to reformulate results on characters of non-compact groups (Vergne [104]).

9

The Index Bundle

Consider a manifold B and a finite dimensional Hermitian superbundle $\mathscr{H} = \mathscr{H}^+ \oplus \mathscr{H}^- \to B$. Let D be an odd endomorphism of $\mathscr{H}$ with components $\mathsf{D}^\pm : \mathscr{H}^\pm \to \mathscr{H}^\mp$, such that $\ker(\mathsf{D})$ has constant rank, so that the family of superspaces $(\ker(\mathsf{D}^z) \mid z \in B)$ forms a superbundle over B, called the index bundle of D. Let $\mathbb{A}$ be a superconnection on $\mathscr{H}$ with zero-degree term equal to the odd endomorphism D of $\mathscr{H}$, and let $\mathscr{F} = \mathbb{A}^2 \in \mathscr{A}(B, \mathrm{End}(\mathscr{H}))$ be the curvature of $\mathbb{A}$; all notations are as in Section 1.4. If we assume that D is self-adjoint and has kernel of constant rank, then at the level of cohomology,

$$\mathrm{ch}(\mathbb{A}) = \mathrm{ch}(\ker(\mathsf{D})).$$

In fact, this equation has a refinement at the level of differential forms.

If $t > 0$, we may define a new superconnection on $\mathscr{H}$, by the formula

$$\mathbb{A}_t = t^{1/2}\mathsf{D} + \mathbb{A}_{[1]} + t^{-1/2}\mathbb{A}_{[2]} + \dots,$$

where $\mathbb{A}_{[1]}$ is the connection associated to $\mathbb{A}$. This superconnection has Chern character form

$$\mathrm{ch}(\mathbb{A}_t) = \mathrm{Str}(e^{-\mathbb{A}_t^2}).$$

Let P_0 be the projection from $\mathscr{H}$ onto the subbundle $\ker(\mathsf{D})$. We may also define a connection on the superbundle $\ker(\mathsf{D})$ by projection,

$$\nabla_0 = P_0 \cdot \mathbb{A}_{[1]} \cdot P_0.$$

In Section 1, we will prove the following result:

$$\lim_{t\to\infty} \mathrm{ch}(\mathbb{A}_t) = \mathrm{ch}(\nabla_0).$$

Intuitively speaking, as t tends to infinity, the supertrace

$$\mathrm{ch}(\mathbb{A}_t) = \mathrm{Str}\big(e^{-\mathbb{A}_t^2}\big) = \mathrm{Str}\big(e^{-t\mathsf{D}^2 + O(t^{1/2})}\big)$$

is pushed onto the sub-bundle $\ker(\mathsf{D})$. The proof, which is given in Section 1, is substantially more complicated than this simple idea.

Our real interest lies in an infinite-dimensional version of this, in which the family D of operators on the finite dimensional bundle $\mathscr{H}$ is replaced by a family of Dirac operators. Our technical tools will be the estimates of Chapter 2 on the heat kernels of generalized Laplacians, and the spectral theorem.

Consider a fibre bundle $\pi : M \to B$ with compact fibres. We denote by $T(M/B)$ the vertical tangent bundle, which is the sub-bundle of TM consisting of vertical vectors, and by $|\Lambda_\pi|$ the vertical density bundle, which is isomorphic to $|\Lambda_M| \otimes \pi^*|\Lambda_B|^{-1}$. If $\mathscr{E}$ is a superbundle on M, we denote by $\pi_*\mathscr{E}$ the infinite-dimensional superbundle on B whose fibre at $z \in B$ is the Fréchet space

$$(\pi_*\mathscr{E})_z = \Gamma(M_z, \mathscr{E}_z \otimes |\Lambda_{M_z}|^{1/2});$$

here, $M_z = \pi^{-1}(z)$ is the fibre over $z \in B$, and $\mathscr{E}_z$ is the restriction of the bundle $\mathscr{E}$ to M_z. The space of sections of $\pi_*\mathscr{E}$ is defined by setting

$$\Gamma(B, \pi_*\mathscr{E}) = \Gamma(M, \mathscr{E} \otimes |\Lambda_\pi|^{1/2}),$$

and the space of differential forms on B with values in $\pi_*\mathscr{E}$ by

$$\mathscr{A}(B, \pi_*\mathscr{E}) = \Gamma(M, \pi^*(\Lambda T^*B) \otimes \mathscr{E} \otimes |\Lambda_\pi|^{1/2}).$$

Let $\mathsf{D} = (\mathsf{D}^z \mid z \in B)$ be a smooth family of Dirac operators. By a superconnection on $\pi_*\mathscr{E}$ adapted to D, we mean a differential operator $\mathbb{A}$ on the bundle $\mathbb{E} = \pi^*(\Lambda T^*B) \otimes \mathscr{E} \otimes |\Lambda_\pi|^{1/2}$ over M, of odd parity, such that

1. (Leibniz's rule) if $v \in \mathscr{A}(B)$ and $s \in \Gamma(M, \mathbb{E})$, then
$$\mathbb{A}(vs) = (d_B v)s + (-1)^{|v|} v\, \mathbb{A}s;$$
2. $\mathbb{A} = \mathsf{D} + \sum_{i=1}^{\dim(B)} \mathbb{A}_{[i]}$, where $\mathbb{A}_{[i]} : \mathscr{A}^\bullet(B, \pi_*\mathscr{E}) \to \mathscr{A}^{\bullet+i}(B, \pi_*\mathscr{E})$.

In the appendix to this chapter, we show that the curvature of the superconnection $\mathbb{A}$ has a smooth heat kernel $\langle x \mid e^{-\mathbb{A}^2} \mid y \rangle$ varying smoothly as a function of $z \in B$ and $x, y \in M_z$. Because of this we can extend the definition of the Chern character of a superconnection to the infinite dimensional bundle $\pi_*\mathscr{E} \to B$.

Suppose that the dimension of $\ker(\mathsf{D}^z)$ is independent of $z \in B$. In this case, the vector spaces $\ker(\mathsf{D}^z)$ combine to form a vector bundle $\operatorname{ind}(\mathsf{D})$, which is the index bundle of the family D; the index of a single Dirac operator is a special case of this construction. In Section 2, we show that, as in the finite-dimensional case, we can define a connection on the index bundle by the formula $\nabla_0 = P_0 \cdot \mathbb{A}_{[1]} \cdot P_0$, and the result of this chapter is a formula due to Bismut for the Chern form $\operatorname{ch}(\nabla_0)$, which is a generalization of the McKean-Singer formula. Just as in the finite-dimensional case, and by the same proof, we will show in Section 3 that

$$\lim_{t \to \infty} \operatorname{ch}(\mathbb{A}_t) = \operatorname{ch}(\nabla_0).$$

In Section 4, we consider the special case where the family D is associated to a principal bundle $P \to B$ with compact structure group G. In this case, it is possible to calculate directly the Chern character form $\mathrm{ch}(\ker(\mathsf{D})) = \mathrm{ch}(\nabla_0)$ using the Kirillov formula of the last chapter, and we obtain the formula

$$\mathrm{ch}(\ker(\mathsf{D})) = (2\pi i)^{-\dim(M/B)/2} \int_{M/B} \hat{A}(M/B)\,\mathrm{ch}(\mathscr{E}/S),$$

where $\dim(M/B) = \dim(M) - \dim(B)$ (we assume that M is connected) and $\hat{A}(M/B)$ is the $\hat{A}$-genus of the vertical tangent bundle $T(M/B)$. This formula is true at the level of cohomology for any family of Dirac operators, as we will show in the next chapter.

Section 5 is devoted to studying the case when the vector spaces $\ker(\mathsf{D}^z)$ vary in dimension. The index bundle $\mathrm{ind}(\mathsf{D})$ is now only locally a vector bundle, but it turns out that it has a well-defined Chern character in de Rham cohomology, and that $\mathrm{ch}(\mathbb{A}_t)$ lies in the same cohomology class for all $t > 0$ if $\mathbb{A}$ is a superconnection on $\pi_*\mathscr{E}$ such that $\mathbb{A}_{[0]} = \mathsf{D}$. In Chapter 10, in the case where the family D is associated to a Clifford connection, we will present Bismut's construction of a very particular superconnection for which $\mathrm{ch}(\mathbb{A}_t)$ possesses a limit $\lim_{t\to 0}\mathrm{ch}(\mathbb{A}_t)$; combined with the results of this chapter, this enables us to calculate the Chern character of the index bundle $\mathrm{ind}(\mathsf{D})$.

In Section 6, we define the zeta-function and zeta-function determinant of a generalized Laplacian. Using this, we define in Section 7 the determinant line bundle $\det(\pi_*\mathscr{E}, \mathsf{D})$ associated to a smooth family of Dirac operators, and, in the case in which the index of D^z is zero, a section $\det(\mathsf{D}^+)$ of $\det(\pi_*\mathscr{E}, \mathsf{D})$. The determinant line bundle has a canonical metric, known as the Quillen metric, and we define a connection on $\det(\pi_*\mathscr{E}, \mathsf{D})$ preserving this metric given the data of a connection on the bundle $M \to B$ and a Hermitian connection on the bundle $\mathscr{E}$ over M.

9.1 The Index Bundle in Finite Dimensions

In this section, we will discuss in detail the finite-dimensional analogue of the index bundle, since it is easier to visualize what is happening here than in the infinite-dimensional case. Notation will be as in Sections 1.4 and 1.5.

Consider a manifold B and a superbundle $\mathscr{H} = \mathscr{H}^+ \oplus \mathscr{H}^- \to B$. Let D be an odd endomorphism of $\mathscr{H}$ with components $\mathsf{D}^\pm : \mathscr{H}^\pm \to \mathscr{H}^\mp$, such that $\ker(\mathsf{D})$ has constant rank, so that the family of superspaces $(\ker(\mathsf{D}^z) \mid z \in B)$ forms a superbundle over B, called the **index bundle** of D. (From now on, points of B will always be denoted by z.) Suppose that $\mathscr{H}^+$ and $\mathscr{H}^-$ have Hermitian structures for which D^- is the adjoint of D^+ (so that D is self-adjoint). If $x \in \mathscr{H}$ and $y \in \ker(\mathsf{D})$, then by the assumption that D is self-adjoint,

$$\langle \mathsf{D}x, y\rangle = \langle x, \mathsf{D}y\rangle = 0.$$

Denote by $\mathscr{H}_0 \subset \mathscr{H}$ the superbundle $\ker(\mathsf{D})$, graded by $\mathscr{H}_0^\pm = \ker(\mathsf{D}^\pm)$, and by $\mathscr{H}_1$ the the image $\mathrm{im}(\mathsf{D}) \subset \mathscr{H}$ of D. The above equation shows that $\mathscr{H}_1$ is the orthogonal complement of $\mathscr{H}_0$,

$$\mathscr{H}^+ = \mathscr{H}_0^+ \oplus \mathscr{H}_1^+,$$
$$\mathscr{H}^- = \mathscr{H}_0^- \oplus \mathscr{H}_1^-,$$

and hence that the bundle $\ker(\mathsf{D}^-)$ is isomorphic to $\operatorname{coker}(\mathsf{D}^+) = \mathscr{H}^- / \operatorname{im}(\mathsf{D}^+)$, and the bundles $\mathscr{H}_1^+$ and $\mathscr{H}_1^-$ are isomorphic.

Let $\mathbb{A} = \mathbb{A}_{[0]} + \mathbb{A}_{[1]} + \mathbb{A}_{[2]} + \dots$ be a superconnection on $\mathscr{H}$, with curvature $\mathscr{F} = \mathbb{A}^2 \in \mathscr{A}(B, \operatorname{End}(\mathscr{H}))$. By (1.34), the Chern character form of $\mathbb{A}$,

$$\operatorname{ch}(\mathbb{A}) = \operatorname{Str}(e^{-\mathscr{F}}),$$

is equal in de Rham cohomology to the difference of the Chern characters of the bundles $\mathscr{H}^+$ and $\mathscr{H}^-$.

Let $\mathbb{A}$ be a superconnection whose zero-degree term $\mathbb{A}_{[0]}$ equals the odd endomorphism D of $\mathscr{H}$. If we assume that D is self-adjoint and has kernel of constant rank, then by the above discussion and (1.34), we have the equality in cohomology

$$\begin{aligned}
\operatorname{ch}(\mathbb{A}) &= \operatorname{ch}(\mathscr{H}^+) - \operatorname{ch}(\mathscr{H}^-) \\
&= \operatorname{ch}(\mathscr{H}_0^+) + \operatorname{ch}(\mathscr{H}_1^+) - \operatorname{ch}(\mathscr{H}_0^-) - \operatorname{ch}(\mathscr{H}_1^-) \\
&= \operatorname{ch}(\mathscr{H}_0^+) - \operatorname{ch}(\mathscr{H}_0^-) = \operatorname{ch}(\ker(\mathsf{D})).
\end{aligned}$$

We will refine this equation to the level of differential forms.

Let P_0 be the orthogonal projection of $\mathscr{H}$ on $\mathscr{H}_0$, and let $P_1 = 1 - P_0$ be the orthogonal projection of $\mathscr{H}$ on $\mathscr{H}_1$. The endomorphism D^+ gives an isomorphism between $\mathscr{H}_1^+$ and $\mathscr{H}_1^-$.

Let $\tilde{\mathbb{A}}$ be the superconnection

$$\tilde{\mathbb{A}} = P_0 \cdot \mathbb{A} \cdot P_0 + P_1 \cdot \mathbb{A} \cdot P_1,$$

which preserves the spaces $\mathscr{A}(B, \mathscr{H}_0)$ and $\mathscr{A}(B, \mathscr{H}_1) \subset \mathscr{A}(B, \mathscr{H})$.

We will make constant use of the following notation: if $K \in \mathscr{A}(B, \operatorname{End}(\mathscr{H}))$, we write

$$K = \begin{vmatrix} \alpha & \beta \\ \gamma & \delta \end{vmatrix},$$

which means simply that

$$\begin{vmatrix} \alpha & \beta \\ \gamma & \delta \end{vmatrix} = \begin{vmatrix} P_0 K P_0 & P_0 K P_1 \\ P_1 K P_0 & P_1 K P_1 \end{vmatrix},$$

with $\alpha \in \Gamma(B, \operatorname{End}(\mathscr{H}_0))$ etc.

Since $\tilde{\mathbb{A}}$ commutes with P_0, we see that its curvature has the form

$$\widetilde{\mathscr{F}} = \tilde{\mathbb{A}}^2 = \begin{vmatrix} R & 0 \\ 0 & S \end{vmatrix}.$$

Denote by ∇_0 the connection on the bundle $\mathscr{H}_0$ given by projecting the connection $\mathbb{A}_{[1]}$ onto the bundle $\mathscr{H}_0$:

$$\nabla_0 = P_0 \cdot \mathbb{A}_{[1]} \cdot P_0.$$

We filter the algebra $\mathscr{M} = \mathscr{A}(B, \mathrm{End}(\mathscr{H}))$ by the subspaces

$$\mathscr{M}_i = \sum_{j \geq i} \mathscr{A}^j(B, \mathrm{End}(\mathscr{H})).$$

Lemma 9.1. *The differential form R lies in $\mathscr{M}_2$, and the curvature of the connection ∇_0 equals $R_{[2]}$.*

Proof. The superconnection $\mathbb{A}_0 = P_0 \cdot \mathbb{A} \cdot P_0$ on the bundle $\mathscr{H}_0$ has curvature $\mathbb{A}_0^2 = R$. Since $P_0 \cdot \mathbb{A}_{[0]} \cdot P_0 = P_0 \cdot D \cdot P_0 = 0$, we see that

$$\mathbb{A}_0 = \nabla_0 + \sum_{i \geq 2} P_0 \cdot \mathbb{A}_{[i]} \cdot P_0,$$

from which the lemma is clear. □

For $t > 0$, let δ_t be the automorphism of $\mathscr{A}(B, \mathscr{H})$ which acts on $\mathscr{A}^i(B, \mathscr{H})$ by multiplication by $t^{-i/2}$. Then $\mathbb{A}_t = t^{1/2} \delta_t \cdot \mathbb{A} \cdot \delta_t^{-1}$ is again a superconnection on $\mathscr{H}$, and the decomposition of $\mathbb{A}_t$ into homogeneous components with respect to the exterior degree is given by the formula

$$\mathbb{A}_t = t^{1/2} \mathbb{A}_{[0]} + \mathbb{A}_{[1]} + t^{-1/2} \mathbb{A}_{[2]} + \dots .$$

The curvature $\mathscr{F}_t = \mathbb{A}_t^2$ of the superconnection $\mathbb{A}_t$ is the operator

$$\mathscr{F}_t = t \delta_t \cdot \mathscr{F} \cdot \delta_t^{-1},$$

and the cohomology class of $\mathrm{ch}(\mathbb{A}_t) = \mathrm{Str}(e^{-\mathscr{F}_t})$ is independent of t, and is equal to the difference of the Chern characters $\mathrm{ch}(\mathscr{H}_0^+) - \mathrm{ch}(\mathscr{H}_0^-)$ for all $t > 0$. We will study the limit of $\mathrm{ch}(\mathbb{A}_t)$ as $t \to \infty$; it is remarkable that the following stronger result holds.

Theorem 9.2. *Let $\mathscr{H} = \mathscr{H}^+ \oplus \mathscr{H}^-$ be a Hermitian super-vector bundle and let D be a Hermitian odd endomorphism of $\mathscr{H}$ whose kernel has constant rank. Let $\mathbb{A}$ be a superconnection on $\mathscr{H}$ with zero-degree term equal to D. For $t > 0$, let*

$$\mathbb{A}_t = t^{1/2} \delta_t \cdot \mathbb{A} \cdot \delta_t^{-1} = t^{1/2} \mathsf{D} + \mathbb{A}_{[1]} + t^{-1/2} \mathbb{A}_{[2]} + \dots$$

be the rescaled superconnection, with curvature $\mathscr{F}_t$. Then for t large,

$$\left\| e^{-\mathscr{F}_t} - e^{-R_{[2]}} \right\|_\ell = O(t^{-1/2})$$

uniformly on compact subsets of B and for all C^ℓ-norms.

Proof. We start by proving the following lemma.

Lemma 9.3. *1. Under the decomposition $\mathscr{H} = \mathscr{H}_0 \oplus \mathscr{H}_1$, the curvature $\mathscr{F}$ can be written as*

$$\mathscr{F} = \begin{vmatrix} X & Y \\ Z & T \end{vmatrix} \in \begin{vmatrix} \mathscr{M}_2 & \mathscr{M}_1 \\ \mathscr{M}_1 & \mathscr{M}_0 \end{vmatrix}.$$

2. *The endomorphism $T_{[0]} \in \Gamma(B, \operatorname{End}(\mathscr{H}_1))$ is equal to $P_1 \cdot D^2 \cdot P_1$ and is positive definite.*
3. *Denote the inverse of $T_{[0]}$ on $\mathscr{H}_1$ by G. The curvature $R_{[2]}$ of the connection ∇_0 on $\mathscr{H}_0$ equals*

$$R_{[2]} = X_{[2]} - Y_{[1]} G Z_{[1]}.$$

Proof. We write $\mathbb{A} = \tilde{\mathbb{A}} + \omega$, where $\omega = P_0 \cdot \mathbb{A} \cdot P_1 + P_1 \cdot \mathbb{A} \cdot P_0 \in \mathscr{M}_1$:

$$\omega = \begin{vmatrix} 0 & \mu \\ \nu & 0 \end{vmatrix}.$$

Expanding the right-hand side of the equation

$$\mathscr{F} = (\tilde{\mathbb{A}} + \omega)^2 = \widetilde{\mathscr{F}} + [\tilde{\mathbb{A}}, \omega] + \omega \wedge \omega,$$

we obtain

$$\begin{aligned} \mathscr{F} &= \begin{vmatrix} R + \mu\nu & P_0[\tilde{\mathbb{A}}, \mu] P_1 \\ P_1[\tilde{\mathbb{A}}, \nu] P_0 & S + \nu\mu \end{vmatrix} \\ &\equiv \begin{vmatrix} R_{[2]} + \mu_{[1]}\nu_{[1]} & \mu_{[1]} D \\ D\nu_{[1]} & D^2 \end{vmatrix} \mod \begin{vmatrix} \mathscr{M}_3 & \mathscr{M}_2 \\ \mathscr{M}_2 & \mathscr{M}_1 \end{vmatrix}. \end{aligned}$$

If we write

$$\mathscr{F} = \begin{vmatrix} X & Y \\ Z & T \end{vmatrix},$$

we see that

$$X_{[2]} - Y_{[1]} G Z_{[1]} = (R_{[2]} + \mu_{[1]}\nu_{[1]}) - (\mu_{[1]} D) G (D\nu_{[1]}) = R_{[2]}. \qquad \square$$

The following lemma is the key step in the proof. We will use the fact that the space of matrices g of the form

$$g = 1 + K, \quad \text{where } K \in \mathscr{M}_1,$$

form a group; to obtain the inverse of such a matrix, we use the formula (which is a finite sum, since B has finite dimension),

$$(1+K)^{-1} = 1 + \sum_{k=1}^{\infty} (-K)^k.$$

Lemma 9.4. *There exists a matrix g with $g - 1 \in \mathscr{M}_1$, such that*

$$g \mathscr{F} g^{-1} = g \begin{vmatrix} X & Y \\ Z & T \end{vmatrix} g^{-1} = \begin{vmatrix} U & 0 \\ 0 & V \end{vmatrix}.$$

Furthermore

$$\begin{aligned} U &\equiv X - YGZ \pmod{\mathscr{M}_3}, \\ V &\equiv T \pmod{\mathscr{M}_1}. \end{aligned}$$

Proof. To construct a matrix g which puts $\mathscr{F}$ in diagonal form, we argue by induction on $\dim(B)-i$. Thus, assume that we have found g_i such that

$$g_i \mathscr{F} g_i^{-1} = \begin{vmatrix} X_i & Y_i \\ Z_i & T_i \end{vmatrix} \in \begin{vmatrix} \mathscr{M}_2 & \mathscr{M}_i \\ \mathscr{M}_i & \mathscr{M}_0 \end{vmatrix},$$

with $T_i \equiv \mathsf{D}^2 \pmod{\mathscr{M}_1}$; in particular,

$$\begin{vmatrix} 0 & 0 \\ 0 & 1-GT_i \end{vmatrix} \in \mathscr{M}_1.$$

We write

$$\begin{vmatrix} 1 & -Y_iG \\ GZ_i & 1 \end{vmatrix} \cdot \begin{vmatrix} X_i & Y_i \\ Z_i & T_i \end{vmatrix} \cdot \begin{vmatrix} 1 & -Y_iG \\ GZ_i & 1 \end{vmatrix}^{-1} = \begin{vmatrix} \tilde{X}_i & \tilde{Y}_i \\ \tilde{Z}_i & \tilde{T}_i \end{vmatrix}.$$

Since $\begin{vmatrix} 0 & -Y_iG \\ GZ_i & 0 \end{vmatrix} \in \mathscr{M}_i$, we see that

$$\begin{vmatrix} 1 & -Y_iG \\ GZ_i & 1 \end{vmatrix}^{-1} - \begin{vmatrix} 1 & Y_iG \\ -GZ_i & 1 \end{vmatrix} \in \mathscr{M}_{2i}.$$

We have the following explicit formulas for $\tilde{X}_i$, $\tilde{Y}_i$, $\tilde{Z}_i$, and $\tilde{T}_i$ mod $\mathscr{M}_{2i}$:

$$\begin{aligned}
\tilde{X}_i &\equiv X_i - 2(Y_iG)Z_i + (Y_iG)T_i(GZ_i) \\
&\equiv X_i \pmod{\mathscr{M}_{2i}}; \\
\tilde{Y}_i &\equiv Y_i(1-GT_i) + (X_i - (Y_iG)Z_i)(Y_iG) \in \mathscr{M}_{i+1}; \\
\tilde{Z}_i &\equiv (1-T_iG)Z_i + (GZ_i)X_i - (GZ_i)Y_i(GZ_i) \in \mathscr{M}_{i+1}; \\
\tilde{T}_i &\equiv T_i + (GZ_i)X_i(Y_iG) + Z_i(Y_iG) + (GZ_i)Y_i \\
&\equiv T_i \pmod{\mathscr{M}_1}.
\end{aligned}$$

Thus, we can continue the induction.

Now, suppose that we have a matrix g of the required form which diagonalizes $\mathscr{F}$. This implies that

$$\begin{vmatrix} 1+K & M \\ N & 1+L \end{vmatrix} \cdot \begin{vmatrix} X & Y \\ Z & T \end{vmatrix} = \begin{vmatrix} U & 0 \\ 0 & V \end{vmatrix} \cdot \begin{vmatrix} 1+K & M \\ N & 1+L \end{vmatrix},$$

for some $\begin{vmatrix} K & M \\ N & L \end{vmatrix} \in \mathscr{M}_1$, from which we obtain the equation

$$\begin{vmatrix} X+KX+MZ & Y+KY+MT \\ NX+Z+LZ & NY+T+LT \end{vmatrix} = \begin{vmatrix} U(1+K) & UM \\ VN & V(1+L) \end{vmatrix}.$$

Since X is in $\mathscr{M}_2$ and K, L, M, N, Y and Z are all in $\mathscr{M}_1$, we see that

(1) $V = (T+LT+NY)(1+L)^{-1} \equiv T \pmod{\mathscr{M}_1}$,
(2) hence $GV \equiv 1 \pmod{\mathscr{M}_1}$,

(3) $U = (X + KX + MZ)(1+K)^{-1} \equiv X + MZ \pmod{\mathscr{M}_3}$,
(4) $Y + MT = UM - KY \in \mathscr{M}_2$,
(5) hence, multiplying on the right by G, $M \equiv -YG \pmod{\mathscr{M}_2}$.

Combining the last two formulas, we see that

$$U \equiv X - YGZ \pmod{\mathscr{M}_3}.$$ □

By this lemma, we may write

$$e^{-t\delta_t(\mathscr{F})} = \delta_t(g)^{-1} \begin{vmatrix} e^{-t\delta_t(U)} & 0 \\ 0 & e^{-t\delta_t(V)} \end{vmatrix} \delta_t(g).$$

Lemma 9.5. *We have the estimate* $|e^{-t\delta_t(V)}| \leq Ce^{-\varepsilon t}$ *for some positive* ε*, as well as for all of its derivatives.*

Proof. Essentially, this is true because $V_{[0]} = T_{[0]} = \mathsf{D}^2$ is positive definite on $\mathscr{H}_1$. The proof uses the Volterra series for $e^{-t\delta_t(V)}$ (see 2.5): if $V = \mathsf{D}^2 + A$, then

$$e^{-t\delta_t(V)} = \sum_{k \geq 0} (-t)^k I_k,$$

where

$$I_k = \int_{\Delta_k} e^{-\sigma_0 t \mathsf{D}^2} \delta_t(A) e^{-\sigma_1 t \mathsf{D}^2} \delta_t(A) \dots e^{-\sigma_{k-1} t \mathsf{D}^2} \delta_t(A) e^{-\sigma_k t \mathsf{D}^2}\, d\sigma_1 \dots d\sigma_k.$$

On the simplex

$$\Delta_k = \{(\sigma_0, \dots, \sigma_k) \in \mathbb{R}^{k+1} \mid \textstyle\sum_{i=0}^k \sigma_i = 1, \sigma_i \geq 0\},$$

one of the σ_i must be greater than $(k+1)^{-1}$. Since the operator $e^{-\sigma_i t \mathsf{D}^2}$ decays exponentially when $t \to \infty$, while the operators $e^{-\sigma_j t \mathsf{D}^2}$, $j \neq i$, are bounded, we see that I_k decays exponentially. The sum is finite, since I_k is a sum of terms of degree at least k with respect to the grading of the differential forms on B. □

Using this lemma and the fact that $t\delta_t(U) = R_{[2]} + O(t^{-1/2})$, which is a consequence of Lemma 9.4, it follows that

$$e^{-t\delta_t(\mathscr{F})} = \delta_t(g)^{-1} \begin{vmatrix} e^{-R_{[2]}} + O(t^{-1/2}) & 0 \\ 0 & 0 \end{vmatrix} \delta_t(g) + O(e^{-\varepsilon t}).$$

Both $\delta_t(g)$ and $\delta_t(g)^{-1} = \delta_t(g^{-1})$ have the form

$$\begin{vmatrix} 1 + O(t^{-1/2}) & O(t^{-1/2}) \\ O(t^{-1/2}) & 1 + O(t^{-1/2}) \end{vmatrix}.$$

It follows that

$$e^{-t\delta_t(\mathscr{F})} = \begin{vmatrix} e^{-R_{[2]}} + O(t^{-1/2}) & O(t^{-1/2}) \\ O(t^{-1/2}) & O(t^{-1}) \end{vmatrix}. \tag{9.1}$$

The argument is similar for the derivatives with respect to the base, using the formula of Theorem 2.48. □

Corollary 9.6. *The limit* $\lim_{t\to\infty} \mathrm{ch}(\mathbb{A}_t)$ *exists, and equals the Chern character of the connection* ∇_0 *on the superbundle* $\mathscr{H}_0 = \ker(\mathsf{D})$.

We will now give an explicit homotopy between $\mathrm{ch}(\mathbb{A}_t)$ and $\lim_{t\to\infty} \mathrm{ch}(\mathbb{A}_t)$. Let us write

$$\alpha(t) = \mathrm{Str}\Big(\frac{d\mathbb{A}_t}{dt} e^{-\mathbb{A}_t^2}\Big) \in \mathscr{A}(B).$$

The transgression formula (1.33) states that

$$\frac{d\,\mathrm{ch}(\mathbb{A}_s)}{ds} = -d\alpha(s).$$

This may also be seen by the following construction, which is valid for any smooth family $\mathbb{A}_s$ of superconnections on a superbundle $\mathscr{E} \to B$. Let $\widetilde{B} = B \times \mathbb{R}_+$, and let $\tilde{\mathscr{E}}$ be the superbundle $\mathscr{E} \times \mathbb{R}_+$ over $\widetilde{B}$, which is the pull-back to $\widetilde{B}$ of $\mathscr{E}$. Define a superconnection $\tilde{\mathbb{A}}$ on $\tilde{\mathscr{E}}$ by the formula

$$(\tilde{\mathbb{A}}\beta)(x,s) = (\mathbb{A}_s\beta(\cdot,s))(x) + ds \wedge \frac{\partial \beta(x,s)}{\partial s}.$$

The curvature $\widetilde{\mathscr{F}}$ of $\tilde{\mathbb{A}}$ is

$$\widetilde{\mathscr{F}} = \mathscr{F}_s - \frac{d\mathbb{A}_s}{ds} \wedge ds,$$

where $\mathscr{F}_s = \mathbb{A}_s^2$ is the curvature of $\mathbb{A}_s$. Since $(ds)^2 = 0$, the Volterra series expansion of $e^{-\widetilde{\mathscr{F}}}$ is

$$e^{-\widetilde{\mathscr{F}}} = e^{-\mathscr{F}_s} + \Big(\int_0^1 e^{-u\mathscr{F}_s} \frac{d\mathbb{A}_s}{ds} e^{-(1-u)\mathscr{F}_s}\,du\Big) \wedge ds.$$

Let

$$\alpha(s) = \mathrm{Str}\Big(\frac{d\mathbb{A}_s}{ds} e^{-\mathscr{F}_s}\Big).$$

Taking the supertrace of the above formula, we see that

$$\mathrm{ch}(\tilde{\mathbb{A}}) = \mathrm{ch}(\mathbb{A}_s) + \alpha(s) \wedge ds,$$

with respect to the decomposition $\widetilde{B} = B \times \mathbb{R}_+$. Since $\mathrm{ch}(\tilde{\mathbb{A}})$ is closed in $\mathscr{A}(\widetilde{B})$ and $\mathrm{ch}(\mathbb{A}_s)$ is closed in $\mathscr{A}(B)$, we infer that

$$\frac{d\,\mathrm{ch}(\mathbb{A}_s)}{ds} = -d\alpha(s).$$

Integrating this formula, we see that

$$\begin{aligned}\mathrm{ch}(\mathbb{A}_t) - \mathrm{ch}(\mathbb{A}_T) &= -\int_t^T \frac{d}{ds}\mathrm{ch}(\mathbb{A}_s)\,ds \\ &= \int_t^T d\,\mathrm{Str}\Big(\frac{d\mathbb{A}_s}{ds} e^{-\mathbb{A}_s^2}\Big)\,ds \\ &= d\int_t^T \alpha(s)\,ds.\end{aligned}$$

Our task is to show that this integral converges as $T \to \infty$. We will prove this by a small modification of the proof of Theorem 9.2.

Theorem 9.7. *The differential form*

$$\alpha(t) = \operatorname{Str}\left(\frac{d\mathbb{A}_t}{dt} e^{-\mathbb{A}_t^2}\right) \in \mathscr{A}(B)$$

satisfies the estimates, for t large,

$$\|\alpha(t)\|_\ell \le C(\ell) t^{-3/2}$$

on compact subsets of B, for all C^ℓ-norms.

For all $t \ge 0$, the integral

$$\int_t^\infty \alpha(s)\, ds \in \mathscr{A}(B)$$

is convergent, and defines a differential form which satisfies the formula

$$\operatorname{ch}(\mathbb{A}_t) - \operatorname{ch}(\nabla_0) = d \int_t^\infty \alpha(s)\, ds.$$

Proof. We have the following formula for $\alpha(t)$:

$$\alpha(t) = \operatorname{Str}\left(\frac{d\mathbb{A}_t}{dt} e^{-t\delta_t(\mathscr{F})}\right).$$

Since

$$\frac{d\mathbb{A}_t}{dt} = \frac{D}{2t^{1/2}} - \sum_{i=2}^{\dim(B)} \frac{(i-1)\mathbb{A}_{[i]}}{2t^{(i+1)/2}},$$

we have

$$\frac{d\mathbb{A}_t}{dt} = \frac{1}{2t^{1/2}} \begin{vmatrix} 0 & 0 \\ 0 & \mathsf{D} \end{vmatrix} + O(t^{-3/2}).$$

Inserting this into (9.1), we see that

$$\frac{d\mathbb{A}_t}{dt} e^{-t\delta_t(\mathscr{F})} = \frac{1}{2t^{1/2}} \left(\begin{vmatrix} 0 & 0 \\ 0 & \mathsf{D} \end{vmatrix} + O(t^{-1}) \right) \begin{vmatrix} e^{-R_{[2]}} + O(t^{-1/2}) & O(t^{-1/2}) \\ O(t^{-1/2}) & O(t^{-1}) \end{vmatrix}$$

$$= \begin{vmatrix} O(t^{-3/2}) & O(t^{-2}) \\ O(t^{-1}) & O(t^{-3/2}) \end{vmatrix}.$$

Since only the diagonal blocks contribute to the supertrace, this proves the theorem. □

In the special case in which $\mathbb{A}$ has the form $\nabla + \mathsf{D}$, so that $\mathbb{A}_{[i]} = 0$ for $i \ge 2$, we see that $\mathbb{A}_t = \nabla + t^{1/2}\mathsf{D}$ equals ∇ at $t = 0$. Furthermore,

$$\alpha(t) = \frac{1}{2t^{1/2}} \operatorname{Str}(\mathsf{D} \cdot e^{-\mathbb{A}_t^2}) = \begin{cases} O(t^{-1/2}) & \text{for } t \to 0, \\ O(t^{-3/2}) & \text{for } t \to \infty. \end{cases}$$

Thus, the integral $\int_0^\infty \alpha(t)\,dt$ defines a differential form on B, and we have

$$\mathrm{ch}(\nabla) - \mathrm{ch}(\nabla_0) = d\int_0^\infty \alpha(t)\,dt. \tag{9.2}$$

This is a refinement of the fact that

$$[\mathrm{ch}(\mathscr{H}^+)] - [\mathrm{ch}(\mathscr{H}^-)] = [\mathrm{ch}(\mathscr{H}_0^+)] - [\mathrm{ch}(\mathscr{H}_0^-)] \in H^\bullet(B)$$

to the level of differential forms.

9.2 The Index Bundle of a Family of Dirac Operators

The goal of this chapter is to give an infinite-dimensional version of the last section, in which the family D of operators on the finite dimensional bundle $\mathscr{H}$ is replaced by a family of Dirac operators. The details are slightly more complicated for two reasons: firstly, because we work in infinite dimensions, and secondly because we will be interested in the more general case in which the dimension of $\ker(\mathsf{D}^z)$ varies as we move about the family.

By a **family** of manifolds $(M_z \mid z \in B)$, we will mean the family of fibres $M_z = \pi^{-1}(z)$ of a smooth fibre bundle $\pi : M \to B$. We assume that the fibres M_z are compact. However, we will not assume that the base is compact. Let $T(M/B) \subset TM$ be the bundle of vertical tangent vectors, with dual $T^*(M/B) \cong T^*M/\pi^*T^*B$. To the family $M \to B$, we associate the vertical density bundle

$$|\Lambda_\pi| = |\Lambda T^*(M/B)| \cong |\Lambda_M| \otimes (\pi^*|\Lambda_B|)^{-1}.$$

When restricted to a fibre M_z of M, $|\Lambda_\pi|$ may be identified with the bundle $|\Lambda_{M_z}|$ of densities along the fibre. We will call a section of $|\Lambda_\pi|$ a **vertical density**.

By a family of vector bundles $(\mathscr{E}_z \mid z \in B)$ we mean a (smooth) vector bundle $\mathscr{E} \to M$, so that $\mathscr{E}_z$ is the restriction of the bundle $\mathscr{E}$ to M_z. For us, $\mathscr{E}$ will always be a complex superbundle with Hermitian metric. To the family $\mathscr{E} \to M$, we associate the infinite-dimensional superbundle $\pi_*\mathscr{E} = \pi_*\mathscr{E}^+ \oplus \pi_*\mathscr{E}^-$ over B, whose fibre at $z \in B$ is the Fréchet space $\Gamma(M_z, \mathscr{E}_z \otimes |\Lambda_{M_z}|^{1/2})$. A smooth section of $\pi_*\mathscr{E}$ over B is defined to be a smooth section of $\mathscr{E} \otimes |\Lambda_\pi|^{1/2}$ over M:

$$\Gamma(B, \pi_*\mathscr{E}) = \Gamma(M, \mathscr{E} \otimes |\Lambda_\pi|^{1/2}). \tag{9.3}$$

The bundle $\pi_*\mathscr{E}$ carries a metric, defined by means of the canonical metric on each fibre $\Gamma(M_z, \mathscr{E}_z \otimes |\Lambda_{M_z}|^{1/2})$. Indeed, if $\phi^z \in \pi_*\mathscr{E}_z = \Gamma(M_z, \mathscr{E}_z \otimes |\Lambda_{M_z}|^{1/2})$, we see that $|\phi^z(x)|^2 \in \Gamma(M_z, |\Lambda_{M_z}|)$, and hence may be integrated:

$$|\phi^z|^2 = \int_{M_z} |\phi^z(x)|^2.$$

The reason for inserting the line bundle $|\Lambda_\pi|^{1/2}$ is precisely in order to have such a canonical metric along the fibres $\pi_*\mathscr{E}$. The fibre bundle M will have a metric along the fibres, which trivializes $|\Lambda_\pi|$. However, unlike in the case where B is a point, it is important to distinguish between the bundles $\pi_*\mathscr{E}$ and $z \mapsto \Gamma(M_z, \mathscr{E}_z)$.

Instead of working with the bundle $\mathrm{End}(\pi_*\mathscr{E})$ of all endomorphisms of $\pi_*\mathscr{E}$, we will restrict our attention to only those endomorphisms of the fibres which are the sum of a differential operator and a smoothing operator. Thus, we will denote by $\mathscr{D}(\mathscr{E})$ the bundle of algebras over B whose fibre at z is the algebra $\mathscr{D}(M_z, \mathscr{E}_z \otimes |\Lambda_{M_z}|^{1/2})$ of differential operators on $\mathscr{E}_z \otimes |\Lambda_{M_z}|^{1/2}$, and whose smooth sections are families of differential operators D^z whose coefficients in a local trivialization of M and $\mathscr{E}$ depend smoothly on the coordinates in B. We denote by $\mathscr{K}(\mathscr{E})$ the bundle of algebras whose fibre at z is the algebra of smoothing operators on the bundle $\mathscr{E}_z \otimes |\Lambda_{M_z}|^{1/2}$, and whose smooth sections are smooth families of smoothing operators K^z. (We refer to reader to the appendix for more details on this bundle.) Since $\mathscr{K}(\mathscr{E})$ is a bundle of modules for $\mathscr{D}(\mathscr{E})$, we may form an algebra from the sums of operators in $\mathscr{D}(\mathscr{E})$ and $\mathscr{K}(\mathscr{E})$.

Definition 9.8. *The* **$\mathscr{P}$-endomorphisms** *of the infinite-dimensional bundle $\pi_*\mathscr{E}$ are smooth sections of the bundle $\mathscr{D}(\mathscr{E}) + \mathscr{K}(\mathscr{E})$, that is, families of operators which may be written in the form*

$$D^z + K^z,$$

where $(D^z \mid z \in B)$ is a smooth family of differential operators acting on the family of bundles $\mathscr{E}_z \otimes |\Lambda_{M_z}|^{1/2}$, and $(K^z \mid z \in B)$ is a smooth family of smoothing operators on the same family of bundles. We write $\mathrm{End}_{\mathscr{P}}(\pi_*\mathscr{E})$ for the space of smooth sections of this bundle.

We will consider a smooth family $\mathsf{D} = (\mathsf{D}^z \mid z \in B) \in \Gamma(B, \mathscr{D}(\mathscr{E}))$ of Dirac operators on $\mathscr{E}$. Thus, D is an odd operator with respect to the $\mathbb{Z}_2$-grading, with components $\mathsf{D}^\pm : \Gamma(B, \pi_*\mathscr{E}^\pm) \to \Gamma(B, \pi_*\mathscr{E}^\mp)$, and $\mathsf{D}^2 \in \Gamma(B, \mathscr{D}(\mathscr{E}))$ is a family of generalized Laplacians. Note that the Dirac operator D^z defines a Riemannian structure on the fibre M_z (that is, an inner product on the vector bundle $T(M/B) \to M$ of vertical tangent vectors), and an action of the Clifford algebra bundle $C(T^*(M/B))$ on $\mathscr{E}$. In particular, we obtain a canonical trivialization of the bundle $|\Lambda_\pi|$.

Now assume that for each $z \in B$, the operator D^z is self-adjoint with respect to the Hermitian structure on the vector bundle $\mathscr{E}_z$. The vector space $\ker(\mathsf{D}^z)$ has finite dimension. The aim of this section is to define the index bundle of the family D as a superbundle on B whose fibre at $z \in B$ is equal to $\ker(\mathsf{D}^z)$ when the dimension of this vector space is independent of z; in Section 5, we will show that the difference bundle $[\ker(\mathsf{D}^+)] - [\ker(\mathsf{D}^-)]$ makes sense even without this condition.

If λ is not an eigenvalue for any operator of the family D^z, then the superbundle $\mathscr{H}_{[0,\lambda)}$ whose fibre at the point z is the spectral subspace of $(\mathsf{D}^z)^2$ associated to the interval $[0, \lambda)$ is a smooth bundle over B, as we will now prove. This superbundle is also a natural candidate for the index bundle, since $(\mathsf{D}^z)^+$ exchanges the even and odd eigenspaces associated to a given non-zero eigenvalue.

Lemma 9.9. *Let $P \in \Gamma(B, \mathscr{K}(\mathscr{E}))$ be a smooth family of smoothing operators on the family of Hermitian vector bundles $\mathscr{E} \to M$, such that P^z is a finite-rank projection for all $z \in B$. Then $\dim(\mathrm{im}(P^z))$ is constant in each component of B, and the spaces $(\mathrm{im}(P^z) \mid z \in B)$ form a smooth bundle $\mathrm{im}(P^z)$ over B.*

Proof. The dimension of the vector space $\operatorname{im}(P^z)$ is an integer, given by the formula

$$\dim(\operatorname{im}(P^z)) = \operatorname{Tr}(P^z) = \int_{M_z} \operatorname{Tr}(\langle x \mid P^z \mid x \rangle),$$

where $\langle x \mid P^z \mid x \rangle \in \Gamma(M_z, \operatorname{End}(\mathscr{E}) \otimes |\Lambda_{M_z}|)$ is the restriction to the diagonal of the kernel of P^z. Since $\langle x \mid P^z \mid x \rangle$ varies smoothly as a function of z by assumption, the trace varies smoothly as well; thus the dimension of the image of P^z is constant.

We now prove that the family of vector spaces $\operatorname{im}(P^z)$ form a smooth superbundle. If $z_0 \in B$, let U be a neighbourhood of z_0 in B such that $\pi : M \to B$ may be trivialized over U. In this way, we may replace the base B by an open ball $U \subset \mathbb{R}^p$, the family M by the product $M_0 \times U$, and the bundle $\mathscr{E}$ by the bundle $\mathscr{E}_0 \times U$, where $\mathscr{E}_0 \to M_0$ is a superbundle over M_0. The space $\pi_* \mathscr{E}_z$ of sections of $\mathscr{E}_z \otimes |\Lambda_{M_z}|^{1/2}$ does not depend on z. This gives a trivialization of $\pi_*\mathscr{E}$ as $\Gamma(M_0, \mathscr{E}_0 \otimes |\Lambda_{M_0}|^{1/2}) \times U$. Thus P^z is a smooth family of finite rank projectors on the fixed space $\Gamma(M_0, \mathscr{E}_0 \otimes |\Lambda_{M_0}|^{1/2})$.

Choose a basis ϕ_j^0, $(1 \leq j \leq m)$, of $\operatorname{im}(P^0)$ at the point $0 \in U$. We may extend these sections smoothly to U, by extending the corresponding sections of $\mathscr{E} \otimes |\Lambda_\pi|^{1/2}$ over M_z to $\pi^{-1}(U) \subset M$, and we obtain m families of sections $(\phi_j)_{j=1}^m \in \Gamma(U, \pi_*\mathscr{E})$. The sections $P\phi_j$ are smooth as a function of z, lie in $\operatorname{im}(P)$, and are linearly independent for $|z|$ sufficiently small. Thus, they form a basis of $\operatorname{im}(P)$, for $|z|$ small. From the smoothness of P with respect to z, we see that $P^z\phi_j^z(y)$ is smooth with respect to $(y,z) \in M_0 \times U$, and from this we deduce the smoothness of the transition maps between two open sets $U \subset B$ with a trivialization of $\pi_*\mathscr{E}$ above them. □

If λ is a real number, let U_λ be the subset of B on which λ is not an eigenvalue of $(\mathsf{D}^z)^2$. Let $(P^z_{[0,\lambda)} \mid z \in U_\lambda)$ be the family of orthogonal projectors onto the spectral subspace $[0,\lambda)$ of $(\mathsf{D}^z)^2$, with respect to the L^2-norm.

Proposition 9.10. *1. The sets U_λ form an open covering of B.*

2. The family $P_{[0,\lambda)}$ lies in $\Gamma(U_\lambda, \mathscr{K}(\mathscr{E}))$; that is, the operators $P^z_{[0,\lambda)}$ form a smooth family of smoothing operators on the bundles $\mathscr{E}_z \otimes |\Lambda_{M_z}|^{1/2}$ for $z \in U_\lambda$.

3. The vector spaces $(\mathscr{H}_{[0,\lambda)})_z = \operatorname{im}(P^z_{[0,\lambda)})$ form a smooth finite-dimensional superbundle $\mathscr{H}_{[0,\lambda)} \subset \pi_*\mathscr{E}$ over U_λ.

Proof. As in the proof of Lemma 9.9, we may assume that the base is an open ball $U \subset \mathbb{R}^p$, that M is a product $M_0 \times U$, and that the bundle $\mathscr{E}$ is the pull-back of a superbundle $\mathscr{E}_0 \to M_0$. The fibre $\Gamma(M_0, \mathscr{E}_0 \otimes |\Lambda_{M_0}|^{1/2})$ of the bundle $\pi_*\mathscr{E}$ carries a metric $(\cdot,\cdot)_z$ which depends on $z \in U$; however, by Lemma 2.31, the Hilbert space completion $\Gamma_{L^2}(M_0, \mathscr{E}_0 \otimes |\Lambda_{M_0}|^{1/2})$ does not depend on the metric.

The family of heat kernels $K^z = e^{-(\mathsf{D}^z)^2}$ at time $t = 1$ is a smooth family of smoothing operators on $\Gamma(M_0, \mathscr{E}_0 \otimes |\Lambda_{M_0}|^{1/2})$ by Theorem 2.48, which extends to a family of operators on the space $\Gamma_{L^2}(M_0, \mathscr{E}_0 \otimes |\Lambda_{M_0}|^{1/2})$ of L^2-sections, self-adjoint with respect to the metric $(\cdot,\cdot)_z$. The projector on the spectral subspace $[0,\lambda)$ of $(\mathsf{D}^z)^2$ is equal to the projector on the spectral subspace $(e^{-\lambda}, 1]$ of K^z. If $e^{-\lambda}$ is not an eigenvalue of K^{z_0}, there exists a neighbourhood U of z_0 such that for $z \in U$, $e^{-\lambda}$

is not an eigenvalue of K^z. This proves that U_λ is an open subset of B. The fact that $\bigcup_\lambda U_\lambda = B$ is clear.

Let C be the circle in the complex plane which crosses the real axis at the points $e^{-\lambda}$ and $1+\varepsilon$. The spectral theorem for K^z shows that, for $z \in U_\lambda$, the projector P^z_λ is given by a contour integral

$$P^z_\lambda = \int_C (K^z - u)^{-1}\, du.$$

Taking derivatives of this formula with respect to z, we obtain a formula for $\partial_z^\alpha P^z_\lambda$, as a sum of contour integrals of the form

$$\int_C (K^z - u)^{-1} (\partial_z^{\alpha_1} K^z) \dots (K^z - u)^{-1} (\partial_z^{\alpha_k} K^z)(K^z - u)^{-1}\, du.$$

The operator $(K^z - u)^{-1}$ is bounded on the space of L^2-sections of $\mathscr{E}_0 \otimes |\Lambda_{M_0}|^{1/2}$ over M_0 (since boundedness is independent of the metric $(\cdot,\cdot)_z$), and commutes with the generalized Laplacian $(\mathsf{D}^z)^2$. It follows from Corollary 2.40 that if Q is a smoothing operator, then both $(K^z - u)^{-1}Q$ and $Q(K^z - u)^{-1}$ are smoothing operators. Since the operator $\partial_z^\alpha K^z$ has a smooth kernel, we see that any derivative $\partial_z^\alpha P^z_\lambda$ has smooth kernel, hence that the family P^z_λ is a smooth family of smoothing operators. □

Corollary 9.11. *Assume that* $\ker(\mathsf{D}^z)$ *has constant dimension for* $z \in B$. *Then the vector spaces* $\ker(\mathsf{D}^z)$ *form a smooth vector bundle* $\ker(\mathsf{D})$ *over* B.

Proof. If $\ker(\mathsf{D}^z)$ has constant dimension for $z \in B$, then by the local compactness of B, there exists for each relatively compact subset $U \subset B$ a small constant $\varepsilon > 0$ such that the projector onto $\ker(\mathsf{D}^z)$ coincides with P^z_ε. □

If $\dim\big(\ker(\mathsf{D}^z)\big)$ is constant, the **index bundle** $\operatorname{ind}(\mathsf{D})$ of the family D is defined to be the superbundle $\ker(\mathsf{D})$. (This is also known as the **analytic index** of the family.)

If $\lambda < \mu$, let $(P^z_{(\lambda,\mu)} \mid z \in U_\lambda \cap U_\mu)$ be the family of orthogonal projectors onto the spectral subspace (λ,μ) of $(\mathsf{D}^z)^2$, so that

$$P_{[0,\mu)} = P_{[0,\lambda)} + P_{(\lambda,\mu)}.$$

Since $P_{(\lambda,\mu)}$ is a smooth family of projections, the family of vector spaces $\mathscr{H}^z_{(\lambda,\mu)} = P^z_{(\lambda,\mu)}$ forms a smooth superbundle over $U_\lambda \cap U_\mu$. It is clear that

$$\mathscr{H}_{[0,\mu)} = \mathscr{H}_{[0,\lambda)} \oplus \mathscr{H}_{(\lambda,\mu)}.$$

9.3 The Chern Character of the Index Bundle

In this section, following Bismut, we will extend Quillen's theory of superconnections to the infinite dimensional bundle $\pi_*\mathscr{E} \to B$, thereby obtaining a formula for the Chern character of $\operatorname{ind}(\mathsf{D})$ in terms of heat kernels. Our treatment is modelled on the finite-dimensional case, which we discussed in Section 1.

The space of sections of $\pi_*\mathscr{E}$ is defined by $\Gamma(B,\pi_*\mathscr{E}) = \Gamma(M,\mathscr{E}\otimes|\Lambda_\pi|^{1/2})$. It is natural to define the space of differential forms on B with values in $\pi_*\mathscr{E}$ by

$$\mathscr{A}(B,\pi_*\mathscr{E}) = \Gamma(M,\pi^*(\Lambda T^*B)\otimes\mathscr{E}\otimes|\Lambda_\pi|^{1/2}).$$

A differential operator on the space $\mathscr{A}(B,\pi_*\mathscr{E})$ is by definition a differential operator on $\Gamma(M,\pi^*(\Lambda T^*B)\otimes\mathscr{E}\otimes|\Lambda_\pi|^{1/2})$. Let

$$\mathscr{A}(B,\mathscr{D}(\mathscr{E})) = \Gamma(B,\Lambda T^*B\otimes\mathscr{D}(\mathscr{E}))$$

be the space of vertical differential operators with differential form coefficients. If a differential operator D on $\mathscr{A}(B,\pi_*\mathscr{E})$ supercommutes with the action of $\mathscr{A}(B)$, then this operator is given by the action of an element of $\mathscr{A}(B,\mathscr{D}(\mathscr{E}))$. Similarly, we write

$$\mathscr{A}(B,\mathscr{K}(\mathscr{E})) = \Gamma(B,\Lambda T^*B\otimes\mathscr{K}(\mathscr{E}))$$

for the space of smooth families of smoothing operators K^z with differential form coefficients. Denote by d_B the exterior differential on B.

Definition 9.12. *Let* D *be a smooth family of Dirac operators on* $\mathscr{E}$*. A* **superconnection** *adapted to the family* D *is a differential operator* $\mathbb{A}$ *on* $\mathscr{A}(B,\pi_*\mathscr{E})$ *of odd parity such that*

1. *(Leibniz's rule) for all* $v\in\mathscr{A}(B)$ *and* $\phi\in\mathscr{A}(B,\pi_*\mathscr{E})$,
$$\mathbb{A}(v\phi) = (d_Bv)\phi + (-1)^{|v|}v\mathbb{A}(\phi);$$
2. $\mathbb{A} = \mathsf{D} + \sum_{i=1}^{\dim(B)}\mathbb{A}_{[i]}$, *where* $\mathbb{A}_{[i]} : \mathscr{A}^\bullet(B,\pi_*\mathscr{E}) \to \mathscr{A}^{\bullet+i}(B,\pi_*\mathscr{E})$.

It is easy to see that $\mathbb{A}_{[i]}$ supercommutes with $\mathscr{A}(B)$ if $i\neq 1$, and hence $\mathbb{A}_{[i]}\in\mathscr{A}^i(B,\mathscr{D}(\mathscr{E}))$ for $i\neq 1$.

Let us show how to construct a superconnection adapted to a family of Dirac operators D. It suffices to define a connection $\nabla^{\pi_*\mathscr{E}}$ on the bundle $\pi_*\mathscr{E}$, that is, a differential operator from $\Gamma(B,\pi_*\mathscr{E}^\pm)$ to $\mathscr{A}^1(B,\pi^*\mathscr{E}^\mp)$ such that

$$\nabla^{\pi_*\mathscr{E}}(f\phi) = df\wedge\phi + f\nabla^{\pi_*\mathscr{E}}\phi$$

for all $f\in C^\infty(B)$ and $\phi\in\Gamma(B,\pi_*\mathscr{E})$. Let us assume that the bundle M/B possesses the additional structure of a connection, that is, a choice of a splitting $TM = T_HM\oplus T(M/B)$, so that the subbundle T_HM is isomorphic to the vector bundle π^*TB. From this, we can define a canonical linear connection on the vertical tangent space $T(M/B)$ using the projection operator

$$P : TM \to T(M/B)$$

with kernel the chosen horizontal tangent space T_HM. If X is a vector field on the base B, denote by X_M its horizontal lift on M, that is, the vector field on M which is a section of $T_H M$ and which projects to X under the pushforward $\pi_* : (T_H M)_x \to T_{\pi(x)}B$.

Furthermore, let us suppose that the bundle $\mathscr{E}$ over M is provided with a connection $\nabla^{\mathscr{E}}$ compatible with its Hermitian structure.

Proposition 9.13. *Let $\alpha \in \Gamma(M, |\Lambda_M|^{1/2})$, $\beta \in \Gamma(B, |\Lambda_B|^{-1/2})$ and $s \in \Gamma(M, \mathscr{E})$. Define the action of $\nabla^{\pi_*\mathscr{E}}_X$, where X is a vector field on B, on*

$$s \otimes \alpha \otimes \pi^*\beta \in \Gamma(M, \mathscr{E} \otimes |\Lambda_\pi|^{1/2}) = \Gamma(B, \pi_*\mathscr{E})$$

by the formula

$$\nabla^{\pi_*\mathscr{E}}_X (s \otimes \alpha \otimes \pi^*\beta) = (\nabla^{\mathscr{E}}_{X_M} s) \otimes \alpha \otimes \pi^*\beta + s \otimes \mathscr{L}(X_M)\alpha \otimes \pi^*\beta + s \otimes \alpha \otimes \pi^*\mathscr{L}(X)\beta.$$

1. *This formula is independent of the tensor product decomposition of a section $s \otimes \alpha \otimes \pi^*\beta \in \Gamma(M, \mathscr{E} \otimes |\Lambda_\pi|^{1/2})$ with $s \in \Gamma(M, \mathscr{E})$, $\alpha \in \Gamma(M, |\Lambda_M|^{1/2})$ and $\beta \in \Gamma(B, |\Lambda_B|^{-1/2})$, and satisfies the formula $\nabla^{\pi_*\mathscr{E}}_{fX} = f\nabla^{\pi_*\mathscr{E}}_X$ for $f \in C^\infty(B)$. Hence, $\nabla^{\pi_*\mathscr{E}}$ is a connection on the bundle $\pi_*\mathscr{E}$ over B.*
2. *The connection $\nabla^{\pi_*\mathscr{E}}$ is compatible with the inner product on $\pi_*\mathscr{E}$.*

Proof. To show the independence of $\nabla^{\pi_*\mathscr{E}}_X$ on the representation of a section of $\pi_*\mathscr{E}$ as a tensor product $s \otimes \alpha \otimes \pi^*\beta$, we must check that if $f \in C^\infty(M)$, then

$$\nabla^{\pi_*\mathscr{E}}((fs) \otimes \alpha \otimes \pi^*\beta) = \nabla^{\pi_*\mathscr{E}}(s \otimes (f\alpha) \otimes \pi^*\beta),$$

and that if $h \in C^\infty(B)$, then

$$\nabla^{\pi_*\mathscr{E}}(s \otimes ((\pi^*h)\alpha) \otimes \pi^*\beta) = \nabla^{\pi_*\mathscr{E}}(s \otimes \alpha \otimes \pi^*(h\beta)).$$

These follow easily from Leibniz's rule.

To show that $\nabla^{\pi_*\mathscr{E}}$ is a connection, we must show that $\nabla^{\pi_*\mathscr{E}}_{fX} = f\nabla^{\pi_*\mathscr{E}}_X$ for $f \in C^\infty(B)$. Using Lemma 1.14, we see that

$$\begin{aligned}
\nabla^{\pi_*\mathscr{E}}_{fX}(s \otimes \alpha \otimes \pi^*\beta) =& \pi^* f(\nabla^{\mathscr{E}}_{X_M} s) \otimes \alpha \otimes \pi^*\beta \\
&+ s \otimes (\pi^* f \mathscr{L}(X_M)\alpha + \tfrac{1}{2}\pi^* X(f)\alpha) \otimes \pi^*\beta \\
&+ s \otimes \alpha \otimes (\pi^* f \pi^* \mathscr{L}(X)\beta - \tfrac{1}{2}\pi^* X(f)\pi^*\beta) \\
=& \pi^* f\big[(\nabla^{\mathscr{E}}_{X_M} s) \otimes \alpha \otimes \pi^*\beta \\
&+ s \otimes \mathscr{L}(X_M)\alpha \otimes \pi^*\beta + s \otimes \alpha \otimes \pi^*\mathscr{L}(X)\beta\big].
\end{aligned}$$

We must now show that $\nabla^{\pi_*\mathscr{E}}$ preserves the inner-product on $\pi_*\mathscr{E}$. This follows from the assumption that $\nabla^{\mathscr{E}}$ is compatible with the Hermitian metric on $\mathscr{E}$, so that

$$\begin{aligned}
\mathscr{L}(X)|s \otimes \alpha \otimes \pi^*\beta|^2 &= \mathscr{L}(X)\Big(\int_{M/B} |s|^2\, \alpha^2 \otimes \pi^*\beta^2\Big) \\
&= \int_{M/B} \mathscr{L}(X_M)(|s|^2\, \alpha^2 \otimes \pi^*\beta^2) \\
&= 2(s \otimes \alpha \otimes \pi^*\beta, \nabla^{\pi_*\mathscr{E}}_X(s \otimes \alpha \otimes \pi^*\beta)).
\end{aligned}$$

□

Thus, associated to a connection on the fibre bundle M/B and a connection on the bundle $\mathscr{E}$, there is a natural superconnection $\mathsf{D}+\nabla^{\pi_*\mathscr{E}}$ for the family of Dirac operators $(\mathsf{D}^z \mid z \in B)$. In Chapter 10, we will also have to consider superconnections $\mathbb{A}$ for which $\mathbb{A}_{[2]}$ does not vanish.

The curvature $\mathscr{F} = \mathbb{A}^2 \in \mathscr{A}(B, \mathscr{D}(\mathscr{E}))$ of a superconnection is a vertical differential operator with differential form coefficients. We have

$$\mathscr{F} = \mathsf{D}^2 + \mathscr{F}_{[+]},$$

where D^2 is a smooth family of generalized Laplacians and

$$\mathscr{F}_{[+]} \in \Gamma(B, \Lambda T^*B \otimes \mathscr{D}(\mathscr{E}))$$

is a smooth family of differential operators with differential form coefficients which raises exterior degree in

$$\Lambda T_z^*B \otimes \Gamma(M_z, \mathscr{E}_z \otimes |\Lambda_{M_z}|^{1/2}).$$

The results of Appendix 1 apply in this situation, and we obtain the existence of a smooth family of heat kernels for $\mathscr{F}$, which we denote by $e^{-t\mathscr{F}} \in \mathscr{A}(B, \mathscr{K}(\mathscr{E}))$:

$$e^{-t\mathscr{F}} = e^{-t\mathsf{D}^2} + \sum_{k>0} (-t)^k I_k,$$

where

$$I_k = \int_{\Delta_k} e^{-\sigma_0 t \mathsf{D}^2} \mathscr{F}_{[+]} e^{-\sigma_1 t \mathsf{D}^2} \mathscr{F}_{[+]} \dots e^{-\sigma_{k-1} t \mathsf{D}^2} \mathscr{F}_{[+]} e^{-\sigma_k t \mathsf{D}^2}.$$

Since I_k vanishes for $k > \dim(B)$, the above sum is finite.

We will show that the Chern character of the index bundle $\operatorname{ind}(\mathsf{D})$ can be computed from a superconnection with $\mathbb{A}_{[0]} = \mathsf{D}$, by the same formula as in the case of a finite dimensional superbundle, namely $\operatorname{ch}(\operatorname{ind}(\mathsf{D})) = \operatorname{Str}(e^{-\mathbb{A}^2})$.

Let $K = (K^z \mid z \in B) \in \mathscr{A}(B, \mathscr{K}(\mathscr{E}))$ be a smooth family of smoothing operators with coefficients in $\mathscr{A}(B)$, given by a kernel

$$\langle x \mid K \mid y \rangle \in \Gamma(M \times_\pi M, \pi^*\Lambda T^*B \otimes (\mathscr{E} \otimes |\Lambda_\pi|^{1/2}) \boxtimes_\pi (\mathscr{E}^* \otimes |\Lambda_\pi|^{1/2})).$$

Here $M \times_\pi M = \{(x,y) \in M \times M \mid \pi(x) = \pi(y)\}$ with projections $\mathrm{pr}_1(x,y) = x$ and $\mathrm{pr}_2(x,y) = y$ to M, and if $\mathscr{E}_1$ and $\mathscr{E}_2$ are two vector bundles on M, the vector bundle $\mathscr{E}_1 \boxtimes_\pi \mathscr{E}_2$ over $M \times_\pi M$ is given by the formula

$$\mathscr{E}_1 \boxtimes_\pi \mathscr{E}_2 = \mathrm{pr}_1^* \mathscr{E}_1 \otimes \mathrm{pr}_2^* \mathscr{E}_2.$$

There is a supertrace on $\mathscr{K}(\mathscr{E}_z)$ over each fibre M_z of M/B, which gives a supertrace

$$\operatorname{Str} : \Gamma(B, \mathscr{K}(\mathscr{E})) \to C^\infty(B).$$

Suppose that $K \in \mathscr{A}(B, \mathscr{K}(\mathscr{E}))$; restricted to the diagonal, its kernel $\langle x \mid K^z \mid x \rangle$ is a smooth section of the bundle $\pi^*\Lambda T^*B \otimes \operatorname{End}(\mathscr{E}) \otimes |\Lambda_\pi|$ over M, and its pointwise supertrace $\operatorname{Str}_{\mathscr{E}} \langle x \mid K^z \mid x \rangle$ is a section in $\Gamma(M, \pi^*\Lambda T^*B \otimes |\Lambda_\pi|)$. This section can be integrated over the fibres to obtain a differential form on B.

Lemma 9.14. *The $\mathscr{A}(B)$-valued supertrace* $\operatorname{Str} : \mathscr{A}(B, \mathscr{K}(\mathscr{E})) \mapsto \mathscr{A}(B)$ *of the family of operators K is the differential form on B*

$$z \longmapsto \int_{M_z} \operatorname{Str}_{\mathscr{E}}\left(\langle x \mid K^z \mid x \rangle\right).$$

If $\mathbb{A}$ is a superconnection on $\pi_* \mathscr{E}$, then $[\mathbb{A}, K]$ is again a family of smoothing operators with differential form coefficients. The following lemma is one of the main steps in the extension of Quillen's construction of Chern characters to the infinite dimensional setting.

Lemma 9.15. $d_B \operatorname{Str}(K) = \operatorname{Str}([\mathbb{A}, K]) \in \mathscr{A}(B)$

Proof. As in the proof of Proposition 9.9, we may assume that the base B is an open neighbourhood $U \subset \mathbb{R}^p$ of $0 \in \mathbb{R}^p$, that the family M is the product $M_0 \times U$, and that the bundle $\mathscr{E}$ is the pull-back to M of the bundle $\mathscr{E}_0 \to M_0$. With respect to the identification $\mathscr{A}(B, \pi_* \mathscr{E}) = \mathscr{A}(U) \otimes \Gamma(M_0, \mathscr{E}_0 \otimes |\Lambda_{M_0}|^{1/2})$, we may write

$$\mathbb{A} = d_U + \sum_I D_I^z \, dz_I,$$

where D_I^z are differential operators on $\Gamma(M_0, \mathscr{E}_0 \otimes |\Lambda_{M_0}|^{1/2})$ depending smoothly on $z \in U$; we also have

$$K^z = \sum_I K_I^z \, dz_I,$$

where K_I^z are smoothing operators on $\Gamma(M_0, \mathscr{E}_0 \otimes |\Lambda_{M_0}|^{1/2})$ depending smoothly on $z \in U$. By Lemma 3.49, we have $\operatorname{Str}([D_I^z, K_J^z]) = 0$, so that we may assume that $\mathbb{A} = d_U$ in this trivialization. We see that

$$[d_U, K] = \sum_{I,i} \frac{\partial K_I^z}{\partial z_i} dz_i \wedge dz_I.$$

On the other hand,

$$\begin{aligned}
d_U \operatorname{Str}(K) &= \sum_{I,i} \frac{\partial}{\partial z_i} \operatorname{Str}(K_I^z) \, dz_i \wedge dz_I \\
&= \sum_{I,i} \frac{\partial}{\partial z_i} \left(\int_{M_0} \operatorname{Str} \langle x \mid K_I^z \mid x \rangle \right) dz_i \wedge dz_I \\
&= \int_{M_0} \sum_{I,i} \operatorname{Str} \left(\frac{\partial}{\partial z_i} \langle x \mid K_I^z \mid x \rangle \right) dz_i \wedge dz_I \\
&= \operatorname{Str}([d_U, K]).
\end{aligned}$$

□

Definition 9.16. *The* **Chern character** *form of a superconnection $\mathbb{A}$ on the bundle $\pi_* \mathscr{E}$, adapted to a family of Dirac operators D, is the differential form on B given by the formula*

$$\operatorname{ch}(\mathbb{A}) = \operatorname{Str}(e^{-\mathbb{A}^2});$$

this is well-defined, since $e^{-\mathbb{A}^2} = e^{-\mathscr{F}} \in \mathscr{A}(B, \mathscr{K}(\mathscr{E}))$.

Theorem 9.17. *Let* $\mathbb{A}$ *be a superconnection on the bundle* $\pi_*\mathcal{E}$ *for the family of Dirac operators* D.

1. *The differential form* $\mathrm{ch}(\mathbb{A})$ *is closed.*
2. *If* $\mathbb{A}_\sigma$ *is a one-parameter family of superconnections on the bundle* $\pi_*\mathcal{E}$ *for the family of Dirac operators* D_σ, *then*

$$\frac{d}{d\sigma}\mathrm{ch}(\mathbb{A}_\sigma) = -d_B \operatorname{Str}\Big(\frac{d\mathbb{A}_\sigma}{d\sigma}e^{-\mathcal{F}_\sigma}\Big).$$

Thus, the class of $\mathrm{ch}(\mathbb{A}_\sigma)$ *in de Rham cohomology is a homotopy invariant of the superconnection* $\mathbb{A}$.

Proof. The fact that $\mathrm{ch}(\mathbb{A})$ is closed follows from Lemma 9.15, since

$$d_B \operatorname{Str}(e^{-\mathcal{F}}) = -\operatorname{Str}([\mathbb{A}, e^{-\mathcal{F}}]) = 0.$$

The proof of the homotopy invariance is similar. As in Proposition 1.41, we argue that

$$\begin{aligned}\frac{d}{d\sigma}\operatorname{Str}(e^{-\mathcal{F}_\sigma}) &= -\operatorname{Str}\Big(\Big[\mathbb{A}_\sigma, \frac{d\mathbb{A}_\sigma}{d\sigma}\Big]e^{-\mathcal{F}_\sigma}\Big)\\ &= -\operatorname{Str}\Big(\Big[\mathbb{A}_\sigma, \frac{d\mathbb{A}_\sigma}{d\sigma}e^{-\mathcal{F}_\sigma}\Big]\Big)\\ &= -d_B \operatorname{Str}\Big(\frac{d\mathbb{A}_\sigma}{d\sigma}e^{-\mathcal{F}_\sigma}\Big),\end{aligned}$$

where all of the steps may be justified by using the fact that $d\mathbb{A}_\sigma/d\sigma$ is a family of vertical differential operators with differential form coefficients.

The fact that the cohomology class of $\mathrm{ch}(\mathbb{A}_\sigma)$ depends only on the homotopy class of D_σ follows as in the finite-dimensional case from the homotopy invariance of $\mathrm{ch}(\mathbb{A})$: if $\mathbb{A}_0$ and $\mathbb{A}_1$ are both superconnections for D, then $\sigma \in [0,1] \mapsto \mathbb{A}_\sigma = \sigma\mathbb{A}_1 + (1-\sigma)\mathbb{A}_0$ is a one-parameter family of superconnections for D. Similarly, when D_σ varies smoothly, we can form the smooth family of superconnections $\mathbb{A}_\sigma = \mathsf{D}_\sigma + \nabla^{\pi_*\mathcal{E}}$, and the cohomology class of $\mathrm{ch}(\mathbb{A}_\sigma)$ will be independent of σ. □

Our goal in the remainder of this section is to prove the following fundamental theorem, which generalizes the results of Section 1 to the case of a family of Dirac operators. Assume that D is a family of Dirac operators such that $\ker(\mathsf{D}^z)$ has constant dimension, so that $\ker(\mathsf{D})$ is a superbundle over B. If P_0^z is the orthogonal projection from $\pi_*\mathcal{E}_z$ to $\ker(\mathsf{D}^z)$, then $P_0 \in \Gamma(B, \mathcal{K}(\mathcal{E}))$ is a smooth family of smoothing operators. The following lemma is clear.

Lemma 9.18. *The operator* ∇_0 *defined by the formula* $\nabla_0 = P_0\mathbb{A}_{[1]}P_0$ *is a connection on the superbundle* $\ker(\mathsf{D})$.

For $t > 0$, let δ_t be the automorphism of $\mathcal{A}(B, \pi_*\mathcal{E})$ which multiplies $\mathcal{A}^i(B, \pi_*\mathcal{E})$ by $t^{-i/2}$. Then $\mathbb{A}_t = t^{1/2}\delta_t \cdot \mathbb{A} \cdot \delta_t^{-1}$ is a superconnection for the family of Dirac operators $t^{1/2}\mathsf{D}$.

Theorem 9.19. *For $t > 0$, let*

$$\mathbb{A}_t = t^{1/2}\delta_t \cdot \mathbb{A} \cdot \delta_t^{-1} = t^{1/2}\mathsf{D} + \mathbb{A}_{[1]} + t^{-1/2}\mathbb{A}_{[2]} + \dots$$

be the rescaled superconnection, with curvature $\mathcal{F}_t = t\delta_t(F)$. Then for t large,

$$\|e^{-\mathcal{F}_t} - e^{-\nabla_0^2}\|_\ell \leq C(\ell)t^{-1/2}$$

uniformly on compact subsets of $M \times_\pi M$, for all C^ℓ-norms.

Proof. The proof is formally almost exactly the same as the proof in the finite-dimensional case given in Section 1, and we give only a few steps. Consider the algebra $\mathscr{M} = \mathscr{A}(B, \mathrm{End}_{\mathscr{P}}(\mathscr{E})) = \Gamma(B, \pi^*\Lambda T^*B \otimes \mathrm{End}_{\mathscr{P}}(\mathscr{E}))$, where $\mathrm{End}_{\mathscr{P}}(\pi_*\mathscr{E})$ is as in Definition 9.8; we filter $\mathscr{M}$ by the subspaces

$$\mathscr{M}_i = \sum_{j \geq i} \mathscr{A}^j(B, \mathrm{End}_{\mathscr{P}}(\mathscr{E})).$$

We will also need the algebra $\mathscr{N} = \mathscr{A}(B, \mathscr{K}(\mathscr{E}))$, filtered in the same way.

Let $G = (G^z \mid z \in B)$ be the family of Green operators G^z of $(\mathsf{D}^z)^2$.

Proposition 9.20. *The action of left and right multiplication by G preserves $\mathscr{N}$.*

Proof. Since this is a local question, we may suppose that the bundles M and $\mathscr{E}$ over B are trivial, by replacing B by an open subset U. We may rewrite the integral representation of G given in (2.11) as follows, where P_0 is the projection onto the kernel of D:

$$G = \int_0^\infty e^{-t(\mathsf{D}^2+P_0)}\,dt - P_0.$$

Thus, we may decompose $\partial^\alpha((G+P_0)K)$, where $K \in \mathscr{N}$, into terms proportional to

$$\int_0^\infty (\partial_z^{\alpha_1} e^{-t(\mathsf{D}^2+P_0)})(\partial_z^{\alpha_2} K)\,dt,$$

where $\alpha_1 + \alpha_2 = \alpha$. Repeated application of Theorem 2.48 shows that this integral has the general form

$$\int_0^\infty \dots \int_0^\infty e^{-t_1(\mathsf{D}^2+P_0)} D_1 e^{-t_2(\mathsf{D}^2+P_0)} D_2 \dots e^{-t_k(\mathsf{D}^2+P_0)} D_k\,dt_1 \dots dt_k,$$

where $k = |\alpha_1| + 1$ and $D_i \in \Gamma(B, \mathrm{End}_{\mathscr{P}}(\pi_*\mathscr{E}))$. This integral may be bounded using the exponential decay of $\langle x \mid e^{-t(\mathsf{D}^2+P_0)} \mid y\rangle$ proved in Proposition 2.37. Using the formula

$$\partial^\alpha(GK) = \partial^\alpha((G+P_0)K) - \partial^\alpha(P_0K),$$

we see that GK is in $\mathscr{N}$. The case where G multiplies on the right is similar. □

Let $P_1 = 1 - P_0$ be the projection onto $\operatorname{im}(\mathsf{D})$. If $K \in \mathscr{M}$, we will write

$$K = \begin{vmatrix} \alpha & \beta \\ \gamma & \delta \end{vmatrix} = \begin{vmatrix} P_0 K P_0 & P_0 K P_1 \\ P_1 K P_0 & P_1 K P_1 \end{vmatrix} \in \begin{vmatrix} \mathscr{N} & \mathscr{N} \\ \mathscr{N} & \mathscr{M} \end{vmatrix}.$$

Let

$$\mathscr{F} = \begin{vmatrix} X & Y \\ Z & T \end{vmatrix}$$

be the curvature of the superconnection $\mathbb{A}$; then X, Y and Z are in $\mathscr{N}$.

Let $R_{[2]}$ be the curvature of the connection ∇_0 on the bundle $\ker(\mathsf{D})$. By the same proof as in Section 1, we have the formula $R_{[2]} = X_{[2]} - Y_{[1]} G Z_{[1]}$.

The space of endomorphisms $g \in \mathscr{M}$ of the form $g = 1 + K$, where $K \in \mathscr{M}_1$, form a group, with

$$(1+K)^{-1} = 1 + \sum_{k=1}^{\infty} (-K)^k.$$

Lemma 9.21. *There exists $g \in \mathscr{M}$ with $g - 1 \in \mathscr{N}_1$, such that*

$$g \begin{vmatrix} X & Y \\ Z & T \end{vmatrix} g^{-1} = \begin{vmatrix} U & 0 \\ 0 & V \end{vmatrix}.$$

Furthermore

$$U \equiv X - YGZ \pmod{\mathscr{N}_3},$$
$$V \equiv T \pmod{\mathscr{N}_1}.$$

The proof is the same as in the finite-dimensional case of Section 1, once we observe that if G is the Green operator of D^2, and if Y and $Z \in \mathscr{N}_i$, then the operators YG and GZ are also in $\mathscr{N}_i$, and $Y(1-GT)$ and $(1-TG)Z$ are in $\mathscr{N}_{i+1}$. Thus, we may construct g as in Section 1 as a product of matrices of the form

$$\begin{vmatrix} 1 & -YG \\ GZ & 1 \end{vmatrix} \in 1 + \begin{vmatrix} 0 & \mathscr{N}_1 \\ \mathscr{N}_1 & 0 \end{vmatrix}.$$

Thus, as before, we may put $\mathscr{F}$ into block diagonal form; there is a family of operators $K \in \mathscr{N}_1$ such that, with $g = 1 + K$,

$$\mathscr{F} = g^{-1} \begin{vmatrix} U & 0 \\ 0 & V \end{vmatrix} g,$$

where $U \equiv R_{[2]} \pmod{\mathscr{N}_3}$ and $V \equiv \mathsf{D}^2 \pmod{\mathscr{N}_1}$. Note that the family of operators $\delta_t(V)$ is for each $t > 0$ the sum of a family of generalized Laplacians D^2 and an element of $P_1 \cdot \mathscr{M}_1 \cdot P_1$. It follows from Appendix 2 that the family of heat kernels $e^{-t\delta_t(V)}$ is a section in $\mathscr{A}(B, \mathscr{K}(\mathscr{E}))$ for each $t > 0$. Furthermore, the uniqueness of solutions of the heat equation implies that

$$e^{-t\delta_t(\mathscr{F})} = \delta_t(g)^{-1} \begin{pmatrix} e^{-t\delta_t(U)} & 0 \\ 0 & e^{-t\delta_t(V)} \end{pmatrix} \delta_t(g).$$

If $A(t):(0,\infty)\to\Gamma(B,\mathscr{K}(\mathscr{E}))$, we will write $A(t)=O(f(t))$ if for all $\varepsilon>0$ and $\ell\in\mathbb{N}$ and each function $\phi\in C_c^\infty(B)$ of compact support, there is a constant $C(\ell,\varepsilon,\phi)$ such that

$$\|\pi^*(\phi)(x)\langle x\,|\,A(t)\,|\,y\rangle\|_\ell\le C(\ell,\varepsilon,\phi)f(t)\quad\text{for all } t>\varepsilon.$$

Let U be a relatively compact open subset of B. If λ is the infimum over U of the lowest non-zero eigenvalue of the operators D^2, then it follows from Proposition 9.49 that over U,

$$P_1e^{-t\delta_t(\mathbb{V})}P_1=O(e^{-t\lambda/2}).$$

Thus, we have

$$e^{-t\delta_t(\mathscr{F})}=\delta_t(g)\begin{vmatrix}e^{-R_{[2]}} & 0\\ 0 & 0\end{vmatrix}\delta_t(g)^{-1}+\delta_t(g)\begin{vmatrix}O(t^{-1/2}) & 0\\ 0 & O(e^{-t\lambda/2})\end{vmatrix}\delta_t(g)^{-1}.$$

Since $\delta_t(g)-1=O(t^{-1/2})$, we see that

$$e^{-t\delta_t(\mathscr{F})}=\begin{vmatrix}e^{-R_{[2]}} & 0\\ 0 & 0\end{vmatrix}+\begin{vmatrix}O(t^{-1/2}) & O(t^{-1/2})\\ O(t^{-1/2}) & O(t^{-1})\end{vmatrix},$$

from which the theorem follows. □

Corollary 9.22. *The limit*

$$\lim_{t\to\infty}\mathrm{ch}(\mathbb{A}_t)=\mathrm{ch}(P_0\mathbb{A}_{[1]}P_0)\in\mathscr{A}(B)$$

holds with respect to each C^ℓ-norm on compact subsets of B.

The above corollary is equivalent to the McKean-Singer theorem when B is a point. In that case, a family of Dirac operators is just a single Dirac operator D on a Clifford bundle $\mathscr{E}\to M$ and the superconnection $\mathbb{A}$ equals D. The Chern character of the index bundle $\mathrm{ind}(\mathsf{D})$ is just a sophisticated name for its dimension, which is the index of D in the usual sense. On the other hand, we see that $\mathrm{Str}\big(e^{-\mathbb{A}_t^2}\big)$ is nothing but $\mathrm{Str}\big(e^{-t\mathsf{D}^2}\big)$, which by the McKean-Singer theorem equals the index of D.

We will now prove an infinite-dimensional version of the transgression formula, which gives an explicit formula for a differential form whose differential is the difference $\mathrm{ch}(\mathbb{A}_t)-\lim_{t\to\infty}\mathrm{ch}(\mathbb{A}_t)$.

Theorem 9.23. *Let D be a family of Dirac operators such that* $\ker(\mathsf{D}^z)$ *has constant dimension. The differential form*

$$\alpha(t)=\mathrm{Str}\Big(\frac{d\mathbb{A}_t}{dt}e^{-\mathbb{A}_t^2}\Big)\in\mathscr{A}(B)$$

satisfies the estimates

$$\|\alpha(t)\|_\ell\le C(\ell)t^{-3/2}$$

on compact subsets of B, for all C^ℓ-norms, and

$$\operatorname{ch}(\mathbb{A}_t) = \lim_{t\to\infty} \operatorname{ch}(\mathbb{A}_t) + d\int_t^\infty \alpha(s)\,ds$$
$$= \operatorname{ch}(P_0\mathbb{A}_{[1]}P_0) + d\int_t^\infty \alpha(s)\,ds.$$

Proof. It is clear that the formula for $\operatorname{ch}(\mathbb{A}_t) - \lim_{t\to\infty}\operatorname{ch}(\mathbb{A}_t)$, known as the transgression formula, is an immediate consequence of the estimate on $\alpha(t)$ as $t\to\infty$.

The proof of the estimate on $\alpha(t)$ is again much the same as that of Theorem 9.7, except that the estimates are all rewritten in terms of the kernels of the operators in question. Since

$$\frac{d\mathbb{A}_t}{dt} - \frac{1}{2t^{1/2}}\mathsf{D} = \sum_{i=2}^{n} \frac{(1-i)\mathbb{A}_{[i]}}{2t^{(i+1)/2}},$$

and $\mathbb{A}_{[i]} \in \mathscr{M}_2$, we see that if $K(t) : (0,\infty) \to \Gamma(B,\mathscr{K}(\mathscr{E}))$ is a family of kernels such that $K(t) = O(t^{-s})$, then

$$\frac{d\mathbb{A}_t}{dt}K(t) - \frac{\mathsf{D}}{2t^{1/2}}K(t) = O(t^{-s-3/2}).$$

Furthermore, the proof of Theorem 9.19 showed that

$$e^{-t\delta_t(\mathscr{F})} = \begin{vmatrix} e^{-R_{[2]}} & 0 \\ 0 & 0 \end{vmatrix} + \begin{vmatrix} O(t^{-1/2}) & O(t^{-1/2}) \\ O(t^{-1/2}) & O(t^{-1}) \end{vmatrix}.$$

It follows that $\alpha(t)$ is the supertrace of

$$\frac{1}{2t^{1/2}}\begin{vmatrix} 0 & 0 \\ 0 & \mathsf{D} \end{vmatrix} \cdot \begin{vmatrix} O(t^{-1/2}) & O(t^{-1/2}) \\ O(t^{-1/2}) & O(t^{-1}) \end{vmatrix} + O(t^{-3/2}) = \begin{vmatrix} O(t^{-3/2}) & O(t^{-3/2}) \\ O(t^{-1}) & O(t^{-3/2}) \end{vmatrix}.$$

Since only the diagonal blocks contribute to the supertrace, this proves the theorem. □

In the next chapter, following Bismut, we will construct a superconnection $\mathbb{A}$ for the family D of Dirac operators associated to a Clifford connection, such that as $t\to 0$,

$$\alpha(t) = O(t^{-1/2}).$$

From this will follow the existence of a limit $\lim_{t\to 0}\operatorname{ch}(\mathbb{A}_t)$, which can in fact be calculated explicitly, and the following analogue of (9.2):

$$\lim_{t\to 0}\operatorname{ch}(\mathbb{A}_t) - \operatorname{ch}(\nabla_0) = d\int_0^\infty \alpha(t)\,dt.$$

Let $R \in \Gamma(B,\mathscr{K}(\mathscr{E}))$ be a smooth family of self-adjoint smoothing operators, such that each operator R^z is odd. Consider the perturbed family of Dirac operators

$$\mathsf{D} + R \in \Gamma(B, \operatorname{End}^-_{\mathscr{P}}(\mathscr{E})). \tag{9.4}$$

If we square $\mathsf{D}+R$, we obtain

$$(\mathsf{D}+R)^2 = \mathsf{D}^2 + (\mathsf{D}R + R\mathsf{D} + R^2).$$

Since D^2 is a smooth family of generalized Laplacians and $\mathsf{D}R + R\mathsf{D} + R^2$ is a smooth family of smoothing operators, we obtain the following lemma.

Lemma 9.24. *The square of the operator* $\mathsf{D}+R$ *has the form* $H = H_0 + K$, *where* H_0 *is a smooth family of generalized Laplacians and* K *is a smooth family of smoothing operators.*

There is no difficulty in extending the results in Sections 2 and 3 to the more general case where the family D of Dirac operators is everywhere replaced by a family $\mathsf{D}+R$ of the form discussed above. In particular, the vector space $\ker((\mathsf{D}+R)^z)$ is finite-dimensional for each $z \in B$, the index of $(\mathsf{D}+R)^z$

$$\operatorname{ind}((\mathsf{D}+R)^z) = \dim(\ker^+((\mathsf{D}+R)^z)) - \dim(\ker^-((\mathsf{D}+R)^z))$$

is independent of $z \in B$ (and equals $\operatorname{ind}(\mathsf{D}^z)$, by the homotopy invariance of the index). As in Proposition 9.20, we can also prove that if the dimension of $\ker((\mathsf{D}+R)^z)$ is constant for all $z \in B$, then the Green operator $G_R = (G_R^z \mid z \in B)$ of $\mathsf{D}+R$ depends smoothly on z, in the sense that if X is a family of smoothing operators, then XG_R and $G_R X$ are families of smoothing operators.

Note that a superconnection $\mathbb{A}$ adapted to $\mathsf{D}+R$ will have $\mathbb{A}_{[0]} = \mathsf{D}+R$, and hence cannot be a differential operator when considered as an operator on $\Gamma(M, \pi^*\Lambda T^*B \otimes \mathscr{E} \otimes |\Lambda_\pi|^{1/2})$. Thus, we will allow a superconnection $\mathbb{A}$ for $\mathsf{D}+R$ to be the sum of a differential operator on $\mathscr{A}(B, \pi_*\mathscr{E}) = \Gamma(M, \pi^*\Lambda T^*B \otimes \mathscr{E} \otimes |\Lambda_\pi|)$ and an element of $\mathscr{A}(B, \operatorname{End}_{\mathscr{P}}(\mathscr{E}))$. This does not cause any change in the wording of the analogues of Theorem 9.17 and 9.19 for $\mathsf{D}+R$, nor in the estimates required in their proof.

First we have the analogue of Theorem 9.17.

Theorem 9.25. *Let* $\mathsf{D}+R$ *be the sum of family of Dirac operators* D *and an odd self-adjoint family of smoothing operators* R. *Let* $\mathbb{A}$ *be a superconnection for the family* $\mathsf{D}+R$.

1. *The differential form* $\operatorname{ch}(\mathbb{A})$ *is closed.*
2. *If* $\mathbb{A}_\sigma$ *is a one-parameter family of superconnections on the bundle* $\pi_*\mathscr{E}$, *then*
$$\frac{d}{d\sigma}\operatorname{ch}(\mathbb{A}_\sigma) = -d_B \operatorname{Str}\Big(\frac{d\mathbb{A}_\sigma}{d\sigma} e^{-\mathscr{F}_\sigma}\Big).$$
Thus, the class of $\operatorname{ch}(\mathbb{A}_\sigma)$ *in de Rham cohomology is a homotopy invariant of the superconnection* $\mathbb{A}$.
3. *In particular, the class of* $\operatorname{ch}(\mathbb{A})$ *in de Rham cohomology depends only on the homotopy class of the Dirac operator* D.

Next, we have the analogue of Theorem 9.19.

Theorem 9.26. *Assume that* $\ker((\mathsf{D}+R)^z)$ *has constant dimension for* $z \in B$.

1. *The vector spaces* $\ker((\mathsf{D}+R)^z)$ *form a smooth vector bundle* $\ker(\mathsf{D}+R)$ *over* B.
2. *The limit*
$$\lim_{t\to\infty} \operatorname{ch}(\mathbb{A}_t) = \operatorname{ch}\big(\ker(\mathsf{D}+R)\big) \in \mathscr{A}(B)$$
holds with respect to each C^ℓ*-norm on compact subsets of* B. □

We could also state the analogue of the transgression formula, but since we will not make use of it, we leave its formulation to the reader.

9.4 The Equivariant Index and the Index Bundle

In this section, we will consider the special case of a family of Dirac operators on a bundle $M \to B$ with compact structure group G. We will see that the Chern character of the index bundle may be calculated using the Kirillov formula (Theorem 8.2). This section is not needed to understand the rest of this chapter or the next.

Let N be a compact Riemannian manifold with an action of a compact Lie group G. Let E be an equivariant Hermitian Clifford bundle over N, with equivariant Clifford connection ∇^E and associated Dirac operator D_N acting on $\Gamma(N,E)$. Then $\ker(\mathsf{D}_N^+)$ and $\ker(\mathsf{D}_N^-)$ are finite-dimensional G-modules; we will denote the actions of G and of its Lie algebra $\mathfrak{g}$ on $\ker(\mathsf{D}_N)$ by ρ_N.

Let $P \to B$ be a principal bundle with structure group G, and let $M = P \times_G N$ be the associated fibre bundle, with base B and typical fibre N. We also form the associated bundle $\mathscr{E} = P \times_G E$, which we consider to be a bundle over M; thus, $\mathscr{E}$ defines a family of bundles $(\mathscr{E}_z \to M_z \mid z \in B)$. A section of $\mathscr{E}$ is a G-invariant section of the pull-back of E to $P \times N$, that is a section $\phi \in \Gamma(P \times N, E)$ such that

$$\phi(pg, g^{-1}x) = g^{-1}\phi(p,x) \quad \text{for } p \in P \text{ and } x \in N.$$

The operator D_N acts fibrewise on $\Gamma(P \times N, E)$ and preserves the subspace of G-invariant sections. Thus D_N induces a family of Dirac operators $\mathsf{D} = (\mathsf{D}^z \mid z \in B)$. It is clear that in this special case the vector space $\ker(\mathsf{D}^z)$ has constant dimension, and that the superbundle $\ker(\mathsf{D})$ over B is induced by the $\mathbb{Z}_2$-graded G-module $\ker(\mathsf{D}_N)$:

$$\ker(\mathsf{D}) = P \times_G \ker(\mathsf{D}_N).$$

Using the Riemannian structure on N, we identify $|\Lambda_N|^{1/2}$ with the trivial line bundle. Let $\theta = \sum_i \theta^i X_i \in \mathscr{A}^1(P) \otimes \mathfrak{g}$ be a connection one-form on the principal bundle $P \to B$, with curvature two-form $\Omega_P \in \mathscr{A}^2(P, \mathfrak{g})_{\mathrm{bas}}$. Since $\pi_*\mathscr{E}$ is an associated vector bundle $P \times_G \Gamma(N,E)$ with infinite dimensional fibre, we deduce from the connection θ on P a natural connection on the bundle $\pi_*\mathscr{E} \to B$. If $\mathscr{L}^E(X_i)$ is the action of the Lie algebra element $X_i \in \mathfrak{g}$ on $\Gamma(N,E)$, we see that

$$\nabla^{\pi_*\mathscr{E}} = d_P + \sum_i \theta^i \mathscr{L}^E(X_i)$$

on G-invariant sections.

The following lemma follows from the obvious fact that $\nabla^{\pi_*\mathscr{E}}$ preserves the bundle $\ker(\mathsf{D})$, which itself follows from the fact that $\mathscr{L}^E(X_i)$ preserves $\ker(\mathsf{D}_N)$.

Lemma 9.27. *The projected connection ∇_0 on $\ker(\mathsf{D})$ may be identified with the connection on $P \times_G \ker(\mathsf{D}_N)$ associated to the connection one-form θ.*

Denote by $\mathrm{ch}(\ker(\mathsf{D}))$ the Chern character of $\ker(\mathsf{D})$ with respect to the connection ∇_0. It follows from this lemma that the Chern character form $\mathrm{ch}(\nabla_0)$ is given in terms of the representation ρ_N by the formula

$$\mathrm{ch}(\nabla_0) = \mathrm{Str}\big(e^{-\rho_N(\Omega_P)}\big).$$

Denote by $I_{\mathfrak{g}}$ the G-invariant analytic function on $\mathfrak{g}$ given by the formula

$$I_{\mathfrak{g}}(X) = \operatorname{ind}_G(e^{-X}, \mathsf{D}_N).$$

The Chern-Weil homomorphism ϕ_θ of Definition 7.33 maps the equivariant de Rham complex $(\mathscr{A}_G(N), d_{\mathfrak{g}})$ to the de Rham complex $(\mathscr{A}(M), d)$. On $\mathbb{C}[\mathfrak{g}]^G$, it coincides with the ordinary Chern-Weil homomorphism, that is, $\alpha \mapsto \alpha(\Omega_P)$, and maps into the space of basic forms on P, which may be identified with forms on B. It follows that

$$\phi_\theta(I_{\mathfrak{g}}) = \operatorname{Str}(e^{-\rho_N(\Omega_P)}) = \operatorname{ch}(\ker(\mathsf{D}))$$

The vertical tangent bundle $T(M/B)$ of M may be identified with the associated bundle $P \times_G TN$. It has a connection $\nabla^{N,\theta}$, obtained by combining the Levi-Civita connection on TN with the connection on the principal bundle P as in Section 7.6. Let $R^{M/B}$ be the curvature of the connection $\nabla^{N,\theta}$. Denote by $\hat{A}(M/B)$ the $\hat{A}$-genus of the connection $\nabla^{N,\theta}$,

$$\hat{A}(M/B) = \det{}^{1/2}\left(\frac{R^{M/B}/2}{\sinh(R^{M/B}/2)}\right).$$

To simplify the discussion, we will assume that N has a spin-structure with spinor bundle $\mathscr{S}(N)$, and that $E = W \otimes \mathscr{S}(N)$ for some G-equivariant Hermitian twisting superbundle W with connection ∇^W. Let $\mathscr{W} = P \times_G W$ be the corresponding Hermitian superbundle over M. Given a connection on W, we obtain a connection $\nabla^{\mathscr{W}}$ on $\mathscr{W}$; denote by $\operatorname{ch}(\mathscr{W})$ the Chern character form of the connection $\nabla^{\mathscr{W}}$:

$$\operatorname{ch}(\mathscr{W}) = \operatorname{Str}(\exp(-F^{\mathscr{W}})).$$

The differential form $\hat{A}(M/B)\operatorname{ch}(\mathscr{W})$ is a closed differential form on M, whose integral over the fibres of M/B is a closed differential form on B.

Proposition 9.28. *The Chern character of the connection ∇_0 on the index bundle* $\ker(\mathsf{D})$ *is given by the formula*

$$\operatorname{ch}(\nabla_0) = (2\pi i)^{-\dim(N)/2} \int_{M/B} \hat{A}(M/B)\operatorname{ch}(\mathscr{W}).$$

Proof. The proof is based on the Kirillov formula of Chapter 8:

$$I_{\mathfrak{g}}(X) = (2\pi i)^{-\dim(N)/2} \int_N \hat{A}_{\mathfrak{g}}(X,N)\operatorname{ch}_{\mathfrak{g}}(X,W),$$

where $\hat{A}_{\mathfrak{g}}(X,N)$ and $\operatorname{ch}_{\mathfrak{g}}(X,W)$ are equivariant characteristic classes given by the formulas

$$\hat{A}_{\mathfrak{g}}(X,N) = \det{}^{1/2}\left(\frac{R^N_{\mathfrak{g}}(X)/2}{\sinh(R^N_{\mathfrak{g}}(X)/2)}\right)$$

$$\operatorname{ch}_{\mathfrak{g}}(X,W) = \operatorname{Str}(\exp(-F^W_{\mathfrak{g}}(X))).$$

By Proposition 7.38 (2), we see that the Chern-Weil map $\phi_\theta : \mathscr{A}_G(N) \to \mathscr{A}(M)$ sends these equivariant characteristic forms to the corresponding characteristic forms of the associated bundles:

$$\phi_\theta(\hat{A}_{\mathfrak{g}}(N)) = \hat{A}(M/B),$$
$$\phi_\theta(\mathrm{ch}_{\mathfrak{g}}(W)) = \mathrm{ch}(\mathscr{W}).$$

The result follows from the formula $\int_{M/B} \circ \phi_\theta = \phi_\theta \circ \int_N$. □

We will see in Chapter 10 that the above formula for $\mathrm{ch}(\ker(\mathsf{D}))$ holds in cohomology for any family of Dirac operators. We will also show that the local version of the family index theorem implies the local Kirillov formula.

9.5 The Case of Varying Dimension

In this section, following Atiyah and Singer, we will show that the index bundle of a family of Dirac operators over a compact base B may be represented as the formal difference of two vector bundles, even without the assumption that $\ker(\mathsf{D})$ has constant dimension.

Let us start by recalling the definition of a difference bundle. This is an equivalence class of superbundles on a manifold B, with respect to the equivalence relation that $\mathscr{E} \sim \mathscr{F}$ if there are vector bundles $\mathscr{G}$ and $\mathscr{H}$ such that

$$\mathscr{E}^+ \oplus \mathscr{G} \cong \mathscr{F}^+ \oplus \mathscr{H};$$
$$\mathscr{E}^- \oplus \mathscr{G} \cong \mathscr{F}^- \oplus \mathscr{H}.$$

The class of $\mathscr{E}$ is denoted by $[\mathscr{E}^+] - [\mathscr{E}^-]$ and is called the **difference bundle** of $\mathscr{E}$. If $\mathscr{G}$ is a vector bundle, we will associate to it the difference bundle $[\mathscr{G}] - [0]$.

If B is compact, the difference bundles forms an abelian ring $K(B)$ called the **K-theory** of B. The sum in this ring is induced by taking the direct sum, and the product by taking the tensor product, of superbundle representatives. It is easily verified that the additive identity is the zero bundle, the multiplicative identity is the class of the trivial line bundle, which we will denote by $[\mathbb{C}]$, and the negative of $[\mathscr{E}^+] - [\mathscr{E}^-]$ is $[\mathscr{E}^-] - [\mathscr{E}^+]$. We will not give any further results about the ring $K(B)$: as explained in the introduction, our emphasis in this book is on explicit representatives, by superbundles or differential forms, rather than the classes that they represent in K-theory $K(B)$ or de Rham cohomology $H^\bullet(B)$. We do note, however, that the Chern character defines a homomorphism of rings from $K(B)$ to the de Rham cohomology $H^\bullet(B)$ of B; this follows from the fact that the Chern character vanishes on superbundles of the form $\mathscr{G} \oplus \mathscr{G}$. In fact, this map may be shown to induce an isomorphism of $K(B) \otimes \mathbb{C}$ with the even de Rham cohomology, by the Atiyah-Hirzebruch spectral sequence.

If $\ker(\mathsf{D})$ has constant dimension, we have defined the index bundle to be the difference bundle represented by $\ker(\mathsf{D})$. Since M is compact, there is a positive ε

such that $U_\lambda = M$ for $0 < \lambda < \varepsilon$. It follows that ind(D) is also represented by $\mathscr{H}_{[0,\lambda)}$ for $0 < \lambda < \varepsilon$. Motivated by this, we would like to define ind(D) in such a way that when restricted to U_λ, it has the superbundle $\mathscr{H}_{[0,\lambda)}$ as representative. The following lemma shows that this approach is consistent on $U_\lambda \cap U_\mu$, for $0 < \lambda < \mu$.

Lemma 9.29. *If $0 < \lambda < \mu$, then the difference bundles represented by $\mathscr{H}_{[0,\lambda)}$ and $\mathscr{H}_{[0,\mu)}$ are equal.*

Proof. By the isomorphism

$$\mathscr{H}_{[0,\mu)} = \mathscr{H}_{[0,\lambda)} \oplus \mathscr{H}_{(\lambda,\mu)},$$

we see that we must prove that on $U_\lambda \cap U_\mu$,

$$\mathscr{H}^+_{(\lambda,\mu)} \cong \mathscr{H}^-_{(\lambda,\mu)}.$$

On the bundle $\mathscr{H}^+_{(\lambda,\mu)}$, the operator $(\mathsf{D}^+)^*\mathsf{D}^+ = \mathsf{D}^-\mathsf{D}^+$ has spectrum lying in the interval (λ,μ), and hence is invertible (an explicit inverse is given by the Green operator G). It follows that the operator

$$\mathsf{D}^+ : \mathscr{H}^+_{(\lambda,\mu)} \to \mathscr{H}^-_{(\lambda,\mu)}$$

induces the desired isomorphism of bundles. □

We will show that if B is compact, there is a difference bundle of the form $[\mathscr{E}] - [\mathbb{C}^N]$ which when restricted to U_λ is equivalent to the difference bundle $\mathscr{H}_{[0,\lambda)}$. This will show that ind(D) may be considered as an element of $K(B)$. We will also generalize Theorem 9.19 to give a formula for the Chern character of ind(D) in the general case.

We make use of the simple observation that if D is a family of Dirac operators such that $\ker(\mathsf{D}^-)$ vanishes, then

$$\dim(\ker(\mathsf{D}^z)) = \operatorname{ind}(\mathsf{D}^z),$$

and is thus independent of z. Thus, one way to perturb a family so that its kernel ker(D) becomes a vector bundle is to modify it in such a way as to ensure that D^+ is surjective for each z, and hence that D^- has vanishing kernel. This is not difficult if we permit ourselves to perturb D^z by a family of smoothing operators.

We introduce the trick of replacing M by an enlarged version $M' = M \cup B$, obtained by adding one point p_z to each fibre M_z of $M \to B$. Similarly, we replace $\mathscr{E}$ by the bundle $\mathscr{E}'$, which coincides with $\mathscr{E}$ over M and whose fibre over the point p_z is a constant vector space V with $V^+ = \mathbb{C}^N$ and $V^- = 0$. Note that $\pi_*\mathscr{E}'$ is isomorphic to $\pi_*\mathscr{E} \oplus (B \times \mathbb{C}^N)$, so that

$$\Gamma(B, \pi_*\mathscr{E}') \cong \Gamma(B, \pi_*\mathscr{E}) \oplus C^\infty(B, \mathbb{C}^N).$$

The family D of Dirac operators corresponds in a natural way to a family of Dirac operators on the family of vector bundles $\mathscr{E}' \to M'$.

Let $\psi : \mathbb{C}^N \to \Gamma(B, \pi_*\mathscr{E}^-)$ be a linear map from $\mathbb{C}^N$ to $\Gamma(B, \pi_*\mathscr{E}^-)$. Let ψ_i be the sections of $\pi_*\mathscr{E}^-$ corresponding to the basis vector $e_i \in \mathbb{C}^N$. We define a family of smoothing operators $R_\psi \in \Gamma(B, \mathscr{K}(\mathscr{E}'))$ on the vector bundles $\mathscr{E}' \to M'$ by the formulas

$$R_\psi \phi = \sum e_i(\psi_i, \phi), \quad \text{for } \phi \in \Gamma(B, \pi_*\mathscr{E}) \subset \Gamma(B, \pi_*\mathscr{E}'),$$
$$R_\psi e_i = \psi_i.$$

It is easy to see that R_ψ is odd and symmetric. We define D_ψ to equal $\mathsf{D} + R_\psi$, given explicitly by the formula

$$\mathsf{D}_\psi^+\Big(\phi \oplus \sum_{i=1}^N u_i e_i\Big) = \mathsf{D}^+\phi + \sum_{i=1}^N u_i \psi_i,$$
$$\mathsf{D}_\psi^-(\phi) = \mathsf{D}^-\phi \oplus \sum_{i=1}^N (\psi_i, \phi) e_i,$$

for $\phi \in \Gamma(B, \pi_*\mathscr{E}^+)$ and $\Gamma(B, \pi_*\mathscr{E}^-)$ respectively, and $u_i \in C^\infty(B)$. We thus have a family $(\mathsf{D}_\psi^+)^z$ of the type described in (9.4).

In order for $\ker(\mathsf{D}_\psi^-)$ to vanish, we must demand that D_ψ^+ is surjective for all $z \in B$. The following lemma shows how to choose suitable ψ_i.

Lemma 9.30. *There exists an integer N and a map $\psi : \mathbb{C}^N \to \Gamma(B, \pi_*\mathscr{E}^-)$ such that the map $(\mathsf{D}_\psi^+)^z$ is surjective at each point in $z \in B$.*

Proof. By Proposition 3.48, we know that

$$\ker(\mathsf{D}_z^-) \oplus \mathsf{D}_z^+\big[\Gamma(M_z, \mathscr{E}_z^+ \otimes |\Lambda_\pi|^{1/2})\big] = \Gamma(M_z, \mathscr{E}_z^- \otimes |\Lambda_\pi|^{1/2})$$

at each point $z \in B$. We need to construct a collection ψ_i of sections of $\pi_*\mathscr{E}^-$ such that at each point $z \in B$, the projection onto the vector space $\ker(\mathsf{D}_z^-)$ of the sections ψ_i^z span it. We proceed as follows.

Given $z_0 \in B$, we can find a $\lambda > 0$ such that $z_0 \in U_\lambda$, a ball $U(z_0) \subset B$ around z_0 and sections $\phi_i \in \Gamma(U(z_0), \pi_*\mathscr{E})$ such that for every $z \in U(z_0)$ the elements ϕ_i^z form a basis of $(\mathscr{H}^-_{[0,\lambda)})_z$. If $\chi \in C_c^\infty(U(z_0))$ is a cut-off function on $U(z_0)$ equal to 1 around z_0, the sections $\chi(z)\phi_i^z$ can be extended to smooth sections of the bundle $\pi_*\mathscr{E}$ on B. Thus, for each $z_0 \in B$, there exists a $\lambda > 0$, an open set $V(z_0) \subset U(z_0)$ and sections $\psi_i \in \Gamma(B, \pi_*\mathscr{E})$ such that ψ_i^z span $(\mathscr{H}^-_{[0,\lambda)})_z$, and hence their projections onto the vector space $\ker((\mathsf{D}^-)^z)$ span this space, for $z \in V(z_0)$. Since B is compact, it is covered by a finite number of sets $V(z_0)$, and we may take for (ψ_i) the union of the sections ψ_i^z for just these sets. □

If λ is a positive number, let $U_{\lambda,\psi}$ be the open subset of U_λ consisting of those points $z \in U_\lambda$ such that the composition of

$$\mathbb{C}^N \xrightarrow{\psi} (\pi_*\mathscr{E}^-)_z \xrightarrow{P_{[0,\lambda)}} (\mathscr{H}_{[0,\lambda)})_z$$

is surjective. If λ_1 is the smallest non-zero eigenvalue of $(\mathsf{D}^{z_0})^2$, then z_0 is in $U_{\lambda,\psi}$ if $\lambda < \lambda_1$. Thus, B is covered by the open sets $\{U_{\lambda,\psi} \mid \lambda \in \mathbb{R}\}$.

Proposition 9.31. *1. If ψ satisfies the conditions of Lemma 9.30, we have the short exact sequence of vector bundles on $U_{\lambda,\psi}$,*

$$0 \to \ker(\mathsf{D}_\psi) \xrightarrow{P_{[0,\lambda)}\oplus 1} \mathscr{H}^+_{[0,\lambda)} \oplus (B \times \mathbb{C}^N) \xrightarrow{P_{[0,\lambda)}\mathsf{D}_\psi} \mathscr{H}^-_{[0,\lambda)} \to 0.$$

Thus, over U_λ, the difference bundles $[\ker(\mathsf{D}_\psi)] - [\mathbb{C}^N]$ and $\mathscr{H}_{[0,\lambda)}$ are equivalent.

2. The difference bundle $[\ker(\mathsf{D}_\psi)] - [\mathbb{C}^N]$ of $K(B)$ is independent of the choice of map $\psi : \mathbb{C}^N \to \Gamma(B, \pi_*\mathscr{E}^-)$ satisfying Lemma 9.30.

Proof. Let us show that the bundle map

$$P_{[0,\lambda)} \oplus 1 : \ker(\mathsf{D}_\psi) \subset \pi_*\mathscr{E}^+ \oplus (B \times \mathbb{C}^N) \to \mathscr{H}^+_{[0,\lambda)} \oplus (B \times \mathbb{C}^N)$$

is injective. Indeed, if $(\phi, u) \in \ker(\mathsf{D}_\psi)$ lies in the kernel of this map, then we see that $u = 0$, and hence that $\phi \in \ker(\mathsf{D})$. However, the projection $P_{[0,\lambda)}$ is the identity on $\ker(\mathsf{D})$ if $\lambda > 0$, and hence $\phi = 0$.

The bundle map

$$P_{[0,\lambda)}\mathsf{D}_\psi : \mathscr{H}^+_{[0,\lambda)} \oplus (B \times \mathbb{C}^N) \to \mathscr{H}^-_{[0,\lambda)}$$

is surjective by the definition of $U_{\lambda,\psi}$.

The composition of the two maps

$$\ker(\mathsf{D}_\psi) \subset \pi_*\mathscr{E}^+ \oplus (B \times \mathbb{C}^N) \xrightarrow{P_{[0,\lambda)}\oplus 1} \mathscr{H}^+_{[0,\lambda)} \oplus (B \times \mathbb{C}^N) \xrightarrow{P_{[0,\lambda)}\mathsf{D}_\psi} \mathscr{H}^-_{[0,\lambda)}$$

is the bundle map from $\ker(\mathsf{D}_\psi)$ to $\mathscr{H}^-_{[0,\lambda)}$ which sends $(\phi, u) \in \ker(\mathsf{D}_\psi)$ to

$$P_{[0,\lambda)}(\mathsf{D}P_{[0,\lambda)}\phi) + P_{[0,\lambda)}\psi(u) = P_{[0,\lambda)}(\mathsf{D}\phi + \psi(u)) = P_{[0,\lambda)}\mathsf{D}_\psi(\phi, u),$$

which vanishes, since $(\phi, u) \in \ker(\mathsf{D}_\psi)$. Thus, we see that

$$0 \to \ker(\mathsf{D}_\psi) \xrightarrow{P_{[0,\lambda)}\oplus 1} \mathscr{H}^+_{[0,\lambda)} \oplus (B \times \mathbb{C}^N) \xrightarrow{P_{[0,\lambda)}\mathsf{D}_\psi} \mathscr{H}^-_{[0,\lambda)} \to 0$$

is a complex. Let us show that it is a short exact sequence. If $\phi \in \left(\mathscr{H}^+_{[0,\lambda)}\right)_z$ and $u \in \mathbb{C}^N$ are such that $P^z_{[0,\lambda)}(\mathsf{D}^z\phi + \psi^z(u)) = 0$, then there exists an $\alpha \in \left(\pi_*\mathscr{E}^+\right)_z$ such that $P^z_{[0,\lambda)}\alpha = 0$ and $(\mathsf{D}^+)^z\phi + \psi^z(u) = (\mathsf{D}^+)^z\alpha$. Thus $(\phi - \alpha, u) \in \ker(\mathsf{D}_\psi)$, and its image by $P_{[0,\lambda)}$ is our element (ϕ, u).

It remains to prove the independence of this construction from the map $\psi : \mathbb{C}^N \to \Gamma(B, \pi_*\mathscr{E}^-)$. Consider another map $\bar{\psi} : \mathbb{C}^{\bar{N}} \to \Gamma(B, \pi_*\mathscr{E}^-)$. The bundle $\ker(\mathsf{D}_\psi)$ may be identified with a subbundle of $\ker(\mathsf{D}_{\psi\oplus\bar{\psi}})$ by sending $(\phi, u) \in \left(\pi_*\mathscr{E}^+\right)_z \oplus \mathbb{C}^N$ to $(\phi, u, 0) \in \left(\pi_*\mathscr{E}^+\right)_z \oplus \mathbb{C}^N \oplus \mathbb{C}^{\bar{N}}$. Consider the sequence of bundles

$$0 \to \ker(\mathsf{D}_\psi) \to \ker(\mathsf{D}_{\psi\oplus\bar{\psi}}) \to (B \times \mathbb{C}^{\bar{N}}) \to 0,$$

where the last arrow maps $(\phi, u, \bar{u}) \in \ker(\mathsf{D}_{\psi\oplus\bar{\psi}})$ to $\bar{u} \in \mathbb{C}^{\bar{N}}$. This sequence is exact at each point $z \in B$. To see this, choose $\bar{u} \in \mathbb{C}^{\bar{N}}$. The surjectivity of D_ψ implies that there exists $u \in \mathbb{C}^N$ and $\phi \in \pi_*\mathscr{E}_z$ such that $\bar{\psi}(\bar{u}) = \mathsf{D}\phi + \psi(u)$. □

Using this proposition, we may define the index bundle of any family of Dirac operators on a compact base.

Definition 9.32. *The* **index bundle** *of a smooth family of Dirac operators over a compact base B is the virtual bundle*

$$\operatorname{ind}(\mathsf{D}) = [\ker(\mathsf{D}_\psi)] - [\mathbb{C}^N],$$

where ψ is chosen as in Proposition 9.31.

We can now prove a generalization of the McKean-Singer formula to superconnections associated to families of Dirac operators.

Theorem 9.33. *Let $\mathbb{A}$ be a superconnection on the bundle $\pi_*\mathscr{E}$ for the family* D. *Then the cohomology class of* $\operatorname{ch}(\mathbb{A})$ *is equal to the Chern character of the index bundle* $\operatorname{ind}(\mathsf{D})$.

Proof. Unlike in the case where $\ker(\mathsf{D})$ had constant dimension, the differential form $\operatorname{Str}(e^{-t\delta_t(\mathbb{A}^2)})$ may have no limit as $t \to \infty$ when the dimension of $\ker(\mathsf{D})$ is not constant. However, we may replace the family D over $M \to B$ by a family D_ψ over $M' \to B$ such that $\ker(\mathsf{D}_\psi)$ is a vector bundle.

Let $\mathbb{A}$ be a superconnection for the family D, and transfer the operators $\mathbb{A}_{[i]}$ to operators on $\mathscr{A}(B, \pi_*\mathscr{E}')$. Let $\tilde{\mathbb{A}}_\varepsilon$ denote the superconnection $\mathbb{A} + \varepsilon R_\psi$ for the family $\mathsf{D} + \varepsilon R_\psi$. Applying Theorem 9.25 to the one-parameter family $\tilde{\mathbb{A}}_\varepsilon$, we see that the cohomology class of the differential form

$$\operatorname{ch}(\tilde{\mathbb{A}}_\varepsilon) = \operatorname{Str}(e^{-\tilde{\mathbb{A}}_\varepsilon^2})$$

does not depend on ε. For $\varepsilon = 1$, we know by Theorem 9.26 that $\operatorname{ch}(\tilde{\mathbb{A}}_\varepsilon)$ represents the cohomology class of $\operatorname{ch}(\ker(\mathsf{D}_\psi))$, the Chern character of the vector bundle $\ker(\mathsf{D}_\psi)$. Taking $\varepsilon = 0$, we see that

$$\operatorname{ch}(\tilde{\mathbb{A}}_0) = \operatorname{ch}(\mathbb{A}) + N \in \mathscr{A}(B).$$

It follows that $\operatorname{ch}(\mathbb{A})$ and $\operatorname{ch}(\ker(\mathsf{D}_\psi)) - N$ represent the same class in cohomology, from which the theorem follows. □

9.6 The Zeta-Function of a Laplacian

In this section, we give an application of the existence of the asymptotic expansion for the heat kernel: the construction of the zeta-function of a generalized Laplacian and the definition of its renormalized determinant. This construction, applied to a family of Dirac operators, is used in the theory of determinant line bundles, which we will describe in the next section. The reader only interested in the family index theorem may omit the rest of this chapter.

If $f \in C^\infty(0,\infty)$ is a function on the positive real line (well-behaved at 0 and ∞ in a sense which we will describe shortly) the **Mellin transform** of f is the function

$$\mathsf{M}[f](s) = \frac{1}{\Gamma(s)} \int_0^\infty f(t) t^{s-1}\, dt.$$

Lemma 9.34. *Let $f \in C^\infty(0,\infty)$ be a function with asymptotic expansion for small t of the form*

$$f(t) \sim \sum_{k \geq -n} f_k t^{k/2} + g \log t,$$

and which decays exponentially at infinity, that is, for some $\lambda > 0$, and t sufficiently large,

$$|f(t)| \leq C e^{-t\lambda}.$$

Let γ be the Euler constant. Then

1. *the Mellin transform $\mathsf{M}[f]$ is a meromorphic function with poles contained in the set $n/2 - \mathbb{N}/2$;*
2. *the Laurent series of $\mathsf{M}[f]$ around $s = 0$ is $-gs^{-1} + (f_0 - \gamma g) + O(s)$.*

Let $h \in C^\infty(0,\infty)$ be a function with asymptotic expansion for small t of the form

$$h(t) \sim \sum_{k \geq -n} h_k t^{k/2},$$

and such that for some $\lambda > 0$, and t large,

$$|h(t)| \leq C e^{-t\lambda}.$$

Then $f(t) = \int_t^\infty h(s)\, ds$ satisfies the above assumptions.

Proof. Choose $\operatorname{Re} s$ large, and split the integration over $[0,\infty)$ into the intervals $[0,1]$ and $(1,\infty)$:

$$\Gamma(s)\mathsf{M}[f](s) = \int_0^1 f(t) t^{s-1}\, dt + \int_1^\infty f(t) t^{s-1}\, dt.$$

The second term, the contribution of the integral at infinity, is clearly entire, by the exponential decay of $f(t)$ as $t \to \infty$. The first term may be expanded in an asymptotic series, and then integrated explicitly, term by term:

$$\begin{aligned}\int_0^1 f(t) t^{s-1}\, dt &= \sum_{k<K} f_k \int_0^1 t^{k/2+s-1}\, dt \\ &\quad + g \int_0^1 (\log t) t^{s-1}\, dt + \int_0^1 O(t^{K/2+s-1})\, dt \\ &= \sum_{k<K} \frac{f_k}{s+k/2} - \frac{g}{s^2} + r(s),\end{aligned}$$

where $r(s)$ is holomorphic in the right-half plane $\operatorname{Re} s > -K/2$. Since the inverse of the Gamma function $\Gamma(s)^{-1}$ is entire, (1) follows.

Since $\Gamma(s)^{-1} = s + \gamma s^2 + O(s^3)$, it follows that $\mathsf{M}[f]$ has the Taylor series around $s = 0$

$$-gs^{-1}+(f_0-\gamma g)+O(s),$$

proving part (2) of the lemma.

If h is as in the statement of the lemma, then to show that $\int_t^\infty h(s)\,ds$ has an asymptotic expansion of the required form, we write, for t small,

$$f(t)=\int_t^1\sum_{k<0}\left(h_k s^{k/2}\right)ds-\int_0^t\Big(h(s)-\sum_{k<0}h_k s^{k/2}\Big)ds$$
$$+\int_0^1\Big(h(s)-\sum_{k<0}h_k s^{k/2}\Big)ds+\int_1^\infty h(s)\,ds$$

from which we obtain

$$f(t)\sim-\sum_{k\neq-2}\frac{h_k}{k/2+1}t^{k/2+1}-h_{-2}\log t+c,$$

where c is the constant

$$c=\sum_{\{k<0|k\neq-2\}}\frac{h_k}{k/2+1}+\int_0^1\Big(h(s)-\sum_{k<0}h_k s^{k/2}\Big)ds+\int_1^\infty h(s)\,ds. \qquad \square$$

If we define $\mathrm{LIM}_{t\to0}f(t)$ to equal $f_0-\gamma g$, where f_0 and g are the coefficients of t^0 and $\log t$ in the asymptotic expansion of f, we see that part (2) of the above lemma implies the formula

$$\frac{d}{ds}\Big|_{s=0}s\mathsf{M}[f]=\underset{t\to0}{\mathrm{LIM}}\,f(t),$$

We will call $\mathrm{LIM}_{t\to0}f(t)$ the **renormalized limit** of $f(t)$ as $t\to0$.

If H is a self-adjoint generalized Laplacian on a compact manifold M and λ is a positive real number, the **zeta-function** of H is the Mellin transform

$$\zeta(s,H,\lambda)=\mathsf{M}\big[\mathrm{Tr}(P_{(\lambda,\infty)}e^{-tH})\big]=\frac{1}{\Gamma(s)}\int_0^\infty\mathrm{Tr}(P_{(\lambda,\infty)}e^{-tH})t^{s-1}\,dt,$$

where $P_{(\lambda,\infty)}$ is the spectral projection associated to H onto the eigenvalues lying in (λ,∞). This function equals $\mathrm{Tr}(P_{(\lambda,\infty)}H^{-s})$ for $s>n/2$, where n is the dimension of the manifold M.

As an example, consider the scalar Laplacian $-d^2/dt^2$ on the circle, with eigenvalues $\{0,1,1,4,4,\dots,n^2,n^2,\dots\}$. If $0<\lambda<1$ and $s>1/2$, we see that

$$\zeta(s,H,\lambda)=\frac{2}{\Gamma(s)}\sum_{n=1}^\infty\int_0^\infty e^{-tn^2}t^{s-1}\,dt=2\sum_{n=1}^\infty n^{-2s}=2\zeta(2s),$$

where $\zeta(s)$ is the Riemann zeta function.

We can understand the behaviour of the zeta-function as a function on the complex plane by means of the Minakshisundaram-Pleijel asymptotic expansion of the heat kernel.

Proposition 9.35. *The function $\zeta(s,H,\lambda)$ possesses a meromorphic extension to the whole complex plane, and is holomorphic at $s=0$.*

Proof. As $t \to 0$, we have

$$\begin{aligned}\operatorname{Tr}\big(P_{(\lambda,\infty)}e^{-tH}\big) &= \operatorname{Tr}(e^{-tH}) - \sum_{k=0}^{m} e^{-t\lambda_k}\\ &\sim \sum_{k=-n/2}^{\infty} t^k a_k,\end{aligned}$$

where $\lambda_0 \le \cdots \le \lambda_m \le \lambda$ are the eigenvalues of H lying in $(-\infty,\lambda]$, enumerated according to multiplicity. The fact that $\operatorname{Tr}\big(P_{(\lambda,\infty)}e^{-tH}\big)$ decays exponentially fast as $t \to \infty$ has been proved in Lemma 2.37. □

Denote by $\zeta'(s,H,\lambda)$ the derivative of $\zeta(s,H,\lambda)$ with respect to s. Following Ray and Singer, we define the **zeta-function determinant** of a generalized Laplacian H for which 0 is not an eigenvalue to be

$$\det(H) = e^{-\zeta'(0,H,0)},$$

This definition is motivated by the fact that if H is a positive endomorphism of a finite dimensional Hermitian vector space V, and

$$\zeta(s,H) = \mathrm{M}\big[\operatorname{Tr}(e^{-tH})\big] = \operatorname{Tr}(H^{-s}),$$

then $\zeta'(0,H) = -\log\det(H)$. The zeta-function determinant of a generalized Laplacian H is a renormalized determinant adapted to the class of generalized Laplacians.

If H is a self-adjoint Laplacian, let $0 < \lambda < \mu$, and let $\lambda < \lambda_1 \le \cdots \le \lambda_m \le \mu$ be an enumeration of the eigenvalues of H lying in the interval $(\lambda,\mu]$. The following result is clear.

Proposition 9.36. *The zeta-function of H satisfies the formula*

$$\zeta(s,H,\lambda) = \zeta(s,H,\mu) + \sum_{i=1}^{m} \lambda_i^{-s},$$

and hence its derivative at $s=0$ satisfies the formula

$$\zeta'(0,H,\lambda) = \zeta'(0,H,\mu) - \sum_{i=1}^{m} \log\lambda_i.$$

In the following proposition, we calculate the variation of the zeta-function $\zeta(s,H^z,\lambda)$ of a family H^z of positive generalized Laplacians on M with respect to the parameter z. We work on the set U_λ where λ is not an eigenvalue of H^z.

Proposition 9.37. *Over the set U_λ, the derivative with respect to the parameter z of the zeta-function of a family of generalized Laplacians is given by the formula*

$$\frac{\partial\zeta(s,H^z,\lambda)}{\partial z} = -s\mathrm{M}\big[\operatorname{Tr}\big(P^z_{(\lambda,\infty)}\partial_z H^z (H^z)^{-1} e^{-tH^z}\big)\big].$$

Proof. Using the method of Lemma 9.34, it is not hard to show that the derivative $\partial\zeta(s,H^z,\lambda)/\partial z$ of the zeta-function is meromorphic. Thus, we may argue at large $\operatorname{Re}s$, where both sides of the formula are given by convergent integrals. If we differentiate the expression

$$\zeta(s,H^z,\lambda) = \frac{1}{\Gamma(s)}\int_0^\infty \operatorname{Tr}\left(P^z_{(\lambda,\infty)}e^{-tH^z}\right)t^{s-1}\,dt$$

for the zeta-function with respect to z and apply Corollary 2.50, we obtain that

$$\frac{\partial}{\partial z}\zeta(s,H^z,\lambda) = -\mathsf{M}\left[t\operatorname{Tr}(P^z_{(\lambda,\infty)}\partial_z H^z e^{-tH^z})\right] + \mathsf{M}\left[\operatorname{Tr}(\partial_z P^z_{(\lambda,\infty)}e^{-tH^z})\right]. \tag{9.5}$$

Since $P = P^z_{(\lambda,\infty)}$ is a projection, Leibniz's rule shows that

$$\partial_z P = P\partial_z P + \partial_z P P;$$

multiplying on the left and right by P, we see that $P\partial_z PP = 0$, so that

$$\partial_z P = P(\partial_z P)(1-P) + (1-P)(\partial_z P)P.$$

Thus the second term on the right-hand side of (9.5) vanishes. On the other hand, integration by parts shows that in general,

$$\mathsf{M}[tf'] = -s\mathsf{M}[f],$$

so that the first term can be rewritten in the desired form. □

The following theorem is the analogue for the zeta-function determinant of the formula

$$\frac{\partial}{\partial z}\log\det(H^z) = \operatorname{Tr}(\partial_z H^z(H^z)^{-1})$$

for the variation of the determinant of a family of operators H_z on a finite-dimensional vector space.

Proposition 9.38. *If H^z is a family of generalized Laplacians on a compact manifold M, then over the set U_λ,*

$$\frac{\partial}{\partial z}\zeta'(s=0,H^z,\lambda) = -\operatorname*{LIM}_{t\to 0}\operatorname{Tr}(P^z_{(\lambda,\infty)}\partial_z H^z(H^z)^{-1}e^{-tH^z}).$$

Proof. To apply Lemma 9.34 (2), we must show that

$$\operatorname{Tr}(P^z_{(\lambda,\infty)}\partial_z H^z(H^z)^{-1}e^{-tH^z})$$

has an asymptotic expansion. Consider the asymptotic expansion

$$\operatorname{Tr}(P^z_{(\lambda,\infty)}e^{-sH^z}) \sim \sum_{k=-n/2}^{\infty} a_k(z)s^k.$$

Recalling from the proof of the previous proposition that

$$\partial_z \operatorname{Tr}\big(P^z_{(\lambda,\infty)} e^{-sH^z}\big) = -s \operatorname{Tr}\big(P^z_{(\lambda,\infty)} \partial_z H^z e^{-sH^z}\big),$$

we see that $\operatorname{Tr}\big(P^z_{(\lambda,\infty)} \partial_z H^z e^{-sH^z}\big)$ has an asymptotic expansion of the form

$$\operatorname{Tr}\big(P^z_{(\lambda,\infty)} \partial_z H^z e^{-sH^z}\big) \sim - \sum_{k=-n/2}^{\infty} \partial_z a_k s^{k-1}.$$

Thus, we can apply Lemma 9.34 to conclude that

$$\int_t^\infty \operatorname{Tr}\big(P^z_{(\lambda,\infty)} \partial_z H^z e^{-sH^z}\big)\, ds = \operatorname{Tr}\big(P^z_{(\lambda,\infty)} \partial_z H^z (H^z)^{-1} e^{-tH^z}\big)$$

has an asymptotic expansion. □

9.7 The Determinant Line Bundle

Using the results of the last section, we will now define the determinant line bundle associated to a family of Dirac operators, its Quillen metric, and its natural connection. It is helpful, however, to see first how the theory looks in finite dimensions.

Let $\mathscr{H} = \mathscr{H}^+ \oplus \mathscr{H}^-$ be a finite-dimensional superbundle over a manifold B, such that that $\dim(\mathscr{H}^+) = \dim(\mathscr{H}^-) = m$. The map

$$\Lambda^m \mathsf{D}^+ : \Lambda^m \mathscr{H}^+ \to \Lambda^m \mathscr{H}^-$$

is a smooth section of the determinant line bundle

$$\det(\mathscr{H}) = (\Lambda^m \mathscr{H}^+)^{-1} \otimes \Lambda^m \mathscr{H}^-,$$

called the **determinant** of D^+, which we denote by $\det(\mathsf{D}^+)$. The section $\det(\mathsf{D}^+)$ vanishes at precisely those points where D^+ is not invertible. In particular, if D^+ is everywhere invertible, the section $\det(\mathsf{D}^+)$ defines a trivialization of the line bundle $\det(\mathscr{H})$, since it is nowhere vanishing.

Let $\nabla^{\mathscr{H}}$ be a connection on the bundle $\mathscr{H}$ which preserves the decomposition $\mathscr{H} = \mathscr{H}^+ \oplus \mathscr{H}^-$; it induces connections $\nabla^{\det(\mathscr{H})}$ and $\nabla^{\operatorname{End}(\mathscr{H})}$ on the bundles $\det(\mathscr{H})$ and $\operatorname{End}(\mathscr{H})$.

Proposition 9.39. *The section* $\det(\mathsf{D}^+)$ *satisfies the differential equation*

$$\nabla^{\det(\mathscr{H})} \det(\mathsf{D}^+) = -\det(\mathsf{D}^+) \operatorname{Tr}_{\mathscr{H}^+}\big((\mathsf{D}^+)^{-1} \nabla^{\operatorname{End}(\mathscr{H})} \mathsf{D}^+\big).$$

Proof. We may assume that B is an open set on which $\mathscr{H}$ has been trivialized, $\mathscr{H}^\pm = U \times H^\pm$, so that $\nabla = d + \omega$, where $\omega = \omega^+ \oplus \omega^-$. In this frame, the connection $\nabla^{\det(\mathscr{H})}$ is given by the formula $d - \operatorname{Str}(\omega)$. If we replace D^+ by a small perturbation $\mathsf{D}^+ + \varepsilon \delta \mathsf{D}^+$, where $\delta \mathsf{D} \in \Gamma(B, \operatorname{Hom}(\mathscr{H}^+, \mathscr{H}^-))$, we see that

$$\begin{aligned}\det(\mathsf{D}^+ + \varepsilon\delta\mathsf{D}^+) &= \det(\mathsf{D}^+(1+\varepsilon(\mathsf{D}^+)^{-1}\delta\mathsf{D}^+)) \\ &= \det(\mathsf{D}^+)\det(1+\varepsilon(\mathsf{D}^+)^{-1}\delta\mathsf{D}^+)) \\ &= \det(\mathsf{D}^+) + \varepsilon\det(\mathsf{D}^+)\operatorname{Tr}((\mathsf{D}^+)^{-1}\delta\mathsf{D}^+) + O(\varepsilon^2),\end{aligned}$$

where we use the fact that

$$\frac{d}{d\varepsilon}\Big|_{\varepsilon=0}\det(1+\varepsilon A) = \operatorname{Tr}(A).$$

From this, applying the sign rule for graded tensor products, we see that

$$\nabla^{\det\mathscr{H}}\det(\mathsf{D}^+) = -\det(\mathsf{D}^+)Tr_{\mathscr{H}^+}((\mathsf{D}^+)^{-1}d\mathsf{D}^+) - \operatorname{Str}(\omega)\det(\mathsf{D}^+).$$

On the other hand,

$$\begin{aligned}\operatorname{Tr}_{\mathscr{H}^+}\big((\mathsf{D}^+)^{-1}\nabla^{\operatorname{End}(\mathscr{H})}\mathsf{D}^+\big) &= \operatorname{Tr}_{\mathscr{H}^+}\big((\mathsf{D}^+)^{-1}(d\mathsf{D}^+ + \omega^-\mathsf{D}^+ + \mathsf{D}^+\omega^+)\big) \\ &= \operatorname{Tr}_{\mathscr{H}^+}\big((\mathsf{D}^+)^{-1}d\mathsf{D}^+\big) + \operatorname{Str}(\omega).\end{aligned}$$ □

We will now show how to define a determinant line bundle using a family of Dirac operators D parametrized by a base B. Recall that if $\mathscr{E}$ and $\mathscr{F}$ are two vector bundles, then

$$\det(\mathscr{E}\oplus\mathscr{F}) \cong \det(\mathscr{E})\otimes\det(\mathscr{F}).$$

Let $0<\lambda<\mu$ be a pair of positive real numbers, and consider the decomposition

$$\mathscr{H}_{[0,\mu)} = \mathscr{H}_{[0,\lambda)} \oplus \mathscr{H}_{(\lambda,\mu)}$$

over the open set $U_\lambda \cap U_\mu$. From this, we see that

$$\det\big(\mathscr{H}_{[0,\mu)}\big) \cong \det\big(\mathscr{H}_{[0,\lambda)}\big)\otimes\det\big(\mathscr{H}_{(\lambda,\mu)}\big).$$

Denote by $\mathsf{D}_{(\lambda,\mu)}$ the Dirac operator restricted to $\mathscr{H}_{(\lambda,\mu)}$. The line bundle $\det\big(\mathscr{H}_{(\lambda,\mu)}\big)$ over $U_\lambda\cap U_\mu$ is trivial, since it has a nowhere-vanishing section $\det\big(\mathsf{D}^+_{(\lambda,\mu)}\big)$. Thus, the line bundles $\det\big(\mathscr{H}_{[0,\lambda)}\big)$ piece together to form a line bundle over all of B, called the **determinant line bundle** associated to the family of Dirac operators D and denoted by $\det(\pi_*\mathscr{E},\mathsf{D})$; to a section $s\in\Gamma\big(U_\lambda\cap U_\mu,\det(\mathscr{H}_{[0,\lambda)})\big)$, we associate the section $s\otimes\det\big(\mathsf{D}^+_{(\lambda,\mu)}\big)$ of $\det\big(\mathscr{H}_{[0,\mu)}\big)$.

We have assumed that the vector bundle $\mathscr{E}$ has a Hermitian metric, and hence that $\pi_*\mathscr{E}$ has a metric too, given on sections s_z and $t_z\in\pi_*\mathscr{E}_z = \Gamma(M_z,\mathscr{E}_z\otimes|\Lambda_\pi|_z^{1/2})$ by

$$\int_{x\in M_z}(s_z(x),t_z(x))_{\mathscr{E}_x}.$$

We would like to define a natural metric on the line bundle $\det(\pi_*\mathscr{E},\mathsf{D})$. From the metric on $\pi_*\mathscr{E}$, we obtain a natural metric on each of the vector bundles $\mathscr{H}_{[0,\lambda)}$, which is known as the **L^2-metric**. It follows that each of the line bundles $\det\big(\mathscr{H}_{[0,\lambda)}\big)$ has a natural metric, which it inherits from the metric on $\mathscr{H}_{[0,\lambda)}$ in the natural way.

However, the metrics on the bundles $\det(\mathcal{H}_{[0,\lambda)})$ and $\det(\mathcal{H}_{[0,\mu)})$ over $U_\lambda \cap U_\mu$ are not equal, and they differ by a factor equal to

$$|\det(\mathsf{D}^+_{(\lambda,\mu)})|,$$

where the metric is the L^2-metric on the trivial line bundle $\det(\mathcal{H}_{(\lambda,\mu)})$.

The family of generalized Laplacians $\mathsf{D}^-\mathsf{D}^+$ on $\pi_*\mathcal{E}^+$ consists of self-adjoint operators, since $(\mathsf{D}^+)^* = \mathsf{D}^-$. The following lemma is clear.

Lemma 9.40. *If $\lambda < \lambda_1 \leq \cdots \leq \lambda_m < \mu$ are the eigenvalues of $\mathsf{D}^-\mathsf{D}^+$ lying between λ and μ, then*

$$|\det(\mathsf{D}^+_{(\lambda,\mu)})| = \prod_{i=1}^m \lambda_i^{1/2}.$$

By choosing local trivializations of the bundles M and $\mathcal{E}$ over B, we see that the zeta-function of the family $\mathsf{D}^-\mathsf{D}^+$ is a meromorphic function of s with smooth dependence on $z \in B$. By Proposition 9.36, its derivative at $s = 0$ satisfies the formula

$$\zeta'(0,\mathsf{D}^-\mathsf{D}^+,\lambda) = \zeta'(0,\mathsf{D}^-\mathsf{D}^+,\mu) - \sum_{i=1}^m \log\lambda_i.$$

From this, we easily see that the metric

$$|\cdot|_Q = e^{-\zeta'(0,\mathsf{D}^-\mathsf{D}^+,\lambda)/2}|\cdot|$$

on the line bundle $\det(\mathcal{H}_{[0,\lambda)})$ agrees with the corresponding metric on the line bundle $\det(\mathcal{H}_{[0,\mu)})$, and hence all of these metrics patch together to form a metric on the line bundle $\det(\pi_*\mathcal{E},\mathsf{D})$, which is the **Quillen metric**.

If the family D has index zero, then $\dim(\mathcal{H}^+_{[0,\lambda)}) = \dim(\mathcal{H}^-_{[0,\lambda)})$ for all $\lambda > 0$, and so that it makes sense to speak of the section $\det(\mathsf{D}^+_{[0,\lambda)})$ of the determinant line bundle $\det(\mathcal{H}_{[0,\lambda)})$ over the set U_λ. Since

$$\det(\mathsf{D}^+_{[0,\mu)}) = \det(\mathsf{D}^+_{[0,\lambda)}) \otimes \det(\mathsf{D}^+_{(\lambda,\mu)}) \in \Gamma(U_\lambda \cap U_\mu, \det(\mathcal{H}_{[0,\mu)})),$$

we see that these sections patch together to give a section $\det(\mathsf{D}^+)$ of the line bundle $\det(\pi_*\mathcal{E},\mathsf{D})$.

Proposition 9.41. *Let D be a family of Dirac operators of index zero on a vector bundle $\mathcal{E}$ over the family of manifolds $M \to B$. There is a canonical smooth section $\det(\mathsf{D}^+)$ of the determinant line bundle $\det(\pi_*\mathcal{E},\mathsf{D})$, which vanishes precisely where D is not invertible. The section $\det(\mathsf{D}^+)$ satisfies the formula*

$$|\det(\mathsf{D}^+)|_Q = \det(\mathsf{D}^-\mathsf{D}^+)^{1/2},$$

where by $\det(\mathsf{D}^-\mathsf{D}^+)$ we mean the zeta-function determinant $e^{-\zeta'(0,\mathsf{D}^-\mathsf{D}^+,0)}$ of the symmetric Laplacian $\mathsf{D}^-\mathsf{D}^+$. (This is the analogue for Dirac operators of the formula $\det(A^*A) = |\det(A)|^2$ in finite dimensions.)

There is a natural connection on the determinant line bundle $\det(\pi_*\mathscr{E}, \mathsf{D})$, compatible with the Quillen metric. To construct this, we need a unitary connection $\nabla^{\pi_*\mathscr{E}}$ on the bundle $\pi_*\mathscr{E}$, for example, the one defined in Section 2 from connections on the bundle M/B and on the Hermitian vector bundle $\mathscr{E}$.

From the connection $\nabla^{[0,\lambda)} = P_{[0,\lambda)}\nabla^{\pi_*\mathscr{E}}P_{[0,\lambda)}$ on $\mathscr{H}_{[0,\lambda)}$, we obtain a connection on $\det(\mathscr{H}_{[0,\lambda)})$ compatible with the L^2-metric, which we denote by $\nabla^{\det(\mathscr{H}_{[0,\lambda)})}$.

The operator $\mathsf{D}_{(\lambda,\infty)} = P_{(\lambda,\infty)}\mathsf{D}$ over the open set U_λ has the property that the bundle $\ker(\mathsf{D}_{(\lambda,\infty)})$ equals $\mathscr{H}_{[0,\lambda)}$. We may define a superconnection associated to the operator $\mathsf{D}_{(\lambda,\infty)}$ by the formula

$$\mathbb{A}_\lambda = \mathsf{D}_{(\lambda,\infty)} + \nabla^{\pi_*\mathscr{E}},$$

and its rescaled version

$$\mathbb{A}_{\lambda,s} = s^{1/2}\mathsf{D}_{(\lambda,\infty)} + \nabla^{\pi_*\mathscr{E}}.$$

Define two differential forms $\alpha^\pm(s,\lambda) \in \mathscr{A}(B)$, given by traces over $\pi_*\mathscr{E}^+$ and $\pi_*\mathscr{E}^-$ respectively:

$$\alpha^\pm(s,\lambda) = \operatorname{Tr}_{\pi_*\mathscr{E}^\pm}\Big(\frac{\partial \mathbb{A}_{\lambda,s}}{\partial s}e^{-\mathbb{A}_{\lambda,s}^2}\Big) = \frac{1}{2s^{1/2}}\operatorname{Tr}_{\pi_*\mathscr{E}^\pm}\Big(\mathsf{D}_{(\lambda,\infty)}e^{-\mathbb{A}_{\lambda,s}^2}\Big).$$

Lemma 9.42. *The one-form components of the differential forms $\alpha^\pm(s,\lambda)$ satisfy*

$$\overline{\alpha^+(s,\lambda)_{[1]}} = \alpha^-(s,\lambda)_{[1]},$$

and have asymptotic expansions for small s of the form

$$\alpha^\pm(s,\lambda)_{[1]} \sim \sum_{k=-N}^{\infty} s^{k/2}a_k^\pm.$$

Proof. To see that $\overline{\alpha^+(s,\lambda)_{[1]}} = \alpha^-(s,\lambda)_{[1]}$, we use the facts that $\mathsf{D}^* = \mathsf{D}$ and that $\nabla^{\pi_*\mathscr{E}}$ respects the metric on $\pi_*\mathscr{E}$.

The one-form component of $\alpha^\pm(s,\lambda)$ is equal to

$$\begin{aligned}\alpha^\pm(s,\lambda)_{[1]} &= \frac{1}{2s^{1/2}}\operatorname{Tr}_{\pi_*\mathscr{E}^\pm}\Big(\mathsf{D}_{(\lambda,\infty)}e^{-s\mathsf{D}_{(\lambda,\infty)}^2 - s^{1/2}[\nabla^{\pi_*\mathscr{E}},\mathsf{D}_{(\lambda,\infty)}]}\Big)_{[1]} \\ &= -\tfrac{1}{2}\operatorname{Tr}_{\pi_*\mathscr{E}^\pm}\Big(\mathsf{D}_{(\lambda,\infty)}[\nabla^{\pi_*\mathscr{E}},\mathsf{D}_{(\lambda,\infty)}]e^{-s\mathsf{D}_{(\lambda,\infty)}^2}\Big) \\ &= -\tfrac{1}{2}\operatorname{Tr}_{\pi_*\mathscr{E}^\pm}\big((1-P_{[0,\lambda)})\mathsf{D}^\mp[\nabla^{\pi_*\mathscr{E}},\mathsf{D}^\pm]e^{-s\mathsf{D}^2}\big),\end{aligned}$$

since $P_{(\lambda,\infty)}(\nabla^{\pi_*\mathscr{E}}P_{(\lambda,\infty)})P_{(\lambda,\infty)} = 0$. The result now follows from Propositions 2.46 and 2.47, which give asymptotic expansions for $\operatorname{Tr}(DP_t)$ and $\operatorname{Tr}(KP_t)$ at small t respectively, where P_t is a heat kernel, D is a differential operator, and K is a smoothing operator. □

By Lemma 9.34, we conclude that the functions $\int_t^\infty \alpha^\pm(s,\lambda)_{[1]}\,ds$ have asymptotic expansions for t small. Let

$$\beta_\lambda^\pm = 2\operatorname{LIM}_{t\to 0} \int_t^\infty \alpha^\pm(s,\lambda)_{[1]}\, ds \in \mathscr{A}^1(B).$$

The sum $\beta_\lambda^+ + \beta_\lambda^-$ is real, while the difference $\beta_\lambda^+ - \beta_\lambda^-$ is imaginary.

Lemma 9.43. $d\zeta'(0,\mathsf{D}^-\mathsf{D}^+,\lambda) = -(\beta_\lambda^+ + \beta_\lambda^-)$

Proof. Using the formula $\int_t^\infty e^{-sH}\, ds = H^{-1}e^{-tH}$, we see that

$$\begin{aligned}\beta_\lambda^+ &= -\operatorname{LIM}_{t\to 0} \int_t^\infty \operatorname{Tr}_{\pi_*\mathscr{E}^+}\big(P_{(\lambda,\infty)}\mathsf{D}^-[\nabla^{\pi_*\mathscr{E}},\mathsf{D}^+]e^{-s\mathsf{D}^2}\big)\, ds \\ &= -\operatorname{LIM}_{t\to 0} \operatorname{Tr}_{\pi_*\mathscr{E}^+}\big(P_{(\lambda,\infty)}\mathsf{D}^-[\nabla^{\pi_*\mathscr{E}},\mathsf{D}^+](\mathsf{D}^-\mathsf{D}^+)^{-1}e^{-t\mathsf{D}^-\mathsf{D}^+}\big).\end{aligned}$$

and

$$\begin{aligned}\beta_\lambda^- &= -\operatorname{LIM}_{t\to 0} \int_t^\infty \operatorname{Tr}_{\pi_*\mathscr{E}^-}\big(P_{(\lambda,\infty)}\mathsf{D}^+[\nabla^{\pi_*\mathscr{E}},\mathsf{D}^-]e^{-s\mathsf{D}^2}\big)\, ds \\ &= \operatorname{LIM}_{t\to 0} \int_t^\infty \operatorname{Tr}_{\pi_*\mathscr{E}^+}\big(P_{(\lambda,\infty)}[\nabla^{\pi_*\mathscr{E}},\mathsf{D}^-]\mathsf{D}^+e^{-s\mathsf{D}^2}\big)\, ds \\ &= \operatorname{LIM}_{t\to 0} \operatorname{Tr}_{\pi_*\mathscr{E}^+}\big(P_{(\lambda,\infty)}[\nabla^{\pi_*\mathscr{E}},\mathsf{D}^-]\mathsf{D}^+(\mathsf{D}^-\mathsf{D}^+)^{-1}e^{-t\mathsf{D}^-\mathsf{D}^+}\big).\end{aligned}$$

Adding the one-forms β_λ^+ and β_λ^- together, we obtain

$$\beta_\lambda^+ + \beta_\lambda^- = \operatorname{LIM}_{t\to 0} \operatorname{Tr}_{\pi_*\mathscr{E}^+}\big(P_{(\lambda,\infty)}[\nabla^{\pi_*\mathscr{E}},\mathsf{D}^-\mathsf{D}^+](\mathsf{D}^-\mathsf{D}^+)^{-1}e^{-t\mathsf{D}^-\mathsf{D}^+}\big).$$

The lemma follows immediately from Proposition 9.38. □

We can now define the connection on $\det(\pi_*\mathscr{E},\mathsf{D})$. On $\det(\mathscr{H}_{[0,\lambda)})$, we take the connection

$$\nabla^{\det(\mathscr{H}_{[0,\lambda)})} + \beta_\lambda^+.$$

Lemma 9.44. *The section* $\det(\mathsf{D}^+_{(\lambda,\mu)})$ *is parallel with respect to the connection* $\nabla^{\det(\mathscr{H}_{(\lambda,\mu)})} + (\beta_\mu^+ - \beta_\lambda^+)$ *on* $\det(\mathscr{H}_{(\lambda,\mu)})$.

Proof. Observe that

$$(\mathsf{D}^+_{(\lambda,\mu)})^{-1}[\nabla^{(\lambda,\mu)},\mathsf{D}^+_{(\lambda,\mu)}] = (\mathsf{D}^-_{(\lambda,\mu)}\mathsf{D}^+_{(\lambda,\mu)})^{-1}\mathsf{D}^-_{(\lambda,\mu)}[\nabla^{(\lambda,\mu)},\mathsf{D}^+_{(\lambda,\mu)}].$$

By Proposition 9.39 applied to the bundle $\mathscr{H} = \mathscr{H}_{(\lambda,\mu)}$, we see that

$$\begin{aligned}\nabla^{\det(\mathscr{H}_{(\lambda,\mu)})}\det(\mathsf{D}^+_{(\lambda,\mu)}) &= -\operatorname{Tr}_{\mathscr{H}^+_{(\lambda,\mu)}}\big(\det(\mathsf{D}^+_{(\lambda,\mu)})(\mathsf{D}^+_{(\lambda,\mu)})^{-1}[\nabla^{(\lambda,\mu)},\mathsf{D}^+_{(\lambda,\mu)}]\big) \\ &= -\det(\mathsf{D}^+_{(\lambda,\mu)})(\beta_\mu^+ - \beta_\lambda^+).\end{aligned}$$

□

The connection $\nabla^{\det(\mathscr{H}_{[0,\mu)})}$ on $\det(\mathscr{H}_{[0,\mu)})$ is the tensor product of the connections $\nabla^{\det(\mathscr{H}_{[0,\lambda)})}$ and $\nabla^{\det(\mathscr{H}_{(\lambda,\mu)})}$:

$$\nabla^{\det(\mathscr{H}_{[0,\mu)})} = \nabla^{\det(\mathscr{H}_{[0,\lambda)})} \otimes 1 + 1 \otimes \nabla^{\det(\mathscr{H}_{(\lambda,\mu)})}.$$

Lemma 9.44 now shows that the connections

$$\nabla^{\det(\mathscr{H}_{[0,\lambda)})} + \beta_\lambda^+ \quad \text{and} \quad \nabla^{\det(\mathscr{H}_{[0,\mu)})} + \beta_\mu^+$$

on the line bundles $\det(\mathscr{H}_{[0,\lambda)})$ and $\det(\mathscr{H}_{[0,\mu)})$ agree over the set $U_\lambda \cap U_\mu$, when we identify these two bundles by multiplying by the section $\det(\mathsf{D}^+_{(\lambda,\mu)})$ of $\det(\mathscr{H}_{(\lambda,\mu)})$. Denote the resulting connection on $\det(\pi_*\mathscr{E}, \mathsf{D})$ by $\nabla^{\det(\pi_*\mathscr{E},\mathsf{D})}$.

Proposition 9.45. *The connection $\nabla^{\det(\pi_*\mathscr{E},\mathsf{D})}$ is compatible with the Quillen metric $|\cdot|_Q$ on $\det(\pi_*\mathscr{E}, \mathsf{D})$.*

Proof. If $|\cdot|$ and $\|\cdot\|$ are two metrics on a line bundle $\mathscr{L}$ such that $\|\cdot\| = e^f|\cdot|$ and ∇ is a connection compatible with the metric $|\cdot|$, then it is easy to see that $\nabla + df + \omega$ will be compatible with the metric $\|\cdot\|$ for any imaginary one-form ω.

The connection $\nabla^{\det(\mathscr{H}_{[0,\lambda)})}$ is compatible with the L^2-metric $|\cdot|$. Thus, we see that any connection of the form

$$\nabla^{\det(\mathscr{H}_{[0,\lambda)})} - \tfrac{1}{2} d\zeta'(0, \mathsf{D}^-\mathsf{D}^+, \lambda) + \omega = \nabla^{\det(\mathscr{H}_{[0,\lambda)})} + \tfrac{1}{2}(\beta_\lambda^+ + \beta_\lambda^-) + \omega$$

where ω is an imaginary one-form, will be compatible with the Quillen metric $|\cdot|_Q = e^{-\zeta'(0,\mathsf{D}^-\mathsf{D}^+,\lambda)/2}|\cdot|$. Choosing $\omega = (\beta_\lambda^+ - \beta_\lambda^-)/2$, we obtain the desired result. □

Appendix. More on Heat Kernels

In this appendix, we will prove a generalization of the construction of heat kernels of Chapter 2 which we need in the course of this chapter.

Let M be a compact manifold, let $\mathscr{E}$ be a vector bundle on M, and let $\mathscr{A} = \sum_{i=0}^n \mathscr{A}^i$ be an finite-dimensional algebra with identity graded by the natural numbers. (In practice, $\mathscr{A}$ will be an exterior algebra.) We will denote by $\mathscr{D}$ the algebra of differential operators $\mathscr{D}(M, \mathscr{E} \otimes |\Lambda|^{1/2})$, by $\mathscr{K}$ the algebra of smoothing operators on $\mathscr{E} \otimes |\Lambda|^{1/2}$, and by $\mathscr{P}$ the algebra of operators acting on $\Gamma(M, \mathscr{E} \otimes |\Lambda|^{1/2})$ generated by $\mathscr{D}$ and by $\mathscr{K}$. We call an element of $\mathscr{P}$ a **$\mathscr{P}$-endomorphism.**

Let $\mathscr{M}$ be the algebra $\mathscr{P} \otimes \mathscr{A}$. We define a decreasing filtration of the algebra $\mathscr{M}$ by setting $\mathscr{M}_i = \sum_{j \geq i} \mathscr{P} \otimes \mathscr{A}^i$. The space $\Gamma(M, \mathscr{E} \otimes |\Lambda|^{1/2}) \otimes \mathscr{A}$ is be made into a module for $\mathscr{M}$, by letting $\mathscr{P}$ act on $\Gamma(M, \mathscr{E} \otimes |\Lambda|^{1/2})$, and letting $\mathscr{A}$ act on itself by left multiplication.

We will prove that an operator of the form

$$\mathscr{F} = H_0 + K + \mathscr{F}_{[+]},$$

where H_0 is a generalized Laplacian, $K \in \mathscr{K}$ and $\mathscr{F}_{[+]} \in \mathscr{M}_1$, has a heat kernel satisfying many of the properties of the heat kernel of a generalized Laplacian. We define a heat kernel for $\mathscr{F}$ to be a continuous map $(t,x,y) \mapsto p_t(x,y) \in (\mathscr{E} \otimes |\Lambda|^{1/2})_x \otimes (\mathscr{E}^* \otimes |\Lambda|^{1/2})_y \otimes \mathscr{A}$ which is C^1 in t, C^2 in x, and satisfies the equation

$$\frac{\partial}{\partial t} p_t(x,y) + \mathscr{F}_x p_t(x,y) = 0,$$

with the boundary condition that for every $s \in \Gamma(M, \mathscr{E} \otimes |\Lambda_M|^{1/2}) \otimes \mathscr{A}$,

$$\lim_{t \to 0} \int_{y \in M} p_t(x,y)\, s(y) = s(x) \quad \text{uniformly in } x \in M.$$

Let us start with the case where $\mathscr{A}$ is the algebra of complex numbers $\mathbb{C}$, so that there is no component $\mathscr{F}_{[+]}$. Thus, we wish to construct the heat kernel of an operator H of the form $H = H_0 + K$, which differs from a generalized Laplacian H_0 by a smoothing operator K. Since we know that H_0 has a smooth heat kernel satisfying certain strong estimates, it is not very surprising that we can prove the same things for H. To construct the heat kernel of H, we use a generalization of the Volterra series of Proposition 2.48.

Proposition 9.46. *The series*

$$Q_t = \sum_{k=0}^{\infty} (-t)^k \int_{\Delta_k} e^{-\sigma_0 t H_0} K e^{-\sigma_1 t H_0} \cdots K e^{-\sigma_k t H_0}\, d\sigma$$

converges in the sense that the corresponding series of kernels converges for $t > 0$, with respect to any C^ℓ-norm, $\ell \geq 0$, to a kernel

$$q_t \in \Gamma\big(M \times M, (\mathscr{E} \otimes |\Lambda|^{1/2}) \boxtimes (\mathscr{E}^* \otimes |\Lambda|^{1/2})\big).$$

The sum is C^∞ with respect to t and is a solution of the heat equation

$$(\partial_t + (H_0)_x + K_x) q_t(x,y) = 0$$

with the following boundary condition at $t = 0$: if ϕ is a smooth section of $\mathscr{E} \otimes |\Lambda|^{1/2}$, then

$$\lim_{t \to 0} Q_t s = s \quad \text{in the uniform norm.}$$

Thus $q_t(x,y)$ is a heat kernel for the operator $H = H_0 + K$. Furthermore the kernel of the difference

$$e^{-tH} - e^{-tH_0} = \sum_{k=1}^{\infty} (-t)^k \int_{\Delta_k} e^{-\sigma_0 t H_0} K e^{-\sigma_1 t H_0} \cdots K e^{-\sigma_k t H_0}\, d\sigma$$

tends to 0 when $t \to 0$.

Proof. Since K is a smoothing operator, the operator $e^{-tH_0}K$ has a smooth kernel for all $t \geq 0$, and

$$\|e^{-tH_0}K\|_\ell \leq C(\ell)\|K\|_\ell,$$

for some constant $C(\ell)$ depending on ℓ. It follows that, for $k \geq 1$,

$$\left\| \int_{\Delta_k} e^{-\sigma_0 t H_0} K e^{-\sigma_1 t H_0} \dots K e^{-\sigma_k t H_0} \right\|_\ell \leq \frac{C(\ell)^{k+1}\|K\|_\ell^k}{k!}.$$

Thus the series $\sum_{k\geq 1}$ converges with respect to the C^ℓ-norm, uniformly for $t \geq 0$, with similar estimates for the derivatives with respect to t. It is easy to verify, as in theorem 2.23, that the complete sum $\sum_{k\geq 0}$ is a solution of the heat equation. The estimate $\|e^{-tH} - e^{-tH_0}\|_\ell = O(t)$ is clear, and implies the boundary condition for $t = 0$. □

If the operator $H = H_0 + K$ is symmetric, then we can extend a number of properties which hold for a generalized Laplacian H_0 to the case where K is non-zero.

1. As in Proposition 2.33, H is essentially self-adjoint on the Hilbert space of L^2-sections of $\mathscr{E} \otimes |\Lambda|^{1/2}$.
2. The unique self-adjoint extension of H has a discrete spectrum bounded below; each eigenspace is finite-dimensional and is contained in the space of smooth sections.
3. The analogue of Proposition 2.37 holds, namely, if $H = H_0 + K$, and if P_1 is the projection onto the eigenfunctions of H with positive eigenvalue, then for each $\ell \in \mathbb{N}$, there exists a constant $C(\ell) > 0$ such that for t sufficiently large,
$$\|\langle x \mid P_1 e^{-tH} P_1 \mid y\rangle\|_\ell \leq C(\ell)e^{-t\lambda_1/2},$$
where λ_1 is the smallest non-zero eigenvalue of H.
4. The Green operator G, which is H^{-1} on $\ker(H)^\perp$ and 0 on $\ker(H)$, operates on the space of smooth sections of $\mathscr{E} \otimes |\Lambda|^{1/2}$, and is given by the formula
$$G = \int_0^\infty P_1 e^{-tH} P_1\, dt.$$

We now turn to the case of an operator $\mathscr{F}$ of the form

$$\mathscr{F} = H_0 + K + \mathscr{F}_{[+]},$$

which differs from a generalized Laplacian by the sum of a smoothing operator $K \in \mathscr{K}$ and an operator $\mathscr{F}_{[+]}$ lying in the ideal $\mathscr{M}_1$. The operators $\mathscr{F}$ and H differ by an operator $\mathscr{F}_{[+]}$ of positive degree in the finite-dimensional graded algebra $\mathscr{A}$. Since we have already constructed the heat kernel of $H = H_0 + K$, the Volterra series once more gives a candidate for the heat kernel of $\mathscr{F}$: for fixed $t > 0$, define the operator $e^{-t\mathscr{F}}$ to equal

$$e^{-t\mathscr{F}} = e^{-tH} + \sum_{k>0} (-t)^k I_k,$$

where
$$I_k = \int_{\Delta_k} e^{-\sigma_0 tH} \mathscr{F}_{[+]} e^{-\sigma_1 tH} \mathscr{F}_{[+]} \dots e^{-\sigma_{k-1} tH} \mathscr{F}_{[+]} e^{-\sigma_k tH} \, d\sigma.$$

The sum is finite, since $I_k \in \mathscr{M}_k$, and for k large, we know that $\mathscr{A}^k = 0$, and hence $\mathscr{M}_k$ is zero. Thus, we only have to make sense of each term in the sum. We need the following lemma.

Lemma 9.47. *Let D be a differential operator of order k. There exists a constant $C > 0$ such that if K is a smoothing operator, one has the following bounds: for $t \in [0,T]$, where T is a positive real number, we have*
$$\|De^{-tH}K\|_\ell \le C\|K\|_{k+\ell}$$
$$\|Ke^{-tH}D\|_\ell \le C\|K\|_{k+\ell}.$$

Proof. There exists a constant $C(\ell)$ such that for $\phi \in \Gamma^\ell(M,\mathscr{E})$ one has for $t \in [0,T]$, $\|e^{-tH}\phi\|_\ell \le C(\ell)\|\phi\|_\ell$. The bound $\|De^{-tH}K\|_\ell \le C(\ell)\|K\|_{k+\ell}$ follows easily. Then we write
$$Ke^{-tH}D = (D^* e^{-tH^*} K^*)^*.$$

The above bound applies to the kernel of $D^* e^{-tH^*} K^*$, yielding $\|Ke^{-tH}D\|_\ell \le C(\ell)\|K\|_{k+\ell}$. □

We can now complete the proof of the following theorem.

Theorem 9.48. *There exists a unique heat kernel $p_t(x,y)$ for $\mathscr{F}$, which is a smooth map from $t \in (0,\infty)$ to $\Gamma\big(M \times M, (\mathscr{E} \otimes |\Lambda|^{1/2}) \boxtimes (\mathscr{E}^* \otimes |\Lambda|^{1/2})\big)$.*

Proof. We must show that each term I_k has a smooth kernel. On the simplex Δ_k, one of the σ_i must be greater than $(k+1)^{-1}$. Since for $(k+1)^{-1} < \sigma \le 1$ and fixed t, the operator $e^{-\sigma tH}$ has uniformly smooth kernel, it follows by iterated application of Lemma 9.47 that the operator
$$e^{-t\sigma_0 H} \mathscr{F}_{[+]} e^{-\sigma_1 tH} \mathscr{F}_{[+]} \dots e^{-\sigma_{k-1} tH} \mathscr{F}_{[+]} e^{-\sigma_k tH}$$

has a smooth kernel which depends continuously on $(\sigma_0,\dots,\sigma_k) \in \Delta_k$. Thus the integral makes sense as an operator with smooth kernel. As in Chapter 2, we see easily that $e^{-tH} + \sum_{k>0}(-t)^k I_k$ satisfies the heat equation for $\mathscr{F}$. The boundary condition as $t \to 0$ holds, since for ϕ smooth, each term $I_k\phi$ has a limit as $t \to 0$; when multiplied by $(-t)^k$, each term contributes zero except the term $I_0 = e^{-tH}$. Uniqueness is proved as in Chapter 2 as a consequence of the existence of the heat kernel for the adjoint operator $\mathscr{F}^*$. □

There is an analogue of Proposition 2.37 for operators of the type $\mathscr{F}$.

Proposition 9.49. *Suppose $\mathscr{F} = H + \mathscr{F}_{[+]}$ where $H = H_0 + K$ is the sum of a symmetric generalized Laplacian H_0 and a symmetric smoothing operator K, and $\mathscr{F}_{[+]} \in \mathscr{M}_1$. If P_1 is the projection onto the positive eigenspace of H and P_1 commutes with $\mathscr{F}_{[+]}$, then there exists an $\varepsilon > 0$ such that the kernel of the operator $e^{-t\mathscr{F}}$ satisfies the estimates*

$$\|\langle x \mid P_1 e^{-t\mathscr{F}} P_1 \mid y\rangle\|_\ell \leq C(\ell) e^{-\varepsilon t}$$

on $M \times M$ as $t \to \infty$.

Proof. Using the Volterra series, we may write

$$e^{-t\mathscr{F}} = \sum_{k=0}^{n} (-t)^k I_k,$$

where

$$I_k = \int_{\Delta_k} e^{-\sigma_0 tH} \mathscr{F}_{[+]} e^{-\sigma_1 tH} \dots \mathscr{F}_{[+]} e^{-\sigma_k tH}.$$

It follows that $P_1 I_k P_1$ equals

$$\int_{\Delta_k} (P_1 e^{-\sigma_0 tH} P_1)(P_1 \mathscr{F}_{[+]} P_1)(P_1 e^{-\sigma_1 tH} P_1) \dots (P_1 \mathscr{F}_{[+]} P_1)(P_1 e^{-\sigma_k tH} P_1).$$

By the analogue of Proposition 2.37 to the operator H, we see that for t sufficiently large, and for each ℓ,

$$\|\langle x \mid P_1 e^{-tH} P_1 \mid y\rangle\|_\ell \leq C(\ell) e^{-t\lambda_1/2}.$$

On the simplex Δ_k, one of the σ_i must be greater than $(k+1)^{-1}$. The estimate now follows easily by induction using Lemma 9.47 combined with the above decay for $e^{-\sigma_i tH}$. □

Let $\pi : M \to B$ be a family of manifolds over a base B. Denote by $M \times_\pi M$ the fibre-product

$$M \times_\pi M = \{(x,y) \in M \times M \mid \pi(x) = \pi(y)\},$$

which is a fibre bundle over B with fibre at $z \in B$ equal to $M_z \times M_z$. Let pr_1 and pr_2 be the two projections from $M \times_\pi M$ to M, and let $|\Lambda_\pi|$ be the vertical density bundle of the fibre bundle $\pi : M \to B$. If $\mathscr{E}_1$ and $\mathscr{E}_2$ are vector bundles on M, let $\mathscr{E}_1 \boxtimes_\pi \mathscr{E}_2$ be the vector bundle over $M \times_\pi M$ given by the formula

$$\mathscr{E}_1 \boxtimes_\pi \mathscr{E}_2 = \mathrm{pr}_1^* \mathscr{E}_1 \otimes \mathrm{pr}_2^* \mathscr{E}_2.$$

Let $\mathscr{E} \to M$ be a family of vector bundles $\mathscr{E}_z \to M_z$. We define a smooth **family of smoothing operators** acting on the bundles $\mathscr{E}_z \to M_z$ along the fibres of $\pi : M \to B$ to be a family of operators with a kernel

$$k \in \Gamma\big(M \times_\pi M, (\mathscr{E} \otimes |\Lambda_\pi|^{1/2}) \boxtimes_\pi (\mathscr{E}^* \otimes |\Lambda_\pi|^{1/2})\big).$$

When restricted to the fibre $M_z \times M_z$, the kernel k may be viewed as a kernel k^z in $\Gamma(M_z \times M_z, (\mathscr{E} \otimes |\Lambda_{M_z}|^{1/2}) \boxtimes_\pi (\mathscr{E}^* \otimes |\Lambda_{M_z}|^{1/2}))$, which defines a smoothing operator K^z on $\pi_* \mathscr{E}_z = \Gamma(M_z, \mathscr{E}_z)$ by the formula

$$(K^z \phi)(x) = \int_{M_z} k^z(x,y)\, \phi(y)$$

for $\phi \in \Gamma(M_z, \mathscr{E}_z \otimes |\Lambda_{M_z}|^{1/2})$. We define the bundle $\mathscr{K}(\mathscr{E})$ over B to be the bundle whose smooth sections are given by

$$\Gamma(B, \mathscr{K}(\mathscr{E})) = \Gamma(M \times_\pi M, (\mathscr{E} \otimes |\Lambda_\pi|^{1/2}) \boxtimes (\mathscr{E}^* \otimes |\Lambda_\pi|^{1/2}));$$

as explained in Definition 9.8, $\mathscr{K}(\mathscr{E})$ is a sub-bundle of $\mathrm{End}_{\mathscr{D}}(\mathscr{E})$.

Suppose we are given a smooth family of generalized Laplacians H^z along the fibres of $M \to B$; in other words, we are given the following data, from which we construct the Laplacians H^z in the canonical way:

1. a smooth family $g_{M/B} \in \Gamma(M, S^2(T(M/B)))$ of metrics along the fibres;
2. a smooth family of connections acting on $\mathscr{E}$ along the fibres, which we may suppose to be the restriction of a smooth connection on the bundle $\mathscr{E} \to M$;
3. a smooth family of potentials $F^z \in \Gamma(M_z, \mathrm{End}(\mathscr{E}_z))$, in other words, an element of $\Gamma(M, \mathrm{End}(\mathscr{E}))$.

We have the following result, which generalizes Theorem 2.48.

Theorem 9.50. *If H^z is a smooth family of generalized Laplacians, then the corresponding heat kernel $p_t(x,y,z)$ defines a smooth family of smoothing operators, that is, a section in $\Gamma(B, \mathscr{K}(\mathscr{E}))$.*

Proof. As usual, around any point $z_0 \in B$, we can find a neighbourhood on which the families M and $\mathscr{E}$ are trivialized. Thus, we may replace B by a ball $U \subset \mathbb{R}^p$ centred at zero, M by the trivial bundle $M_0 \times U$, and $\mathscr{E}$ by the bundle $\mathscr{E}_0 \times U$, where $\mathscr{E}_0$ is a bundle over M_0. Since the changes of coordinates used to obtain this trivialization are smooth, as are their inverses, we see that the data used to define the family of generalized Laplacians H^z give a smooth family of data for defining generalized Laplacians on the bundle $\mathscr{E}_0 \otimes |\Lambda_{M_0}|^{1/2}$, parametrized by the ball U. By Theorem 2.48, we know that for any $t > 0$, the derivative $\partial_z^k p_t(x,y,H^z)$ is a smooth family of smooth kernels on $\mathscr{E}_0$, from which the theorem follows. □

The manifold M embeds inside $M \times_\pi M$ as the diagonal $M \subset M \times_\pi M$. If we restrict a kernel $k \in \Gamma(M \times_\pi M, (\mathscr{E} \otimes |\Lambda_\pi|^{1/2}) \boxtimes_\pi (\mathscr{E}^* \otimes |\Lambda_\pi|^{1/2}))$ to the diagonal, it becomes a section of $\mathrm{End}(\mathscr{E}) \otimes |\Lambda_\pi|$, and its pointwise trace $\mathrm{Tr}(k)$ becomes a section of the bundle of vertical densities $|\Lambda_\pi|$. Such a section may be integrated over the fibres to give a function on B; we will denote this integral by $\int_{M/B} \mathrm{Tr}(k(x,x))$. In this language, the formula for the trace along the fibres of the family of operators K^z becomes

$$\mathrm{Tr}(K^z) = \int_{M/B} \mathrm{Tr}(\langle x \mid K^z \mid x \rangle) \in C^\infty(B).$$

We may also state the family version of this theorem. Let $\mathscr{A}$ be a bundle of finite-dimensional graded algebras with identity over B (in practice, $\mathscr{A} = \Lambda T^*B$) and let $\mathscr{M}$ be the bundle of filtered algebras $\mathscr{M} = \mathscr{A} \otimes \mathrm{End}_{\mathscr{P}}(\mathscr{E})$. Let $\mathscr{F} \in \Gamma(B, \mathscr{A} \otimes \mathrm{End}_{\mathscr{P}}(\mathscr{E}))$ be a family of $\mathscr{P}$-endomorphism with coefficients in $\mathscr{A}$, of the form

$$\mathscr{F} = H_0 + K + \mathscr{F}_{[+]},$$

where $H_0 \in \Gamma(B, \mathscr{D}(\mathscr{E}))$ and $K \in \Gamma(B, \mathscr{K}(\mathscr{E}))$ are smooth families of generalized Laplacians and smoothing operators, and $\mathscr{F}_{[+]}$ is an element of $\Gamma(B, \mathscr{M}_1) = \sum_{i=1}^n \Gamma(B, \mathscr{A}^i \otimes \mathrm{End}_{\mathscr{P}}(\mathscr{E}))$.

Theorem 9.51. *For each $t > 0$, the kernel of the operator $e^{-t\mathscr{F}}$ is a smooth family of smoothing operators with coefficients in $\mathscr{A}$, that is, a smooth section in $\Gamma(B, \mathscr{A} \otimes \mathscr{K}(\mathscr{E}))$.*

Bibliographic Notes

The main theorem of this chapter, Theorem 9.33, is proved in the first part of Bismut's article on the family index theorem [30]. This theorem is a synthesis of the heat kernel approach of McKean and Singer towards the local index theorem with Quillen's theory of superconnections. Our other main reference for this chapter is Berline and Vergne [25], where Theorem 9.2 is proved. The proof of our refinement Theorem 9.19 of Theorem 9.33 is new.

Sections 1, 2 and 3

Atiyah and Singer first defined the family index of a continuous family of elliptic pseudodifferential operators, as an element of the K-theory of the base [16]. They then derived a formula for this family index, generalizing the Riemann-Roch theorem of Grothendieck [40]. For expositions of topological K-theory, see the books of Atiyah [3] and Karoubi [72].

Bismut, by contrast, obtained a formula for the Chern character of the family index in the special case of a family of Dirac operators; in this, he was inspired by Quillen's definition of the Chern character of a superconnection [94].

The results on the convergence of the transgressed Chern character (Theorems 9.7 and 9.23) are new, but were inspired by an analogue of Theorem 9.23 for complex manifolds due to Gillet and Soulé [68].

Section 4

The example presented in this section is from Section 5 of Atiyah and Singer [16]. However, since we have the Kirillov formula of Chapter 8 at our disposal, we are able to give a formula for the Chern character form of the index bundle.

Section 5

The techniques of this section are those of Atiyah and Singer [16].

Section 6

The zeta-function of the Laplace-Beltrami operator was introduced by Minakshisundaram and Pleijel [87]. This work was extended to elliptic pseudodifferential operators by Seeley [100], and applied in the original proof of the Atiyah-Bott fixed point formula [5] (see Section 6.2). The definition of the zeta-function determinant of a Laplacian was given by Ray and Singer [95].

Section 7

Our definition of the determinant line bundle is taken from an article by Bismut and Freed [35]; the construction is based on the ideas of Quillen [93]. This theory began with the work of physicists, in particular Alvarez-Gaumé and Witten on anomalies of quantum field theories (which is in fact precisely the theory of the Chern class of the determinant line bundle of a family of Dirac operators); see for example [107].

10

The Family Index Theorem

Let $\pi : M \to B$ be a family of oriented Riemannian manifolds $(M_z \mid z \in B)$, and let $\mathscr{E}$ be a bundle on M such that $\mathscr{E}_z = \mathscr{E}|_{M_z}$ is a Clifford module for each z; suppose in addition that there is a connection $\nabla^{\mathscr{E}}$ given on $\mathscr{E}$ whose restriction to each bundle $\mathscr{E}_z$ is a Clifford connection. Let $\pi_*\mathscr{E}$ be the infinite-dimensional bundle over B whose fibre at $z \in B$ is the space $\Gamma(M_z, \mathscr{E}_z)$; let $\mathsf{D} = (\mathsf{D}^z \mid z \in B)$ be the family of Dirac operators acting on the fibres of $\pi_*\mathscr{E}$, constructed from the Clifford module structure and Clifford connection on $\mathscr{E}$. The aim of this chapter is to calculate the Chern character of the index bundle $\operatorname{ind}(\mathsf{D}) \in K(B)$, by introducing a superconnection for the family of Dirac operators D whose Chern character is explicitly calculable. (Note that because the fibres of the bundle M/B have Riemannian metrics, the line bundle $|\Lambda_\pi|^{1/2}$ has a canonical trivialization; this reconciles the above definition of $\pi_*\mathscr{E}$ with that in Chapter 9.) This theorem is a generalization of Theorem 4.1 of Chapter 4; as in that chapter, we must assume that the Dirac operators D_z are associated to Clifford connections on $\mathscr{E}_z$.

We introduce the bundle $\mathbb{E} = \pi^*\Lambda T^*B \otimes \mathscr{E}$, which has the property that

$$\mathscr{A}(B, \pi_*\mathscr{E}) \cong \Gamma(M, \mathbb{E}).$$

Recall that a superconnection for the family D is a differential operator $\mathbb{A}$ on the bundle $\mathbb{E} \to M$, which is odd with respect to the total $\mathbb{Z}_2$-grading of the bundle $\mathbb{E}$, such that

1. (Leibniz's rule) if $v \in \mathscr{A}(B)$ and $s \in \Gamma(M, \mathbb{E})$, then
$$\mathbb{A}(vs) = (d_B v)s + (-1)^{|v|} v\,\mathbb{A}s;$$
2. $\mathbb{A} = \mathsf{D} + \sum_{i=1}^{\dim(B)} \mathbb{A}_{[i]}$, where $\mathbb{A}_{[i]} : \mathscr{A}^{\bullet}(B, \pi_*\mathscr{E}) \to \mathscr{A}^{\bullet+i}(B, \pi_*\mathscr{E})$.

Let $\mathbb{A}_t = t^{1/2}\delta_t \cdot \mathbb{A} \cdot \delta_t = \sum_{i=0}^{\dim(B)} t^{(1-i)/2}\mathbb{A}_{[i]}$ be the rescaled superconnection corresponding to $\mathbb{A}$. In this chapter, we will describe a particular superconnection $\mathbb{A}$ for the family of Dirac operators D, for which the limit

$$\lim_{t\to 0} \operatorname{ch}(\mathbb{A}_t) = \lim_{t\to 0} \operatorname{Str}(e^{-\mathbb{A}_t^2})$$

exists in the space $\mathscr{A}(B)$. Note that there is no reason *a priori* why the limit $\lim_{t\to 0}\operatorname{ch}(\mathbb{A}_t)$ should exist; this is analogous to the fact that the local index theorem is not true for arbitrary Dirac operators, but only those associated to an ordinary Clifford connection. Since the superconnection $\mathbb{A}$ was constructed by Bismut, we will call it the Bismut superconnection.

Recall that in the last chapter, we proved that $\operatorname{ch}(\mathbb{A}_t)$ is a differential form which lies in the de Rham cohomology class of the index bundle $\operatorname{ind}(\mathsf{D}) \in K(B)$ of the family D. It follows that the limit $\lim_{t\to 0}\operatorname{ch}(\mathbb{A}_t)$ lies in the same cohomology class. The formula for this limit, explained below, gives a way of calculating the Chern character of the index bundle. Thus, Bismut's theorem implies a local version of the family index theorem of Atiyah and Singer for the family D.

To define the Bismut superconnection, we need a connection for the fibration M/B. In other words, if $T(M/B)$ is the bundle of vertical vectors, we need a decomposition of TM into a direct sum of $T(M/B)$ and a horizontal tangent bundle $T_H M$, isomorphic to π^*TB. This allows us to construct a canonical connection $\nabla^{M/B}$ on $T(M/B)$, with curvature $R^{M/B}$; we do this in Section 1. The bundle $\mathscr{E}$ is assumed to be a Clifford module for the Clifford bundle $C(M/B) = C(T^*(M/B))$. We assume the connection $\nabla^{\mathscr{E}}$ is compatible with the connection $\nabla^{M/B}$. Denote by $F^{\mathscr{E}/S}$ the twisting curvature of $\nabla^{\mathscr{E}}$: if we are given a spin-structure on the bundle $T(M/B)$ with spinors $\mathscr{S}_{M/B}$, so that $\mathscr{E} = \mathscr{S}_{M/B} \otimes \mathscr{W}$ for the bundle $\mathscr{W} = \operatorname{Hom}_{C(M/B)}(\mathscr{S}_{M/B}, \mathscr{E})$, the twisting curvature $F^{\mathscr{E}/S}$ is the curvature $F^{\mathscr{W}}$ of the twisting bundle $\mathscr{W}$. Let

$$\hat{A}(M/B) = \det{}^{1/2}\left(\frac{R^{M/B}/2}{\sinh(R^{M/B}/2)}\right)$$

be the $\hat{A}$-genus of the connection $\nabla^{M/B}$, and let

$$\operatorname{ch}(\mathscr{E}/S) = \operatorname{Str}_{\mathscr{E}/S}(e^{-F^{\mathscr{E}/S}})$$

be the relative Chern character of the Clifford module $\mathscr{E}$. We will prove Bismut's theorem,

$$\lim_{t\to 0}\operatorname{ch}(\mathbb{A}_t) = (2\pi i)^{-n/2}\int_{M/B}\hat{A}(M/B)\operatorname{ch}(\mathscr{E}/S),$$

where n is the dimension of the fibre.

Now make the stronger assumption that the dimension of the kernel $\ker(\mathsf{D})$ is constant, so that $\ker(\mathsf{D})$ is a finite-dimensional subbundle of $\pi_*\mathscr{E}$. If P_0 is the projection from $\pi_*\mathscr{E}$ onto the finite dimensional bundle $\ker(\mathsf{D})$, we may form the projected connection $\nabla_0 = P_0\mathbb{A}_{[1]}P_0$ on $\operatorname{ind}(\mathsf{D})$. The Chern character $\operatorname{ch}(\mathbb{A}_t)$ has a limit as $t\to\infty$ as well, equal to the Chern character of ∇_0:

$$\lim_{t\to\infty}\operatorname{ch}(\mathbb{A}_t) = \operatorname{ch}(\nabla_0).$$

We will prove that the integral

$$\int_0^\infty \operatorname{Str}\left(\frac{d\mathbb{A}_t}{dt}e^{-\mathbb{A}_t^2}\right)dt$$

converges in $\mathscr{A}(B)$ under the hypothesis that $\ker(\mathsf{D})$ is constant in dimension, and thus derive the transgression formula for the Chern character associated to the Bismut connection,

$$\operatorname{ch}(\nabla_0) = (2i\pi)^{-n/2} \int_{M/B} \hat{A}(M/B) \operatorname{ch}(\mathscr{E}/S) + d \int_0^\infty \operatorname{Str}\Big(\frac{d\mathbb{A}_t}{dt} e^{-\mathbb{A}_t^2}\Big)\, dt.$$

In Sections 1–3, we study the geometry of a family M/B with a given horizontal tangent bundle. Thinking of M as a Riemannian manifold with degenerate metric $g = g_{M/B}$ on T^*M, defined to vanish on the horizontal cotangent vectors, then we can construct a natural Clifford module structure and Clifford connection on $\mathbb{E}$ using one more piece of data, a metric on the base B.

Having constructed the degenerate Clifford module structure on $\mathbb{E}$, we define the Bismut superconnection $\mathbb{A}$ to be the associated Dirac operator; we then prove the important result that $\mathbb{A}$ does not depend on the horizontal metric g_B used in the definition of the Clifford module structure and Clifford connection on $\mathbb{E}$. It is interesting to observe that $\mathbb{A}$ has in general terms up to degree two:

$$\mathbb{A} = \mathbb{A}_{[0]} + \mathbb{A}_{[1]} + \mathbb{A}_{[2]}.$$

Let $\mathscr{F} = \mathbb{A}^2$ be the curvature of the Bismut superconnection. In Section 3, we will prove Bismut's explicit formula for $\mathscr{F}$, and explain how it is a generalization of the Lichnerowicz formula.

By the results of Chapter 9, for each $t > 0$, the heat operator $e^{-t\mathscr{F}}$ acting on $\Gamma(M, \mathbb{E})$ has a kernel

$$k_t \in \Gamma(M \times_\pi M, \pi^* \Lambda T^*B \otimes (\mathscr{E} \boxtimes_\pi \mathscr{E}^*)).$$

If we restrict the kernel k_t to the diagonal, we obtain a section $k_t(x,x)$ of the bundle of algebras $\pi^* \Lambda T^*B \otimes \operatorname{End}(\mathscr{E})$ over M. The bundle $\operatorname{End}(\mathscr{E})$ is isomorphic to $C(M/B) \otimes \operatorname{End}_{C(M/B)}(\mathscr{E})$, and applying the symbol map to $C(M/B)$, we obtain a section of $\pi^* \Lambda T^*B \otimes \Lambda T^*(M/B) \otimes \operatorname{End}_{C(M/B)}(\mathscr{E})$. Using the connection on the fibre bundle $M \to B$, we obtain an isomorphism of $\pi^* \Lambda T^*B \otimes \Lambda T^*(M/B)$ with ΛT^*M, so that we obtain a differential form $k_t(x,x)$ on M, with values in the bundle $\operatorname{End}_{C(M/B)}(\mathscr{E})$. We will calculate $\lim_{t\to 0} \operatorname{ch}(\mathbb{A}_t)$ by showing that $k_t(x,x)$ has an asymptotic expansion

$$k_t(x,x) \sim (4\pi t)^{-n/2} \sum_{i=0}^\infty t^i k_i(x)$$

with coefficients $k_i \in \sum_{j \le 2i} \mathscr{A}^j(M, \operatorname{End}_{C(M/B)}(\mathscr{E}))$, such that

$$\sum_{i=0}^{\dim(M)/2} (k_i)_{[2i]} = \hat{A}(M/B) \exp(-F^{\mathscr{E}/S}).$$

From this result, we immediately deduce that $\lim_{t\to 0} \operatorname{ch}(\mathbb{A}_t)$ exists.

In Section 5, we prove the transgression formula, and use this in Section 6 to calculate the "anomaly" formula of Bismut-Freed for the curvature of the connection on the determinant line bundle introduced in Section 9.7. Finally, in Section 7, we return to the case studied in Section 9.5 in which the bundle M/B is associated to a principal bundle $P \to B$ with compact structure group G, and show that in this case the local family index theorem is equivalent to the local Kirillov formula of Chapter 8.

10.1 Riemannian Fibre Bundles

Let $\pi : M \to B$ be a fibre bundle with Riemannian metrics on the fibres. In this section, we will describe two different families of connections that may be constructed on the tangent bundle TM of the total space M. This material is used in the rest of this chapter in the proof of the family index theorem.

Let $\pi : M \to B$ be a fibre bundle. We will denote by $T(M/B)$ the bundle of vertical tangent vectors. Let us assume that the bundle M/B possesses the following additional structures:

1. a connection, that is, a choice of a splitting $TM = T_H M \oplus T(M/B)$, so that the subbundle $T_H M$ is isomorphic to the vector bundle $\pi^* TB$;
2. a connection $\nabla^{M/B}$ on $T(M/B)$.

We denote by P the projection operator

$$P : TM \to T(M/B)$$

with kernel the chosen horizontal tangent space $T_H M$. If X is a vector field on the base B, denote by X_M its horizontal lift on M, that is, the vector field on M which is a section of $T_H M$ and which projects to X under the pushforward $\pi_* : (T_H M)_x \to T_{\pi(x)} B$. We will often make use of a local frame e_i of the vertical tangent bundle $T(M/B)$, and of a local frame f_α of TB, with dual frames e^i and f^α; using the connection, we obtain a local frame of the tangent bundle TM.

There are three tensors, S, k and Ω, canonically associated to a family of Riemannian manifolds with connection M/B. These are defined as follows:

1. The tensor S (the **second fundamental form**) is the section of the bundle

$$\operatorname{End}(T(M/B)) \otimes T_H^* M \cong T^*(M/B) \otimes T(M/B) \otimes T_H^* M$$

defined by

$$(S(X,\theta),Z) = \langle \nabla_Z^{M/B} X - P[Z,X], \theta \rangle$$

for $Z \in \Gamma(M, T_H M)$, $X \in \Gamma(M, T(M/B))$ and $\theta \in \Gamma(M, T^*(M/B))$. It is clear that this is a tensor, since P vanishes on $T_H M$.

2. The one-form $k \in \mathscr{A}^1(M)$ (the **mean curvature**) is the trace of S:

$$k(Z) = \operatorname{Tr}(S(Z)) = \sum (S(e_i, e^i), Z).$$

3. The tensor Ω is the section of the bundle $\mathrm{Hom}(\Lambda^2 T_H M, T(M/B))$ over M defined by the formula
$$\Omega(X,Y) = -P[X,Y] \quad \text{for } X \text{ and } Y \text{ in } \Gamma(M, T_H M).$$
This is clearly antisymmetric in X and Y, and is a tensor since if we replace Y by fY, for $f \in C^\infty(M)$, then
$$\Omega(X, fY) = -fP[X,Y] - X(f)PY = f\Omega(X,Y).$$
By comparison with (1.6), we see that Ω may be identified with the **curvature** of the fibre bundle M/B.

The total exterior differential d_M may be expressed in terms of $\nabla^{M/B}$, S, Ω and the vertical exterior differential $d_{M/B}$ of the fibre bundle $M \to B$. We extend $d_{M/B}$ to an operator on $\Gamma(M, \Lambda T_H^* M \otimes \Lambda T^*(M/B))$ by the formula
$$d_{M/B}(\pi^* \nu \otimes \beta) = (-1)^{|\nu|} \pi^* \nu \otimes d_{M/B} \beta$$
for $\nu \in \mathscr{A}(B)$ and $\beta \in \Gamma(M, \Lambda T^*(M/B))$. We cannot define a horizontal differential on $\Gamma(M, \Lambda T_H^* M \otimes \Lambda T^*(M/B))$, but we may define an operator δ_B by means of the connection $\nabla^{M/B}$ on $\Gamma(M, \Lambda T^*(M/B))$:
$$\delta_B(\pi^* \nu \otimes \beta) = \pi^*(d_B \nu) \wedge \beta + (-1)^{|\nu|} \pi^* \nu \otimes \sum_\alpha f^\alpha \wedge \nabla^{M/B}_{f_\alpha} \beta.$$

If $\theta \in \Gamma(M, T^*(M/B))$, we define the contraction
$$\langle S, \theta \rangle \in \Gamma(M, T_H^* M \otimes T^*(M/B)) \subset \mathscr{A}^2(M)$$
by the formula
$$\langle S, \theta \rangle (e_i, f_\alpha) = \langle S(e_i, \theta), f_\alpha \rangle,$$
and the contraction $\langle \Omega, \theta \rangle \in \Gamma(M, \Lambda^2 T_H^* M) \subset \mathscr{A}^2(M)$.

Proposition 10.1. $d_M = d_{M/B} + \delta_B - \sum_i \langle S, e^i \rangle \iota(e_i) + \sum_i \langle \Omega, e^i \rangle \iota(e_i)$

Proof. It is easy to check that both sides are derivations of $\mathscr{A}(M)$, and that they agree on horizontal forms. Thus, we only need to check that they agree on the vertical one forms e^i. We start by checking this when both sides are evaluated on $e_k \wedge f_\alpha$: on the one hand,
$$(d_M e^i)(e_k, f_\alpha) = (e^i, [f_\alpha, e_k]),$$
while by the definition of S, the right-hand side equals
$$-\langle \nabla^{M/B}_{f_\alpha} e^i, e_k \rangle - \langle \nabla^{M/B}_{f_\alpha} e_k, e^i \rangle + ([f_\alpha, e_k], e^i) = ([f_\alpha, e_k], e^i).$$
We must also evaluate both sides on $f_\alpha \wedge f_\beta$: but the equality comes down to
$$(d_M e^i)(f_\alpha, f_\beta) = \langle \Omega(f_\alpha, f_\beta), e^i \rangle. \qquad \square$$

Let us now suppose that $M \to B$ is a family of Riemannian manifolds, thus, on each fibre $M_z = \pi^{-1}(z)$, $z \in B$, there is given a Riemannian metric, which we will denote by $g_{M/B}$; in other words, we are given an inner product on the bundle of vertical tangent vectors $T(M/B)$. There is then a canonical connection $\nabla^{M/B}$ on $T(M/B)$, constructed as follows.

If we choose a Riemannian metric g_B on the base B and pull it up to $T_H M$ by means of the identification $T_H M \cong \pi^* TB$ given by the connection on the bundle M/B, we obtain an inner product on the bundle $T_H M$, which we will call a horizontal metric. We may now form the total metric $g = g_B \oplus g_{M/B}$ on the tangent bundle TM of M, by means of the identification $TM \cong T_H M \oplus T(M/B)$. Let ∇^g be the Levi-Civita connection on TM with respect to this metric, and define a connection $\nabla^{M/B}$ on the bundle $T(M/B)$ by projecting this connection

$$\nabla^{M/B} = P \cdot \nabla^g \cdot P.$$

The following proposition shows that the choice of metric g_B on the base is irrelevant in constructing $\nabla^{M/B}$. In what follows, we will denote by $g(X,Y)$ the inner product of two vector fields X and Y with respect to the metric g, or simply (X,Y) if the metric g is evident from the context.

Proposition 10.2. *The connection $\nabla^{M/B}$ on $T(M/B)$ is independent of the metric g_B on TB used in its definition.*

Proof. This follows easily from the formula (1.18) for the Levi-Civita connection ∇^g, which we will use constantly in this chapter:

$$\begin{aligned} 2(\nabla^g_X Y, Z) = & ([X,Y],Z) - ([Y,Z],X) + ([Z,X],Y) \\ & + X(Y,Z) + Y(Z,X) - Z(X,Y). \end{aligned}$$

If the vectors X, Y and Z are all vertical, that is, sections of $T(M/B)$, then the right-hand side reduces to the Levi-Civita connection on the fibres for the vertical metric $g_{M/B}$. On the other hand, if X is horizontal but Y and Z are vertical, then $[Y,Z]$ is vertical, so that $([Y,Z],X)$ vanishes, and we see that

$$2(\nabla^{M/B}_X Y, Z) = (P[X,Y],Z) + (P[Z,X],Y) + X(Y,Z).$$

From this formula, it is clear that only the vertical metric $g_{M/B}$ and the vertical projection P are used to define $\nabla^{M/B}_X$ for X horizontal. □

The vertical metric allows us to think of the tensor S as a section of $T^*(M/B) \otimes T^*(M/B) \otimes T^*_H M$. Let us give a more explicit formula for the tensor S.

Lemma 10.3. *If $\nabla^{M/B}$ is the connection on $T(M/B)$ associated to a vertical Riemannian metric $g_{M/B}$, then S is given by the formula*

$$2(S(X,Y),Z) = Z(X,Y) - (P[Z,X],Y) - (P[Z,Y],X).$$

Proof. This follows easily from the explicit formula for the Levi-Civita connection. □

It is evident from this lemma that $(S(X,Y),Z)$ is symmetric in X and Y. In fact, it represents the horizontal covariant derivative of the metric.

Let n be the dimension of the fibres of M/B, and let $\nu_{M/B} \in \mathscr{A}^n(M)$ be the Riemannian volume form along the fibres, where we identify a section of the bundle $\Lambda^n T^*(M/B)$ with a section of the bundle $\Lambda^n T^*M$ by means of the connection on the fibre bundle M/B.

Lemma 10.4. $d\nu_{M/B} = k \wedge \nu_{M/B} + \sum_{i=1}^{n} \langle \Omega, e^i \rangle \iota(e_i) \nu_{M/B}$

Proof. We use the formula of Proposition 10.1. Since $\nabla^{M/B}$ preserves the metric $g_{M/B}$ on $T(M/B)$, and hence $\nabla^{M/B} \nu_{M/B} = 0$, we see that

$$\begin{aligned} d\nu_{M/B} = &- \sum_i \langle S, e^i \rangle \iota(e_i) \nu_{M/B} \\ &+ \sum_i \langle \Omega, e^i \rangle \iota(e_i) \nu_{M/B}. \end{aligned}$$

Furthermore,

$$\begin{aligned} \sum_i \langle S, e^i \rangle \iota(e_i) \nu_{M/B} &= \sum_{i,j,\alpha} \langle S(e_j, e^i), f^\alpha \rangle e^j \wedge f^\alpha \iota(e_i) \nu_{M/B} \\ &= -k(f_\alpha) \nu_{M/B}. \end{aligned}$$

□

If we choose a metric g_B on B, we obtain a Levi-Civita connection on TB, which may be pulled back to give a connection on the bundle $T_H M \cong \pi^* TB$. We can form a new connection on TM taking the direct sum of the connection $\nabla^{M/B}$ on $T(M/B)$ and $\pi^* \nabla^B$ on $T_H M$ (which we will write simply as ∇^B). We will denote this connection by $\nabla^\oplus$:

$$\nabla^\oplus = \nabla^B \oplus \nabla^{M/B}.$$

Note that if we replace g by the rescaled metric $u g_B + g_{M/B}$, where $u > 0$, then ∇^B does not change, and so neither does $\nabla^\oplus$.

The connection $\nabla^\oplus$ preserves the metric g, but has non-vanishing torsion in general. Our next task is to compare the connections ∇^g and $\nabla^\oplus$ on TM. In order to do this, we introduce a three-tensor ω on M which only depends on the data that we are given for the fibre bundle $M \to B$, namely the vertical metrics $g_{M/B}$ and the choice of connection on M/B.

Definition 10.5. *Let $\omega \in \mathscr{A}^1(M, \Lambda^2 T^*M)$ be the $\Lambda^2 T^*M$-valued one-form on M defined by the formula*

$$\begin{aligned} \omega(X)(Y,Z) = S(X,Z)(Y) &- S(X,Y)(Z) \\ &+ \tfrac{1}{2}(\Omega(X,Z),Y) - \tfrac{1}{2}(\Omega(X,Y),Z) + \tfrac{1}{2}(\Omega(Y,Z),X). \end{aligned}$$

The above formula is antisymmetric in Y and Z, so that ω takes its values in Λ^2T^*M, as claimed.

The following result, due to Bismut, is the main result of this section.

Proposition 10.6. *The Levi-Civita connection ∇^g is related to the connection $\nabla^\oplus$ by the following formula:*

$$g(\nabla^g_X Y,Z) = g(\nabla^\oplus_X Y,Z) + \omega(X)(Y,Z).$$

Proof. We will make use of the following lemma

Lemma 10.7. *If Y is the horizontal lift of a vector field on B, and if X is vertical, then $[X,Y]$ is vertical.*

Proof. If $\phi \in C^\infty(B)$, then $Y(\pi^*\phi) = \pi^*(\pi_* Y\phi)$, while, since X is vertical, $X(\pi^*\phi) = 0$ and $(X\cdot Y)\pi^*\phi = 0$. This shows that $[X,Y](\pi^*\phi) = 0$, and hence that $[X,Y]$ is vertical. □

Now, observe that $(\nabla^g_X Y,Z) - (\nabla^\oplus_X Y,Z)$ is antisymmetric in Y and Z, because each of the connections ∇^g and $\nabla^\oplus$ preserves the total metric g on M. The proof of the theorem now consists on a case-by-case examination of the different situations in which X, Y and Z are horizontal or vertical.

1. If X, Y and Z are all horizontal lifts of vector fields on the base or all vertical, it is easy to see that $(\nabla^g_X Y,Z)$ equals $(\nabla^\oplus_X Y,Z)$ while $\omega(X)(Y,Z)$ is zero.
2. If both Y and Z are vertical, then $(\nabla^g_X Y,Z)$ and $(\nabla^\oplus_X Y,Z)$ are both equal to $(\nabla^{M/B}_X Y,Z)$, and once more $\omega(X)(Y,Z)$ is zero.
   In each of the remaining cases, $(\nabla^\oplus_X Y,Z)$ is equal to zero.
3. If X and Y are vertical and Z is horizontal, then $(\nabla^g_X Y,Z)$ equals
$$-(S(X,Y),Z) = \omega(X)(Y,Z).$$
4. If X and Y are horizontal lifts of vector fields on B and Z is vertical, then $(\nabla^g_X Y,Z)$ equals
$$\tfrac{1}{2}([X,Y],Z) + \tfrac{1}{2}([Z,Y],X) + \tfrac{1}{2}([Z,X],Y) - \tfrac{1}{2}Z(X,Y) = \tfrac{1}{2}([X,Y],Z),$$
as follows from the above lemma. But this equals $-\frac{1}{2}(\Omega(X,Y),Z)$ by the definition of Ω.
5. If X is vertical and Y and Z are the horizontal lifts of vector fields on B, then $(\nabla^g_X Y,Z)$ is equal to
$$\tfrac{1}{2}([X,Y],Z) - \tfrac{1}{2}([X,Z],Y) - \tfrac{1}{2}([Y,Z],X)$$
which equals $\frac{1}{2}(\Omega(Y,Z),X)$ by the same argument as for (4). □

□

10.2 Clifford Modules on Fibre Bundles

A fibre bundle M with the data that we considered in the last section may be thought of as a Riemannian manifold with a degenerate metric g_0 on its cotangent bundle T^*M, which vanishes in the horizontal cotangent directions $T(M/B)^\perp \subset T^*M$. This metric explodes in the horizontal directions of the tangent bundle TM of M, so that in contrast to Riemannian geometry, we distinguish between T^*M and TM. Since the Clifford algebra of a vector space V with vanishing metric is isomorphic to the exterior algebra ΛV, we see that the Clifford bundle $C(T^*M)$ of M with respect to the degenerate metric g^0, which we will denote by $C_0(M)$, may be identified with the bundle of algebras $\pi^*\Lambda T^*B \otimes C(M/B)$, where $C(M/B)$ denotes the Clifford bundle $C(T^*(M/B))$.

In this section, we will generalize the discussion of Clifford modules and Clifford connections of Chapter 3 to the setting of the Clifford bundle $C_0(M)$.

Fix a metric g_B on the base. If $u \in (0,1]$ is a small positive number, let g^u be the metric on T^*M equal to the vertical metric $g_{M/B}$ on the vertical cotangent vectors, and ug_B on the horizontal ones:

$$g^u = g_{M/B} \oplus ug_B.$$

In particular, $g^0 = \lim_{u\to 0} g^u$ is the degenerate metric that we introduced above. The family of metrics g^u is only a tool in investigating the geometry of g^0. Let g_u be the dual metric on TM,

$$g_u = g_{M/B} \oplus u^{-1} g_B,$$

which explodes in the horizontal directions as $u \to 0$.

Let $C_u(M) = C(T^*M, g^u)$ be the Clifford algebra bundle of the bundle of inner-product spaces (T^*M, g^u), and denote the canonical quantization map of Proposition 3.5 from ΛT^*M to $C_u(M)$ by $\mathbf{c}_u$. Thus, the Clifford bundle $C_0(M)$ is the limit of the one-parameter family of algebras bundles $C_u(M)$.

Let τ^u denote the bundle map from $\Lambda^2 T^*M$ to $\operatorname{End}(T^*M)$ which is defined implicitly by

$$\mathbf{c}_u(\tau^u(\alpha)\xi) = [\mathbf{c}_u(\alpha), \mathbf{c}_u(\xi)],$$

where $\alpha \in \Lambda^2 T_x^*M$ and $\xi \in T_x^*M$; then

$$\tau^u(v_1 \wedge v_2)\xi = 2\big(g^u(v_1,\xi)v_2 - g^u(v_2,\xi)v_1\big).$$

In order to understand this definition better, we will write it out explicitly in terms of an orthogonal frame of T_x^*M of the form $\{e^i\} \cup \{f^\alpha\}$, where e^i is an orthonormal frame for $T^*(M/B)$ and f^α is an orthonormal frame for T^*B, with respect to the metric g_B. In terms of this frame,

$$
\begin{aligned}
\tfrac{1}{2}\tau^u(e^ie^j)e^k &= \delta^{ik}e^j - \delta^{jk}e^i \\
\tfrac{1}{2}\tau^u(e^if^\alpha)e^j &= \delta^{ij}f^\alpha \\
\tau^u(f^\alpha f^\beta)e^i &= 0 \\
\tfrac{1}{2}\tau^u(e^if^\alpha)f^\beta &= -u\delta^{\alpha\beta}e^i \\
\tfrac{1}{2}\tau^u(f^\alpha f^\beta)f^\gamma &= u\left(\delta^{\alpha\gamma}f^\beta - \delta^{\beta\gamma}f^\alpha\right) \\
\tau^u(e^ie^j)f^\alpha &= 0.
\end{aligned} \tag{10.1}
$$

It is clear from these formulas that the family of actions τ^u has a well-defined limit as $u \to 0$, which we denote by τ^0:

$$
\begin{aligned}
\tfrac{1}{2}\tau^0(e^ie^j)e^k &= \delta^{ik}e^j - \delta^{jk}e^i \\
\tfrac{1}{2}\tau^0(e^if^\alpha)e^j &= \delta^{ij}f^\alpha \\
\tau^0(f^\alpha f^\beta)e^i &= 0 \\
\tau^0(a)f^\alpha &= 0 \quad \text{for all } a \in \Lambda^2T^*M.
\end{aligned} \tag{10.2}
$$

It is important to note that $\tau^0(a)$ vanishes on T^*B.

We will denote the negative of the adjoint of $\tau^u(a) \in \operatorname{End}(T^*M)$, where $a \in \Lambda^2T^*M$, by $\tau_u(a) \in \operatorname{End}(TM)$, and the negative of the adjoint of $\tau^0(a)$ by $\tau_0(a)$. Note the formula

$$\tfrac{1}{2}g_u(\tau_u(a)X,Y) = \langle a, X\wedge Y\rangle,$$

for $X,Y \in \Gamma(M,TM)$ and $a \in \Lambda^2T^*M$.

It is clear that Λ^2T^*M has a family of Lie brackets induced on it by the maps τ^u, defined by the formula

$$\tau^u[\alpha_1,\alpha_2]_u = [\tau^u(\alpha_1),\tau^u(\alpha_2)].$$

We denote the limit of $[\alpha_1,\alpha_2]_u$ as $u \to 0$ by $[\alpha_1,\alpha_2]_0$, which satisfies

$$\tau^0[\alpha_1,\alpha_2]_0 = [\tau^0(\alpha_1),\tau^0(\alpha_2)].$$

Let $\nabla^{M,u}$, $u > 0$, be the Levi-Civita connection on TM corresponding to the metric g_u, and let $\nabla^\oplus$ be the direct-sum connection introduced in the last section. In the notation of this section, we may restate Proposition 10.6 as

$$\nabla^{M,u} = \nabla^\oplus + \tfrac{1}{2}\tau_u(\omega).$$

where ω is the element of $\mathscr{A}^1(M,\Lambda^2T^*M)$ of Definition 10.5. It follows that the family $\nabla^{M,u}$ of connections has a well-defined limit as $u \to 0$, which we will denote by $\nabla^{T^*M,0}$, and which is given by the formula

$$\nabla^{M,0} = \nabla^\oplus + \tfrac{1}{2}\tau_0(\omega).$$

We denote by $\nabla^{T^*M,u}$ and $\nabla^{T^*M,0}$ the dual connections on TM, that is,

$$
\begin{aligned}
\nabla^{T^*M,u} &= \nabla^\oplus + \tfrac{1}{2}\tau^u(\omega) \\
\nabla^{T^*M,0} &= \nabla^\oplus + \tfrac{1}{2}\tau^0(\omega)
\end{aligned} \tag{10.3}
$$

Proposition 10.8. *1. The connection $\nabla^{M,u}$ is torsion-free.*
2. The projection of $\nabla^{M,u}$ to the sub-bundle $T(M/B)$, equals $\nabla^{M/B}$.
3. Restricted to each fibre, the connection $\nabla^{M,0}$ depends only on the vertical metric $g_{M/B}$ and the connection on the fibre bundle M/B.

Proof. For $u>0$, the connection $\nabla^{M,u}$ is a Levi-Civita connection, so is torsion-free. Since $\lim_{u\to 0}\nabla^{M,u}=\nabla^{M,0}$, it follows that $\nabla^{M,0}$ is torsion-free. The connection $\nabla^{M/B}$ on the bundle $T(M/B)$ is by definition equal to the projection of $\nabla^{M,u}$ onto $T(M/B)$, and taking the limit $u\to 0$, we see that this is true for $\nabla^{M,0}$. Finally, the fact that the restriction of $\nabla^{M,0}$ to a fibre of the bundle M/B depends only on the vertical metric $g_{M/B}$ and the connection on the bundle M/B follows from (10.3), since $\pi^*\nabla^B=0$ on vertical vectors. □

Consider the section of the bundle $\Lambda^2T^*M\otimes\Lambda^2T^*(M/B)$,

$$R^{M/B}(W,X)(Y,Z)=(R^{M/B}(W,X)Y,Z)_{M/B},$$

where $R^{M/B}\in\mathscr{A}^2(M,\operatorname{End}(T(M/B)))$ is the curvature of the connection $\nabla^{M/B}$ on $T(M/B)$. Using the connection on the bundle $M\to B$, we may extend $R^{M/B}$ to a four-tensor on M by setting $R^{M/B}(W,X)(Y,Z)=0$ if Y or Z is horizontal. Let $R^u\in\mathscr{A}^2(M,\operatorname{End}(TM))$ be the Riemannian curvature of the metric g_u, and let $R^B\in\mathscr{A}^2(B,\operatorname{End}(TB))$ be the Riemannian curvature of the metric g_B on the base B.

Proposition 10.9. *Let $\mathsf{R}\in\Gamma(M,\Lambda^2T^*M\otimes\Lambda^2T^*M)$ be the four-tensor defined by the formula*

$$\mathsf{R}(W,X,Y,Z)=\lim_{u\to 0}\Big(g_u(R^u(W,X)Y,Z)-u^{-1}(R^B(W,X)Y,Z)_B\Big).$$

1. The above limit exists and equals $R^{M/B}+\nabla^{\oplus}\omega+\frac14[\omega,\omega]_0$.
2. When Y and Z are vertical, $\mathsf{R}(W,X,Y,Z)$ equals $R^{M/B}(W,X,Y,Z)$.
3. If W, X, Y and Z are vector fields on M, then

$$\mathsf{R}(W,X,Y,Z)+\mathsf{R}(X,Y,W,Z)+\mathsf{R}(Y,W,X,Z)=0.$$

4. $\mathsf{R}(W,X,Y,Z)=\mathsf{R}(Y,Z,W,X)$

Proof. Written as a four-tensor $g_u(R^u(W,X)Y,Z)$, the curvature of the metric g_u is a polynomial in u and u^{-1}, and we must show that R is the constant term in this expansion, by proving that

$$\lim_{u\to 0}u\,g_u(R^u(W,X)Y,Z)=(R^B(W,X)Y,Z)_B.$$

From the formula $\nabla^{M,u}=\nabla^{\oplus}+\frac12\tau_u(\omega)$, we see that

$$\begin{aligned}(\nabla^{M,u})^2&=(\nabla^{\oplus})^2+\tfrac12[\nabla^{\oplus},\tau_u(\omega)]+\tfrac18[\tau_u(\omega),\tau_u(\omega)]\\&=(\nabla^{\oplus})^2+\tfrac12\tau_u(\nabla^{\oplus}\omega)+\tfrac12\tau_u\big(\tfrac14[\omega,\omega]_u\big).\end{aligned}$$

Since $(\nabla^{\oplus})^2 = (\pi^*\nabla^B)^2 + (\nabla^{M/B})^2$, it follows that

$$\begin{aligned} g_u(R^u(W,X)Y,Z) = & u^{-1}(\pi^* R^B(W,X)Y,Z)_B + (R^{M/B}(W,X)Y,Z) \\ & + (\nabla^{\oplus}\omega)(W,X)(Y,Z) + \tfrac{1}{4}[\omega,\omega]_u(W,X)(Y,Z), \end{aligned}$$

where we recall that $[\omega,\omega]_u = (\tau^u)^{-1}([\tau^u(\omega),\tau^u(\omega)])$. The existence of the limit as $u \to 0$ now follows from the fact that $[\omega,\omega]_u = [\omega,\omega]_0 + O(u)$, and we have

$$\begin{aligned} \mathsf{R}(W,X,Y,Z) = & (R^{M/B}(W,X)Y,Z)_{M/B} \\ & + (\nabla^{\oplus}\omega)(W,X)(Y,Z) + \tfrac{1}{4}[\omega,\omega]_0(W,X)(Y,Z). \end{aligned}$$

The other properties of R are clear from its definition: for example, (3) follows from the fact that both R^u and R^B, being Riemannian curvatures, satisfy the desired symmetry. □

Let $\mathscr{E}$ be a Clifford module along the fibres of the bundle M/B, that is, a Hermitian vector bundle over M with a skew-adjoint action

$$c : C(T^*(M/B)) \to \operatorname{End}(\mathscr{E})$$

of the vertical Clifford bundle of M/B, and a Hermitian connection $\nabla^{\mathscr{E}}$ compatible with this action,

$$[\nabla^{\mathscr{E}}_X, c(\alpha)] = c(\nabla^{M/B}_X \alpha),$$

for $X \in \Gamma(M,TM)$ and $\alpha \in \Gamma(M,T^*(M/B))$.

Denote by $\mathbb{E}$ the vector bundle over M defined by the formula

$$\mathbb{E} = \pi^*\Lambda T^*B \otimes \mathscr{E}.$$

This bundle carries a natural action m_0 of the degenerate Clifford module $C_0(M)$. In order to define this action, it suffices to define the actions of $T^*_H M \subset C_0(M)$ and $T^*(M/B) \subset C_0(M)$. The Clifford action of a horizontal cotangent vector $\alpha \in \Gamma(M,T^*_H M)$ is given by exterior multiplication

$$m_0(\alpha) = \varepsilon(\alpha)$$

acting on the first factor $\Lambda T^*_H M$ in $\mathbb{E}$, while the Clifford action of a vertical cotangent vector simply equals its Clifford action on $\mathscr{E}$. This Clifford module will be the one of the main tools in calculating the family index of the family D. In order to study $\mathbb{E}$, we will write it as the limit of a family of Clifford modules for the bundles of Clifford algebras $C_u(M)$, all constructed on the same underlying vector bundle as $\mathbb{E}$.

In order to define the Clifford action

$$m_u : C_u(M) \to \operatorname{End}(\mathbb{E}) \cong \operatorname{End}(\Lambda T^*_H M \otimes \mathscr{E}),$$

it suffices to define the actions of $T^*_H M \subset C_u(M)$ and $T^*(M/B) \subset C_u(M)$. The Clifford action of a horizontal cotangent vector $\alpha \in \Gamma(M,T^*_H M)$ is given by the formula

$$m_u(\alpha) = \varepsilon(\alpha) - u\iota(\alpha),$$

acting on the first factor $\Lambda T^*_H M$ in $\mathbb{E}$, while the Clifford action of a vertical cotangent vector simply equals its Clifford action on $\mathscr{E}$. It is a straightforward task to check that

$$m_u(\alpha)^2 = -g^u(\alpha,\alpha) \quad \text{for } \alpha \in \Gamma(M, T^*M).$$

In particular, we see that the limiting Clifford module action $\lim_{u\to 0} m_u$ is just the degenerate action m_0 introduced above.

There are connections $\nabla^{\mathbb{E},\oplus}$ and $\nabla^{\mathbb{E},u}$ on the Clifford module $\mathbb{E}$ analogous to the connections $\nabla^{\oplus}$ and $\nabla^{T^*M,u}$ that we have constructed on the bundle T^*M. As before, to construct these connections, we must choose a horizontal metric g_B on B. The connection $\nabla^{\mathbb{E},\oplus}$ on $\mathbb{E} \cong \Lambda\pi^*T^*B \otimes \mathscr{E}$ is defined by taking the sum of the Levi-Civita connection $\pi^*\nabla^B$ on the bundle $\Lambda\pi^*T^*B$ with the connection $\nabla^{\mathscr{E}}$ on $\mathscr{E}$,

$$\nabla^{\mathbb{E},\oplus} = \pi^*\nabla^B \otimes 1 + 1 \otimes \nabla^{\mathscr{E}};$$

the connection $\nabla^{\mathbb{E},u}$ is defined by the formula, inspired by Proposition 10.6,

$$\nabla^{\mathbb{E},u} = \nabla^{\mathbb{E},\oplus} + \tfrac{1}{2} m_u(\omega),$$

where ω is the one-form defined in Section 1.

Proposition 10.10. *The connection $\nabla^{\mathbb{E},u}$ is a Clifford connection for the Clifford action m_u of $C_u(M)$ on $\mathbb{E}$, in other words,*

$$[\nabla^{\mathbb{E},u}_X, m_u(\alpha)] = m_u(\nabla^{T^*M,u}_X \alpha),$$

for $X \in \Gamma(M, TM)$ and $\alpha \in \Gamma(M, T^*M)$. In particular, the connection

$$\nabla^{\mathbb{E},0} = \lim_{u\to 0} \nabla^{\mathbb{E},u} = \nabla^{\mathbb{E},\oplus} + \tfrac{1}{2} m_0(\omega)$$

is a Clifford connection for the Clifford action m_0 of the Clifford algebra bundle $C_0(M)$ on $\mathbb{E}$,

$$[\nabla^{\mathbb{E},0}_X, m_0(\alpha)] = m_0(\nabla^{T^*M,0}_X \alpha).$$

The restriction of $\nabla^{\mathbb{E},u}$, and in particular of $\nabla^{\mathbb{E},0}$, to each fibre of the bundle M/B is independent of the horizontal metric g_B used in its definition.

Proof. The first step of the proof is to show that $\nabla^{\mathbb{E},\oplus}$ is a Clifford connection with respect to any of the Clifford actions m_u, if $C_u(M)$ is given the connection $\nabla^{\oplus}$:

$$[\nabla^{\mathbb{E},\oplus}_X, m_u(\alpha)] = m_u(\nabla^{\oplus}_X \alpha).$$

This formula follows from considering the two cases, in which X respectively horizontal and vertical. The proof that $\nabla^{\mathbb{E},u}$ is a Clifford connection for the action m_u and the connection $\nabla^{T^*M,u}$ on $C_u(M)$ follows by observing that, almost by definition, $[m_u(\omega(X)), m_u(\alpha)] = m_u(\tau^u(\omega(X))\alpha)$. Finally, the independence of the horizontal metric g_B of the restriction of $\nabla^{\mathbb{E},u}$ to a fibre of the bundle M/B follows as it did for $\nabla^{M,u}$ from the fact that $\pi^*\nabla^B$ vanishes on vertical vectors. □

In the rest of this chapter, we will make use of the frame $\{e^i\} \cup \{f^\alpha\}$ of T^*M introduced in the last section, as well as the dual frame $\{e_i\} \cup \{f_\alpha\}$ of TM. We adopt the convention for indices that $i, j, \ldots$, label vertical vectors, $\alpha, \beta, \ldots$, label horizontal vectors, while $a, b, \ldots$, label all vectors, horizontal or vertical. With this convention, we denote by e^a any one element of the cotangent frame $\{e^i\} \cup \{f^\alpha\}$, and the Clifford action $m_u(e^a)$ by m_u^a.

In the next section, we will need a formula for the curvature of the Clifford connection $\nabla^{\mathbb{E},0}$. First, observe that an easy generalization of Proposition 3.43 shows that the curvature $(\nabla^{\mathscr{E}})^2$ of the connection $\nabla^{\mathscr{E}}$ on $\mathscr{E}$ equals

$$-\tfrac{1}{2} \sum_{i<j;a<b} R^{M/B}_{ijab} c^i c^j \varepsilon^a \varepsilon^b + \sum_{a<b} F^{\mathscr{E}/S}_{ab} \varepsilon^a \varepsilon^b.$$

where $R^{M/B}_{ijab} = (R^{M/B}(e_a, e_b)e_j, e_i)$ and $F^{\mathscr{E}/S} \in \mathscr{A}^2(M, \mathrm{End}_{C(M/B)}(\mathscr{E}))$ is by definition the twisting curvature of the Clifford connection $\nabla^{\mathscr{E}}$ on the Clifford module $\mathscr{E}$ over $C(M/B)$. In the special case in which the fibres M/B have a spin-structure with spinor bundle $\mathscr{S}(M/B)$, we may write $\mathscr{E}$ as $\mathscr{E} = \mathscr{S}(M/B) \otimes \mathscr{W}$, where the twisting bundle $\mathscr{W}$ is a Hermitian vector bundle with connection $\nabla^{\mathscr{W}}$, and the twisting curvature $F^{\mathscr{E}/S}$ is then nothing but the curvature of the connection $\nabla^{\mathscr{W}}$.

Let λ be the natural action of $\mathrm{End}(TB)$ on ΛTB defined in (1.26). Then $\lambda(R^B)$ is given by the formula (3.15).

Proposition 10.11. *The curvature of the connection* $\nabla^{\mathbb{E},0}$ *on* $\mathbb{E}$ *equals*

$$\lambda(R^B) + \tfrac{1}{2} m_0(\mathsf{R}) + F^{\mathscr{E}/S}.$$

Proof. We have

$$\left(\nabla^{\mathbb{E},\oplus} + \tfrac{1}{2} m_0(\omega)\right)^2 = (\nabla^{\mathbb{E},\oplus})^2 + \tfrac{1}{2}[\nabla^{\mathbb{E},\oplus}, m_0(\omega)] + \tfrac{1}{8}[m_0(\omega), m_0(\omega)].$$

The proposition follows from the formulas

1. $(\nabla^{\mathbb{E},\oplus})^2 = \lambda((\nabla^B)^2) + (\nabla^{\mathscr{E}})^2 = \lambda(R^B) + \tfrac{1}{2} m_0(R^{M/B}) + F^{\mathscr{E}/S}$,
2. $[\nabla^{\mathbb{E},\oplus}, m_0(\omega)] = m_0(\nabla^{\oplus}\omega)$, and
3. $[m_0(\omega), m_0(\omega)] = m_0([\omega, \omega]_0)$. □

□

If the base B has a spin-structure, with associated spinor bundle $\mathscr{S}(B)$, there is another Clifford module on M, this time considered with the non-degenerate metric $g = g_B + g_{M/B}$, namely $\pi^*\mathscr{S}(B) \otimes \mathscr{E}$. Indeed, there is a natural decomposition

$$C(M) \cong \pi^*C(B) \otimes C(M/B)$$

of Clifford algebra bundles over M, corresponding to the decomposition $T^*M \cong \pi^*T^*B \oplus T^*(M/B)$. We may introduce two natural connections in this situation, which are respectively Clifford connections on $\mathscr{S}(B) \otimes \mathscr{E}$ with respect to the connections $\nabla^{\oplus}$ and ∇^g on $C(M)$. We will not give the details of the proof, since it is very similar to that of Proposition 10.10.

Proposition 10.12. *1. The connection*

$$\nabla^{\mathscr{S}(B)\otimes\mathscr{E},\oplus} = \nabla^B \otimes 1 + 1 \otimes \nabla^{\mathscr{E}}$$

on $\mathscr{S}(B)\otimes\mathscr{E}$ is a Clifford connection with respect to the connection $\nabla^{\oplus}$ on $C(M)$.

2. Denote by $c(\omega) \in \mathscr{A}^1(M, C(M))$ the one-form defined by the formula

$$c(\omega) = \tfrac{1}{2}\sum_{abc} \omega(e_a)(e_b, e_c) e^a \otimes c(e^b)c(e^c).$$

The connection

$$\nabla^{\mathscr{S}(B)\otimes\mathscr{E}} = \nabla^{\mathscr{S}(B)\otimes\mathscr{E},\oplus} + \tfrac{1}{2}c(\omega)$$

on the bundle $\mathscr{S}(B)\otimes\mathscr{E}$ is a Clifford connection with respect to the Levi-Civita connection ∇^g on $C(M)$.

3. If M is a spin manifold with spinor bundle $\mathscr{S}_M$, and $\mathscr{E}$ is the relative spinor bundle $\mathscr{S}(M/B) \cong \mathrm{Hom}_{C(B)}(\pi^*\mathscr{S}(B), \mathscr{S}_M)$, then $\mathscr{S}(B)\otimes\mathscr{E}$ is naturally isomorphic to $\mathscr{S}_M$, and $\nabla^{\mathscr{S}(B)\otimes\mathscr{E}}$ may be identified with the Levi-Civita connection on $\mathscr{S}(M)$.

To finish this section, we give a lemma that we will need later.

Lemma 10.13. *If $\mathbb{E}$ is any Clifford module for the Clifford algebra bundle $C_u(M)$, with action m_u, then we have the equality*

$$\sum_{abc} \omega(e_a)(e_b, e_c) m_u^a m_u^b m_u^c = 2\sum_{\alpha} k(f_\alpha) m_u^\alpha - \tfrac{1}{2}\sum_{\alpha\beta i} (\Omega(f_\alpha, f_\beta), e_i) m_u^\alpha m_u^\beta m_u^i.$$

Proof. Using the fact that $\omega(e_a)(e_b, e_c)$ vanishes when the three indices a, b and c are all horizontal and also when both b and c are vertical, the left-hand side is seen to equal

$$\sum_{ij\alpha} 2\omega(e_i)(e_j, f_\alpha) m_u^i m_u^j m_u^\alpha + \sum_{i\alpha\beta} \omega(e_i)(f_\alpha, f_\beta) m_u^i m_u^\alpha m_u^\beta + \sum_{\alpha\beta i} 2\omega(f_\alpha)(e_i, f_\beta) m_u^\alpha m_u^i m_u^\beta.$$

Writing this ought explicitly, we get

$$-2\sum_{ij} S(e_i, e_j)(f_\alpha) m_u^i m_u^j m_u^\alpha - \tfrac{1}{2}\sum_{i\alpha\beta} (\Omega(f_\alpha, f_\beta), e_i) m_u^\alpha m_u^\beta m_u^i.$$

Since $S(e_i, e_j)(f_\alpha)$ is symmetric in i and j, only the trace of the tensor S contributes to the first sum, and we obtain our formula. □

10.3 The Bismut Superconnection

Let $\pi : M \to B$ be a fibre bundle with data as in the first section, namely vertical Riemannian metrics and a connection. Recall that given a vector bundle over M, $\pi_*\mathscr{E}$ is the infinite-dimensional bundle over B whose fibre at $z \in B$ is the space $\Gamma(M_z, \mathscr{E})$. If $\mathscr{E}$ is a Clifford module for the vertical Clifford algebra bundle $C(M/B)$, the Dirac operators D^z along the fibres $(M_z \mid z \in B)$, associated to the Clifford module $\mathscr{E}_z$ over M_z, combine to give a smooth family of Dirac operators $\mathsf{D} = (\mathsf{D}^z \mid z \in B)$, which is a smooth section of $\operatorname{End}_{\mathscr{P}}(\pi_*\mathscr{E}) \subset \operatorname{End}(\pi_*\mathscr{E})$ over B. In this section, we will describe a superconnection, which we call the Bismut superconnection, on the bundle $\pi_*\mathscr{E}$, with zero-form component D, whose Chern character form is particularly well-behaved. In view of its fundamental character, we will present the Bismut superconnection from two different viewpoints which are not obviously equivalent.

The following definition is as in Section 9.3, except that supress the factor $|\Lambda_\pi|^{1/2}$, since it is canonically trivialized by the vertical Riemannian metric on M.

Definition 10.14. *The space $\mathscr{A}(B, \pi_*\mathscr{E})$ of differential forms on B with coefficients in $\pi_*\mathscr{E}$ is the space of sections of the Clifford module $\mathbb{E} \cong \pi^*(\Lambda T^*B) \otimes \mathscr{E}$ over M.*

We introduce the Bismut superconnection

$$\mathbb{A} : \mathscr{A}(B, \pi_*\mathscr{E}) \to \mathscr{A}(B, \pi_*\mathscr{E})$$

as a Dirac operator for the Clifford module $\mathbb{E} \to M$. This is the motivation for introducing the whole apparatus of the last section.

Proposition 10.15. *The Dirac operator on $\mathbb{E}$, given by the formula*

$$\mathbb{A} = \sum_a m_0^a \nabla_a^{\mathbb{E},0},$$

is a superconnection when thought of as an operator on the space $\mathscr{A}(B, \pi_*\mathscr{E})$; we call $\mathbb{A}$ the **Bismut superconnection**. *More explicitly, the restriction of $\mathbb{A}$ to a map $\Gamma(M, \mathscr{E}) \to \Gamma(M, \mathbb{E})$ is given by the formula*

$$\sum_i c^i \nabla_i^{\mathscr{E}} + \sum_\alpha \varepsilon^\alpha (\nabla_\alpha^{\mathscr{E}} + \tfrac{1}{2} k(f_\alpha)) - \tfrac{1}{4} \sum_{\alpha<\beta} \sum_i \varepsilon^\alpha \varepsilon^\beta c^i (\Omega(f_\alpha, f_\beta), e_i).$$

The superconnection $\mathbb{A}$ is independent of the horizontal metric g_B used in its definition.

Proof. It is clear that $\mathbb{A}$ satisfies the first condition for it to be a superconnection on the bundle $\pi_*\mathscr{E}$, namely it is a differential operator on the bundle $\mathbb{E} = \pi^*\Lambda T^*B \otimes \mathscr{E}$ over M which is odd with respect to the total $\mathbb{Z}_2$-grading.

Next, we check that $\mathbb{A}$ satisfies the formula $\mathbb{A}(v s) = (d_B v)s + (-1)^{|v|} v \mathbb{A} s$, for $v \in \mathscr{A}(B)$ and $s \in \Gamma(M, \mathbb{E})$, where d_B is the exterior differential on B. It is easily seen that $\mathbb{A}(v s) = \sum_\alpha \varepsilon^\alpha \nabla_\alpha^B v + (-1)^{|v|} v \mathbb{A} s$. Since the Levi-Civita connection is torsion free, we see that $\sum_\alpha \varepsilon^\alpha \nabla_\alpha^B v = d_B v$, as required.

In order to check the explicit formula for $\mathbb{A}$ on $\Gamma(M,\mathscr{E})$, observe that

$$\mathbb{A} = \sum_i c^i \nabla_i^{\mathrm{E},\oplus} + \sum_\alpha \varepsilon^\alpha \nabla_\alpha^{\mathrm{E},\oplus} + \tfrac{1}{4}\sum_{abc} \omega(e_a)(e_b,e_c) m_0^a m_0^b m_0^c.$$

The proof of the formula is completed by Lemma 10.13. In particular, we see that $\mathbb{A}_{[0]} = \mathsf{D}$. □

The decomposition of the Bismut superconnection with respect to degree $\mathbb{A} = \mathbb{A}_{[0]} + \mathbb{A}_{[1]} + \mathbb{A}_{[2]}$ has in general terms up to degree two. Indeed, this superconnection is a natural example of a superconnection having non-zero terms of degree higher than one.

Proposition 10.16. *The connection part $\mathbb{A}_{[1]}$ of the Bismut superconnection is the natural unitary connection $\nabla^{\pi_*\mathscr{E}}$ on the bundle $\pi_*\mathscr{E}$ of Proposition 9.13, bearing in mind the identification of $\mathscr{E}\otimes|\Lambda_\pi|^{1/2}$ with $\mathscr{E}$ coming from the vertical metric $g_{M/B}$ on M/B.*

Proof. Choose a metric g_B on B, and let $|v_B|$ and $|v_M|$ be the Riemannian densities on B and M with respect to the metrics g_B and $g_M = g_B \oplus g_{M/B}$; clearly, $|v_{M/B}| = |v_M| \otimes \pi^*|v_B|^{-1}$. It will suffice to check that if X is a vector field on B, then

$$\mathscr{L}(X_M)|v_M|\otimes\pi^*|v_B|^{-1} + |v_M|\otimes\pi^*\mathscr{L}(X)|v_B|^{-1} = k(X)|v_M|\otimes|v_B|^{-1},$$

since this will imply that

$$\nabla_X^{\pi_*\mathscr{E}}(s\otimes|v_M|^{1/2}\otimes\pi^*|v_B|^{-1/2}) = (\nabla_{X_M}^{\mathscr{E}} + \tfrac{1}{2}k(X))s\otimes|v_M|^{1/2}\otimes\pi^*|v_B|^{-1/2}.$$

But this is a local result, so we may replace the vertical density $|v_M|\otimes\pi^*|v_B|^{-1}$ by the vertical volume form $v_{M/B}$ in the calculation. By (1.4), we see that

$$\mathscr{L}(X_M)v_{M/B} = \iota(X_M)dv_{M/B} + d(\iota(X_M)v_{M/B}).$$

The second term vanishes, since $v_{M/B}$ is vertical, while the second equals $k(X)v_{M/B}$ by Lemma 10.4. □

In order to compute the Chern character of the superconnection $\mathbb{A}$, we need the formula for its curvature $\mathbb{A}^2$ which we referred to above. Let us introduce the smooth family $\Delta^{M/B} = (\Delta^z \mid z\in B) \in \Gamma(B,\mathrm{End}(\pi_*\mathscr{E}))$ of generalized Laplacians along the fibres M_z of the bundle M/B; each Laplacian Δ^z is just the generalized Laplacian acting on the bundle $\mathscr{E}_z \to M_z$ corresponding to the connection $\nabla^{\mathrm{E},0}|_{M_z}$ and with zero potential.

Theorem 10.17. *Let $F^{\mathscr{E}/S} \in \mathscr{A}(M,\mathrm{End}_{C(M/B)}(\mathscr{E}))$ be the twisting curvature of the Clifford module $\mathscr{E}$, and let $r_{M/B}$ be the scalar curvature of the fibres of M/B. Then*

$$\mathbb{A}^2 = \Delta^{M/B} + \tfrac{1}{4}r_{M/B} + \sum_{a<b} m_0^a m_0^b F^{\mathscr{E}/S}(e_a,e_b).$$

Proof. The proof of this theorem is similar to the proof of Lichnerowicz's formula, Theorem 3.52. To begin with, denoting the covariant derivative ∇_{e_a} in the direction e_a by ∇_a for any connection ∇ on a bundle over M, we have

$$\begin{aligned}\mathbb{A}^2 &= \tfrac{1}{2}\sum_{ab}[m_0^a\nabla_a^{\mathbb{E},0}, m_0^b\nabla_b^{\mathbb{E},0}] \\ &= \tfrac{1}{2}\sum_{ab}[m_0^a, m_0^b]\nabla_a^{\mathbb{E},0}\nabla_b^{\mathbb{E},0} \\ &\quad + \sum_{ab} m_0^a[\nabla_a^{\mathbb{E},0}, m_0^b]\nabla_b^{\mathbb{E},0} + \tfrac{1}{2}\sum_{ab} m_0^a m_0^b[\nabla_a^{\mathbb{E},0}, \nabla_b^{\mathbb{E},0}].\end{aligned}$$

Since our frame e^a is the union of an orthonormal basis e^i of $T^*(M/B)$ and a basis f^α of T^*B, we see that the first term of this sum is equal to $-\sum_i(\nabla_i^{\mathbb{E},0})^2$. Using the fact that $\nabla^{\mathbb{E},0}$ is a Clifford connection, proved in Proposition 10.10, the second term is equal to

$$\begin{aligned}\sum_{ab} m_0^a m_0(\nabla_a^{T^*M,0}e^b)\nabla_b^{\mathbb{E},0} &= \sum_{abc} m_0^a m_0^c\langle\nabla_a^{T^*M,0}e^b, e_c\rangle\nabla_b^{\mathbb{E},0} \\ &= -\tfrac{1}{2}\sum_{ac} m_0^a m_0^c \nabla_{[e_a,e_c]}^{\mathbb{E},0} + \sum_i \nabla_{\nabla_i^{M/B}e_i}^{\mathbb{E},0}.\end{aligned}$$

Here we have used the facts that $\nabla^{M,0}$ agrees with $\nabla^{M/B}$ when restricted to a fibre M_z, that the connection $\nabla^{M,0}$ on TM is torsion-free, and the adjunction formula

$$\nabla_a^{T^*M,0}e^b = \sum_c\langle\nabla_a^{T^*M,0}e^b, e_c\rangle e^c = -\sum_c\langle e^b, \nabla_a^{M,0}e_c\rangle e^c,$$

It follows that

$$\mathbb{A}^2 = \Delta^{M/B} + \tfrac{1}{2}\sum_{ab} m_0^a m_0^b\left([\nabla_a^{\mathbb{E},0}, \nabla_b^{\mathbb{E},0}] - \nabla_{[e^a,e^b]}^{\mathbb{E},0}\right).$$

By Proposition 10.11, $\sum_{ab} m_0^a m_0^b R^{\mathbb{E},0}(e_a, e_b)$ equals

$$\sum_{ab} m_0^a m_0^b \lambda(R^B(e_a, e_b)) - \tfrac{1}{4}\sum_{abcd} m_0^a m_0^b m_0^c m_0^d \mathsf{R}_{cdab} + \sum_{ab} m_0^a m_0^b F^{\mathcal{E}/S}(e^a, e^b),$$

where $\mathsf{R}_{abcd} = \mathsf{R}(e_c, e_d)(e_b, e_a)$. The first term of this sum equals

$$-\tfrac{1}{2}\sum_{\alpha\beta\gamma\delta}\varepsilon^\alpha\varepsilon^\beta\varepsilon^\gamma\iota^\delta R^B_{\gamma\delta\alpha\beta},$$

which vanishes since the antisymmetrization of R^B over any three indices is zero. Furthermore, we have

$$\begin{aligned}\sum_{abcd} m_0^a m_0^b m_0^c m_0^d \mathsf{R}_{abcd} &= \sum_{abd} m_0^a m_0^b m_0^a m_0^d \mathsf{R}_{abad} + \sum_{abd} m_0^a m_0^b m_0^b m_0^d \mathsf{R}_{abbd} \\ &= 2\sum_{adi} m_0^a m_0^d \mathsf{R}_{aidi} = -2\sum_{ij}\mathsf{R}_{ijij},\end{aligned}$$

where we have used for the first equality Proposition 10.9 and antisymmetrization over the first three indices, and for the second equality the symmetry $\mathsf{R}_{aidi} = \mathsf{R}_{diai}$. But the four-tensor R coincides on vertical tangent vectors with the Riemannian curvature tensor of the fibre. □

We will finish this section by giving another definition of the Bismut superconnection. Locally, we may assume that the base B is a spin manifold, with spinor bundle $\mathscr{S}(B)$. Let D_M be the Dirac operator on $\Gamma(M, \mathscr{E} \otimes \pi^* \mathscr{S}(B))$ corresponding to the connection $\nabla^{\mathscr{E}\otimes\mathscr{S}(B)} = \nabla^{\mathscr{S}(B)\otimes\mathscr{E},\oplus} + \frac{1}{2} c(\omega)$ introduced in the last section.

If $\mathscr{W}$ is a finite-dimensional bundle over B with superconnection $\mathbb{A}$, we can construct a Clifford superconnection $\mathbb{B} = \mathbb{A} \otimes 1 + 1 \otimes \nabla^{\mathscr{S}(B)}$ on the twisted Clifford module $\mathscr{W} \otimes \mathscr{S}(B)$ over B. Let $\mathsf{D}_{\mathbb{B}}$ be the Dirac operator on $\Gamma(B, \mathscr{W} \otimes \mathscr{S}(B))$ associated to this superconnection. We proved in Proposition 3.42 that the map $\mathbb{A} \mapsto \mathsf{D}_{\mathbb{B}}$ defined a one-to-one correspondence between superconnections on $\mathscr{W}$ and Dirac operators on $\Gamma(B, \mathscr{W} \otimes \mathscr{S}(B))$, that is, odd differential operators D satisfying the identity $[\mathsf{D}, f] = c(df)$, for $f \in C^\infty(B)$. We will say that $\mathsf{D}_{\mathbb{B}}$ is the Dirac operator associated to $\mathbb{A}$.

We may identify the spaces of sections $\Gamma(B, \mathscr{S}(B) \otimes \pi_* \mathscr{E})$ with $\Gamma(M, \mathscr{E} \otimes \pi^* \mathscr{S}(B))$. We define a Dirac operator on the Clifford module $\pi_* \mathscr{E} \otimes \mathscr{S}(B)$ to be an odd differential operator D on the space of sections $\Gamma(B, \pi_* \mathscr{E} \otimes \mathscr{S}(B)) \cong \Gamma(M, \mathscr{E} \otimes \pi^* \mathscr{S}(B))$ such that $[\mathsf{D}, \pi^* f] = c(\pi^*(df))$ for all $f \in C^\infty(B)$. The following result is an infinite-dimensional extension of Proposition 3.42.

Proposition 10.18. *There is a one-to-one correspondence between superconnections $\mathbb{A}$ on the bundle $\pi_* \mathscr{E}$ and Dirac operators on the Clifford module $\pi_* \mathscr{E} \otimes \mathscr{S}(B)$.*

The total Dirac operator D_M on $\Gamma(M, \mathscr{E} \otimes \pi^* \mathscr{S}(B))$, when considered as an operator on the isomorphic space $\Gamma(B, \mathscr{S}(B) \otimes \pi_* \mathscr{E})$, is a Dirac operator for the Clifford module $\pi_* \mathscr{E} \otimes \mathscr{S}(B)$ over B. Indeed, the fact that $[\mathsf{D}_M, \pi^* f] = c(\pi^*(df))$ for $f \in C^\infty(B)$ is just a special case of the formula $[\mathsf{D}_M, f] = c(df)$ for $f \in C^\infty(M)$. Thus, there is a superconnection $\mathbb{A}$ on the bundle $\pi_* \mathscr{E}$ whose associated Dirac operator may be identified with D_M. In fact, this superconnection on $\pi_* \mathscr{E}$ is just the Bismut superconnection.

Theorem 10.19. *The Dirac operator on*

$$\Gamma(B, \pi_* \mathscr{E} \otimes \mathscr{S}(B)) \cong \Gamma(M, \mathscr{E} \otimes \pi^* \mathscr{S}(B))$$

associated to the Bismut superconnection $\mathbb{A}$ is the total Dirac operator D_M.

Proof. Let $\mathbb{B} = \nabla^{\mathscr{S}(B)} \otimes 1 + 1 \otimes \mathbb{A}$. If $s \in \Gamma(B, \mathscr{S}(B))$ and $t \in \Gamma(M, \mathscr{E})$, we see that

$$\begin{aligned}\mathsf{D}_{\mathbb{B}}(s \otimes t) &= s \otimes \sum_i c^i \nabla_i^{\mathscr{E}} t + \sum_\alpha c^\alpha s \otimes \nabla_\alpha^{\mathscr{E}} t \\ &\quad + \sum_\alpha (c^\alpha \nabla_\alpha^B s + \tfrac{1}{2} k(f_\alpha) c^\alpha s) \otimes t - \tfrac{1}{4} \sum_{\alpha<\beta} \sum_i (\Omega(f_\alpha, f_\beta), e_i) c^\alpha c^\beta s \otimes c^i t.\end{aligned}$$

Thus, we see that $\mathsf{D}_{\mathbf{B}} = \sum_a c^a \nabla_a^{\mathscr{S}(B)\otimes\mathscr{E},\oplus} + \frac{1}{4}h$, where h is the section of $C(M)$ given by the formula

$$h = 2\sum_{\alpha} k(f_\alpha)c^\alpha - \tfrac{1}{2}\sum_{\alpha\beta i}(\Omega(f_\alpha, f_\beta), e_i)c^\alpha c^\beta c^i.$$

By Lemma 10.13, h equals $\sum_{abc} \omega(e_a)(e_b, e_c)c^a c^b c^c$.

Since $\nabla^{\mathscr{E}\otimes\mathscr{S}(B)} = \nabla^{\mathscr{S}(B)\otimes\mathscr{E},\oplus} + \frac{1}{2}c(\omega)$, the operator D_M equals

$$\mathsf{D}_M = \sum_a c^a \nabla_a^{\mathscr{S}(B)\otimes\mathscr{E},\oplus} + \tfrac{1}{4}\sum_{abc}\omega(e_a)(e_b, e_c)c^a c^b c^c,$$

which we see is the same as $\mathsf{D}_{\mathbf{A}}$. □

10.4 The Family Index Density

In this section, we use the method of Chapter 4 to prove the local family index theorem of Bismut, that is, to calculate $\lim_{t\to 0}\mathrm{ch}(\mathbb{A}_t)$. Let us summarize the data that we are given:

1. A fibre bundle $\pi : M \to B$ with a vertical metric $g_{M/B}$ and connection, that is, a decomposition of the tangent bundle of M into the direct sum of the vertical tangent bundle $T(M/B)$ and the horizontal tangent bundle π^*TB; from this, we obtain a connection $\nabla^{M/B}$ on the vertical tangent bundle $T(M/B)$.
2. a Clifford module $\mathscr{E}$ for the vertical Clifford bundle $C(M/B)$, with Clifford connection $\nabla^{\mathscr{E}}$ compatible with $\nabla^{M/B}$.

Using just this data, we can construct a family of twisted Dirac operators $\mathsf{D} = (\mathsf{D}^z \mid z \in B)$, and the Bismut superconnection $\mathbb{A}$, which satisfies $\mathbb{A}_{[0]} = \mathsf{D}$. Let n denote the dimension of the fibres M_z.

Let $\mathscr{A}_z$ be the finite dimensional algebra ΛT_z^*B. The curvature $\mathscr{F} = \mathbb{A}^2$ of $\mathbb{A}$ acts on the space $\mathscr{A}_z \otimes (\pi_*\mathscr{E})_z$ of sections of the bundle $\mathbb{E} = \Lambda\pi^*T^*B \otimes \mathscr{E}$ along the fibre M_z.

The following lemma is a restatement of Theorem 9.48.

Lemma 10.20. *For each $t > 0$, the heat operator $e^{-t\mathscr{F}}$ acting on $\Gamma(M, \mathbb{E})$ has a kernel*

$$\langle x \mid e^{-t\mathscr{F}} \mid y\rangle \in \Gamma(M \times_\pi M, \pi^*\mathscr{A} \otimes \mathscr{E} \boxtimes_\pi \mathscr{E}^*),$$

in the sense that if $\phi \in \Gamma(M, \mathbb{E})$, we have

$$(e^{-t\mathscr{F}}\phi)(x) = \int_{M_z} \langle x \mid e^{-t\mathscr{F}} \mid y\rangle \phi(y)\, dy,$$

where dy is the Riemannian volume form of the fibre M_z and $z = \pi(x)$.

The manifold M embeds inside $M \times_\pi M$ as the diagonal $M \subset M \times_\pi M$. If we restrict the heat kernel of $\mathscr{F}$ to the diagonal, it becomes a section of the bundle of algebras $\pi^*\Lambda T^*B \otimes \operatorname{End}(\mathscr{E})$. Since $\mathscr{E}$ is a Clifford module over the Clifford algebra $C(M/B)$, we see that

$$\begin{aligned}\pi^*\Lambda T^*B \otimes \operatorname{End}(\mathscr{E}) &\cong \pi^*\Lambda T^*B \otimes C(M/B) \otimes \operatorname{End}_{C(M/B)}(\mathscr{E}) \\ &\cong \pi^*\Lambda T^*B \otimes \Lambda T^*(M/B) \otimes \operatorname{End}_{C(M/B)}(\mathscr{E}) \\ &= \Lambda T^*M \otimes \operatorname{End}_{C(M/B)}(\mathscr{E}).\end{aligned}$$

Thus, the restriction to the diagonal $k_t(x,x)$ of the heat kernel of $\mathscr{F}$ is identified with a differential form on M with values in the bundle $\operatorname{End}_{C(M/B)}(\mathscr{E})$.

Let $\hat{A}(M/B)$ be the $\hat{A}$-genus of the bundle $T(M/B)$ for the connection $\nabla^{M/B}$,

$$\hat{A}(M/B) = \det{}^{1/2}\left(\frac{R^{M/B}/2}{\sinh(R^{M/B}/2)}\right),$$

and let $F^{\mathscr{E}/S} \in \Gamma(M, \operatorname{End}_{C(M/B)}(\mathscr{E}))$ be the twisting curvature of the bundle $\mathscr{E}$. The aim of this section is to prove the following generalization of Theorem 4.1.

Theorem 10.21. *Consider the asymptotic expansion of $k_t(x,x)$*

$$k_t(x,x) \sim (4\pi t)^{-n/2} \sum_{i=0}^{\infty} t^i k_i(x).$$

1. *The coefficient k_i lies in $\sum_{j \le 2i} \mathscr{A}^j(M, \operatorname{End}_{C(M/B)}(\mathscr{E}))$.*
2. *The full symbol of $k_t(x,x)$, defined by $\sigma(k) = \sum_{i=0}^{\dim(M)/2} \sigma_{2i}(k_i)$, is given by the formula*

$$\sigma(k) = \hat{A}(R^{M/B}) \exp(-F^{\mathscr{E}/S}) \in \mathscr{A}(M, \operatorname{End}_{C(M/B)}(\mathscr{E}))$$

Before proving the theorem, let us show how it allows us to compute the Chern character of the index bundle $\operatorname{ind}(\mathsf{D})$ of the family of vertical Dirac operators D. Recall the operator δ_t^B on $\mathscr{A}(B)$ which multiplies differential forms of degree i by $t^{-i/2}$. Let

$$\mathbb{A}_t = t^{1/2}\delta_t^B \cdot \mathbb{A} \cdot (\delta_t^B)^{-1} = t^{1/2}\mathbb{A}_{[0]} + \mathbb{A}_{[1]} + t^{-1/2}\mathbb{A}_{[2]}$$

be the rescaled Bismut superconnection, with curvature $\mathscr{F}_t = t\delta_t^B \cdot \mathscr{F} \cdot (\delta_t^B)^{-1}$. By Theorem 9.19, the Chern character of the bundle $\operatorname{ind}(\mathsf{D})$ is represented, for every $t > 0$, by the closed differential form $\operatorname{ch}(\mathbb{A}_t) = \operatorname{Str}(e^{-\mathscr{F}_t})$ on B.

Lemma 10.22. *Extend the map* $\operatorname{Str} : \Gamma(M_z, \operatorname{End}(\mathscr{E}_z)) \to C^\infty(M_z)$ *to a map* $\operatorname{Str} : \Gamma(M_z, \mathscr{A}_z \otimes \operatorname{End}(\mathscr{E}_z)) \to \mathscr{A}_z \otimes C^\infty(M_z)$. *Then*

$$\operatorname{ch}(\mathbb{A}_t) = \int_{M_z} \delta_t^B\Big(\operatorname{Str}_{\mathscr{E}_x}(k_t(x,x))\Big)\, dx.$$

Proof. Using the formula $\mathscr{F}_t = t\delta_t^B \cdot \mathscr{F} \cdot (\delta_t^B)^{-1}$, we see that

$$\begin{aligned}\operatorname{ch}(\mathbb{A}_t) = \operatorname{Str}(e^{-\mathscr{F}_t}) &= \operatorname{Str}(e^{-t\delta_t^B \cdot \mathscr{F} \cdot (\delta_t^B)^{-1}}) \\ &= \operatorname{Str}\Big(\delta_t^B(e^{-t\mathscr{F}})\Big) \\ &= \delta_t^B\Big(\operatorname{Str}(e^{-t\mathscr{F}})\Big)\end{aligned}$$

Now, the kernel of the operator $e^{-t\mathscr{F}}$ is by definition equal to $k_t(x,y)$, and the lemma follows now from the fact that

$$\operatorname{Str}(e^{-t\mathscr{F}_z}) = \int_{M_z} \operatorname{Str}_{\mathscr{E}_x}(k_t(x,x))\,dx \in \Lambda T_z^*B. \qquad \square$$

Let us show that the local family index theorem of Bismut is a corollary of Theorem 10.21. Let $\operatorname{ch}(\mathscr{E}/S) = \operatorname{Str}_{\mathscr{E}/S}(e^{-F^{\mathscr{E}/S}})$ be the relative Chern character of the Clifford module $\mathscr{E}$. Let

$$T_{M/B} : \mathscr{A}(M) \to \Gamma(M, \pi^*\Lambda T^*B)$$

be the map given by decomposing the bundle ΛT^*M as a tensor product $\Lambda T^*(M/B) \otimes \pi^*\Lambda T^*B$ and applying the Berezin integral map to the first factor, that is, projecting onto $\Lambda^n T^*(M/B) \otimes \pi^*\Lambda T^*B$ and then dividing by the vertical Riemannian volume form $\nu_{M/B}$ in $\Gamma(M, \Lambda^n T^*(M/B))$.

Theorem 10.23 (Bismut). *The section*

$$\delta_t^B\Big(\operatorname{Str}_{\mathscr{E}_x}(k_t(x,x))\Big) \in \Gamma(M, \pi^*\Lambda T^*B)$$

has a limit when $t \to 0$ equal to

$$(2\pi i)^{-n/2} T_{M/B}(\hat{A}(M/B)\operatorname{ch}(\mathscr{E}/S)) \in \Gamma(M, \pi^*\Lambda T^*B).$$

Consequently, the differential form $\operatorname{ch}(\mathbb{A}_t) \in \mathscr{A}(B)$ *has a limit when $t \to 0$ given by the formula*

$$\lim_{t\to 0} \operatorname{ch}(\mathbb{A}_t) = (2\pi i)^{-n/2} \int_{M/B} \hat{A}(R^{M/B})\operatorname{ch}(\mathscr{E}/S).$$

Proof. Consider the bigrading on ΛT_x^*M given by

$$\Lambda^{p,q} T_x^*M = \sum_{p,q} \mathscr{A}_z^p \otimes \Lambda^q T_x^*(M/B).$$

If we decompose the symbol of $a \in \mathscr{A}_z \otimes \operatorname{End}(\mathscr{E}_x)$ according to this bigrading,

$$\sigma(a) = \sum_{p,q} \sigma_{[p,q]}(a)$$

where

$$\sigma_{[p,q]}(a) \in \mathscr{A}_z^p \otimes \Lambda^q T_x^*(M/B) \otimes \operatorname{End}_{C(M)}(\mathscr{E}_x),$$

we have the formula

$$\mathrm{Str}_{\mathscr{E}_x}(a) = (-2i)^{n/2} \sum_p \mathrm{Str}_{\mathscr{E}/S}(\sigma_{[p,n]}(a)).$$

Applying this formula to the supertrace of

$$k_t(x,x) \sim (4\pi t)^{-n/2} \sum_i t^i k_i(x),$$

we see that

$$\delta_t^B(\mathrm{Str}_{\mathscr{E}_x}(k_t(x,x))) \sim (2\pi i)^{-n/2} \sum_{j,p} t^{j-(n+p)/2} \mathrm{Str}_{\mathscr{E}/S}(\sigma_{[p,n]}(k_j(x))).$$

Since $\sigma_{[p,n]}(k_j) = 0$ if $2j \leq n+p$, we see that there is no singular term in the asymptotic expansion of $\delta_t^B(\mathrm{Str}_{\mathscr{E}_x}(k_t(x,x)))$ as $t \to 0$. The explicit formula of Theorem 10.21 for $\sum_j \sigma_{2j}(k_j)$ implies the formula for the leading order in the asymptotic expansion. □

Bearing in mind the family McKean-Singer theorem of Chapter 9, we obtain the cohomological form of the Atiyah-Singer family index theorem.

Corollary 10.24 (Atiyah-Singer).

$$\mathrm{ch}(\mathrm{ind}(\mathsf{D})) = (2\pi i)^{n/2} \int_{M/B} \hat{A}(M/B)\, \mathrm{ch}(\mathscr{E}/S) \in H^{2\bullet}(B).$$

We now turn to the proof of Theorem 10.21, which is a fairly straightforward generalization of Theorem 4.1, the case in which B is a point. We start by choosing a point $z \in B$, but instead of working with the generalized Laplacian D^2 on the manifold M as in Chapter 4, we will work with the curvature $\mathscr{F}^z$ of the Bismut superconnection $\mathbb{A}$ restricted to the fibre M_z, which we may think of as a generalized Laplacian with coefficients in the exterior algebra $\mathscr{A}_z$.

Fix a point $x_0 \in M_z$, and let $V = T_{x_0}(M/B)$ and $H = T_z B$ be the vertical and horizontal tangent spaces at x_0; thus

$$T = T_{x_0} M = V \oplus H.$$

Let $U = \{\mathbf{x} \in V \mid \|\mathbf{x}\| < \varepsilon\}$, where ε is smaller than the injectivity radius of the fibre M_z. We identify U by the exponential map $\mathbf{x} \mapsto \exp_{x_0} \mathbf{x}$ with a neighborhood of x_0 in M_z. Let $\tau^{M/B}(x_0,x)$ be the parallel transport map in the bundle $T(M/B)$ along the geodesic from x to x_0, defined with respect to the connection $\nabla^{M/B}$ defined in Section 1; since we are working on a single fibre M_z, this connection is nothing but the Levi-Civita connection of M_z. Using this map, we can identify the fibre $T_{\mathbf{x}}(M/B)$ with the space V, so that the space of differential forms $\mathscr{A}(U)$ is identified with $C^\infty(U, \Lambda V^*)$.

Choose an orthonormal basis $d\mathbf{x}_i$ of $V^* = T^*_{x_0}(M/B)$, and let

$$e^i \in \Gamma(U, T^*(M/B)) = \Gamma(U, V^*)$$

be the orthonormal frame of $T^*(M/B)$ over U obtained by parallel transport of $d\mathbf{x}_i$ along geodesics by the Levi-Civita connection on M_z. We denote by e^a a local frame of T^*M on U consisting of the union of the cotangent frame e^i and of a fixed basis f^α of T_z^*B.

Let $E = \mathscr{E}_{x_0}$ be the fibre of the Clifford module $\mathscr{E}$ at x_0, let S_V be the spinor space of V^*, and let $W = \mathrm{Hom}_{C(V^*)}(S_V, E)$, so that E is naturally isomorphic to $S_V \otimes W$. Recall the Clifford action m_0 of T^*M on $\Lambda\pi^*T^*B \otimes \mathscr{E}$, for which vertical cotangent vectors act by Clifford multiplication on $\mathscr{E}$, and horizontal tangent vectors act by exterior multiplication on $\mathscr{A} = \Lambda\pi^*T^*B$. Let $\tau^{\mathbb{E}}(x_0, x)$ be the parallel transport map in the bundle $\mathscr{A}_z \otimes \mathscr{E}$ along the geodesic from x to x_0, defined with respect to the Clifford connection $\nabla^{\mathbb{E},0}$ of Section 2. Using this map, we can identify the fibre $\mathscr{A}_z \otimes \mathscr{E}_x$ of $\mathbb{E}$ at x with the space $\Lambda H^* \otimes S_V \otimes W$, and the space of sections $\Gamma(U, \mathscr{A}_z \otimes \mathscr{E})$ with $C^\infty(U, \Lambda H^* \otimes S_V \otimes W)$.

Although the bundle $\mathbb{E}$ over M_z is the tensor product of the bundle $\mathscr{E}$ with the trivial bundle $\mathscr{A}_z$, the term $\frac{1}{2} m_0(\omega)$ in the definition of the connection $\nabla^{\mathbb{E},0}$ means that parallel transport with respect to $\nabla^{\mathbb{E},0}$ is not equal simply to the tensor product of the identity on $\mathscr{A}_z$ with parallel transport in $\mathscr{E}$ by the connection $\nabla^{\mathscr{E}}$. Using the parallel transport map $\tau^{\mathbb{E}}(x_0, x)$, we can analyse the Clifford action m_0 of T_x^*M on $\mathbb{E}_x$ by transporting it back to x_0, where it becomes an $\mathbf{x}$-dependent action of $T^* = H^* \oplus V^*$ on $\mathscr{A}_z \otimes E$.

Lemma 10.25. *If c^i is the Clifford action of the cotangent vector $d\mathbf{x}_i$ on E, and ε^α is multiplication by f^α in the exterior algebra ΛH^*, we have*

$$m_0(e^i) = c^i + \sum_\alpha u^i_\alpha \varepsilon^\alpha$$
$$m_0(f^\alpha) = \varepsilon^\alpha,$$

where u^i_α are smooth functions on U satisfying $u^i_\alpha(\mathbf{x}) = O(|\mathbf{x}|)$.

Proof. Let $\mathscr{R}$ be the radial vector field on U. The frame e^i satisfies $\nabla^{M/B}_{\mathscr{R}} e^i = 0$. It follows that

$$\nabla^{M,0}_{\mathscr{R}} e^i = \tfrac{1}{2} \tau^0(\omega(\mathscr{R})) e^i = \sum_\alpha \omega(\mathscr{R})(e_i, f_\alpha)\varepsilon^\alpha,$$

by (10.3), and the fact that $\omega(X,Y) = 0$ if both X and Y are vertical. Now, we use the fact, proved in Proposition 10.10, that $\nabla^{\mathbb{E},0}$ is a Clifford connection: $[\nabla^{\mathbb{E},0}_X, m_0(\alpha)] = m_0(\nabla^{M,0}_X \alpha)$. This shows that

$$[\nabla^{\mathbb{E},0}_{\mathscr{R}}, m_0(e^i)] = m_0(\nabla^{M,0}_{\mathscr{R}} e^i) = \sum_\alpha \omega(\mathscr{R})(e_i, f_\alpha) f^\alpha.$$

Integrating this ordinary differential equation with the initial condition that at $\mathbf{x} = 0$, $m_0(e^i) = c^i$, we obtain the first formula.

The second formula is clear: since $\tau^0(a) f^\alpha = 0$ for all $a \in \Lambda^2 T_x^*M$, the connection $\nabla^{M,0}$ coincides with $\pi^*\nabla^B$ on the sub-bundle π^*TB of horizontal vectors, so that $\nabla^{M,0}_{\mathscr{R}} f^\alpha = \nabla^B_{\mathscr{R}} f^\alpha = 0$. □

We need another simple lemma.

Lemma 10.26. *In the trivialization of* $\mathbb{E}$ *over* U *induced by the parallel transport map* $\tau^{\mathbb{E}}(x_0,x)$, *the connection* $\nabla^{\mathbb{E},0}$ *equals* $d+\Theta$, *where*

$$\Theta(\partial_i) = -\tfrac{1}{4}\sum_{j;a<b}(\mathsf{R}(\partial_i,\partial_j)e_a,e_b)m^a m^b \mathbf{x}^j + \sum_{a<b} f_{iab}(\mathbf{x})m^a m^b + g_i(\mathbf{x});$$

here $f_{iab}(\mathbf{x}) = O(|\mathbf{x}|^2) \in C^\infty(U)$ *and* $g_i(\mathbf{x}) = O(|\mathbf{x}|) \in C^\infty(U)\otimes \mathrm{End}(W)$.

Proof. Let $\mathscr{R}$ be the radial vector field on U. Since $\Theta(\mathscr{R}) = 0$, the one-form Θ is determined (see (1.12)) by the equation

$$\begin{aligned}\mathscr{L}(\mathscr{R})\Theta &= \iota(\mathscr{R})\left(\nabla^{\mathbb{E},0}\right)^2\\ &= \iota(\mathscr{R})(\lambda(R^B) + \tfrac{1}{2}m_0(\mathsf{R}) + F^{\mathscr{E}/S}) \quad \text{by Proposition 10.11,}\\ &= \tfrac{1}{2}\iota(\mathscr{R})m_0(\mathsf{R}) + \iota(\mathscr{R})F^{\mathscr{E}/S};\end{aligned}$$

the last line holds because $\iota(\mathscr{R})\lambda(R^B) = 0$, $\mathscr{R}$ being a vertical vector field and R^B being horizontal. □

Let Δ^z be the Laplacian on M_z associated to the connection $\nabla^{\mathbb{E},0}$, so that the curvature of the Bismut superconnection equals

$$\mathscr{F}^z = \Delta^z + \tfrac{1}{4}r_{M_z} + \sum_{a<b} m_0^a m_0^b F^{\mathscr{E}/S}(e_a,e_b).$$

We can transfer this operator to $C^\infty(U,\Lambda T^*\otimes \mathrm{End}(W))$ using the quantization map, that is, by replacing the Clifford action $m_0(e^a)$ at $\mathbf{x}=0$ by the action

$$m^i = \varepsilon^i - \iota^i\,,\; m^\alpha = \varepsilon^\alpha.$$

The resulting operator L is given by the explicit formula

$$L = -\sum_i\left((\nabla_i^{\mathbb{E},0})^2 - \nabla^{\mathbb{E},0}_{\nabla_i e_i}\right) + \tfrac{1}{4}r_{M_z} + \sum_{a<b} F^{\mathscr{E}/S}_{ab} m^a m^b.$$

Next, we introduce the rescaling operator

$$\delta_u : C^\infty(U,\Lambda T^*\otimes \mathrm{End}(W)) \to C^\infty(U,\Lambda T^*\otimes \mathrm{End}(W)),$$

which is basic to the proof; if $a\in C^\infty(U,\Lambda^i T^*\otimes \mathrm{End}(W))$, then

$$\delta_u(a)(\mathbf{x}) = u^{-i/2}a(u^{1/2}\mathbf{x}).$$

Furthermore, if a is an element of $C^\infty(U\times\mathbb{R},\Lambda^i T^*\otimes\mathrm{End}(W))$, we define

$$\delta_u(a)(t,\mathbf{x}) = u^{-i/2}a(ut,u^{1/2}\mathbf{x}).$$

By Lemma 10.25,

$$\lim_{u\to 0} u^{1/2}\delta_u\cdot m_0(\theta)\cdot\delta_u^{-1} = \varepsilon(\theta) \quad \text{for } \theta\in C^\infty(U,T^*).$$

In the next lemma, we calculate the leading term of the asymptotic expansion of $\delta_u\nabla^{\mathbb{E},0}\delta_u^{-1}$. The following lemma follows immediately from Lemma 10.26.

Lemma 10.27. *In the trivialization of $\mathscr{A}_z \otimes \mathscr{E}_z$ over U induced by the parallel transport map $\tau^{\mathbb{E}}(x_0,x)$,*

$$u^{1/2}\delta_u \cdot \nabla^{\mathbb{E},0}_{\partial_i} \cdot \delta_u^{-1} = \partial_i - \tfrac{1}{4}\sum_{j;a<b} \mathsf{R}(\partial_i,\partial_j,e_a,e_b)_{x_0} m^a m^b \mathbf{x}^j + O(u^{1/2}).$$

We can now show that the family of differential operators $u\delta_u L \delta_u^{-1}$ acting on $C^\infty(U,\Lambda T^* \otimes \mathrm{End}(W))$ has a limit as $u \to 0$; this is the most important step in the proof. Since the twisting curvature $\mathsf{F} = F^{\mathscr{E}/S}_{x_0}$ of the Clifford module $\mathscr{E}$ at x_0 is an element of $\Lambda^2 T^* \otimes \mathrm{End}(W)$, it acts on the space $C^\infty(U,\Lambda T^* \otimes \mathrm{End}(W))$ by multiplication. Similarly, the matrix-coefficients

$$a_{ij} = \left(R^{M/B}\partial_i,\partial_j\right)_{x_0} = \sum_{a<b}\left(R^{M/B}(e_a,e_b)\partial_i,\partial_j\right)_{x_0}\varepsilon^a\varepsilon^b$$

of the curvature of the bundle $T(M/B)$ belong to $\Lambda^2 T^*$ and act on the space $C^\infty(U,\Lambda T^* \otimes \mathrm{End}(W))$.

Proposition 10.28. *The differential operator $u\delta_u \cdot L \cdot \delta_u^{-1}$ on $C^\infty(U,\Lambda T^* \otimes \mathrm{End}(W))$ has a limit K when u tends to 0, equal to*

$$K = -\sum_i \Big(\partial_i - \tfrac{1}{4}\sum_j a_{ij}\mathbf{x}_j\Big)^2 + \mathsf{F}.$$

Proof. By Lemma 10.27, the differential operator $u^{1/2}\delta_u \nabla^{\mathbb{E},0}_{\partial_i}\delta_u^{-1}$ equals

$$\partial_i - \tfrac{1}{4}\sum_{j;a<b} (\mathsf{R}(\partial_i,\partial_j)_{x_0} e_a, e_b)\varepsilon^a\varepsilon^b\mathbf{x}_j + O(u^{1/2}).$$

The fundamental symmetry of the four-tensor R (Proposition 10.9 (3)) shows that

$$\mathsf{R}(\partial_i,\partial_j,e_a,e_b) = \mathsf{R}(e_a,e_b,\partial_i,\partial_j) = (R^{M/B}(e_a,e_b)\partial_i,\partial_j),$$

from which it follows that

$$u^{1/2}\delta_u \nabla^{\mathbb{E},0}_{e_i}\delta_u^{-1} = \partial_i - \tfrac{1}{4}a_{ij}\mathbf{x}_j + O(u^{1/2}).$$

The operator $u\delta_u \cdot L \cdot \delta_u^{-1}$ is equal to

$$-\sum_i \left(u^{1/2}\delta_u \cdot \nabla^{\mathbb{E},0}_{e_i} \cdot \delta_u^{-1}\right)^2 + \sum_{a<b} u\delta_u \cdot m^a m^b \cdot \delta_u^{-1} F^{\mathscr{E}/S}(e^a,e^b)(u^{1/2}\mathbf{x})$$
$$- u\sum_i \delta_u \cdot \nabla^{\mathbb{E},0}_{\nabla_{e_i}e_i} \cdot \delta_u^{-1} + \tfrac{u}{4} r_{M_z}(u^{1/2}\mathbf{x}).$$

It is clear that both the third and fourth terms behave as $O(u^{1/2})$ as $u \to 0$, while

$$\sum_{a<b} u\delta_u m^a m^b \delta_u^{-1} F^{\mathscr{E}/S}(e^a,e^b)(u^{1/2}\mathbf{x}) = \sum_{a<b}\varepsilon^a\varepsilon^b \mathsf{F}_{ab} + O(u^{1/2}).$$

Putting all of these equations together, and using the fact that $e_i = \partial_i$ at $\mathbf{x} = 0$, we obtain the theorem. □

Consider the time-dependent function on U given by transferring the heat kernel $\langle x \mid e^{-t\mathcal{F}^z} \mid x_0 \rangle$ to the open set $U \subset T_{x_0}M_z$ by means of the parallel transport map $\tau^{\mathbb{E}}(x_0,x)$:

$$k(t,\mathbf{x}) = \tau^{\mathbb{E}}(x_0,x)\langle x \mid e^{-t\mathcal{F}^z} \mid x_0 \rangle, \quad \text{where } x = \exp_{x_0} \mathbf{x}.$$

Then $k(t,\mathbf{x})$ lies in $C^\infty(U, \Lambda H^* \otimes \mathrm{End}(S_V) \otimes \mathrm{End}(W))$; using the symbol map

$$\mathrm{End}(S_V) \to C(V^*) \to \Lambda V^*,$$

pwe may think of $k(t,\mathbf{x})$ as a map from U to $\Lambda V^* \otimes \Lambda H^* \otimes \mathrm{End}(W) \cong \Lambda T^* \otimes \mathrm{End}(W)$, and is a solution to the heat equation

$$(\partial_t + L)k(t,\mathbf{x}) = 0$$

with initial condition $\lim_{t\to 0} k(t,\mathbf{x}) = \delta(\mathbf{x})$. As in Chapter 4, we rescale the heat kernel as follows.

Definition 10.29. *The rescaled heat kernel $r(u,t,\mathbf{x})$ equals*

$$r(u,t,\mathbf{x}) = \sum_{i=0}^{n} u^{(n-i)/2} k(ut, u^{1/2}\mathbf{x})_{[i]}.$$

The choice of the particular rescaling operator δ_u is motivated by the fact that

$$\lim_{u\to 0} r(u,t,0) = \lim_{u\to 0} \delta_u(k_t(x_0)) \in \Lambda T^*_{x_0}M,$$

and the right-hand side of this equation is precisely what we must calculate to prove Theorem 10.21.

It is clear that $r(u,t,\mathbf{x})$ satisfies the differential equation

$$(\partial_t + u\delta_u \cdot L \cdot \delta_u^{-1}) r(u,t,\mathbf{x}) = 0.$$

The rest of the proof proceeds in exactly the same way as the proof of Theorem 4.1. Lemma 4.19 applies to the operator $\mathbb{A}^2$ on M_z just as well as it does when the base B is a point, and we obtain $\Lambda T^* \otimes \mathrm{End}(W)$-valued polynomials $\gamma_i(t,\mathbf{x})$ on $\mathbb{R} \times V$ such that for every integer N, the function $r^N(u,t,\mathbf{x})$ defined by

$$q_t(\mathbf{x}) \sum_{i=-n}^{2N} u^{i/2} \gamma_i(t,\mathbf{x})$$

approximates $r(u,t,\mathbf{x})$ in the following sense: for $N > j + |\alpha|/2$,

$$\|\partial_t^j \partial_{\mathbf{x}}^\alpha (r(u,t,\mathbf{x}) - r^N(u,t,\mathbf{x}))\| \le C(N,j,\alpha) u^N,$$

for $0 < u \le 1$ and $(t,\mathbf{x}) \in (0,1) \times U$. Just as in Chapter 4, $\gamma_i(0,0) = 0$ if $i \neq 0$, while $\gamma_0(0,0) = 1$. Using the fact that $L(u) = K + O(u^{1/2})$, we can show in the same way as when the base B is a point that there are no poles in the Laurent series expansion of $r(u,t,\mathbf{x})$, that is, that $\gamma_i(t,\mathbf{x}) = 0$ for $i < 0$ so that

$$r(u,t,\mathbf{x}) \sim q_t(\mathbf{x}) \sum_{i=0}^{\infty} u^{i/2} \gamma_i(t,\mathbf{x}).$$

The limit $\lim_{u \to 0} r(u,t,\mathbf{x}) = q_t(\mathbf{x}) \gamma_0(t,\mathbf{x})$ is a formal solution to the heat equation

$$(\partial_t + K) q_t(\mathbf{x}) \gamma_0(t,\mathbf{x}) = 0$$

with initial condition $\gamma_0(0,0) = 1$. The following theorem is a consequence of Theorem 4.13.

Theorem 10.30. *The limit* $\lim_{u \to 0} r(u,t,\mathbf{x})$ *exists and is given by the formula*

$$(4\pi t)^{-n/2} \det{}^{1/2}\Big(\frac{tR^{M/B}/2}{\sinh(tR^{M/B}/2)}\Big) \exp\Big(-t\mathsf{F} - \frac{1}{4t}\Big\langle \mathbf{x} \Big| \frac{R^{M/B}/2}{\coth(tR^{M/B}/2)} \Big| \mathbf{x} \Big\rangle\Big).$$

Theorem 10.21 follows from this lemma when we set $t = 1$ and $\mathbf{x} = 0$.

10.5 The Transgression Formula

In this section, we will show that the transgression formula

$$\operatorname{ch}(\mathbb{A}_s) - \operatorname{ch}(\mathbb{A}_T) = d \int_s^T \operatorname{Str}\Big(\frac{\partial \mathbb{A}_t}{\partial t} e^{-\mathbb{A}_t^2}\Big)\, dt$$

has a limit as $s \to 0$, when $\mathbb{A}$ is the Bismut superconnection; of course, for an arbitrary superconnection for the family D, this is not true.

Let us start with an analogue in finite dimensions of the technique which we will use. Recall the construction of Section 9.1, in which we associated to a smooth family of superconnections $\mathbb{A}_s$ on a superbundle $\mathscr{E}$ over a manifold B parametrized by $\mathbb{R}_+$ a superconnection $\tilde{\mathbb{A}}$ on the superbundle $\tilde{\mathscr{E}} = \mathscr{E} \times \mathbb{R}_+$ over the manifold $\widetilde{B} = B \times \mathbb{R}_+$. Consider the special case in which $\mathbb{A}_s$ is the rescaled superconnection

$$\mathbb{A}_s = s^{1/2} \delta_s^B \cdot \mathbb{A} \cdot (\delta_s^B)^{-1},$$

and let $\tilde{\mathbb{A}}_t$ be the dilation of the corresponding superconnection $\tilde{\mathbb{A}}$ on the superbundle $\tilde{\mathscr{E}}$:

$$\tilde{\mathbb{A}}_t = \mathbb{A}_{st} + ds \wedge \frac{\partial}{\partial s},$$

with curvature $\widetilde{\mathscr{F}}_t = \tilde{\mathbb{A}}_t^2$.

Lemma 10.31. *We have*

$$\operatorname{Str}\Big(e^{-\widetilde{\mathscr{F}}_t}\Big)\Big|_{s=1} = \operatorname{Str}\Big(e^{-\mathscr{F}_t}\Big) - t\alpha(t) \wedge ds,$$

where $\alpha(t)$ *equals*

$$\alpha(t) = \operatorname{Str}\Big(\frac{d\mathbb{A}_t}{dt} e^{-\mathbb{A}_t^2}\Big).$$

Proof. Since $\tilde{\mathbb{A}}_t = \mathbb{A}_{ts} + d_{\mathbb{R}_+}$, it follows that at $s = 1$,

$$\widetilde{\mathscr{F}}_t = \mathscr{F}_t + t\frac{d\mathbb{A}_t}{dt} \wedge ds.$$

The lemma now follows from the Volterra expansion and the definition of $\alpha(t)$. □

The following theorem is proved by applying the family index theorem to a superconnection over $\widetilde{B} = B \times \mathbb{R}_+$ constructed from the Bismut superconnection in the same way as above.

Theorem 10.32. *1. The differential form*

$$\alpha(t) = \operatorname{Str}\Big(\frac{d\mathbb{A}_t}{dt} e^{-\mathbb{A}_t^2}\Big)$$

has an asymptotic expansion as $t \to 0$ of the form

$$\alpha(t) = \sum_{j=1}^{\infty} t^{j/2-1} \alpha_{j/2},$$

where $\alpha_{j/2} \in \mathscr{A}(B)$.

2. The Chern form $\mathrm{ch}(\mathbb{A}_t)$ is given by the transgression formula

$$\mathrm{ch}(\mathbb{A}_t) = (2\pi i)^{-n/2} \int_{M/B} \hat{A}(M/B)\,\mathrm{ch}(\mathscr{E}/S) - d\int_0^t \alpha(s)\,ds.$$

3. If $\dim(\ker(\mathsf{D}^z))$ is independent of $z \in B$, so that $\mathrm{ind}(\mathsf{D})$ is a vector bundle with connection $\nabla_0 = P_0 \mathbb{A}_{[1]} P_0$, then

$$\begin{aligned}\mathrm{ch}(\nabla_0) &= \lim_{s\to\infty} \mathrm{ch}(\mathbb{A}_s) \\ &= (2\pi i)^{-n/2} \int_{M/B} \hat{A}(M/B)\,\mathrm{ch}(\mathscr{E}/S) - d\int_0^\infty \alpha(s)\,ds.\end{aligned}$$

Proof. It is clear that Parts (2) and (3) are an immediate consequence of Part (1) and the explicit formula for $\lim_{t\to 0} \mathrm{ch}(\mathbb{A}_t)$ which comes from the local family index theorem.

Let $\mathbb{R}_+$ be the positive real line $(0,\infty)$, and consider the family of manifolds $\widetilde{M} = M \times \mathbb{R}_+$ over the base $\widetilde{B} = B \times \mathbb{R}_+$. To prove that the function $s \mapsto \alpha(s)$ is integrable as $s \to 0$, we will apply the family index theorem to this family, with respect to the following additional data:

1. the vertical metric $g_{\widetilde{M}/\widetilde{B}}$ on $\widetilde{M}$ equals $s^{-1} g_{M/B}$;
2. the connection on the family $\widetilde{M} \to \widetilde{B}$ equals the natural extension of the given connection on the family M/B in the directions tangential to B, and the trivial one in the direction tangential to $\mathbb{R}_+$, for which the horizontal tangent space is spanned at each point by $\partial/\partial s$;
3. the Clifford bundle over $\widetilde{M}$ is $\mathscr{E} \times \mathbb{R}_+$, with Clifford connection $\nabla^{\mathscr{E}} + d_{\mathbb{R}_+}$.

We denote by $\tilde{S}$, $\tilde{k}$ and $\tilde{T}$, the metric and tensors corresponding to the above data on $\widetilde{M}$, and by $\tilde{\mathbb{A}}$ and $\widetilde{\mathscr{F}}$ the corresponding Bismut superconnection and its curvature. Let $n = \dim(M) - \dim(B)$ be the dimension of the fibres of the family $M \to B$.

Lemma 10.33. *1. If X and Y are vertical vector fields on M, then the tensor $\tilde{S}(X,Y)$ equals*

$$\tilde{S}(X,Y) = \frac{1}{s} S(X,Y) - \frac{1}{2s^2}(X,Y)\,ds.$$

2. The one-form $\tilde{k}$ equals $\tilde{k} = k - \frac{n}{2s} ds$.

3. The tensor $\tilde{\Omega}$ equals Ω.

4. The Bismut superconnection for the above data on the family $\widetilde{M} \to \widetilde{B}$ equals

$$\tilde{\mathbb{A}} = \mathbb{A}_s + d_{\mathbb{R}_+} - \frac{n}{4s} ds.$$

Proof. Given vector fields on M, we may extend them to vector fields on all of $\widetilde{M}$ by taking the extension which is independent of $s \in \mathbb{R}_+$. Denote by ∂_s the vector field on $\widetilde{M}$ which generates translation along $\mathbb{R}_+$. If X and Y are vertical vector fields on M and Z is horizontal, $\tilde{S}(X,Y)$ may be calculated as follows:

$$\begin{aligned}
(\tilde{S}(X,Y),\partial_s) &= \tfrac{1}{2}\frac{\partial}{\partial s}(X,Y)_{\widetilde{M}} \\
&= \tfrac{1}{2}\frac{\partial}{\partial s} s^{-1}(X,Y) = -\tfrac{1}{2}s^{-2}(X,Y), \\
(\tilde{S}(X,Y),Z) &= \tfrac{1}{2}(Z\cdot(X,Y)_{\widetilde{M}} - (P[Z,X],Y)_{\widetilde{M}} - (P[Z,Y],X)_{\widetilde{M}}) \\
&= s^{-1}(S(X,Y),Z).
\end{aligned}$$

The formula for the one-form $\tilde{k}$ follows by taking the trace of $\tilde{S}$.

It is easy to see that $\tilde{\Omega} = \Omega$, since $[\partial_s, X] = 0$ if X is any vector field on M.

To compare the Bismut superconnections $\tilde{\mathbb{A}}$ and $\mathbb{A}$, we use the above formulas, the explicit formula for $\mathbb{A}$ given in Proposition 10.15, and the fact that if $\alpha \in \mathscr{A}^1(M)$, the Clifford action $\tilde{c}(\alpha)$ on $\tilde{\mathscr{E}}$ equals $s^{1/2}c(\alpha)$. □

If we take the square of the above formula for the total Bismut superconnection, we see that

$$\widetilde{\mathscr{F}} = \tilde{\mathbb{A}}^2 = \mathbb{A}_s^2 + \frac{\partial \mathbb{A}_s}{\partial s} ds,$$

which has Chern character

$$\mathrm{ch}(\tilde{\mathbb{A}}) = \mathrm{ch}(\mathbb{A}_s) - \alpha(s)\,ds.$$

The theorem will follow from the local family index theorem applied to the Bismut superconnection $\tilde{\mathbb{A}}$. By this theorem, we know that $\delta_t^{\widetilde{B}} \operatorname{Str}_{\mathscr{E}}(e^{-t\widetilde{\mathscr{F}}}) \in \mathscr{A}(\widetilde{B})$ has an asymptotic expansion in t as $t \to 0$ of the form

$$\delta_t^{\widetilde{B}} \operatorname{Str}(e^{-t\widetilde{\mathcal{F}}}) = \operatorname{Str}(e^{-\mathcal{F}_{st}}) - \operatorname{Str}\Big(\frac{\partial \mathbb{A}_{st}}{\partial s} e^{-\mathcal{F}_{st}}\Big)\, ds$$
$$\sim \sum_{j=0}^{\infty} t^{j/2}(\phi_{j/2} - \alpha_{j/2}\, ds).$$

where $\phi_{j/2} \in \mathcal{A}(\widetilde{B})$ and $\alpha_{j/2} \in \mathcal{A}(\widetilde{B})$ are differential forms on $\widetilde{B}$ which do not involve ds. Using the fact that

$$\frac{\partial \mathbb{A}_{st}}{\partial s}\Big|_{s=1} = t\frac{\partial \mathbb{A}_t}{\partial t},$$

we may set $s = 1$, and we obtain the asymptotic expansion

$$\operatorname{Str}\Big(\frac{\partial \mathbb{A}_t}{\partial t} e^{-\mathcal{F}_t}\Big) \sim \sum_{j=0}^{\infty} t^{j/2-1} \alpha_{j/2}(x, s=1).$$

To complete the proof, we must show that $\alpha_0 = 0$. We use the explicit formula for $\phi_0 - \alpha_0\, ds$ furnished by the family index theorem:

$$\phi_0 - \alpha_0\, ds = (2\pi i)^{-n/2} \int_{\widetilde{M}/\widetilde{B}} \hat{A}(\widetilde{M}/\widetilde{B}) \operatorname{ch}(\mathcal{E}/S).$$

But the differential form $\hat{A}(\widetilde{M}/\widetilde{B})\operatorname{ch}(\mathcal{E}/S) \in \mathcal{A}(\widetilde{M})$ is the pull-back to $\widetilde{M}$ of the differential form $\hat{A}(M/B)\operatorname{ch}(\mathcal{E}/S) \in \mathcal{A}(M)$; hence, it does not involve ds, and we see that α_0 vanishes. □

10.6 The Curvature of the Determinant Line Bundle

In Section 9.7, we defined the determinant line bundle $\det(\pi_*\mathcal{E}, \mathsf{D})$, its Quillen metric $|\cdot|_Q$, whose definition involves the zeta-function determinant of the operators $\mathsf{D}^-\mathsf{D}^+$ restricted to the eigenspace (λ, ∞), and its natural connection. In this section, following Bismut and Freed, we will calculate the curvature of the natural connection on $\det(\pi_*\mathcal{E}, \mathsf{D})$, when D is a family of Dirac operators associated to a connection on $\mathcal{E}$.

If $\mathcal{E}$ is a finite-dimensional superbundle, there is a relationship between the Chern character of $\mathcal{E}$ with respect to a connection $\nabla^{\mathcal{E}}$ and the curvature of its determinant line bundle $\det(\mathcal{E})$ with respect to the corresponding connection $\nabla^{\det(\mathcal{E})}$:

$$(\nabla^{\det(\mathcal{E})})^2 = -\operatorname{Str}((\nabla^{\mathcal{E}})^2) = \operatorname{ch}(\nabla^{\mathcal{E}})_{[2]}.$$

We will see a similar relationship in infinite dimensions; the curvature of the connection that we described in Section 9.7 on the line bundle $\det(\pi_*\mathcal{E}, \mathsf{D})$ will be shown to equal

$$\lim_{t\to 0} \operatorname{ch}(\mathbb{A}_t)_{[2]} = (2\pi i)^{-n/2}\Big(\int_{M/B} \hat{A}(M/B) \operatorname{ch}(\mathcal{E}/S)\Big)_{[2]}.$$

Let us first consider a special case, in which $\ker(\mathsf{D})$ is a vector bundle, so $\det(\pi_*\mathscr{E}, \mathsf{D})$ and $\det(\ker(\mathsf{D}))$ may be identified. In this case, the connection $\nabla_0 = P_0 \mathbb{A}_{[1]} P_0$ on $\ker(\mathsf{D})$ is compatible with the L^2-metric, and Theorem 10.32 shows that

$$\operatorname{ch}(\nabla_0) = (2\pi i)^{-n/2} \int_{M/B} \hat{A}(M/B) \operatorname{ch}(\mathscr{E}/S) - d \int_0^\infty \alpha(s)\, ds.$$

If we take the two-form component of this formula, we see that

$$\left(\nabla^{\det(\ker(\mathsf{D}))}\right)^2 = (2\pi i)^{-n/2} \left(\int_{M/B} \hat{A}(M/B) \operatorname{ch}(\mathscr{E}/S) \right)_{[2]} - d \int_0^\infty \alpha(s)_{[1]}\, ds,$$

from which we see immediately that

$$\nabla^{\det(\ker(\mathsf{D}))} + \int_0^\infty \alpha(s)_{[1]}\, ds$$

is a connection on the bundle $\det(\ker(\mathsf{D}))$ with curvature

$$(2\pi i)^{-n/2} \left(\int_{M/B} \hat{A}(M/B) \operatorname{ch}(\mathscr{E}/S) \right)_{[2]}.$$

This connection is compatible with the Quillen metric. It would be very interesting to have similar geometric interpretations of the other components of the transgression formula.

Recall that the Quillen metric is defined on the representative $\det(\mathscr{H}_{[0,\lambda)})$ of the line bundle $\det(\pi_*\mathscr{E}, \mathsf{D})$ over U_λ as

$$|\cdot|_Q = e^{-\zeta'(0,\mathsf{D}^-\mathsf{D}^+,\lambda)/2} |\cdot|,$$

where $|\cdot|$ is the L^2-metric on $\det(\mathscr{H}_{[0,\lambda)})$.

The operator $\mathsf{D}_{(\lambda,\infty)} = P_{(\lambda,\infty)} \mathsf{D}$ over the open set U_λ has the property that the bundle $\ker(\mathsf{D}_{(\lambda,\infty)})$ equals $\mathscr{H}_{[0,\lambda)}$. We may define a superconnection $\mathbb{A}_\lambda$ by the formula

$$\mathbb{A}_\lambda = \mathsf{D}_{(\lambda,\infty)} + \nabla^{\pi_*\mathscr{E}}.$$

Similarly, we define the rescaled version of this superconnection

$$\mathbb{A}_{\lambda,s} = s^{1/2} \mathsf{D}_{(\lambda,\infty)} + \nabla^{\pi_*\mathscr{E}}.$$

Define two one-forms $\alpha^\pm(s,\lambda)_{[1]} \in \mathscr{A}^1(B)$, given by traces over $\pi_*\mathscr{E}^+$ and $\pi_*\mathscr{E}^-$ respectively:

$$\alpha^\pm(s,\lambda)_{[1]} = \operatorname{Tr}_{\pi_*\mathscr{E}^\pm} \left(\frac{\partial \mathbb{A}_{\lambda,s}}{\partial s} e^{-\mathbb{A}_{\lambda,s}^2} \right)_{[1]}.$$

By Lemma 9.42, we see that $\overline{\alpha^+(s,\lambda)_{[1]}} = \alpha^-(s,\lambda)_{[1]}$.

As in Section 9.7, we denote the one-forms $2\operatorname{LIM}_{t\to 0} \int_t^\infty \alpha^\pm(s,\lambda)_{[1]}\, ds$ by $\beta_\lambda^\pm$; these satisfy $\overline{\beta_\lambda{}^+} = \beta_\lambda^-$. Thus, the sum $\beta_\lambda^+ + \beta_\lambda^-$ is real, while the difference $\beta_\lambda^+ - \beta_\lambda^-$ is imaginary. Also, we proved that

$$-d\zeta'(0,\mathsf{D}^-\mathsf{D}^+,\lambda) = \beta_\lambda^+ + \beta_\lambda^-.$$

Thus, we see that the sum of differential forms $\beta_\lambda^+ + \beta_\lambda^-$ is an exact real one-form.

We defined the connection $\nabla^{\det(\pi_*\mathscr{E})}$ on $\det(\pi_*\mathscr{E},\mathsf{D})$ as follows: on the bundle $\det(\mathscr{H}_{[0,\lambda)})$, we took the connection

$$\nabla^{\det(\mathscr{H}_{[0,\lambda)})} + \beta_\lambda^+.$$

Our goal is to calculate the curvature of this connection. In the following lemma, we calculate the exterior differential of the imaginary one-form $\beta_\lambda^+ - \beta_\lambda^-$.

Lemma 10.34. *The difference $\alpha^+(s,\lambda)_{[1]} - \alpha^-(s,\lambda)_{[1]}$ is bounded by a multiple of $s^{-1/2}$ in $\mathscr{A}(U_\lambda)$ for small s, and hence*

$$\beta_\lambda^+ - \beta_\lambda^- = 2\int_0^\infty \big(\alpha^+(s,\lambda) - \alpha^-(s,\lambda)\big)_{[1]}\, ds.$$

The exterior differential of this differential form is given by the formula

$$\tfrac{1}{2} d(\beta_\lambda^+ - \beta_\lambda^-) + \operatorname{ch}(\nabla^{[0,\lambda)})_{[2]} = (2\pi i)^{-n/2}\Big(\int_{M/U_\lambda} \hat{A}(M/B)\operatorname{ch}(\mathscr{E}/S)\Big)_{[2]}$$

Proof. We have seen that

$$\alpha^\pm(s,\lambda)_{[1]} = -\tfrac{1}{2}\operatorname{Tr}_{\pi_*\mathscr{E}^\pm}\Big(\mathsf{D}[\nabla^{\pi_*\mathscr{E}},\mathsf{D}]e^{-s\mathsf{D}^2}P_{(\lambda,\infty)}\Big).$$

Since the Bismut superconnection $\mathbb{A}$ coincides with $\mathsf{D} + \nabla^{\pi_*\mathscr{E}}$ up to terms of degree 2, we see that

$$\begin{aligned}\alpha^\pm(s)_{[1]} &= \operatorname{Tr}_{\pi_*\mathscr{E}^\pm}\Big(\frac{\partial \mathbb{A}_s}{\partial s} e^{-\mathbb{A}_s^2}\Big)_{[1]}\\ &= -\tfrac{1}{2}\operatorname{Tr}_{\pi_*\mathscr{E}^\pm}\Big(\mathsf{D}[\nabla^{\pi_*\mathscr{E}},\mathsf{D}]e^{-s\mathsf{D}^2}\Big).\end{aligned}$$

From this, it follows that

$$\operatorname{Tr}_{\pi_*\mathscr{E}^\pm}\Big(\frac{\partial \mathbb{A}_s}{\partial s} e^{-\mathbb{A}_s^2} - \frac{\partial \mathbb{A}_{\lambda,s}}{\partial s} e^{-\mathbb{A}_{\lambda,s}^2}\Big)_{[1]} = -\operatorname{Tr}_{\pi_*\mathscr{E}^\pm}\big(\mathsf{D}[\nabla^{\pi_*\mathscr{E}},\mathsf{D}]e^{-s\mathsf{D}^2}P_{[0,\lambda)}\big).$$

The operator $P_{[0,\lambda)}$ is a smoothing operator, so that by Lemma 9.47, for s small,

$$\Big\|\alpha^\pm(s,\lambda)_{[1]} - \operatorname{Tr}_{\pi_*\mathscr{E}^\pm}\Big(\frac{\partial \mathbb{A}_s}{\partial s} e^{-\mathbb{A}_s^2}\Big)_{[1]}\Big\|_\ell \le C(\ell).$$

The result now follows from Theorem 10.32. □

We now arrive at the main result of this section (this is what is known to physicists as the "anomaly formula").

Theorem 10.35. *The curvature of the connection* $\nabla^{\det(\pi_*\mathscr{E},\mathsf{D})}$ *equals*

$$(2\pi i)^{-n/2}\Big(\int_{M/U_\lambda}\hat{A}(M/B)\operatorname{ch}(\mathscr{E}/S)\Big)_{[2]}.$$

Proof. We have already done all of the work. Indeed,

$$\begin{aligned}(\nabla^{\det(\pi_*\mathscr{E},\mathsf{D})})^2 &= (\nabla^{\det(\mathscr{H}_{[0,\lambda)})}+\beta_\lambda^+)^2\\ &= (\nabla^{\det(\mathscr{H}_{[0,\lambda)})})^2+d\beta_\lambda^+\\ &= (\nabla^{\det(\mathscr{H}_{[0,\lambda)})})^2+\tfrac{1}{2}d(\beta_\lambda^+-\beta_\lambda^-),\end{aligned}$$

since the differential form $\beta_\lambda^+ + \beta_\lambda^-$ is closed. The theorem now follows from Lemma 10.34. □

10.7 The Kirillov Formula and Bismut's Index Theorem

In this section, we will show that the local family index theorem, in the case in which the family $M \to B$ is associated to a principal bundle $P \to B$ with compact structure group G, is equivalent to the Kirillov formula of Chapter 8. All of our notations are as in Section 9.4.

We start by calculating the Bismut superconnection $\mathbb{A}$. The vertical tangent bundle $T(M/B) = P \times_G TN$ has a connection $\nabla^{N,\theta}$, obtained by combining the Levi-Civita connection ∇^N on TN with the connection θ on the principal bundle $P \to B$, as explained in Section 7.6. On the other hand, the connection θ determines a connection on the fibre bundle $M \to B$, and hence, by the construction of Section 1, a connection $\nabla^{M/B}$ on $T(M/B)$.

Lemma 10.36. *The two connections* $\nabla^{N,\theta}$ *and* $\nabla^{M/B}$ *on* $T(M/B)$ *are equal.*

Proof. Since this formula is local, we assume that P is the trivial bundle $P = B \times G$, with connection form $\theta = \sum_{k=1}^m \theta^k X_k \in \mathscr{A}(B,\mathfrak{g})$, so that $M = B \times N$.

If V is a vector field on B, its horizontal lift is $V + \theta(V)_N$. If Y and Z are vector fields on N, then by the definition of $\nabla^{M/B}$, we see that

$$\begin{aligned}2(\nabla^{M/B}_{V+\theta(V)_N}Y,Z) &= ([\theta(V)_N,Y],Z)+([Z,\theta(V)_N],Y)+\theta(V)_N(Y,Z)\\ &= 2([\theta(V)_N,Y],Z),\end{aligned}$$

where we have used the fact that the vertical metric on M is independent of the point z in the base B, and that $\theta(V)_N$ is a Killing vector field. On the other hand, Lemma 7.37 shows that

$$\begin{aligned}\nabla^{N,\theta}_{V+\theta(V)_N}Y &= \nabla_{\theta(V)_N}Y+\mu^N(\theta(V)_N)Y\\ &= [\theta(V)_N,Y].\end{aligned}$$

□

In the same way, the bundle $\mathscr{E} = P \times_G E$ has a connection $\nabla^{E,\theta}$, obtained by combining the Clifford connection ∇^E on $E \to N$ with the connection θ. Using the fact that for any $Y \in \mathfrak{g}$ and $a \in \Gamma(N, T^*N)$,

$$[\nabla^E(Y), c(a)] = c(\mu^N(Y)a),$$

we see that $\nabla^{E,\theta}$ is a Clifford connection on $\mathscr{E}$ compatible with $\nabla^{N,\theta} = \nabla^{M/B}$. We denote this connection by $\nabla^{\mathscr{E}}$.

The equivariant Chern-Weil homomorphism ϕ_θ maps the equivariant de Rham complex $(\mathscr{A}_G(N), d_\mathfrak{g})$ to the de Rham complex $(\mathscr{A}(M), d)$. Using the fact that $\nabla^{N,\theta}$ and $\nabla^{E,\theta}$ equal $\nabla^{M/B}$ and $\nabla^{\mathscr{E}}$ respectively, we see that

$$\phi_\theta(\hat{A}_\mathfrak{g}) = \hat{A}(M/B),$$
$$\phi_\theta(\operatorname{ch}_\mathfrak{g}(E/S)) = \operatorname{ch}(\mathscr{E}/S),$$

where $\hat{A}_\mathfrak{g}$ and $\operatorname{ch}_\mathfrak{g}(E/S)$ are the equivariant $\hat{A}$-genus of N and the equivariant relative Chern character of E.

In the fibre bundle $M \to B$, the tensor S is equal to zero, and hence its trace k is also zero. To calculate the Bismut superconnection, we must describe the curvature tensor Ω of M.

Lemma 10.37. *If V and W are two vector fields on B, then*

$$\Omega(V, W) = -\Omega_P(V_P, W_P)_N,$$

where V_P and W_P are the horizontal lifts of V and W to P, $\Omega_P \in \mathscr{A}^2(P, \mathfrak{g})$ is the curvature of P, and for any element $X \in \mathfrak{g}$, X_N is the corresponding vector field on N, as well as the induced vertical vector field on $M = P \times_G N$.

Since the bundle $\pi_*\mathscr{E}$ is the associated bundle to the representation of G on $\Gamma(N, E)$, it has a canonical connection given on the space

$$\mathscr{A}(B, \pi_*\mathscr{E}) \cong (\mathscr{A}(P) \otimes \Gamma(N, E))_{\text{bas}}$$

by the formula

$$\nabla^{\pi_*\mathscr{E}} = d_P + \sum_{k=1}^m \theta^k \mathscr{L}^E(X_k).$$

This connection has curvature

$$(\nabla^{\pi_*\mathscr{E}})^2 = \sum_{k=1}^m \Omega_P^k \mathscr{L}^E(X_k).$$

There is a linear map $c : \mathfrak{g} \to \Gamma(N, \operatorname{End}(E))$, defind by sending $X \in \mathfrak{g}$ to $c(X_N)$, the operator of Clifford multiplication by the vector field $X_N \in \Gamma(N, TN)$ associated to X. By means of the formula

$$c(\Omega_P) = \sum_{k=1}^m \Omega_P^k c(X_k),$$

we obtain an action of the curvature Ω_P of P on $\mathscr{A}(P) \otimes \Gamma(N, E)$.

Theorem 10.38. *1. The Bismut superconnection of the bundle $P \times_G N$ is the restriction to $\big(\mathscr{A}(P) \otimes \Gamma(N,E)\big)_{\mathrm{bas}}$ of the operator*

$$1 \otimes \mathsf{D}_N + \nabla^{\pi_*\mathscr{E}} + \tfrac{1}{4} c(\Omega_P).$$

2. The Chern character $\mathrm{ch}(\mathbb{A}_t) = \mathrm{Str}\big(e^{-\mathbb{A}_t^2}\big) \in \mathscr{A}(B)$ is independent of t.

Proof. From the formula for $\mathbb{A}$ given in Proposition 10.15, we see that $\mathbb{A}_{[0]} = \mathsf{D}_N$ and $\mathbb{A}_{[2]} = \frac{1}{4} c(\Omega_P)$. To check that

$$\sum_\alpha \varepsilon^\alpha \nabla^{\mathscr{E}}_\alpha = \nabla^{\pi_*\mathscr{E}},$$

we may assume that P is trivial, as in the proof of Lemma 10.36. We see that

$$\begin{aligned}
\sum_\alpha \varepsilon^\alpha \nabla^{\mathscr{E}}_\alpha &= \sum_\alpha \varepsilon^\alpha \nabla^{E,\theta}_{\partial_\alpha + \theta(\partial_\alpha)_N} \\
&= \sum_\alpha \varepsilon^\alpha \Big(\nabla^E_{\theta(\partial_\alpha)_N} + \mu^N\big(\theta(\partial_\alpha)_N\big)\Big) \\
&= \sum_\alpha \varepsilon^\alpha \mathscr{L}^E(\theta(\partial_\alpha)_N) \\
&= \sum_k \theta^k \mathscr{L}^E(X_k).
\end{aligned}$$

The connection $\nabla^{\pi_*\mathscr{E}}$ commutes with D_N, since D_N is G-invariant. Furthermore, $\nabla^{\pi_*\mathscr{E}}$ and $c(\Omega_P)$ commute, by the Bianchi identity $\nabla\Omega_P = 0$. If we let Q_t be the operator

$$Q_t = \mathbb{A}_t - \nabla^{\pi_*\mathscr{E}} = t^{1/2} D_N + \frac{1}{4t^{1/2}} c(\Omega) \in \Gamma(B, \mathrm{End}_{\mathscr{D}}(\pi_*\mathscr{E})),$$

we see that

$$\frac{d\mathbb{A}_t^2}{dt} = \left[\frac{d\mathbb{A}_t}{dt}, \mathbb{A}_t\right] = \left[\frac{d\mathbb{A}_t}{dt}, Q_t\right],$$

and hence that

$$\frac{d}{dt}\mathrm{Str}\big(e^{-\mathbb{A}_t^2}\big) = -\mathrm{Str}\left(\left[\frac{d\mathbb{A}_t}{dt}, Q_t e^{-\mathbb{A}_t^2}\right]\right) = 0,$$

since the supertrace vanishes by Lemma 3.49. □

We have seen in Section 9.4 that Kirillov's formula for $\mathrm{ind}_G(e^{-X}, \mathsf{D}_N)$ implies that

$$\lim_{t\to\infty} \mathrm{Str}\big(e^{-\mathbb{A}_t^2}\big) = (2\pi i)^{-n/2} \int_{M/B} \hat{A}(M/B)\, \mathrm{ch}(\mathscr{E}/S).$$

We see from the above theorem that the formula

$$\lim_{t\to 0} \mathrm{Str}\big(e^{-\mathbb{A}_t^2}\big) = (2\pi i)^{-n/2} \int_{M/B} \hat{A}(M/B)\, \mathrm{ch}(\mathscr{E}/S),$$

is, in the case of a compact structure group, a consequence of the Kirillov formula for the equivariant index of D_N. We will strengthen this result by showing that the the local family index theorem of Bismut is implied in this context by the local Kirillov formula for the operator $K(X) = \left(D + \frac{1}{4}c(X)\right)^2 + \mathcal{L}^E(X)$ of Section 8.3.

The operator $K(X)$ is a differential operator depending polynomially on X, and hence acts on the space $\mathbb{C}[\mathfrak{g}] \otimes \Gamma(N,E)$, by the formula

$$(Ks)(X) = K(X)s(X).$$

If p is a point in the principal bundle P lying in the fibre of $z \in B$, we may specialize the operator $K(X)$ to an operator $K(\Omega_p)$ acting on $\Lambda^2 T^*{}_{\pi(p)}B \otimes \Gamma(N,E)$, by replacing $X = \sum_{k=1}^m a^k X_k \in \mathfrak{g}$, $a^k \in \mathbb{C}$, by

$$\Omega_p = \sum_{i=1}^{m} a^k (\Omega_P^k)_p X_k \in \Lambda^2 T^*{}_{\pi(p)} B \otimes \mathfrak{g}.$$

On the other hand, the point $p \in P$ gives an identification of the fibres M_z with N, so that we may also consider the curvature $\mathbb{A}^2$ of the Bismut superconnection to be a differential operator $\mathbb{A}_p^2$ acting on $\Lambda^2 T^*{}_{\pi(p)}B \otimes \Gamma(N,E)$.

Proposition 10.39. *The two operators $K(\Omega_p)$ and $\mathbb{A}_p^2$ are equal.*

Proof. This follows from Theorem 10.38 and the fact that $\nabla^{\pi_*\mathcal{E}}$ supercommutes with D_N and $c(\Omega)$:

$$\begin{aligned}\mathbb{A}^2 &= \left(D + \tfrac{1}{4}c(\Omega_P)\right)^2 + \left(\nabla^{\pi_*\mathcal{E}}\right)^2 \\ &= \left(D + \tfrac{1}{4}c(\Omega_P)\right)^2 + \mathcal{L}(\Omega_P).\end{aligned}$$

□

From this lemma, we see that the coefficients of the asymptotic expansion of $\langle p \mid e^{-t\mathbb{A}^2} \mid p \rangle$ depend polynomially on Ω_p^k, and are obtained from the coefficients of the asymptotic expansion of $\langle p \mid e^{-tK(X)} \mid p \rangle$ by replacing X by Ω_p.

The converse result is also true: the local family index theorem implies the local Kirillov formula. To show this, we use the following lemma from the theory of classifying spaces.

Lemma 10.40. *There exist fibre bundles $P_N \to B_N$ for which the Chern-Weil map $\mathbb{C}[\mathfrak{g}]^G \to \mathcal{A}(B_N)$ is an injection up to polynomials of degree N, for any $N > 0$.*

Proof. Let V be a faithful unitary representation of G. From the embedding

$$U(V \otimes \mathbb{C}^N) \times U(V \otimes \mathbb{C}^N) \subset U(V \otimes \mathbb{C}^{2N}),$$

we see that $U(V \otimes \mathbb{C}^N)$ acts freely on $P_N = U(V \otimes \mathbb{C}^{2N}) / U(V \otimes \mathbb{C}^N)$. There is an embedding of G in $U(V \otimes \mathbb{C}^N)$, in which we map $g \in G$ to $g \otimes I$. Let $B_N = P_N / G$. The result follows from the fact that if $m > n$, $\pi_i(U(m)/U(n)) = 0$ for $i \leq 2n$. □

The local index theorem for families for the bundle $P_N \to B_N$ implies the local Kirillov formula for $K(X)$ up to order N. Since N is arbitrary, we obtain the local Kirillov formula to all orders.

Bibliographic Notes

Sections 1–4

The results of the first four sections of this chapter are due to Bismut [30], although the proof of the main result (Theorem 10.21), the local family index theorem, follows our proof of the local index theorem in Chapter 4 of this book. In his work, Bismut uses the term Levi-Civita superconnection where, for obvious reasons, we prefer the term Bismut superconnection. Theorem 10.19 is new, and gives a natural interpretation of the Bismut superconnection. For closely related approaches to the family index theorem, see the articles of Berline-Vergne [25], Donnelly [53] and Zhang [109].

Sections 5 and 6

The "anomaly" formula for the curvature of the natural connection on the determinant line bundle was proved by Witten in the special case where there are no zero modes (see for example [107]), based on the work of many theoretical physicists. Further special cases were proved by Atiyah-Singer [17] and Quillen [93], and the theorem is stated and proved in full generality in the article of Bismut and Freed [36]. In this chapter, it appears in a generalized form inspired by the article of Gillet and Soulé [68].

Section 7

The relationship between the family index theorem for a family with compact structure group and the equivariant index theorem is in Atiyah and Singer [16]. The remarkable observation that such a relationship holds at the level of differential operators, as in Proposition 10.39, is due to Bismut [30].

For an extension of the results of this chapter in the direction of algebraic geometry, see Bismut, Gillet and Soulé [37].

References

1. Alvarez-Gaumé, L.: Supersymmetry and the Atiyah-Singer index theorem. Comm. Math. Phys., **90** , 161–173 (1983).
2. Atiyah, M. F.: Collected works. Clarendon Press, Oxford (1988).
3. Atiyah, M. F.: K-theory. Benjamin, New York (1967).
4. Atiyah, M. F.: Circular symmetry and stationary phase approximation. Astérisque, **131**, 43–59 (1985).
5. Atiyah, M. F. and Bott, R.: A Lefschetz fixed point formula for elliptic complexes, I. Ann. Math., **86**, 374–407 (1967).
6. Atiyah, M. F. and Bott, R.: A Lefschetz fixed point formula for elliptic complexes, II. Applications. Ann. Math., **88**, 451–491 (1968).
7. Atiyah, M. F. and Bott, R.: The moment map and equivariant cohomology. Topology, **23**, 1–28 (1984).
8. Atiyah, M. F., Bott, R. and Patodi, V. K.: On the heat equation and the index theorem. Invent. Math., **19**, 279–330 (1973). Errata, **28**, 277–280 (1975).
9. Atiyah, M. F., Bott, R. and Shapiro, A.: Clifford modules. Topology, **3**, Suppl. 1, 3–38 (1964).
10. Atiyah, M. F. and Hirzebruch, F.: Riemann-Roch theorems for differentiable manifolds. Bull. Amer. Math. Soc., **65**, 276–281 (1959).
11. Atiyah, M. F., Patodi, V. K. and Singer, I. M.: Spectral assymetry and Riemannian geometry, I. Math. Proc. Camb. Phil. Soc., **77**, 43–69 (1975).
12. Atiyah, M. F. and Segal, G. B.: The index of elliptic operators, II. Ann. Math., **87**, 531–545 (1968).
13. Atiyah, M. F. and Singer, I. M.: The index of elliptic operators on compact manifolds. Bull. Amer. Math. Soc., **69**, 422–433 (1963).
14. Atiyah, M. F. and Singer, I. M.: The index of elliptic operators, I. Ann. Math., **87**, 484–530 (1968).
15. Atiyah, M. F. and Singer, I. M.: The index of elliptic operators, III. Ann. Math., **87**, 546–604 (1968).
16. Atiyah, M. F. and Singer, I. M.: The index of elliptic operators IV. Ann. Math., **93**, 119–138 (1971).
17. Atiyah, M. F. and Singer, I. M.: Dirac operators coupled to vector potentials. Proc. Nat. Acad. Sci. U.S.A., **81**, 2597–2600 (1984).
18. Baum, P. and Cheeger, J.: Infinitesimal isometries and Pontryagin numbers. Topology, **8**, 173–193 (1969).

19. Berger, M., Gauduchon, P. and Mazet, E.: Le spectre d'une variété riemannienne. Lecture Notes in Mathematics **194**, Springer, Berlin New York (1971).
20. Berline, N. and Vergne, M.: Classes caractéristiques équivariantes. Formules de localisation en cohomologie équivariante. C.R. Acad. Sci. Paris, **295**, 539–541 (1982).
21. Berline, N. and Vergne, M.: Fourier transforms of orbits of the coadjoint representation. In: Trombi, P. C. (ed) Representation theory of reductive groups (Park City, Utah, 1982). Progress in Mathematics, **40**, Birkäuser, Boston (1983), pp. 53–67.
22. Berline, N. and Vergne, M.: Zéros d'un champ de vecteurs et classes charactéristiques équivariantes. Duke Math. J., **50**, 539–549 (1983).
23. Berline, N. and Vergne, M.: The equivariant index and Kirillov character formula. Amer. J. Math., **107**, 1159–1190 (1985).
24. Berline, N. and Vergne, M.: A computation of the equivariant index of the Dirac operator. Bull. Soc. Math. France, **113**, 305–345 (1985).
25. Berline, N. and Vergne, M.: A proof of the Bismut local theorem for a family of Dirac operators. Topology, **26**, 435–463 (1987).
26. Berline, N. and Vergne, M.: L'indice équivariant des opérateurs transversalement elliptique. Invent. Math., **124**, 51–101 (1996).
27. Berezin, F. A.: The method of second quantization. Pure and applied physics, **24**, Academic Press, New York London (1966).
28. Bismut, J.-M.: The Atiyah-Singer theorems: a probabilistic approach. J. Funct. Anal., **57**, 56–99 (1984).
29. Bismut, J.-M.: The Atiyah-Singer theorems: a probabilistic approach, II. J. Funct. Anal., **57**, 329–348 (1984).
30. Bismut, J.-M.: The index theorem for families of Dirac operators: two heat equation proofs. Invent. Math., **83**, 91–151 (1986).
31. Bismut, J.-M.: Localization formulas, superconections, and the index theorem for families. Comm. Math. Phys., **103**, 127–166 (1986).
32. Bismut, J.-M.: The infinitesimal Lefschetz formulas: a heat equation proof. J. Funct. Anal., **62**, 435–457 (1985).
33. Bismut, J.-M.: Local index theory and higher analytic torsion. In: Proceedings of the International Conference of Mathematicians (Berlin 1998), Doc. Math., Extra Vol. **I**, 143–162 (1998).
34. Bismut, J.-M. and Cheeger, J.: η-invariants and their adiabatic limits. J. Amer. Math. Soc., **2**, 33–70 (1989).
35. Bismut, J.-M. and Freed, D. S.: The analysis of elliptic families: Metrics and connections on determinant bundles. Comm. Math. Phys., **106**, 159–176 (1986).
36. Bismut, J.-M. and Freed, D. S.: The analysis of elliptic families: Dirac operators, eta invariants, and the holonomy theorem of Witten. Comm. Math. Phys., **107**, 103–163 (1986).
37. Bismut, J.-M., Gillet, H. and Soulé, C.: Analytic torsion and holomorphic determinant bundles. Comm. Math. Phys., **115**, 49–78, 79–126, 301–351 (1988).
38. Block, J. and Getzler, E.: Equivariant cyclic homology and equivariant differential forms. Ann. Sci. Ecole Norm. Sup. (4), **27**, 493–527 (1994).
39. Borel, A. and Hirzebruch, F.: Characteristic classes and homogeneous spaces. Amer. J. Math., **80**, 458–538 (1958).
40. Borel, A. and Serre, J.-P.: Le théorème de Riemann-Roch (d'après Grothendieck). Bull. Soc. Math. France, **86**, 97–136 (1958).
41. Bost, J. B.: Fibrés déterminants, déterminants régularisés et mesures sur les espaces de modules de courbes complexes, In: Séminaire Bourbaki, Astérisque, **152–153**, 113–149 (1987).

42. Bott, R.: Homogeneous vector bundles. Ann. Math., **66**, 203–248 (1957).
43. Bott, R.: Vector fields and characteristic numbers. Michigan Math. J., **14**, 231–244 (1967).
44. Bott, R.: A residue formula for holomorphic vector fields. J. Diff. Geom., **1**, 311–330 (1967).
45. Bott, R. and Tu, L.: Differential forms in algebraic topology. Graduate Texts in Mathematics, **82**, Springer, New York Berlin (1982).
46. Duflo, M.: Opérateurs transversalement elliptiques et formes différentielles équivariantes (d'après N. Berline et M. Vergne). Séminaire Bourbaki, Astérisque, **237**, 29–45 (1996).
47. Cartan, H.: La transgression dans un groupe de Lie et dans un espace fibré principal. In: Colloque de Topologie (espaces fibrés), Bruxelles, 1950, 57–71. Georges Thone, Liége (1951).
48. Cartan, H.: Théorème d'Atiyah-Singer sur l'indice d'un operateur differentiel. Séminaire Cartan-Schwartz, 16e année: 1963/64, Sécrétariat mathématique, Paris, 1965.
49. Chern, S. S.: A simple intrinsic proof of the Gauss-Bonnet formula for closed Riemannian manifolds. Ann. Math., **45**, 741–752 (1944).
50. Chern, S. S. and Simon, J.: Characteristic forms and geometric invariants. Ann. Math., **99**, 48–69 (1974).
51. Chevalley, C.: The algebraic theory of spinors. Columbia Univ. Press, New York (1954).
52. Connes, A. and Moscovici, H.: The L^2-index theorem for homogeneous spaces of Lie groups. Ann. Math., **115**, 291–330 (1982).
53. Donnelly, H.: Local index theorem for families. Michigan Math. J., **35**, 11–20 (1988).
54. Donnelly, H. and Patodi, V. K.: Spectrum and the fixed point set of isometries. Topology, **16**, 1–11 (1977).
55. Duflo, M., Heckman, G. and Vergne, M.: Projection d'orbites, formule de Kirillov et formule de Blattner. Mem. Soc. Math. France, **15**, 65–128 (1984).
56. Duflo, M. and Vergne, M.: Orbites coadjointes et cohomologie équivariante. In: Duflo, M. et al. (eds) The orbit method in representation theory, Progress in Mathematics, **82**, Birkhäuser, Boston, MA (1990).
57. Duflo, M. and Vergne, M.: Cohomologie équivariante et descente. Astérisque, **215**, 5–108 (1993).
58. Duistermaat, J. J. and Heckman, G.: On the variation in the cohomology of the symplectic form of the reduced phase space. Invent. Math., **69**, 259–268 (1982). Addendum, **72**, 153–158 (1983).
59. Erdélyi, A.: Asymptotic expansions. Dover, New York (1956).
60. Friedan, D. and Windey, P.: Supersymmetric derivation of the Atiyah-Singer index theorem. Nucl. Phys., **B235**, 394–416 (1984).
61. Getzler, E.: Pseudodifferential operators on supermanifolds and the index theorem. Comm. Math. Phys., **92**, 163–178 (1983).
62. Getzler, E.: A short proof of the Atiyah-Singer index theorem. Topology, **25**, 111–117 (1986).
63. Getzler, E.: The Bargmann representation, generalized Dirac operators and the index of pseudodifferential operators on $\mathbb{R}^n$. In: Maeda, Y. et al. (eds) Symplectic geometry and quantization. Contemporary Mathematics, **179**, Amer. Math. Soc., Providence, RI (1994).
64. Gilkey, P. B.: Curvature and the eigenvalues of the Laplacian for elliptic complexes. Advances in Math., **10**, 344–382 (1973).
65. Gilkey, P. B.: Curvature and the eigenvalues of the Dolbeault complex for Kaehler manifolds. Advances in Math., **11**, 311–325 (1973).

66. Gilkey, P. B.: Lefschetz fixed point formulas and the heat equation. In: Byrnes, C. (ed) Partial differential equations and geometry, 91–147. Lecture Notes in Pure and Applied Mathematics, **48**, Marcel Dekker, New York (1979).
67. Gilkey, P. B.: Invariance theory, the heat equation and the Atiyah-Singer index theorem. Mathematics Lecture Series, **11**. Publish or Perish, Wilmington, DE (1984).
68. Gillet, H. and Soulé, C.: Analytic torsion and the arithmetic Todd genus. Topology, **30**, 21–54 (1991).
69. Harish-Chandra: Fourier transforms on a Lie algebra, I. Amer. J. Math., **79**, 193–257 (1957).
70. Hirzebruch, F.: Topological methods in algebraic geometry, 3rd ed. Grundlehren der mathematische Wissenschaften, **131**. Springer, New York (1966).
71. Hirzebruch, F.: The signature theorem: reminiscences and recreation. In: Prospects in mathematics, 3–31. Ann. of Math. Studies, **70**, Princeton Univ. Press, Princeton, NJ (1971).
72. Karoubi, M.: K-theory. An introduction. Grundlehren der mathematische Wissenschaften, **226**, Springer, Berlin New York (1978).
73. Kirillov, A. A.: Method of orbits in the theory of unitary representations of Lie groups. Funktsional Anal. i Prilozhen., **2**, 96–98 (1968).
74. Kirillov, A. A.: Elements of the theory of representations. Grundlehren der mathematische Wissenschaften **220**, Springer, Berlin New York (1976).
75. Kobayashi, S. and Nomizu, K.: Foundations of differential geometry. Vol. I. John Wiley and Sons, New York (1963).
76. Kostant, B.: Quantization and unitary representations. Prequantization. In: Lectures in Modern Analysis and Applications, III, 87–208. Lecture Notes in Mathematics, **170**, Springer, Berlin (1970).
77. Kotake, T.: The fixed-point formula of Atiyah-Bott via parabolic operators. Comm. Pure Appl. Math., **22**, 789–806 (1969).
78. Kotake, T.: An analytical proof of the classical Riemann-Roch theorem. In: Global analysis (Proc. Sympos. Pure Math., **16**, Berkeley, Calif., 1968), 137–146. Amer. Math. Soc., Providence, R.I. (1970).
79. Lafferty, J. D., Yu, Y. L. and Zhang, W. P.: A direct geometric proof of the Lefschetz fixed point formulas. Trans. Amer. Math. Soc., **329**, 571–583 (1992).
80. Lawson, H. B., Jr. and Michelson, M.-L.: Spin geometry. Princeton Mathematical Series, **38**, Princeton Univ. Press, Princeton, NJ (1989).
81. Lichnerowicz, A.: Spineurs harmoniques. C.R. Acad. Sci. Paris. Sér. A, **257**, 7–9 (1963).
82. Mathai, V. and Quillen, D.: Superconnections, Thom classes and equivariant differential forms. Topology, **25**, 85–110 (1986).
83. McKean, H. and Singer, I. M.: Curvature and the eigenvalues of the Laplacian. J. Diff. Geom., **1**, 43–69 (1967).
84. Milnor, J.: Morse theory, Ann. of Math. Studies, **51**. Princeton Univ. Press, Princeton (1963).
85. Milnor, J. W.: Remarks concerning spin manifolds. In: Cairns, S. S. (ed) Differential and combinatorial topology (A symposium in honor of Marston Morse), 55–62. Princeton Univ. Press, Princeton, NJ (1965).
86. Milnor, J. W. and Stasheff, J. D.: Characteristic classes. Ann. of Math. Studies, **76**, Princeton Univ. Press, Princeton, NJ (1974).
87. Minakshisundaram, S. and Pleijel, A.: Some properties of the eigenfunctions of the Laplace operator on Riemannian manifolds. Canad. J. Math., **1**, 242–256 (1949).
88. Palais, R. S.: Seminar on the Atiyah-Singer index theorem., Ann. of Math. Studies, **57**, Princeton Univ. Press, Princeton (1965).

89. Patodi, V. K.: Curvature and the eigenforms of the Laplace operator. J. Diff. Geom., **5**, 233–249 (1971).
90. Patodi, V. K.: An analytic proof of the Riemann-Roch-Hirzebruch theorem. J. Diff. Geom., **5**, 251–283 (1971).
91. Patodi, V. K.: Holomorphic Lefschetz fixed point formula. Bull. Amer. Math. Soc., **79**, 825–828 (1973).
92. Quillen, D.: The spectrum of an equivariant cohomology ring, I and II. Ann. Math., **94**, 549–602 (1971).
93. Quillen, D.: Determinants of Cauchy-Riemann operators over a Riemann surface. Funktsional Anal. i Prilozhen., **19**, no. 1, 37–41.
94. Quillen, D.: Superconnections and the Chern character. Topology, **24**, 89–95 (1985).
95. Ray, D. B. and Singer, I. M.: R-torsion and the Laplacian on Riemannian manifolds. Adv. Math., **7**, 145–210 (1971).
96. Ray, D. B. and Singer, I. M.: Analytic torsion for complex manifolds. Ann. Math., **98**, 154–177 (1973).
97. Roe, J.: Elliptic operators, topology and asymptotic methods. Second Edition. Pitman Research Notes in Mathematics Series, **395**. Longman, Harlow (1998).
98. Rossmann, W.: Kirillov's character formula for reductive groups. Invent. Math., **48**, 207–220 (1978).
99. Rossmann, W.: Equivariant multiplicities on complex varieties. Astérisque, **173-174**, 313–330 (1989).
100. Seeley, R.: Complex powers of an elliptic operator. In: Singular Integrals (Proc. Sympos. Pure Math., **10**), 288–307. Amer. Math. Soc., Providence, RI (1967).
101. Seeley, R.: A proof of the Atiyah-Bott-Lefschetz fixed point formula. An. Acad. Brasil Cience **41** (1969), 493–501.
102. Segal, G. B.: Equivariant K-theory. Inst. Hautes Études Sci. Publ. Math., **34**, 129–151 (1968).
103. Vergne, M.: Formule de Kirillov et indice de l'opérateur de Dirac. In: Proceedings of the International Congress of Mathematicians (Warszawa, 1983), 921–934. North-Holland, Amsterdam (1984).
104. Vergne, M.: Geometric quantization and equivariant cohomology. In: First European Congress of Mathematics, Vol. I (Paris, 1992), 249–295. Progr. Math., **119**, Birkhaüser, Basel (1994).
105. Wells, Jr, R. O.: Differential analysis on complex manifolds. Second Edition. Graduate Texts in Mathematics, **65**. Springer, New York Berlin (1980).
106. Witten, E.: Supersymmetry and Morse theory. J. Diff. Geom., **17**, 661–692 (1982).
107. Witten, E.: Global gravitational anomalies. Comm. Math. Phys., **100**, 197–229 (1985).
108. Yu, Y. L.: Local index theorem for Dirac operator. Acta Math. Sinica (N.S.), **3**, 152–169 (1987).
109. Zhang, W. P.: The local Atiyah-Singer index theorem for families of Dirac operators. In: Jiang, B. J. et al. (eds) Differential geometry and topology, 351–366. Lecture Notes in Mathematics, **1369**. Springer, Berlin (1989).

List of Notation

$\Gamma(M,\mathscr{E})$	space of smooth sections of $\mathscr{E}$	13
$\Gamma_c(M,\mathscr{E})$	space of compactly supported smooth sections of $\mathscr{E}$	13
$\mathrm{GL}(\mathscr{E})$	frame bundle of $\mathscr{E}$	15
$\mathscr{L}(X)$	Lie derivative acting on vector fields	16
$\mathscr{L}^{\mathscr{E}}(X)$	Lie derivative acting on sections of $\mathscr{E}$	16
$\mathscr{A}(M)$	space of differential forms on M	16
$\mathscr{A}_c(M)$	space of compactly supported differential forms on M	16
$\alpha_{[i]}$	i-form component of α	16
$\lvert\alpha\rvert$	degree of α	16
$\iota(v)$	contraction	17, 67
$\varepsilon(\alpha)$	exterior multiplication by α	17
$H^\bullet(M)$	de Rham cohomology of M	18
$H_c^\bullet(M)$	compactly supported de Rham cohomology of M	18
$\mathscr{A}(M,\mathscr{E})$	differential forms on manifold M with values in $\mathscr{E}$	19, 41
$\nabla^{\mathscr{E}_1\otimes\mathscr{E}_2}$	tensor product of covariant derivatives	24
$\tau_\gamma(t)$	parallel translation along smooth path γ	25
$\mathscr{R}$	Euler (radial) vector field	25, 50
$\lvert\Lambda\rvert^{1/2}$	bundle of half-densities	29
$\pi_*\alpha$	integral of α along the fibres of a fibre bundle	30
$\int_{M/B}\alpha$	another notation for $\pi_*\alpha$	30
$\mathrm{O}(M)$	orthonormal frame bundle of Riemannian manifold M	31
$\mathfrak{so}(TM)$	$\mathrm{O}(M)\times_{O(n)}\mathfrak{so}(n)$	32
$R(X,Y)$	Riemannian curvature	32
Ric	Ricci curvature	33
r_M	scalar curvature	33
$\lvert dx\rvert$	Riemannian density	33
grad f	gradient of the function f	33
$\mathbf{x}$	point in the normal coordinate system around x_0	34
$j(\mathbf{x})$	Jacobian of exponential map around x_0	37

$[a,b]$	supercommutator 37
Str	supertrace 38
$\det(\mathscr{E})$	determinant line-bundle of $\mathscr{E}$ 39
$T(a)$	Berezin integral 40
$\mathrm{Pf}(A)$	Pfaffian of A 41
$\det^{1/2}(A)$	another notation for $\mathrm{Pf}(A)$ 41
$\mathbb{A}$	superconnection 42
$\mathbb{A}_{[i]}$	degree i component of $\mathbb{A}$ 43
$\mathrm{ch}(\mathbb{A})$	Chern character form of $\mathbb{A}$ 48, 280
$\mathrm{ch}(\mathscr{E})$	Chern character of $\mathscr{E}$ 48
$\hat{A}(M)$	$\hat{A}$-genus of M 49
$\mathrm{Td}(M)$	Todd genus of the almost-complex manifold M 50
$\chi(M)$	Euler class of the oriented manifold M 56
$\mathrm{Eul}(M)$	Euler number of M 58, 125
$\Delta^{\mathscr{E}}$	Laplacian associated to $\mathscr{E}$ 64
D^*	formal adjoint of D 65
$d^*\alpha$	divergence of α 66, 124
$q_t(\mathbf{x},\mathbf{y})$	heat kernel of Euclidean space 70
$\int_{\mathfrak{g}}^{\text{asymp}}$	asymptotic expansion of a Gaussian integral 71, 170
$\Gamma^{-\infty}(M,\mathscr{E})$	space of generalized sections of $\mathscr{E}$ 71
$\Gamma^{\ell}(M,\mathscr{E})$	space of C^{ℓ}-sections of $\mathscr{E}$ 71
$\mathscr{E}_1 \boxtimes \mathscr{E}_2$	external tensor product of $\mathscr{E}_1$ and $\mathscr{E}_2$ 72
$\langle x \mid P \mid y\rangle$	kernel of the operator P at (x,y) 72
$p_t(x,y,H)$	heat kernel of Laplacian H 73, 84
Δ_k	k-simplex 74
$\tau(x,y)$	parallel translation along geodesic joining x and y 80
$\Vert A\Vert_{\mathrm{HS}}$	Hilbert-Schmidt norm of A 85
$\Gamma_{L^2}(M,\mathscr{E}\otimes\lvert\Lambda\rvert^{1/2})$	space of square-summable sections of $\mathscr{E}\otimes\lvert\Lambda\rvert^{1/2}$ 85
$\mathrm{Tr}(K)$	trace of K 86
G^k	k-th Green kernel of H 89
$C(V)$	Clifford algebra of V 100
$c(v)$	action of $v\in V$ on a Clifford module 101
$\mathrm{End}_{C(V)}(E)$	endorphisms of Clifford module E over $C(V)$ 101
$\sigma(a)$	symbol map from $C(V)$ to ΛV 101
$C^2(V)$	degree 2 subspace of Clifford algebra 101
τ	canonical homomorphism from $C^2(V)$ to $\mathfrak{so}(V)$ 102
$\mathrm{Spin}(V)$	double cover of $\mathrm{SO}(V)$ 103
$j_V(X)$	$\det\left(\frac{\sinh(X/2)}{X/2}\right)$ 104
Γ	chirality element of $C(V)$ 106
$S=S^+\oplus S^-$	spinor representation of $C(V)$ 106
$\mathrm{Str}_{E/S}$	relative supertrace for Clifford module E 109
$\lambda(X)$	action of $\mathrm{U}(V)$ on $\Lambda T^{1,0}V$ 109
$C(M)$	Clifford bundle of Riemannian manifold M 110
$\mathrm{Spin}(M)$	principal $\mathrm{Spin}(n)$-bundle of spin-manifold M 111

$\pi_*\mathscr{E}$	vector bundle of sections along fibres of $\mathscr{E}$	274
$(\mathsf{D}^z \mid z \in B)$	family of Dirac operators	274
$\mathscr{H}_{[0,\lambda)}$	subbundle of $\pi_*\mathscr{E}$ on which $\mathsf{D}^2 < \lambda$	275
$P_{[0,\lambda)}$	spectral projection onto $\mathsf{D}^2 < \lambda$	275
U_λ	subset of B on which λ is not an eigenvalue of D^2	275
$\mathrm{ind}(\mathsf{D})$	family index of D	276, 293
$P_{(\lambda,\mu)}$	spectral projection onto $\lambda < \mathsf{D}^2 < \mu$	276
$\mathscr{H}_{(\lambda,\mu)}$	subbundle of $\pi_*\mathscr{E}$ on which $\lambda < \mathsf{D}^2 < \mu$	276
$\nabla^{\pi_*\mathscr{E}}$	natural connection on $\pi_*\mathscr{E}$	277
$R^{M/B}$	vertical Riemannian curvature	288
$\hat{A}(M/B)$	$\hat{A}$-genus of $T(M/B)$	288, 331
$\mathsf{M}[f]$	Mellin transform of f	293
$\mathop{\mathsf{LIM}}_{t\to 0} f(t)$	renormalized limit of f	295
$\zeta(s,H,\lambda)$	zeta function of H	295
$\det(H)$	zeta function determinant of H	296
$\det(\pi_*\mathscr{E},\mathsf{D})$	determinant line bundle associated to the family D	299
$\det(\mathsf{D}^+)$	determinant of D^+ (section of $\det(\pi_*\mathscr{E},\mathsf{D})$)	300
$\lvert\cdot\rvert_Q$	Quillen metric	300
$\nabla^{\det(\pi_*\mathscr{E},\mathsf{D})}$	natural connection on $\det(\pi_*\mathscr{E},\mathsf{D})$	303

Index

Druck und Bindung: Strauss Offsetdruck GmbH